High and Low Erucic Acid Rapeseed Oils

**PRODUCTION, USAGE, CHEMISTRY, AND
TOXICOLOGICAL EVALUATION**

D1678199

CONTRIBUTORS

R. G. Ackman	B. E. McDonald
D. H. C. Beach	S. V. Pande
G. S. Boulter	W. J. Pigden
A. H. Corner	M. R. Pollard
J. K. Daun	A. D. Roden
R. K. Downey	F. D. Sauer
E. R. Farnworth	H. Sprecher
H. C. Grice	B. R. Stefansson
H. A. Heggtveit	P. K. Stumpf
J. K. G. Kramer	B. F. Teasdale
T. K. Mag	B. K. Thompson
M. Vaisey-Genser	

High and Low Erucic Acid Rapeseed Oils

PRODUCTION, USAGE, CHEMISTRY, AND TOXICOLOGICAL EVALUATION

Edited by

JOHN K. G. KRAMER

FRANK D. SAUER

Animal Research Centre
Research Branch, Agriculture Canada
Ottawa, Ontario, Canada

WALLACE J. PIGDEN

F. A. W. Consultants Ltd.
Ottawa, Ontario, Canada

1983

ACADEMIC PRESS

A Subsidiary of Harcourt Brace Jovanovich, Publishers

Toronto   New York   London
Paris   San Diego   San Francisco   São Paulo   Sydney   Tokyo

ACADEMIC PRESS CANADA
55 Barber Greene Road, Don Mills, Ontario   M3C 2A1

United States Edition published by ACADEMIC PRESS, INC.
111 Fifth Avenue, New York, New York 10003

United Kingdom Edition published by ACADEMIC PRESS, INC. (LONDON) LTD.
24/28 Oval Road, London NW1   7DX

Library of Congress Cataloging in Publication Data
Main entry under title:

High and low erucic acid rapeseed oils.

   Includes bibliographies and index.
   1. Rape-oil. 2. Rape (Plant) I. Kramer
J. K. G. (John K. G.) II. Sauer, F. D.
(Frank D.) III. Pigden, W. J. (Wallace J.)
TP684.R3H53 1983   665'.35   82-13805
ISBN 0-12-425080-7

Canadian Cataloguing in Publication Data

Main entry under title:
High and low erucic acid rapeseed oils

Includes bibliographies and index.
ISBN 0-12-425080-7

1. Rape-oil. 2. Rape (Plant). I. Kramer, John
K. G. II. Sauer, Frank D. III. Pigden, W. J.
(Wallace James), Date

TP684.R3H53   665'.35   C83-098053-9

PRINTED IN THE UNITED STATES OF AMERICA

83 84 85 86   9 8 7 6 5 4 3 2 1

Contents

3   The History and Marketing of Rapeseed Oil in Canada

G. S. BOULTER

4   Chemical Composition of Rapeseed Oil

R. G. ACKMAN

Contents

5 Pathways of Fatty Acid Biosynthesis in Higher Plants with Particular Reference to Developing Rapeseed

P. K. STUMPF and M. R. POLLARD

6 The Development of Improved Rapeseed Cultivars

B. R. STEFANSSON

7 The Introduction of Low Erucic Acid Rapeseed Varieties into Canadian Production

J. K. DAUN

8   Rapeseed Crushing and Extraction

D. H. C. BEACH

9   The Commercial Processing of Low and High Erucic Acid Rapeseed Oils

B. F. TEASDALE and T. K. MAG

10   Current Consumption of Low Erucic Acid Rapeseed Oil by Canadians

M. VAISEY-GENSER

11   The Problems Associated with the Feeding of High Erucic Acid Rapeseed Oils and Some Fish Oils to Experimental Animals

F. D. SAUER and J. K. G. KRAMER

12 Cardiopathology Associated with the Feeding of Vegetable and Marine Oils

A. H. CORNER

13 The Composition of Diets Used in Rapeseed Oil Feeding Trials

E. R. FARNWORTH

14 The Metabolism of Docosenoic Acids in the Heart

F. D. SAUER and J. K. G. KRAMER

15   The Regulation of Long-Chain Fatty Acid Oxidation

S. V. PANDE

16   The Mechanisms of Fatty Acid Chain Elongation and Desaturation in Animals

H. SPRECHER

20   Studies with High and Low Erucic Acid Rapeseed Oil in Man

B. E. McDONALD

21   The Relevance to Humans of Myocardial Lesions Induced in Rats by Marine and Rapeseed Oils

H. C. GRICE and H. A. HEGGTVEIT

22 Some Recent Innovations in Canola Processing Technology

A. D. RODEN

Contributors

Numbers in parentheses indicate the pages on which the authors' contributions begin.

R. G. *Ackman* (85), Fisheries Research and Technology Laboratory, Technical University of Nova Scotia, Halifax, Nova Scotia, Canada B3J 2X4

D. H. C. *Beach* (181), Beach Doodchenko & Associates, Prince Albert, Saskatchewan, Canada S6V 5S9

G. S. *Boulter*[1] (61), Rapeseed Association of Canada, Vancouver, British Columbia, Canada V6N 2E2

A. H. *Corner* (293), Animal Diseases Research Institute, Agriculture Canada, Nepean, Ontario, Canada K2H 8P9

J. K. *Daun* (161), Grain Research Laboratory Division, Canadian Grain Commission, Agriculture Canada, Winnipeg, Manitoba, Canada R3C 3G8

R. K. *Downey* (1), Research Station, Research Branch, Agriculture Canada, Saskatoon, Saskatchewan, Canada S7N 0X2

E. R. *Farnworth* (315), Animal Research Centre, Research Branch, Agriculture Canada, Ottawa, Ontario, Canada K1A OC6

H. C. *Grice*[2] (551), F.D.C. Consultants Inc., Nepean, Ontario, Canada K2G 2X7

H. A. *Heggtveit* (551), Department of Pathology, McMaster University Medical Centre, Hamilton, Ontario, Canada L8N 3Z5

J. K. G. *Kramer* (253, 335, 413, 475), Animal Research Centre, Research Branch, Agriculture Canada, Ottawa, Ontario, Canada K1A OC6

T. K. *Mag* (197), Research Centre, Canada Packers Inc., Toronto, Ontario, Canada M6N 1K4

B. E. *McDonald* (535), Department of Foods and Nutrition, University of Manitoba, Winnipeg, Manitoba, Canada R3T 2N2

S. V. *Pande* (355), Laboratory of Intermediary Metabolism, Clinical Research Institute of Montreal, Montreal, Quebec, Canada H2W 1R7

W. J. *Pigden*[3] (21), F.A.W. Consultants Ltd., Ottawa, Ontario, Canada K2B 5P6

M. R. *Pollard*[4] (131), Department of Biochemistry and Biophysics, University of California, Davis, California 95616

[1]Present address: 4 Semana Crescent, Vancouver, British Columbia, Canada V6N 2E2.
[2]Present address: 71 Norice Drive, Nepean, Ontario, Canada K2G 2X7.
[3]Present address: 850 Norton Avenue, Ottawa, Ontario, Canada K2B 5P6.
[4]Present address: ARCO-Plant Cell Research Institute, 6560 Trinity Court, Dublin, California 94566.

A. D. Roden (563), Quality Control and Engineering, CSP Foods Ltd., Dundas, Ontario, Canada L9H 2E7

F. D. Sauer (253, 335, 413, 475), Animal Research Centre, Research Branch, Agriculture Canada, Ottawa, Ontario, Canada K1A OC6

H. Sprecher (385), Department of Physiological Chemistry, The Ohio State University, Columbus, Ohio 43210

B. R. Stefansson (143), Department of Plant Science, University of Manitoba, Winnipeg, Manitoba, Canada R3T 2N2

P. K. Stumpf (131), Department of Biochemistry and Biophysics, University of California, Davis, California 95616

B. F. Teasdale (197), Research Centre, Canada Packers Inc., Toronto, Ontario, Canada M6N 1K4

B. K. Thompson (515), Engineering and Statistical Research Institute, Research Branch, Agriculture Canada, Ottawa, Ontario, Canada K1A OC6

M. Vaisey-Genser (231), Department of Foods and Nutrition, University of Manitoba, Winnipeg, Manitoba, Canada R3T 2N2

Foreword

It is with great pleasure that I accept the invitation to write about the role of the Research Branch of Agriculture Canada in helping to establish rapeseed as a major crop and a valuable export commodity for Canada. Scientists from the Research Branch of Agriculture Canada were involved in the earliest stages of introducing rapeseed into Canada starting with the testing work done at the Dominion Forage Laboratories of the Experimental Farm Services at Saskatoon. From this modest beginning, a large multidisciplinary Research Branch Program was developed which included experts in the area of plant breeding, nutrition, biochemistry, physiology and industrial processing. Much of the work was carried out in close collaboration with scientists from industry, universities, and other government departments.

Through the efforts of the Research Branch of Agriculture Canada, rapeseed has been introduced into areas where other oilseed crops do not thrive, as for example, the regions north of the 52nd parallel in Canada's Western Provinces. The combined effort of these scientists has helped to propel rapeseed oil to fourth place in the world's production of edible vegetable oils and to firmly establish it in global commerce.

Since the introduction of rapeseed into Canada in the 1940s, a new industry has been developed in Western Canada which in 1981 attained the status of a billion dollar industry, second only to wheat in importance. The development of high quality oil and meal from rapeseed must be attributed to a large degree to the research and development effort devoted to this crop by the scientific community and which has paid off handsomely.

At this time it is my pleasure to thank the many individuals and institutions who have made contributions to the achievement of these important goals. Lastly I wish to acknowledge the outstanding contribution and key leadership of the three editors, Drs. Kramer, Sauer, and Pigden for drawing together and making available to the public the fascinating developments and evolution of rapeseed into a vital and important modern crop.

E. J. LeRoux
ASSISTANT DEPUTY MINISTER
RESEARCH BRANCH
AGRICULTURE CANADA
OTTAWA, ONTARIO, CANADA

Preface

Rapeseed oil, although used since antiquity, is a relatively new food product. In the Western World it was almost unknown before World War II. Thus, it is not surprising that this oil underwent approximately two decades of intensive testing, particularly since it contained new and unfamiliar compounds peculiar to the *Brassica* family. In particular there was the long chain monounsaturated fatty acid, i.e., erucic acid, plus some sulfur-containing compounds, i.e., the glucosinolates which have antithyroid activity. The presence of glucosinolates in the oil were never of great concern since they are water soluble and stay in the meal. Erucic acid was a different story. From the beginning it was clear that this long chain monoenoic fatty acid was poorly metabolized, at least by the rat, and that in this species it caused a host of problems which ranged from fat accumulation in heart muscle to fatty deposits in the adrenal gland and ovarian tissue. As more and more studies were initiated, it became clear that there were significant species differences in erucic acid metabolism and that not all species were as poorly equipped as the rat to metabolize this acid. Nevertheless, responding to demands from health regulatory agencies, the industry began changing to new rapeseed cultivars over a decade ago and today the older high erucic acid cultivars have been phased out and have been replaced by new rapeseed plants, the seed of which is almost devoid of both glucosinolates and erucic acid. These cultivars are called "canola" to distinguish them from the older rapeseed varieties.

It should be mentioned that the heavy investment in research on rapeseed not only yielded valuable information about the crop itself but also produced a body of scientific information which greatly helped in the understanding of fat metabolism and interspecies differences in metabolic pathways. The role of the chain shortening process and peroxisomal oxidation was made clearer by studies on docosenoic acid metabolism. Valuable information was obtained from experiments which dealt with the effects of erucic acid on adrenal gland activity and prostaglandin biosynthesis. Some excellent research was done also on the effect of erucic acid on cardiac mitochondrial respiratory activity. Moreover, some interesting theories were formulated as to the role of saturated and unsaturated long chain fatty acids on the development of heart lesions in rats fed high fat diets. These lesions that developed in rat hearts were a cause of considerable concern until recent studies revealed that larger mammalian species such as swine and monkeys were relatively immune to these lesions and, that when these lesions were present, they were not diet or fat related.

This volume covers a wide range of subjects related to rapeseed, i.e., from plant breeding to industrial processing to nutrition and biochemistry. The editors felt that this was appropriate since for many people rapeseed oil is a totally new food and that therefore all aspects of its production and use are of interest. Furthermore, rapeseed oil, or more appropriately, canola oil, as all vegetable oils, has its own unique fatty acid composition and its extraction, processing, and refining techniques are not exactly like those of other vegetable oils.

The editors would like to express sincere thanks to the many contributors who made this book possible and who so willingly shared their expert knowledge with the rest of us.

John K. G. Kramer
Frank D. Sauer
Wallace J. Pigden

1

The Origin and Description of the *Brassica* Oilseed Crops

R. K. DOWNEY

I. INTRODUCTION

Brassica oilseed crops annually occupy over 11 million hectares of the world's agricultural lands and provide over 8% of the world's edible vegetable oil. Because of their ability to survive and grow at relatively low temper-

High and Low Erucic Acid Rapeseed Oils
Copyright © 1983 by Academic Press Canada
All rights of reproduction in any form reserved.
ISBN 0-12-425080-7

atures, they are one of very few edible oil sources that can be successfully produced in the extremes of the temperate regions. This characteristic also makes them well adapted to cultivation at high elevations and as winter crops in the subtropics. In general, far less heat units are required for growth and development of the oilseed *Brassicas* than for soybean and sunflower production.

In temperate regions, oilseed rape (*Brassica napus* L.) and turnip rape (*Brassica campestris* L.) predominate, while in the semitropics of Asia *B. campestris* and Indian mustard [*Brassica juncea* (L.) Czern.] are major vegetable oil sources. The English word rape, as it applies to the oilseed forms of *B. campestris* and *B. napus*, has arisen from the Latin word rapum, meaning turnip. The word mustard was derived from the European practice of mixing the sweet "must" of old wine with crushed seeds of black mustard [*Brassica nigra* (L.) Koch.] to form a hot paste, "hot must" or "mustum ardens", hence the modern term mustard (Hemingway, 1976).

The small round seeds of these crops contain over 40% oil and upon oil extraction yield a meal, on a dry matter basis, with over 40% high quality protein. In many Asian countries the meal is prized as an organic fertilizer, but in the Western World it is used exclusively as a high quality protein feed supplement for livestock and poultry.

The consuming public is frequently unaware of the dietetic importance of *Brassica* oilseed crops since their oil and meal products are usually processed and blended as they enter the food chain. On the other hand, almost every consumer is familiar with their close relatives, the cole vegetables such as cabbage, cauliflower, broccoli, and Brussels sprouts, the condiment mustards, and the root crops of turnips, rutabagas and radishes. Many farmers directly feed forage rapes and kales while others attempt to rid their fields of weedy species such as black and wild mustard.

II. DOMESTICATION AND SPECIES RELATIONSHIPS

Domestication of *Brassica* vegetables and oilseeds undoubtedly occurred whenever and wherever the economic value of the locally adapted weed was recognized. Indeed, they were probably among the earliest plants domesticated by man since there is evidence that some vegetable types were in wide-scale use in the neolithic age (Chang, 1968; Hyams, 1971), and the ancient Indian Sanskrit writings of 2000–1500 B.C. make direct references to the oilseed rapes and mustard (Singh, 1958). Similarly ancient Greek, Roman, and Chinese writings of the period 500–200 B.C. specifically mention these crops and describe their medicinal values (Prakash and Hinata, 1980).

In the domestication of plants of the *Brassica* genus man has utilized and modified, through selection, almost every plant part including roots, stems, leaves, terminal and axillary buds, and seeds. Since early botanists attempt-

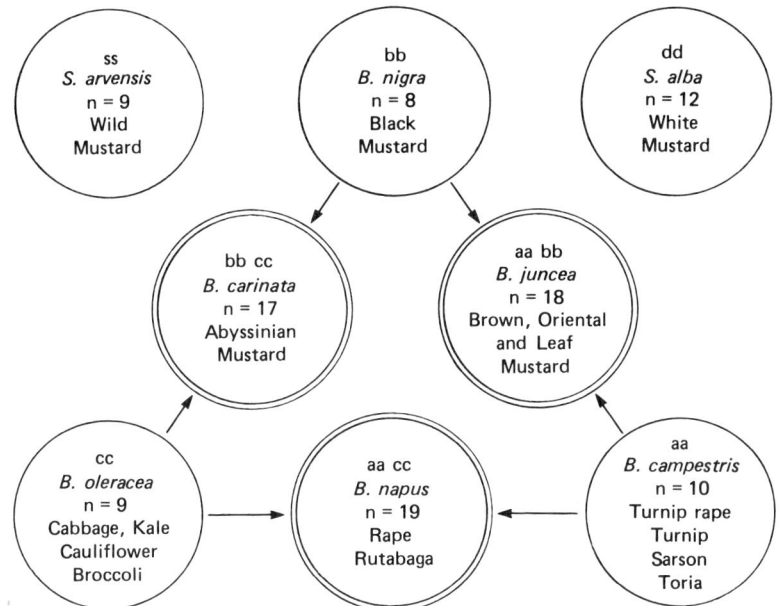

Fig. 1.   Genome and chromosome relationships of some economically important *Brassica* and *Sinapis* species. After U (1935).

ed to separate plant groups on the basis of morphological characteristics, the occurrence of similar or identical plant forms in more than one *Brassica* species resulted in much confusion and misclassification. However, some excellent cytogenetic studies by Morinaga and his students in the early 1930s established the true relationship between the two rape species and their close relatives (see Prakash and Hinata, 1980, for references). By making interspecific crosses and cytologically analyzing chromosome conjugation in the progeny, these researchers demonstrated that the three species with the higher chromosome number, *B. napus*, *B. juncea*, and *B. carinata*, were amphidiploids derived from the monogenomic or basic species *B. nigra*, *B. campestris*, and *B. oleracea* (Fig. 1). The accuracy of this scheme was subsequently corroborated by the synthesis of the existing amphidiploid species.

In addition, there is cytological evidence that the three basic species are themselves secondary polyploids, probably originating from a common ancestor with a basic chromosome number of 5 or 6. It has also been suggested that the genus *Sinapis*, which includes the mustard species *S. alba*, our source of hot dog or cream salad mustard, the radish genus *Raphanus* and the genus *Eruca* may also have evolved from this same progenitor (for references, see Downey, 1966; Prakash and Hinata, 1980).

III. ORIGIN AND DISTRIBUTION

A. *Brassica campestris*

As with many other plant groups of the temperate regions, the tribe *Brassicae* appears to have originated near the Himalayan region (Hedge, 1976). Of the *Brassica* crops, *B. campestris* appears to have had the widest distribution, with secondary centers of diversity in Europe, Western Russia, Central Asia, and the Near East (Sinskaia, 1928; Vavilov, 1949; Mizushima and Tsunoda, 1967; Zhukovsky and Zeven, 1975). The oilseed form, as distinct from the leafy vegetables, appears to have evolved separately in the European-Mediterranean area, with a second center of origin in Asia. The Asian types known as brown and yellow sarson and toria are distinct from the European oleiferous types, not only in their general morphology but also in the chemical makeup of their seeds, tending to have a distinct pattern of fatty acid and glucosinolate composition. Protein analyses also support the separate European and Asian origins (Denford, 1975), although it is clear from chemotaxonomy that the Indian subcontinent forms of sarson and toria are conspecific with European turnip rape as well as tame and wild turnips (Vaughan *et al.*, 1976).

There is evidence to suggest that at least 2000 years ago *B. campestris* was distributed from the Atlantic islands in the west to the eastern shores of China and Korea, and from northernmost Norway, south to the Sahara, and on into India (Sinskaia, 1928). It should be noted, however, that none of the *Brassicas* was native to the Americas.

B. *Brassica napus*

In contrast, development of the oilseed and root forms of *B. napus* appears to be relatively recent. The Greeks and Romans knew of the *B. napus* swede or rutabaga root crops; however, reference to these forms does not appear in the ancient literature, and wild populations of *B. napus* have not been found (Prakash and Hinata, 1980). Because the species is the result of an interspecific cross between *B. campestris* and *B. oleracea*, it could only arise where the two parental species were growing in close proximity. Since the distribution of wild *B. oleracea* was confined to the Mediterranean area, it is generally agreed that *B. napus* originated in Southern Europe.

It is only in relatively recent times that *B. napus* forms have been introduced to Japan, China, and the Indian subcontinent. In the Far East the *B. napus* form has been more productive than indigenous oilseed forms of *B. campestris*. Today, most of the rapeseed produced in China, Korea and Japan is harvested from *B. napus* type plants that have been bred from interspecific crosses between introduced *B. napus* and older indigenous *B. cam-*

pestris varieties (Shiga, 1970). On the Indian subcontinent introduced *B. napus* material has not been nearly as successful, and Asian forms of oilseed *B. campestris* and *B. juncea* still predominate.

C. *Brassica juncea*

The oilseed and condiment forms of *B. juncea* have enjoyed almost as wide a geographical distribution as *B. campestris*. However, there is some uncertainty about their origin. After reviewing the evidence, Prakash and Hinata (1980) concluded that *B. juncea* first evolved in the Middle East where the putative parent species, *B. campestris* and black mustard (*B. nigra*), would have originally come together. However, since black mustard was a valuable spice of very early times, it soon spread over Europe, Africa, Asia, India, and the Far East (Hemingway, 1976). Thus, *B. juncea* may have arisen more than once with different progenitors and in different localities, accounting for centers of diversity in China, Eastern India, and the Caucasus (Hemingway, 1976; Prakash and Hinata, 1980).

IV. CROP FORMS AND CULTIVATION

In Europe and the Americas the crop is normally sown in drill rows at a seeding rate of 5–8 kg/ha. However, in India, Pakistan, and other countries of the Indian subcontinent rape and mustard seed are usually broadcast on the soil surface and then buried by drawing a heavy plank over the fields. In such circumstances, unless the crop is to be irrigated, it is usually sown mixed with cereal grains to reduce the risk of a complete crop failure. In China and Japan the traditional planting method has been to sow beds containing a dense stand of seedlings and transplant the seedlings into the field immediately following the rice harvest.

The small seed of these oilseed crops must be sown shallow, not more than 2–3 cm, into firm, moist soil to ensure a uniform germination and stand. Under good growing conditions emergence of the two cotyledons or seed leaves usually occurs within 4 or 5 days, followed quickly by the development of the first true leaves. The crop quickly establishes a rosette of broad leaves which tend to shade the surrounding area and reduce weed competition.

Subsequent growth patterns differ, depending on the climate of the production zone and on the form and species being grown. In Northern European and some South American countries, notably Chile, the winter or biennial form of *B. napus* dominates the production area. This form remains in the rosette or vegetative stage until it undergoes a long (40 days) vernalization period at temperatures near freezing (Andersson and Olsson, 1961). Thus, in

Europe and the Far East, *B. napus*, normally sown in August–September, remains in the rosette stage during the winter, flowers toward the end of May, and is harvested in July.

The biennial form of *B. campestris* is also grown and although it is more winter hardy and requires a shorter growing season than *B. napus*, the yield of seed and oil is less under favorable growing conditions. Consequently, the area of winter *B. campestris* production is limited to the more rigorous climates of central Sweden and Finland (Lööf, 1972). In general, the winter forms of both species are less winter hardy than winter barley; thus, their distribution is restricted to maritime climates in the temperate zones.

The biennial form of *B. juncea* does not appear to have evolved in nature although such types have apparently been synthesized (Voskresenskaya and Shpota, 1973).

The summer form of *B. napus* is sown in Europe and Canada in April and May, respectively. The crop flowers in June and July and is harvested in September, resulting in a growing season of about 160 days in Europe and 105 days in Canada. In Europe, summer rape is grown to a very limited extent, usually to reseed winter-damaged rape fields. In Canada, however, this form and species makes up about 55% of the Canadian production, with the percentage of acres sown to *B. napus* gradually increasing as earlier maturing varieties become available (see Chapter 6).

Summer turnip rape, *B. campestris*, is an important crop in northern Sweden, Finland, and Canada where the short growing season requires a crop that can withstand late spring frosts and mature before fall frosts occur. Normally, the crop is sown in May and harvested in August, requiring about 95 days from seeding to maturity. Although having a lower potential seed and oil yield than *B. napus* varieties, the short growing season requirements of summer turnip rape provide growers with a low risk, high return crop for the most northerly areas.

On the Indian subcontinent, *B. juncea* and the sarson form of *B. campestris* are usually sown in October or November and harvested in March or April. However, the toria form can be sown in September and harvested in December.

In Sweden, where the summer and winter forms of both species are grown, average 1961–1968 yields for the winter forms of *B. napus* and *B. campestris* were 2700 and 1900 kg/ha, respectively, while the respective yield of the summer forms of the two species were 1700 and 1300 kg/ha (Loof, 1972). In Canada, yields of summer *B. napus* and *B. campestris* are lower, approximately 1200 and 900 kg/ha, respectively. *Brassica juncea* in Canada normally yields about 10% above *B. napus*. In Pakistan and India, average yields of rape and mustard are about 400–600 kg/ha. However, under irrigation *B. juncea* yields of 2000–2500 kg/ha are common (Prakash, 1980).

V. PLANT AND SEED DEVELOPMENT

Depending on form, species, and climate, the plant may remain in the rosette stage for as short a period as 30 or as long as 210 days. However, once the day length and temperature trigger floral initiation, the plant bolts rapidly. The development stages of the summer turnip rape have been documented by Harper and Berkenkamp (1975) (Fig. 2). Summer rape follows a similar pattern (Fig. 3).

The inflorescence is racemose with no terminal flower (Figs. 4-1 and 4-2). Flowering begins at the lowest bud on the main raceme and continues upward with three to five or more flowers opening per day. In *B. napus* and *B. campestris,* flowering at base of secondary racemes is initiated about 3 days after floral initiation on the main raceme (McGregor, 1981). In *B. campestris,* however, apical dominance is not nearly so pronounced, making the identification of a primary raceme difficult and leading to a less structured appearance in the mature plant.

The flower is radial with four erect, prominent sepals, and four petals which alternate with the sepals in the form of a cross (Fig. 4-3). Note the receptive surface of the stigma centered within the four inner stamens, the two outer stamens, and the four petals forming a cross from which the Cruciferae family derives its name. The petal color is normally pale yellow, but several shades of yellow have been identified and numerous genes have been reported to affect flower color (Morice, 1960; Alam and Aziz, 1954). There are six stamens, the two outer stamens being distinctly shorter than the inner four which surround the stigma. There are four nectaries spaced equidistant and between the two whorls of stamens. Two of the nectaries are at the base of the two outer stamens.

Flowers of *B. campestris* are normally smaller and darker yellow than those of *B. napus*. The two species are more clearly distinguished at flowering by examining the position of the buds to the open flowers which surround them (Figs. 4-1 and 4-2). In *B. napus* the buds are normally borne above the open flowers while in *B. campestris* and *B. juncea* the buds are held below the uppermost open flowers. The shape of the leaves on the flowering stock can also be used to distinguish the two rape species, *B. juncea* and other mustard species (Fig. 5). In *B. campestris* the leaf blade clasps the stem completely, while in *B. napus* the leaf only partially clasps the stem. In *B. juncea* the leaf blade does not reach the stem and terminates well up the petiole.

Following pollination by wind or insect, the petals are shed and the pistil elongates to form a pod (silique) with two carpels separated by a false septum. A single row of seeds develops within each of the two loculi (Fig. 6). The number of seeds per pod varies with the species, form, and environment, but normally a pod contains between 15 and 40 seeds. At maturity the

Fig. 5. Distinguishing characteristics of rapeseed and mustard leaves. (a) The blade of the upper leaves of *B. campestris* fully clasp the stem; (b) *B. napus* leaves partially clasp the stem; while (c) the leaf blade of *B. juncea* terminates well up the petiole.

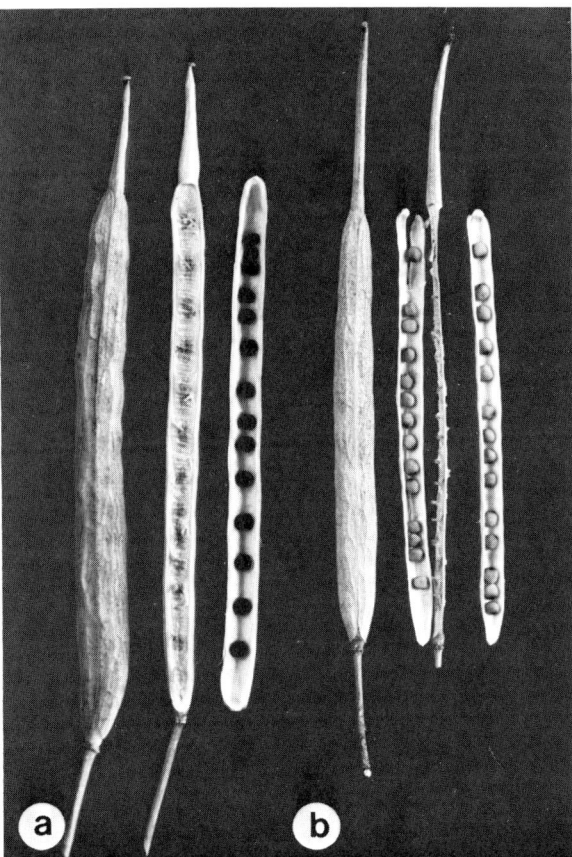

Fig. 6.   Typical pods of (a) *B. napus* showing an intact and opened pod with the seeds of the upper locule exposed, while those of the lower locule are obscured by the central lamella; (b) an intact and opened pod of *B. campestris*.

pod is easily split along the false septum, and the seed may be shattered and lost. Of the three oilseed brassicas, *B. juncea* is the most resistant to shattering, and *B. napus* the most susceptible.

Rape and mustard seeds are predominantly embryo tissue as opposed to cereal grains which are largely endosperm. The seed coat consists of an outer epidermis, a palisade layer of thick-walled columnar-shaped cells, and a layer of crushed parenchyma (Fig. 7). The endosperm is made up of a single row of aleurone cells with a thin layer of crushed parenchyma separating it from the embryo. The developing seed is uniformly green throughout. However, as the seed matures pigments begin to accumulate in the palisade cells, and the seed coat gradually changes from dark green to

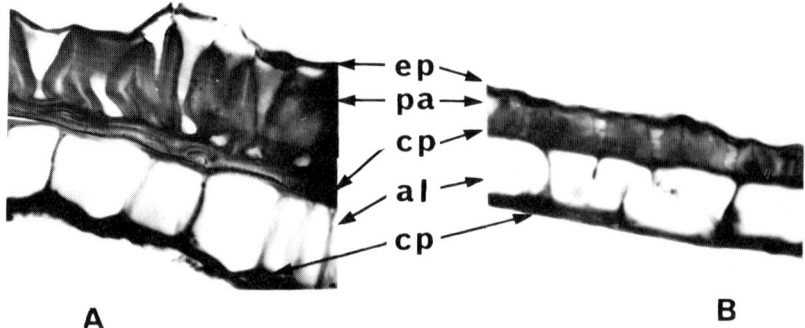

A   **B**

Fig. 7. Seed coats of rapeseed. (A) Brown seed; (B) yellow seed. ×375. ep, epidermis; pa, palisade; cp, crushed parenchyma; al, aleurone (endosperm). From Stringam *et al.*, 1974.

black, brown, reddish-brown, or yellow, depending on the genetic makeup of the mother plant. In some species the palisade cells tend to vary in height, resulting in characteristic reticulations on the seed surface (Vaughan, 1970). Undamaged embryos also lose their green color and are a uniform bright yellow at maturity.

In *B. napus* all present varieties are black seeded while in *B. juncea* both brown and pure yellow seeded varieties are grown. In *B. campestris* the majority of present varieties are dark to reddish-brown, but pure yellow varieties of yellow sarson and yellow-brown Canadian varieties, Candle and Tobin, are in commercial production (Fig. 4-4).

Yellow coated seeds have recently been recognized as having several advantages over brown hulled forms. Yellow seeded strains of all the *Brassica* oilseeds contain a higher oil and protein content and a lower fiber level than brown seeded material with the same genetic background (Stringam *et al.*, 1974; Jonsson and Bengtsson, 1970). All these changes are economically desirable, but, in addition, the oil from yellow seed is generally lighter in color and the yellow seed coat blends with other feedstuffs so that the feed

TABLE I

Normal Ranges in Seed Size of *Brassica* Oilseeds by Species and Form

Species and form	g/1000 seeds
B. napus winter	4.5–5.5
B. campestris winter	3.0–4.0
B. napus summer	3.5–4.5
B. campestris summer	2.0–3.0
B. juncea summer	2.8–3.5
B. campestris sarson	4.0–4.5

TABLE II

Effect of Seed Color and Size on Oil, Protein, and Fiber Levels in *B. campestris*

	Seed		Seed oil (%)[b]	Percentage in meal of	
Variety	Size[a]	Color		Protein[c]	Fiber[c]
Torch	2.5	Brown	39.9	41.6	14.2
Candle	2.5	Yellow brown	42.2	42.2	10.9
Yellow sarson	4.4	Pure yellow	44.5	47.0	8.2

[a] g/1000 seeds.
[b] Moisture-free basis.
[c] Percentage in oil- and moisture-free meal.

manufacturer has greater freedom to modify his formulas without visually altering the appearance of the finished feed. The increased oil and protein percentage and the reduced fiber content of yellow seed occur because yellow seed coats are significantly thinner than brown seed coats (Fig. 7). As a result, the yellow seed contains a higher proportion of oil and protein-rich embryo and a lower proportion of fiber-rich hull.

Seed size is also an important consideration in reducing hull percentage. In general, the winter forms produce larger seeds than those found in the summer types, while seed of *B. napus* varieties is normally larger than that of *B. campestris* (Table I). The *B. campestris* Indian sarson types, which have a larger seed than many *B. napus* varieties, are the exception to the rule. By combining the large seed and yellow seed coat characteristics very significant improvement is possible in oil, protein, and fiber levels (Table II).

A very large proportion of the embryo consists of an inner and a larger outer cotyledon arranged in a conduplicate fashion. The cotyledons are attached to the short hypocotyl immediately below the epicotyl or growing tip from which will emerge the first true leaf and meristem. At the opposite end of the hypocotyl is the radicle or root. The position of the radicle within the seed can usually be observed as a distinct ridge on the surface of the seed.

VI. RAPESEED OIL

A. Oil Content

The oil contained in the seed is its most valuable component. Normally a kilogram of oil is worth about twice that of a kilogram of the high protein meal. About 80% of the seed oil is concentrated in lipid droplets in the cells of the cotyledons. The oil levels of the hypocotyl and root are lower while the seed coat and the adhering endosperm layer contain only 7–12% of the total seed oil (Stringam et al., 1974).

In Sweden and countries of the European Economic Community, rapeseed is purchased from producers on an oil percentage basis. However, the usual practice in other countries is to purchase the seed on the basis of seed weight delivered. Oil content is influenced by many factors, including temperature and moisture during seed development, nitrogen fertilization, and the crop species and form being grown. In general, cool, moist growing conditions favor high oil contents, while increasing rates of nitrogen fertilization usually reduce oil percentage but increase oil yield per acre. Normally the winter forms and the larger seeded B. napus species yield the highest oil content seed.

B. Oil Composition

The makeup of a vegetable oil is at least of equal importance to the quantity of oil that can be extracted since it is the fatty acid composition of an oil that determines its value to the processor. Initially, rapeseed oil was used primarily for industrial purposes such as lamp oil and later as a steam engine lubricant. It was well known that rapeseed oil and other Brassica seed oils differed from other vegetable oils in containing significant quantities of the long chain 20- and 22-carbon monoenoic acids, eicosenoic and erucic. Indeed, the presence of erucic acid gives rapeseed oil a superior ability to cling to steam- and water-washed metal surfaces. This fact led to the beginning of the rapeseed growing and processing industry in Canada during World War II when the allied navy's supply of rapeseed lubricating oil was cut off. Although these and other early industrial uses are no longer significant markets, the special properties of high erucic acid rapeseed and mustard oils now have other important industrial applications (see Chapter 9).

The other fatty acids of rapeseed oil are also contained in most other edible vegetable oils in greater or lesser amounts (Table III). It should be noted, however, that the levels of the saturated fatty acids, palmitic and stearic, are the lowest among the major edible oils and that the level of linolenic acid is comparable to that found in soybean oil. This latter fact tends to make these two oils competitors in the marketplace.

The fatty acid composition of rapeseed oil is largely determined by the genetic makeup of the developing embryo rather than the maternal parent (Downey and Harvey, 1963; Thomas and Kondra, 1973). However, the levels of the polyunsaturated fatty acids, linoleic and linolenic, are strongly influenced by the environment during oil deposition and seed maturation. As in other vegetable oils, higher temperatures during oil deposition tend to reduce the level of polyunsaturated fatty acids. Such large genotype X environmental interactions have hampered breeding efforts to reduce the level

TABLE III
Fatty Acid Composition of High and Low Erucic Acid Oilseed *Brassica* Crops and Common Edible Vegetable Oils

Species, crop or variety	Ref.[a]	Fatty acid composition in percent[b]												
		14:0	16:0	16:1	18:0	18:1	18:2	18:3	20:0	20:1	22:0	22:1	24:0	24:1
B. napus rape														
Victor winter	1	0.0	3.0	0.3	0.8	9.9	13.5	9.8	0.6	6.8	0.7	53.6	0.0	1.0
Jet Neuf winter	2	0.0	4.9	0.4	1.4	56.4	24.2	10.5	0.7	1.2	0.3	0.0	0.0	0.0
Target summer	2	0.0	3.0	0.3	1.5	20.9	13.9	9.1	0.5	12.2	0.3	38.3	0.0	0.0
Tower summer	2	0.0	3.9	0.3	1.1	59.7	23.3	8.6	0.8	1.8	0.2	0.3	0.0	0.0
B. campestris turnip rape														
Duro winter	1	0.0	2.0	0.2	1.0	12.9	13.4	9.1	0.7	9.6	0.2	49.8	0.0	1.1
Yellow sarson	2	0.0	1.8	0.2	0.9	13.1	12.0	8.2	0.9	6.2	0.0	55.5	0.0	1.2
Echo summer	2	0.0	2.5	0.2	1.0	32.5	18.8	8.9	0.6	12.0	0.0	23.5	0.0	0.0
Tobin summer	2	0.0	3.8	0.1	1.2	58.6	24.0	10.3	0.6	1.0	0.1	0.3	0.0	0.0
B. juncea mustard														
Indian origin	3	0.0	2.5	0.3	1.2	8.0	16.4	11.4	1.2	6.4	1.2	46.2	0.7	1.9
Leth 22A	2	0.0	2.8	0.3	1.0	20.9	22.4	15.6	0.8	13.4	0.0	22.8	0.0	0.0
Zem. 1	2	tr[c]	3.6	0.4	2.0	45.0	33.9	11.8	0.7	1.5	0.3	0.1	0.2	0.5
Glycine max soybean														
Group 1 var.	4	0.0	15.3	0.0	4.2	23.6	48.2	8.7	0.0	0.0	0.0	0.0	0.0	0.0
Helianthus annuus sunflower														
Peredovik	5	0.1	5.8	0.1	5.2	16.0	71.5	0.2	0.2	0.1	0.7	0.0	0.1	0.0
Carthamus tinctorius safflower														
US-10	6	0.0	7.6	0.0	2.0	10.8	79.6	0.0	0.0	0.0	0.0	0.0	0.0	0.0
Zea maus corn														
U.S. sources	7	tr	11.5	0.0	2.2	26.6	58.7	0.8	0.2	0.0	0.0	0.0	0.0	0.0
Gossupium hirsutum cotton	8	1.0	23.4	0.8	2.5	17.9	54.2	0.0	0.0	0.0	0.0	0.0	0.1	0.0
Archis hyogaea peanut														
Virginia Bunch	9	9.2	0.0	0.0	3.1	57.2	23.4	0.0	1.4	1.4	2.6	0.0	1.8	0.0
Cook Jumbo	9	6.7	0.0	0.0	4.3	71.4	11.1	0.0	1.6	1.0	2.7	0.0	1.3	0.0

[a] References: 1, Appelqvist (1969); 2, Downey (unpublished data); 3, Appelqvist (1970); 4, Hymowitz et al. (1972); Collins and Sedgwick (1959); 5, Earle et al. (1968); 6, Knowles (1968); 7, Beadle et al. (1965); 8, Anderson and Worthington (1971); 9, Worthington and Hammons (1971).

[b] Fatty acids represented by the number of carbon atoms and double bonds.

[c] tr = Trace amounts.

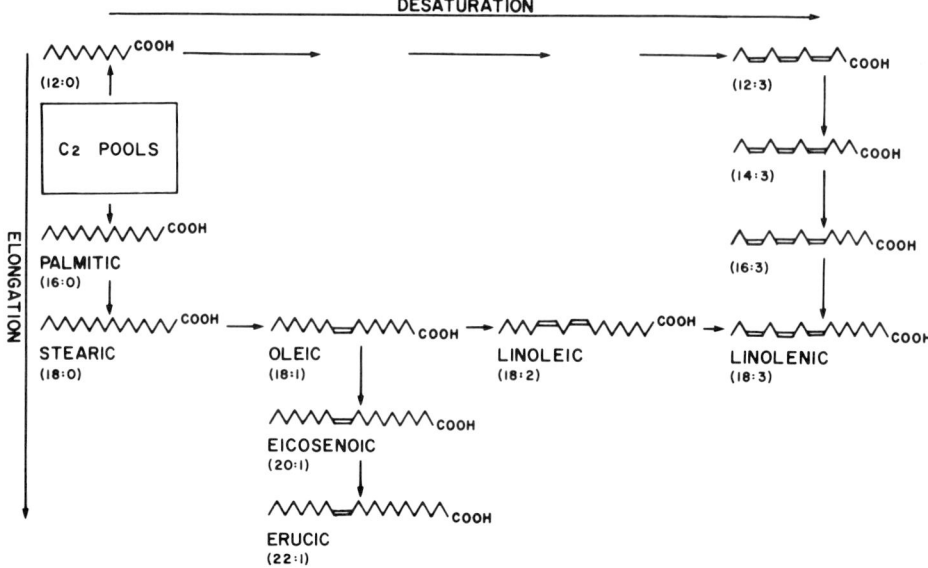

Fig. 8.   Suggested biosynthetic pathway of major rapeseed fatty acids.

of linolenic acid and increase linoleic acid in rapeseed varieties (see Chapter 6).

The generally accepted biosynthetic pathway for rapeseed fatty acids is given in Fig. 8. Erucic acid biosynthesis has been shown to result from the addition of a two-carbon fragment to the carboxyl end of oleic acid to form eicosenoic, followed by a second two carbon addition to form erucic acid (Downey and Craig, 1964; Jonsson, 1977). In the breeding of low erucic acid plants of *B. napus* (Stefansson *et al.*, 1961), *B. campestris* (Downey, 1964), and *B. juncea* (Kirk and Oram, 1981) the embryo's genetic ability for carbon chain elongation of the fatty acids has been blocked, resulting in an accumulation of the precursor, oleic acid (Table III). Physical limitations of positioning the long chain fatty acids on the glycerol molecule tend to prevent erucic acid levels of over 65% from being realized (Tallent, 1972; Calhoun *et al.*, 1975). A series of alleles have been identified in *B. napus* and *B. campestris* which now makes it possible to breed strains containing almost any level of erucic acid from less than 1 to over 60% of the total fatty acids (Krzymanski and Downey, 1969). However, for nutritional and industrial reasons only very low or very high erucic acid oils are likely to find a place in today's competitive markets.

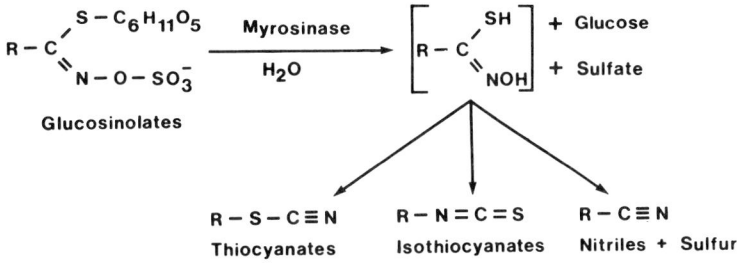

Fig. 9. General structure of glucosinolates and enzymatic hydrolysis products.

VII. RAPESEED MEAL

Rapeseed meal is used exclusively in Canada and Europe as a high protein feed supplement for livestock and poultry. However, until recently the value and marketability of this major oilseed processing by-product has been limited by the presence of glucosinolates in the seed. Glucosinolates, of which some 90 are known, are characteristic of the Cruciferae family (Fenwick *et al.*, 1982). These compounds, when hydrolyzed by the myrosinase enzyme present in *Brassica* seed and vegetative tissues, yield isothiocyanates, oxazolidinethiones, thiocyanates, or nitriles (Fig. 9). The characteristic odor and flavor of *Brassica* vegetables and condiments are largely due to the kind and amounts of these compounds present in their tissues. However, these compounds are also active goitrogens and so limit the amount of rapeseed or mustard meal that can be fed to nonruminant animals, such as swine and poultry, if reduced feed efficiency and weight gains as well as reproductive difficulties are to be avoided (Bell, 1977). To prevent such adverse effects and to improve meal palatability, the myrosinase enzyme is normally heat inactivated as the first step in the oil extraction process (see Chapter 8). However, the ultimate solution has been to breed varieties of rapeseed which have very low levels of glucosinolates combined with low erucic acid oil and high seed, oil, and protein yield. This has now been accomplished in Canadian summer *B. napus* and *B. campestris*.

VIII. CHANGING QUALITY

The development of low erucic acid rapeseed oil created an entirely new natural vegetable oil with physical and nutritional properties distinct from all previously known vegetable oils. To distinguish this new oil in the scientific and commercial literature, terms such as LEAR (*low erucic acid rape-seed oil*), and Canbra (*Canadian Brassica*) were coined in Canada to identify rapeseed oil containing less than 5% erucic acid. In West Germany the term "Sinola" is used to indicate certified seed and oil containing less than 2% erucic acid. Although Sinola is still used in West Germany, the other terms have been superseded in Canada by the name "canola." This term has now been registered and adopted to describe low erucic, low glucosinolate seed and its products. Canola seed contains less than 5% erucic acid as a percentage of the total fatty acids in the oil, and less than 30 μmoles of glucosinolates in the oil-free meal. It is expected that the name and definition of canola will be accepted worldwide since it is presently registered in Canada and Australia, and accepted in Japan and several other consuming countries.

The changeover to low erucic acid oil followed by the introduction of low glucosinolate seed has had a tremendous positive effect on the Canadian rapeseed-processing industry. Eventually this impact will be felt worldwide as Canadian exports become exclusively canola and as other producing countries develop varieties of similar quality. Europe already grows low erucic acid varieties almost exclusively, and high yielding, hardy canola varieties should soon be available from European breeding programs (Röbbelen and Thies, 1980). Australia already has some canola varieties while Pakistan, India, and China are incorporating these qualities into adapted varieties. Unfortunately, no low glucosinolate plants of *B. juncea* have yet been identified.

ACKNOWLEDGMENTS

The suggestions and assistance of my colleagues Drs. Klassen, McGregor, and Woods are gratefully acknowledged, and special thanks are extended to Mr. R. E. Underwood, Saskatoon Research Station photographer, for the photographic work in this chapter.

REFERENCES

Alam, Z., and Aziz, M. H. (1954). *Pak. J. Sci. Res.* **6**, 27–36.
Anderson, O. E., and Worthington, R. E. (1971). *Agron. J.* **63**, 566–569.
Andersson, G., and Olsson, G. (1961). *In* "Manual of Plant Breeding" (H. Kappert and W. Rudorf, eds.), Vol. 5, pp. 1–66. Parey, Berlin.
Appelqvist, L.-A. (1969). *Hereditas* **61**, 9–44.
Appelqvist, L.-A. (1970). *Fette, Seifen, Anstrichm.* **72**, 783–792.
Beadle, J. B., Just, D. E., Morgan, R. E., and Reiners, R. A. (1965). *J. Am. Oil Chem. Soc.* **42**, 90–95.

Bell, J. M. (1977). _Proc. Symp. Rapeseed Oil, Meal, By-Prod. Util., 1977_, Rapeseed Assoc. Canada, Publ. No. 45, pp. 137–194.

Calhoun, W., Crane, J. M., and Stamp, D. L. (1975). _J. Am. Oil Chem. Soc._ **52**, 363–365.

Chang, K. (1968). _Science_ **162**, 519–526.

Collins, F. I., and Sedgwick, V. E. (1959). _J. Am. Oil Chem. Soc._ **36**, 641–644.

Denford, K. E. (1975). _Bot. Notiser_ **128**, 455–462.

Downey, R. K. (1964). _Can. J. Plant Sci._ **44**, 295.

Downey, R. K. (1966). _Can. Dep. Agric. Publ._ **1257**, 7–23.

Downey, R. K., and Craig, B. M. (1964). _J. Am. Oil Chem. Soc._ **41**, 475–478.

Downey, R. K., and Harvey, B. L. (1963). _Can. J. Plant Sci._ **43**, 271–275.

Earle, F. R., Van Etten, C. H., Clark, T. F., and Wolff, I.A. (1968). _J. Am. Oil Chem. Soc._ **45**, 876–879.

Fenwick, G. R., Heaney, R. K., and Mullin, W. J. (1982). _CRC Crit. Rev._ (in press).

Harper, F. R., and Berkenkamp, B. (1975). _Can. J. Plant Sci._ **55**, 657–658.

Hedge, I. C. (1976). _In_ "The Biology and Chemistry of the Cruciferae" (J. G. Vaughan, A. J. MacLeod, and B. M. G. Jones, eds.), pp. 1–35. Academic Press, New York.

Hemingway, J. S. (1976). _In_ "Evolution of Crop Plants" (N. W. Simmonds, ed.), pp. 19–21. Longmans, Green, New York.

Hyams, E. (1971). "Plants in the Service of Man," pp. 33–61. J. M. Dent & Sons, London.

Hymowitz, T., Palmer, R. G., and Hadley, H. H. (1972). _Trop. Agric. (Trinidad)_ **49**, 245–250.

Jonsson, R. (1977). _Hereditas_ **87**, 207–218.

Jonsson, R., and Bengtsson, L. (1970). _Sver. Utsaedesfoeren. Tidskr._ **80**, 149–155.

Kirk, J. T., and Oram, R. N. (1981). _J. Aust. Inst. Agric. Sci._ **47**, 51–52.

Knowles, P. F. (1968). _Crop Sci._ **8**, 641.

Krzymanski, J., and Downey, R. K. (1969). _Can. J. Plant Sci._ **49**, 313–319.

Lööf, B. (1972). _In_ "Rapeseed; Cultivation, Composition, Processing and Utilization" (L.-A. Appelqvist and R. Ohlson, eds.), pp. 49–59. Amer. Elsevier, New York.

McGregor, D. I. (1981). _Can. J. Plant Sci._ **61**, 275–282.

Mizushima, U., and Tsunoda, S. (1967). _Tohoku J. Agric. Res._ **17**, 249–276.

Morice, J. (1960). _Ann. Inst. Natl. Rech. Agron., Ser. B_ **10**, 155–168.

Prakash, S. (1980). _In_ "Brassica Crops and Wild Allies" (S. Tsunoda, K. Hinata, and C. Gomez-Campo, eds.), pp. 151–163. Jpn. Sci. Soc. Press, Tokyo.

Prakash, S., and Hinata, K. (1980). _Opera Bot._ **55**, 1–57.

Röbbelen, G., and Thies, W. (1980). _In_ "Brassica Crops and Wild Allies" (S. Tsunoda, K. Hinata, and C. Gomez-Campo, eds.), pp. 285–299. Jpn. Sci. Soc. Press, Tokyo.

Shiga, T. (1970). _Jpn.. Agric. Res. Q._ **5**, 5–10.

Singh, D. (1958). "Rape and Mustard." Indian Central Oilseed Committee, Bombay.

Sinskaia, E. N. (1928). _Bull. Appl. Bot., Genet. Plant Breed._ **19**, 1–648.

Stefansson, B. R., Hougen, F. W., and Downey, R. K. (1961). _Can. J. Plant Sci._ **41**, 218–219.

Stringam, G. R., McGregor, D. I., and Pawlowski, S. H. (1974). _Proc. Int. Rapskongr., 4th, 1974_ pp. 99–108.

Tallent, W. H. (1972). _J. Am. Oil Chem. Soc._ **49**, 15–19.

Thomas, P. M., and Kondra, Z. P. (1973). _Can. J. Plant Sci._ **53**, 221–225.

U, N. (1935). _Jpn. J. Bot._ **7**, 389–452.

Vaughan, J. G. (1970). "The Structure and Utilization of Oil Seeds." Chapman & Hall, London.

Vaughan, J. G., Phelan, J. R., and Denford, K. E. (1976). _In_ "The Biology and Chemistry of the Cruciferae" (J. G. Vaughan, A. J. MacLeod, and B. M. G. Jones, eds.), pp. 119–137. Academic Press, New York.

Vavilov, N. I. (1949). _Chron. Bot._ **13**, 1–364.

Voskresenskaya, G. S., and Shpota, V. I. (1973). *In* "Handbook of Selection and Seed Growing of Oil Plants" (V. S. Pustovoit, ed.), pp. 149–205. Israel Program for Scientific Translations, Jerusalem.

Worthington, R. E., and Hammons, R. O. (1971). *Oleagineux* **26**, 695–700.

Zhukovsky, P.M., and Zeven, A. C., (1975). "Dictionary of Cultivated Plants and Their Centres of Diversity," pp. 29–30. Pudoc, Wageningen.

2

World Production and Trade of Rapeseed and Rapeseed Products

W. J. PIGDEN

I. INTRODUCTION

Rapeseed is a major source of food and feed throughout the world. It ranks fourth in the world production of edible vegetable oils and fifth in seed

High and Low Erucic Acid Rapeseed Oils
Copyright © 1983 by Academic Press Canada
All rights of reproduction in any form reserved.
ISBN 0-12-425080-7

production, this difference in ranking being due to its comparatively high oil content.

It has a long history of human use in the Orient, dating back to 1000 B.C. Until 1938, China was the world's principal producer supplying 2.5 million tonnes of the world's approximately 4 million tonnes output of rapeseed and mustard (Editorial, 1981). Since 1945, production and use has greatly increased in the Western World, especially in Canada and Europe. In Canada for example, the value of the crop has grown from zero in the early 1940s to the billion dollar level in 1981. Canadian rapeseed is now not only an important source of food and feed but is also a major export crop, second only to wheat.

Since the late 1960s to the present date, plant breeders have diligently selected rapeseed to minimize or eliminate the erucic acid from the oil and glucosinolates from the meal. Much of the recent expansion in the rapeseed industry in Canada and Europe is attributable to the dramatic improvements in the quality of oil and meal over the past 20 years. Another important factor is the wide adaptability of different types of rapeseed to different environments and their ability to thrive in harsh climates. In Canada and Europe most rapeseed is grown north of the 45° parallel. However, in China and India production is comparatively close to the equator, approximately 25°–40°N. This combination of improved quality and wide adaptability makes the crop very attractive to many developed and developing countries which are otherwise deficient in edible oil and in high protein meal.

Canadian rapeseed oil for edible use is now called canola oil, which is derived from rapeseed low in both erucic acid and glucosinolates. A second oil which is high in erucic acid is produced in limited quantities for industrial use only, and is marketed separately from the edible oils. In the Orient, rapeseed and mustard oils, which are both high in erucic acid, are used extensively for food. Mustard [*B. juncea* (L.)] is grown and processed for oil in the same manner as rapeseed.

Rapeseed meal is a by-product of the production of rapeseed oil. In the past, in the Orient the meal was used principally as a fertilizer with limited amounts fed to animals. In Canada and Europe, the meal was fed to livestock but with severe restrictions on its use because of its antinutritional (goitrogenic) effects in animals. Since 1974, however, with the availability of low glucosinolate low erucic acid rapeseed (double low) in Canada (i.e., canola), the production and use of the meal has increased substantially. The same trend is now occurring in Europe with the availability of double low rapeseed. The meal derived from these two types of rapeseed, i.e., high and low in glucosinolates, is handled separately in commerce.

For the earlier detailed history of production and trade in rapeseed from 1934 to 1970, the reader is referred to an excellent review by Ohlson (1972).

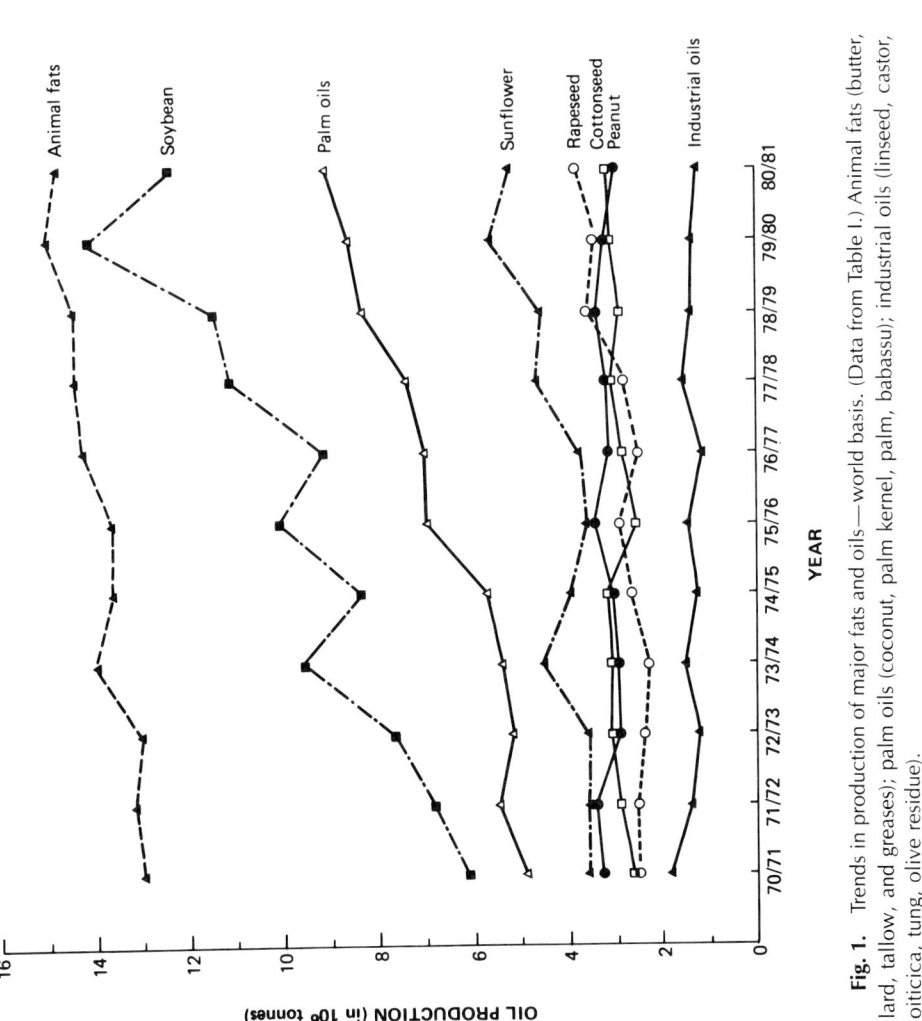

Fig. 1. Trends in production of major fats and oils—world basis. (Data from Table I.) Animal fats (butter, lard, tallow, and greases); palm oils (coconut, palm kernel, palm, babassu); industrial oils (linseed, castor, oiticica, tung, olive residue).

TABLE I

Total World Production of Oils and Fats by Commodity[a] (in 1000 tonnes)

Commodity	1970/ 1971	1971/ 1972	1972/ 1973	1973/ 1974	1974/ 1975	1975/ 1976	1976/ 1977	1977/ 1978	1978/ 1979	1979/ 1980	1980/ 1981
Edible vegetable oils											
Cottonseed	2,654	2,929	3,111	3,168	3,219	2,687	2,804	3,185	3,012	3,200	3,247
Peanut	3,377	3,535	2,921	3,091	3,182	3,594	3,193	3,131	3,357	3,128	3,110
Soybean	6,266	6,846	7,588	9,542	8,325	10,168	8,843	10,123	11,696	14,386	12,431
Sunflower	3,612	3,638	3,548	4,521	3,989	3,669	3,744	4,732	4,710	5,608	4,724
Rapeseed	2,508	2,596	2,493	2,475	2,713	2,912	2,485	2,698	3,673	3,432	3,863
Sesame	721	656	612	635	603	623	600	621	655	626	612
Safflower	226	300	239	210	217	329	217	275	326	342	255
Olive[b]	1,437	1,559	1,437	1,526	1,419	1,788	1,334	1,591	1,574	1,425	1,840
Corn	289	292	309	311	297	408	410	460	466	512	518
Palm oils											
Coconut	2,514	2,878	2,445	2,237	2,922	3,400	3,099	3,181	2,794	3,030	3,325
Palm kernel	465	457	434	488	507	505	542	560	616	670	696
Palm	1,937	2,146	2,230	2,610	2,916	3,059	3,388	3,784	4,295	4,812	5,034
Babassu kernel	72	107	105	105	105	122	132	143	145	128	130

24

Industrial oils										
Linseed	1,234	860	720	755	745	737	684	907	744	806	696
Castor	346	320	404	493	339	304	262	327	392	377	396
Oiticica	20	14	1	11	11	15	14	14	14	14	14
Tung	143	136	110	96	108	111	100	95	101	100	90
Olive residue[c]	131	132	139	146	132	192	153	172	146	136	172
Animal fats											
Butter[d]	4,115	4,389	4,533	4,502	4,572	4,800	4,944	4,981	4,987	4,989	4,972
Lard	4,315	4,259	4,134	4,379	4,330	3,380	3,571	3,871	3,941	3,962	3,827
Tallow and greases	4,672	4,584	4,428	5,121	4,723	5,471	5,815	5,527	5,814	6,063	6,047
Marine oils											
Whale	70	65	55	40	45	15	15	8	10	10	10
Sperm whale	135	125	125	120	119	76	64	58	58	58	58
Fish	1,173	934	809	1,001	1,003	979	1,004	1,236	1,285	1,283	1,222
Total	42,432	43,757	42,930	47,583	46,541	49,344	47,418	52,413	54,810	59,097	57,245

[a] From Canada Grains Council (1981).
[b] Excludes olive residue oil.
[c] Includes quantities of refined oil for edible purposes.
[d] Fat content.

TABLE II

Area under Rapeseed Production by Country[a] (in 1000 hectares)

Country	1971	1972	1973	1974	1975	1976	1977	1978	1979	1980	1981
Argentina	5	1	1	1	1	1	3	7	20	30	7
Australia	42	87	77	17	12	16	8	19	22	40	24
Austria	4	4	3	2	2	2	2	2	3	2	4
Bangladesh	215	189	191	176	196	193	194	206	190	195	214
Belgium-Lux.	1	1	1	1	1	—	1	1	1	1	1
Brazil	—	—	—	—	—	—	—	—	—	3	20
Canada	1,639	2,147	1,323	1,275	1,279	1,829	720	1,453	2,825	3,406	2,080
Chile	49	56	31	25	45	60	54	34	54	50	24
China	1,750	2,050	2,400	2,350	2,300	2,750	2,346	2,225	2,600	2,761	2,841
Czechoslovakia	35	52	53	57	45	63	63	73	79	55	91
Denmark	13	25	31	46	48	72	44	39	47	65	104
Ethiopia	14	15	15	50	50	50	50	25	25	25	25
Finland	7	7	5	10	13	17	14	23	30	33	55
France	311	322	314	327	318	261	282	275	253	223	395
Germany, East	103	104	111	122	123	132	130	125	124	113	124
Germany, West	85	95	106	108	110	90	95	105	121	127	138
Hungary	29	44	50	50	33	46	52	60	70	33	51
India	3,323	3,613	3,319	3,457	3,680	3,339	3,128	3,584	3,544	3,475	3,600
Ireland	—	—	—	—	—	—	—	—	1	1	1
Italy	3	3	3	5	1	1	1	1	1	1	1
Japan	14	11	8	5	4	4	3	3	3	3	3

Country											
Korea, Rep. of	23	29	17	13	26	27	25	27	19	12	15
Mexico	6	6	6	5	6	6	5	6	6	3	2
Netherlands	7	10	15	15	14	14	12	11	10	7	8
Norway	2	2	4	4	5	6	5	5	5	5	6
Pakistan	510	562	534	538	452	470	519	412	433	409	450
Poland	298	362	276	315	258	309	398	400	337	180	320
Romania	3	3	3	7	13	13	7	4	8	10	10
Spain	—	—	—	—	—	—	2	4	4	8	30
Sweden	90	114	150	165	150	164	140	120	151	155	172
Switzerland	9	102	10	9	10	10	11	11	12	12	13
Taiwan	2	2	1	1	3	5	3	3	2	2	2
Turkey	3	3	2	1	1	—	4	9	9	27	30
United Kingdom	4	5	7	14	25	39	48	55	64	74	92
U.S.S.R.	4	6	5	15	11	11	14	14	14	11	20
Yugoslavia	6	9	9	5	5	7	11	20	35	42	32
Total	8,609	10,041	9,081	9,191	9,240	10,007	8,394	9,361	11,122	11,599	11,005

[a] From U.S. Department of Agriculture (1981).

TABLE III

Yields of Rapeseed and Mustard Seed by Country[a] (in kg/hectare)

Country	1971	1972	1973	1974	1975	1976	1977	1978	1979	1980	1981
Argentina	800	2,000	2,000	2,000	2,000	2,000	667	714	750	767	714
Australia	810	632	377	647	750	750	1,125	842	1,045	1,000	750
Austria	2,250	1,750	2,333	2,500	2,500	3,000	2,000	2,500	2,000	1,500	2,000
Bangladesh	642	603	565	563	607	554	588	650	721	605	701
Belgium-Lux.	1,000	1,000	3,000	2,000	1,000	—	—	1,000	2,000	2,000	2,000
Brazil	—	—	—	—	—	—	—	—	—	1,000	1,000
Canada	999	1,004	983	947	910	1,005	1,163	1,358	1,238	1,001	1,194
Chile	1,673	1,393	1,129	1,320	1,244	1,083	1,537	1,529	1,204	1,460	1,167
China	629	649	667	655	609	600	575	528	718	870	839
Czechoslovakia	1,800	1,942	2,019	2,053	2,089	2,079	2,127	2,219	2,101	1,455	2,352
Denmark	2,077	2,000	1,452	2,000	2,333	1,819	1,841	1,974	1,936	2,308	2,038
Ethiopia	429	400	400	400	400	400	400	400	400	400	400
Finland	1,571	1,429	1,400	1,500	1,308	1,471	1,571	1,391	1,500	1,667	1,491
France	1,871	2,059	2,264	1,930	2,066	1,816	1,901	1,411	2,245	2,287	2,762
Germany, East	1,796	1,894	2,108	2,016	2,423	2,750	2,462	2,464	2,565	1,779	2,379
Germany, West	2,176	2,400	2,349	2,056	2,736	2,211	2,337	2,686	2,736	2,528	2,732
Hungary	1,586	1,614	1,040	1,360	1,364	1,413	1,269	1,483	1,529	1,242	1,922
India	595	397	545	493	612	580	496	460	525	412	597
Ireland	—	—	—	—	—	—	—	—	1,000	1,000	1,000
Italy	3,667	6,000	5,333	6,600	4,000	5,000	2,000	1,000	2,000	2,000	2,000

Japan	1,643	1,455	1,625	1,800	1,750	1,500	1,667	1,667	1,667	1,333	2,000
Korea, Rep. of	1,087	1,276	1,235	1,385	1,308	1,259	1,400	1,296	1,158	2,250	1,933
Mexico	1,000	1,000	1,000	1,000	1,000	1,000	1,000	1,000	1,000	1,000	1,000
Netherlands	3,143	3,300	3,000	2,733	3,214	2,643	2,833	2,727	2,300	2,571	3,625
Norway	2,500	1,000	500	1,750	1,400	1,000	1,800	2,000	1,800	1,800	1,667
Pakistan	527	527	528	535	540	568	563	575	573	604	556
Poland	1,899	1,644	1,558	1,625	2,027	2,350	2,462	1,770	2,050	1,294	1,769
Romania	1,667	1,667	1,667	1,714	1,231	1,308	1,571	1,500	1,375	1,000	1,800
Spain	—	—	—	—	—	—	2,000	1,750	1,750	1,250	1,200
Sweden	1,889	1,965	1,920	1,806	2,067	1,768	1,764	1,967	1,914	1,703	1,657
Switzerland	2,111	235	2,400	2,222	2,800	1,800	2,182	2,000	2,333	2,583	2,615
Taiwan	1,000	1,000	1,000	2,000	1,333	1,000	1,000	1,000	1,000	1,000	1,000
Turkey	1,000	1,000	1,000	2,000	2,000	—	1,500	1,556	1,444	1,593	1,667
United Kingdom	2,000	2,000	2,000	2,214	2,200	1,564	2,313	2,582	2,422	2,676	3,272
U.S.S.R.	1,000	1,000	1,400	800	727	818	1,143	1,071	1,071	727	400
Yugoslavia	1,667	2,000	1,556	1,600	2,400	2,000	2,182	2,000	2,086	2,214	2,125

[a] From U.S. Department of Agriculture (1981).

The present chapter discusses only the production and trade of rapeseed and its products from 1971 to 1981.

II. PRODUCTION OF RAPESEED AND MUSTARD—WORLD BASIS

A. All Fats and Oils

Total world production of fats and oils is presented in Table I and trends are shown graphically in Fig. 1. During the period 1970–1971 to 1980–1981, total production of fats and oils has increased by 35% from 42,432 to 57,245 million tonnes. The major contributors to this expansion, expressed as percentage increases, are soybean oil 98%, palm oils 84%, rapeseed oil 54% and sunflower oil 30%. Production of most other major oil sources has either remained relatively static, recorded small increases, or actually declined. For example, animal fats, the largest single source, registered an increase of only 13%, marine oils 4% and industrial oils declined by 37% (Table I).

Table I shows that rapeseed oil now ranks fourth in world production of edible oils after soybean, palm, and sunflower oils. In recent years rapeseed oil has moved ahead of such major contributors as cottonseed, peanut, and coconut oils.

B. Rapeseed and Mustard

1. AREAS HARVESTED

There has been about a 28% expansion in total areas of rapeseed grown over the last decade (Table II). Much of this is attributable to substantially increased areas planted to rapeseed in China and Canada. Considering the major producers in 1981, India had by far the largest area (3,600,000 hectares) followed by China (2,841,000), Canada (2,080,000), Pakistan (450,000), Poland (320,000), and France (395,000).

The area of rapeseed harvested in Canada was down by about 39% in 1981 from 1980 because of the increased area planted to grains. There is reason to believe the Canadian producer made the decision to reduce plantings to rapeseed because of high carry over of stocks in 1979–1980 and 1980–1981, and the higher initial payments on grain to Canadian producers in 1981 (Duncker, 1981).

2. YIELDS

The average yield of rapeseed and mustard by country is given in Table III. In the major producing areas important improvements in yield appear to have been achieved principally by China, Canada, and among some of the large producers in Europe, e.g., France, West and East Germany and the United Kingdom. In India, Pakistan, and Bangladesh, yields have remained essentially static.

Yields per hectare vary tremendously throughout the world, ranging from high yields of 2000–3500 kg/ha in such countries as the Netherlands, France, West and East Germany and the United Kingdom, to 400–800 kg/ha in India and China. High yields have been recorded for Italy between 1971 and 1976 (Table III). These are probably not representative of large scale production, but rather may indicate very intensive farming practices on a small area.

3. SEED PRODUCTION

There are four major rapeseed producing areas, namely, Canada, China, Europe and the Indian subcontinent. Total world production of rapeseed increased by about 4,089,000 tonnes or 57% over the 1971–1981 period as shown in Table IV. The increased world production is mainly due to very large increases in China and Canada, which more than doubled over the 1971–1980 period. However, while China's production was maintained in 1981, Canada's crop dropped by 27% (900,000 tonnes), for reasons given above. In Europe, production by the four major producers, France, East Germany, West Germany, and Poland has increased steadily by some 811,000 tonnes from 1971 to 1981, an improvement of 53%. The fairly substantial increase in 1981 is chiefly due to the recent doubling of the French crop, as a result of the availability of new low erucic acid rapeseed varieties, and favorable market conditions.

4. OIL PRODUCTION

The total world production of rapeseed oil increased by 93%, or 2,105,000 tonnes, over the 1971–1981 period (Table V). The production of rapeseed oil does not necessarily reflect the production of rapeseed since some countries are large importers of rapeseed, which shows up as increased oil production, notably, Japan and West Germany. Conversely, some countries are large exporters of rapeseed which shows up as decreased production of oil, i.e., Canada, Sweden, and Denmark.

The increases in production of rapeseed oil by the major producers, expressed as tonnes of oil and percent increase per country over the 1971–1981 period were China 404,000 (116%), Canada 341,000 (443%), Japan 308,000 (179%), West Germany 264,000 (254%), United Kingdom 137,000 (489%), India 120,000 (23%), France 76,000 (41%), and East Germany 57,000 (89%). Many smaller producers, e.g., Czechoslovakia, Finland, Hungary, Netherlands, and Yugoslavia have also recorded large percentage increases. Production of rapeseed oil in Poland, which accounts for

TABLE IV

Production of Rapeseed and Mustard Seed by Country[a] (in 1000 tonnes)

Country	1971	1972	1973	1974	1975	1976	1977	1978	1979	1980	1981
Argentina	4	2	2	2	2	2	2	5	15	23	5
Australia	34	55	29	11	9	12	9	16	23	40	18
Austria	9	7	7	5	5	6	4	5	6	3	8
Banglacesh	138	114	108	99	119	107	114	134	137	118	150
Belgium-Lux.	1	1	3	2	1	2	—	1	2	2	2
Brazil	—	—	—	—	—	—	—	—	—	3	20
Canada	1,638	2,155	1,300	1,207	1,164	1,839	837	1,973	3,497	3,411	2,483
Chile	82	78	35	33	56	65	83	52	65	73	28
China	1,100	1,330	1,600	1,540	1,400	1,650	1,350	1,175	1,868	2,402	2,384
Czechoslovakia	63	101	107	117	94	131	134	162	166	80	214
Denmark	27	50	45	92	112	131	81	77	91	150	212
Ethiopia	6	6	6	20	20	20	20	10	10	10	10
Finland	11	10	7	15	17	25	22	32	45	55	82
France	582	663	711	631	657	474	536	388	568	510	1,091
Germany, East	185	197	234	246	298	363	320	308	318	201	295
Germany, West	185	228	249	222	301	199	222	282	331	321	377
Hungary	46	71	52	68	45	65	66	89	107	41	98
India	1,976	1,433	1,808	1,704	2,252	1,936	1,551	1,650	1,860	1,433	2,150
Ireland	—	—	—	—	—	—	—	—	1	1	1
Italy	11	18	16	33	4	5	2	1	2	2	2
Japan	23	16	13	9	7	6	5	5	5	4	6

Korea, Rep. of	25	37	21	18	34	34	35	35	22	27	29
Mexico	6	6	6	5	6	6	5	6	6	3	2
Netherlands	22	33	45	41	45	37	34	30	23	18	29
Norway	5	2	2	7	7	6	9	10	9	9	10
Pakistan	269	296	282	288	244	267	292	237	248	247	250
Poland	566	595	430	512	523	726	980	708	691	233	566
Romania	5	5	5	12	16	17	11	6	11	10	18
Spain	—	—	—	—	—	—	4	7	7	10	36
Sweden	170	224	288	298,	310	290	247	236	289	264	285
Switzerland	19	24	24	20	28	18	24	22	28	31	34
Taiwan	2	2	1	2	4	5	3	3	2	2	2
Turkey	3	3	2	2	2	—	6	14	13	43	50
United Kingdom	8	10	14	31	55	61	111	142	155	198	301
U.S.S.R.	4	6	7	12	8	9	16	15	15	8	8
Yugoslavia	10	18	14	8	12	14	24	40	73	93	68
Total	7,235	7,796	7,473	7,312	7,857	8,528	7,159	7,876	10,709	10,079	11,324

a From U.S. Department of Agriculture (1981).

TABLE V

Production of Rapeseed and Mustard Seed Oil by Country[a] (in 1000 tonnes)

Country	1971	1972	1973	1974	1975	1976	1977	1978	1979	1980	1981
Algeria	18	24	18	17	38	25	34	31	18	21	23
Argentina	1	1	1	1	1	1	1	2	6	9	2
Australia	13	6	15	5	5	5	3	6	10	16	7
Austria	2	1	2	1	1	1	2	1	1	1	—
Bangladesh	70	47	51	37	50	44	40	44	48	46	53
Belgium-Lux.	2	—	2	1	1	2	1	2	1	10	12
Brazil	—	—	—	—	—	—	—	—	—	37	8
Canada	77	106	134	126	108	141	226	259	296	365	418
Chile	66	67	58	63	23	26	33	21	26	30	11
China	347	419	504	485	441	520	425	370	588	757	751
Czechoslovakia	48	39	41	50	46	61	60	65	70	34	85
Denmark	1	3	4	7	6	6	6	2	4	6	5
Ethiopia	3	3	3	8	8	8	8	4	4	6	4
Finland	6	6	6	6	8	10	8	12	17	20	30
France	184	261	223	217	190	167	184	182	229	218	260
Germany, East	64	68	84	85	102	125	102	113	112	82	121
Germany, West	104	102	97	166	149	121	181	151	214	303	368
Hungary	8	24	19	24	13	19	17	18	43	18	34
India	525	523	561	522	599	598	564	578	558	430	645
Ireland	—	—	—	—	—	—	—	—	—	1	1
Italy	162	155	112	30	7	12	7	9	17	26	85
Japan	172	233	288	275	304	304	324	352	434	406	480

Korea, Rep. of	9	13	12	11	12	13	14	12	18	16	17
Mexico	2	2	11	13	2	2	2	2	2	4	8
Morocco	5	6	6	6	—	7	—	—	8	5	8
Netherlands	23	18	31	26	55	34	34	27	30	35	63
Pakistan	84	70	75	77	65	70	78	63	64	78	78
Poland	175	181	145	159	183	219	270	236	259	87	208
Romania	1	1	2	2	4	5	3	2	4	3	6
Spain	—	—	—	—	1	1	2	4	6	5	13
Sweden	45	55	61	62	62	77	73	69	82	71	65
Switzerland	10	10	8	11	7	10	9	11	13	14	14
Taiwan	1	1	10	—	2	2	1	1	1	1	1
Turkey	1	1	1	1	1	—	2	4	4	9	10
United Kingdom	28	42	43	10	38	66	110	130	131	127	165
United States	—	1	—	—	2	1	—	—	1	—	—
U.S.S.R.	2	2	3	5	3	4	6	5	5	2	2
Yugoslavia	4	7	5	3	5	9	9	15	9	13	34
Total	2,263	2,498	2,636	2,522	2,542	2,716	2,803	3,334	3,310	4,095	4,368

[a] From U.S. Department of Agriculture (1981).

TABLE VI

Production of Rapeseed Meal[a] (in 1000 tonnes)

Country	1971	1972	1973	1974	1975	1976	1977	1978	1979	1980	1981
Algeria	27	37	27	26	57	38	51	47	28	31	35
Argentina	2	1	1	1	1	1	1	3	8	12	3
Australia	17	7	19	20	7	6	5	8	12	22	9
Austria	2	1	3	1	1	2	2	1	1	1	—
Bangladesh	139	93	101	73	97	86	80	86	82	78	91
Belgium–Lux.	4	1	3	2	2	2	1	2	2	15	18
Brazil	—	—	—	—	—	—	—	—	—	53	11
Canada	113	163	204	194	158	197	315	357	417	521	574
Chile	85	87	76	82	27	32	40	25	32	36	14
China	594	718	864	832	756	891	729	635	1,009	1,297	1,287
Czechoslovakia	68	55	58	71	65	87	86	92	99	48	121
Denmark	2	4	5	10	9	8	8	3	6	6	7
Ethiopa	3	3	3	11	11	11	11	6	6	6	6
Finland	10	10	10	10	12	15	13	19	27	32	48
France	249	352	301	293	245	215	237	234	294	295	349
Germany, East	83	88	109	110	133	162	133	147	145	94	139
Germany, West	155	—	147	259	220	181	266	233	316	455	560
Hungary	11	35	27	34	18	27	25	26	61	26	48
India	1,065	1,060	1,140	1,059	1,216	1,215	1,146	1,172	1,132	872	1,309
Ireland	—	—	—	—	—	—	—	—	1	1	1
Italy	213	204	153	40	9	15	10	12	23	34	110
Japan	252	344	415	391	405	398	426	484	616	558	684

Korea, Rep. of	15	21	19	17	19	21	22	20	30	26	29
Mexico	4	4	17	2	4	4	4	4	4	6	12
Morocco	7	9	9	10	—	11	—	13	13	8	12
Netherlands	34	27	46	38	80	49	50	39	30	52	92
Pakistan	198	166	158	162	137	148	163	133	136	140	140
Poland	242	249	200	220	254	293	374	335	346	125	296
Romania	2	1	2	4	6	7	6	2	5	5	9
Spain	—	—	—	—	1	1	4	6	9	7	19
Sweden	63	77	99	97	81	104	108	98	111	108	98
Switzerland	10	13	13	11	16	10	13	12	15	17	19
Taiwan	1	1	16	1	2	3	2	2	1	1	1
Turkey	2	2	1	1	1	—	3	6	6	15	17
United Kingdom	39	60	62	15	54	94	156	98	188	181	239
United States	—	1	1	—	3	1	—	—	1	—	—
U.S.S.R.	2	3	4	7	5	5	9	9	7	3	3
Yugoslavia	6	11	8	5	7	14	14	22	11	17	46
Total	3,719	3,908	4,321	4,109	4,119	4,354	4,513	4,371	5,230	5,206	6,456

[a] From U.S. Department of Agriculture (1981).

about 50% of Eastern European production, has remained essentially static.

There are major differences in the composition of the oil and the meal produced in different regions. In India, Pakistan, Bangladesh, and China most of the rapeseed and mustard is of the high erucic acid (about 42%), high glucosinolate type. In Canada and Europe rapeseed is mainly of the low erucic low glucosinolate type achieved through extensive plant breeding programs since 1960. By 1978, the average level of erucic acid in Canadian rapeseed oil was only 1.3% (Chapter 7). Similar progress is being achieved in Western Europe. In the Eastern European countries progress appears to be somewhat slower. The oil and meal from the new rapeseed varieties are preferred products and are extensively traded in international markets.

In Canada and Europe the percentage of oil yield from rapeseed is 41–42% based on an 8.5% moisture content of the seed, which is much greater than the 33–35% oil content reported for rapeseed and mustard from India and China. These large differences are attributable partly to varietal differences, but probably more importantly to relatively inefficient oil extraction procedures generally used in that part of the world which leaves about 12% of the oil with the meal. In the Asian countries of India, China, Pakistan, and Bangladesh, oil extracted from the seed is used domestically for human consumption and very little is exported. In these countries and in Japan, rapeseed oil has traditionally been preferred as a cooking oil.

In the Orient, both rapeseed and mustard oils are used extensively for food and large quantities of mustard are produced for oil in India, Pakistan and Bangladesh. In Canada and Europe there is no production of mustard oil for edible oil purposes. All edible Canadian rapeseed oil is now from canola varieties, i.e., derived from seed low in erucic acid and low in glucosinolates. High erucic acid rapeseed oil is used exclusively for industrial purposes, such as lubricants, slip agents for molds, and polymers. In Canada these high erucic acid oils are produced and marketed separately from canola oil. Rapeseed oils in the Western World are either very low in erucic acid for edible use or very high (over 40%) in erucic acid for industrial use.

5. MEAL PRODUCTION

The world production of rapeseed meal is shown in Table VI. In Asia essentially all the locally produced meal is of the high glucosinolate type. It is used principally for fertilizer with small amounts fed to ruminants. However, most countries importing rapeseed now prefer the Canadian canola seed because of the superior nutritional value of the low glucosinolate meal. The Chinese are very interested in developing low glucosinolate cultivars for their own use for similar reasons.

In Canada, all of the meal now being produced by the crushing plants is of the low glucosinolate (canola) type and this accounts for the substantially increased domestic and export demand for Canadian meal. In Europe the changeover to low glucosinolate seed is well under way.

TABLE VII

Exports of Rapeseed[a] (in 1000 tonnes)

Country	1971	1972	1973	1974	1975	1976	1977	1978	1979	1980	1981
Australia	3	40	7	2	1	1	—	—	—	—	—
Austria	5	4	2	3	3	3	—	4	4	2	7
Belgium-Lux.	3	2	2	1	—	—	—	1	1	1	1
Canada	1,062	966	1,226	889	593	683	1,018	1,014	1,721	1,743	1,372
Denmark	19	36	43	81	95	107	72	76	86	133	171
France	224	248	227	122	234	105	131	11	87	23	478
Germany, East	1	4	10	5	3	25	50	11	23	7	15
Germany, West	79	82	48	22	62	20	27	26	25	26	47
Hungary	26	7	2	6	11	17	24	42	7	8	22
Netherlands	12	26	12	16	16	14	9	8	8	12	13
Poland	44	52	—	52	—	100	190	25	—	—	3
Romania	2	3	1	5	5	4	2	1	1	1	1
Sweden	59	79	134	147	139	122	67	58	88	88	136
United Kingdom	—	—	—	3	—	1	5	1	1	1	1
Yugoslavia	—	—	—	—	—	—	—	—	30	60	5
Total	1,539	1,549	1,714	1,354	1,162	1,202	1,595	1,278	2,082	2,105	2,272

[a] From U.S. Department of Agriculture (1981). World exports will not equal imports because of differing market years and because some minor countries are not included in the totals.

TABLE VIII

Imports of Rapeseed[a] (in 1000 tonnes)

Country	1971	1972	1973	1974	1975	1976	1977	1978	1979	1980	1981
Algeria	45	62	46	44	97	64	86	80	47	53	60
Australia	1	1	14	29	5	—	—	—	—	—	—
Austria	1	—	—	—	—	—	—	1	—	1	—
Bangladesh	104	50	68	30	52	43	27	18	13	12	14
Belgium-Lux.	8	2	4	2	2	2	2	3	2	24	30
Brazil	—	—	—	—	—	—	—	—	—	90	—
Chile	91	100	121	132	—	—	—	—	—	—	—
Czechoslovakia	62	1	15	15	26	30	25	8	17	7	10
Denmark	—	1	5	3	—	—	—	—	3	2	—
Finland	6	7	9	2	3	—	—	—	—	—	—
France	103	237	74	36	33	28	34	64	54	98	15
Germany, West	209	133	101	251	176	162	286	174	291	786	623
Hungary	—	—	—	—	—	3	3	1	13	14	12
India	45	30	62	32	18	3	100	250	—	—	—
Ireland	—	—	—	—	—	—	—	—	1	1	1
Italy	376	356	261	70	12	23	16	20	39	60	200

Japan	407	604	687	672	659	718	776	832	1,129	1,067	1,200
Korea, Rep. of	2	—	14	12	—	3	5	—	34	16	20
Mexico	—	—	23	29	—	—	—	—	—	7	18
Morocco	12	15	15	16	—	18	—	—	21	14	20
Netherlands	53	42	51	45	118	67	67	51	65	90	155
Norway	20	13	16	11	14	22	13	6	6	5	6
Spain	—	—	—	—	2	2	2	3	9	3	—
Taiwan	—	—	25	—	—	—	—	—	—	—	—
United Kingdom	65	103	101	—	45	115	186	43	187	138	141
United States	—	2	—	—	5	2	—	—	2	—	—
Yugoslavia	—	—	—	—	—	10	—	—	—	—	15
Total	1,610	1,759	1,697	1,431	1,267	1,315	1,628	1,554	1,933	2,488	2,540

[a] From U.S. Department of Agriculture (1981). World exports will not equal imports because of differing market years and because some minor countries are not included in the totals.

III. PRODUCTION, IMPORT, AND EXPORT OF RAPESEED BY COUNTRIES

This section covers the production, imports, and exports of rapeseed, rapeseed oil, and rapeseed meal by country. As is implied in Tables VII to XI, rapeseed can be transported as the intact seed for processing at the destination, or it can be processed at the place of origin and shipped as oil or meal. In the discussion to follow, the major rapeseed producing countries are discussed separately.

A. Canada

Canada is the world's largest producer and exporter of rapeseed, a position achieved largely within the last decade. Except for a small area in British Columbia, all commercial production is in the three Prairie Provinces, Alberta, Saskatchewan, and Manitoba, where it is well adapted to the northern climate, north of the 49° latitude (Table XII). Varieties grown are of the summer rape type; no winter rape varieties suited to the Canadian climate are commercially available.

Canada has led the world in development and production of low erucic, low glucosinolate (canola) varieties of rapeseed. About 87% of the 1981 crop and essentially all the rapeseed crushed is now canola so that Canadian rapeseed and its products have achieved a high quality standard.

The data in Table XIII indicate the very rapid increase in crushing capacity developed over the period 1975–1981. Present crushing capacity is about one million tonnes of seed per year but the new plants now under construction will increase this capacity to 1.3–1.4 million tonnes. Nearly all the crushing plants are located in Western Canada and use the prepress solvent or solvent process (Chapter 8). The increasing proportion of the crop crushed in Canada has meant that new markets for oil and for meal had to be developed.

In world trade, Japan has for many years been Canada's best customer for rapeseed, importing about 1.0 million tonnes of Canadian seed per year since 1979 (Table VIII). The Netherlands, West Germany, Algeria, Bangladesh and France have also imported considerable quantities of Canadian seed in recent years and other new markets are being developed.

Exports of Canadian rapeseed oil have been increasing in line with the increases in crushing capacity, moving from zero in 1972 to 199,000 tonnes in 1981 and surpassed only by France and West Germany. The Canadian crushing capacity is now being further enlarged to accommodate the expanding markets for canola oil and meal.

Canada has been importing large quantities of U.S. soybeans or soybean meal to provide a protein supplement for livestock. Canola meal is now

TABLE IX

Exports of Rapeseed Oil[a] (in 1000 tonnes)

Country	1971	1972	1973	1974	1975	1976	1977	1978	1979	1980	1981
Belgium-Lux.	3	2	1	1	1	2	2	1	1	3	5
Brazil	—	—	—	—	—	—	—	—	—	1	—
Canada	—	—	25	34	19	33	92	74	111	152	199
Denmark	—	—	3	5	5	3	4	2	3	4	1
Finland	—	—	—	—	—	2	2	3	4	2	4
France	52	79	130	100	125	133	125	140	182	154	208
Germany, East	—	—	—	1	2	2	4	5	6	10	10
Germany, West	60	56	63	110	96	62	96	164	99	181	235
Hungary	8	9	9	3	6	6	9	11	2	5	7
India	—	—	1	—	—	—	—	—	—	—	—
Italy	—	—	10	3	—	—	5	6	8	7	10
Japan	13	11	3	3	2	1	2	2	1	1	1
Korea, Rep. of	—	—	—	—	—	—	—	—	13	5	6
Netherlands	20	22	33	24	33	34	35	25	27	27	34
Pakistan	4	—	—	—	—	—	—	—	—	—	—
Poland	30	47	44	36	44	80	75	72	67	7	5
Sweden	18	38	43	37	32	51	47	43	46	40	38
United Kingdom	—	—	—	—	—	3	35	14	25	15	10
Total	208	264	365	357	365	412	533	562	595	614	773

[a] From U.S. Department of Agriculture (1981). World exports will not equal imports because of differing market years and because some minor countries are not included in the totals.

TABLE X

Imports of Rapeseed Oil[a] (in 1000 tonnes)

Country	1971	1972	1973	1974	1975	1976	1977	1978	1979	1980	1981
Algeria	40	29	24	51	59	59	82	82	122	40	70
Australia	2	—	—	1	2	2	4	2	4	6	6
Austria	9	13	21	18	8	6	8	5	6	9	9
Bangladesh	104	50	19	30	52	43	27	18	10	3	4
Belgium-Lux.	5	5	6	7	9	15	7	5	4	9	12
Chile	3	3	1	1	14	6	4	1	7	2	1
Cyprus	2	2	2	2	1	2	4	—	—	—	—
Czechoslovakia	—	1	—	1	1	1	1	2	2	—	2
Denmark	—	—	—	—	—	—	—	—	—	1	—
France	9	10	4	10	3	—	4	2	3	5	6
Germany, West	13	13	15	19	23	17	34	38	44	39	74
Hong Kong	28	29	29	28	24	28	23	24	30	29	30
Hungary	—	—	1	3	5	6	4	1	3	3	3
India	—	—	27	39	6	11	225	260	167	150	150
Ireland	—	2	1	2	1	2	1	2	3	3	3
Italy	45	57	36	10	14	41	37	34	72	71	50
Japan	—	3	17	7	15	14	8	14	11	8	30

Malaysia	1	1	—	—	—	—	—	—	—	—	—
Morocco	11	31	39	50	84	29	6	6	24	40	20
Netherlands	10	10	4	1	11	10	13	9	30	24	32
Nigeria	—	—	—	—	—	—	1	2	43	87	120
Pakistan	—	—	1	3	1	—	—	—	—	—	—
Poland	1	3	—	—	—	1	—	—	—	8	7
Spain	—	—	7	4	3	15	5	4	10	13	—
Tunisia	—	—	14	7	10	1	—	—	14	—	—
United Kingdom	6	9	13	11	7	2	1	3	6	10	10
United States	5	5	6	8	3	6	7	6	—	6	6
U.S.S.R.	—	—	—	—	—	8	—	4	32	—	—
Yugoslavia	—	—	—	—	1	24	24	5	—	—	—
Total	294	276	286	313	357	349	527	529	647	568	645

[a] From U.S. Department of Agriculture (1981). World exports will not equal imports because of differing market years and because some minor countries are not included in the totals.

45

TABLE XI

Exports and Imports of Rapeseed Meal[a] (in 1000 tonnes)

Country	1971	1972	1973	1974	1975	1976	1977	1978	1979	1980	1981
Exports											
Argentina	1	1	—	—	—	1	1	—	3	2	—
Australia	8	4	6	—	—	—	—	—	1	—	—
Belgium-Lux.	—	—	1	1	—	—	2	3	5	2	2
Brazil	—	—	—	—	—	—	—	—	—	49	1
Canada	—	—	19	48	11	28	107	156	170	176	204
Chile	10	13	—	20	20	18	16	15	13	15	—
Denmark	—	1	2	2	4	1	2	1	2	1	1
Ethiopa	3	3	3	3	10	10	4	3	6	4	4
France	100	155	104	81	72	73	75	117	164	129	172
Germany, West	62	—	48	148	114	93	70	60	65	87	109
Hungary	—	—	—	—	—	—	1	—	4	2	2
India	15	15	14	14	10	15	12	14	10	5	10
Italy	74	88	40	18	3	5	4	2	6	3	30
Japan	—	—	—	—	—	—	1	—	—	—	—
Morocco	9	7	9	6	2	9	—	—	11	8	10
Netherlands	5	4	14	13	20	19	35	36	29	32	38
Pakistan	—	5	43	62	20	37	60	74	32	22	30
Poland	15	29	—	—	—	—	7	9	4	—	—
Sweden	—	—	—	—	—	—	—	—	—	—	5
United Kingdom	—	—	—	2	2	17	29	23	29	33	45
Total	302	325	303	398	288	326	426	513	554	570	663

Imports											
Algeria	—	—	—	2	—	—	2	—	—	—	1
Austria	8	10	12	12	6	1	2	6	2	4	5
Belgium–Lux.	64	69	68	69	48	71	74	56	59	41	35
Denmark	23	43	24	47	45	62	67	72	107	116	115
France	11	5	16	14	4	7	24	3	—	2	2
Germany, West	67	—	56	55	33	49	119	189	161	202	207
Hungary	—	—	—	—	—	—	1	—	1	2	2
Ireland	—	—	—	—	—	3	5	6	5	5	5
Italy	—	—	—	4	2	5	5	—	1	—	—
Japan	13	7	11	11	—	14	24	25	5	7	—
Korea, Rep. of	—	—	—	—	—	—	17	11	29	6	35
Netherlands	67	111	77	92	85	98	130	95	129	110	135
Norway	31	12	21	34	39	50	59	52	65	84	90
Spain	—	—	—	—	23	9	5	—	—	—	—
Sweden	2	—	2	—	1	—	—	—	—	—	—
United Kingdom	96	96	91	55	49	23	38	50	61	56	36
United States	—	—	—	—	—	—	—	—	—	15	—
Total	382	353	378	395	335	392	572	565	625	650	668

[a] From U.S. Department of Agriculture (1981). World exports will not equal imports because of differing market years and because some minor countries are not included in the totals.

TABLE XII

Production of Rapeseed in Canada by Province (1978–1981)[a] (in tonnes)

Province	1978	1979	1980	1981
Manitoba	578,336	657,000	317,500	340,000
Saskatchewan	1,451,510	1,281,400	997,900	680,000
Alberta	1,383,471	1,485,500	1,134,000	648,000
British Columbia	62,236	136,100	56,700	26,100

[a] From Statistics Canada (1978–1982).

replacing a considerable part of the imported soybean meal and this trend will no doubt continue. In fact, canola meal exports reached a volume of 204,000 tonnes in 1981 (Table XI). The main customers in recent years have been West Germany, Netherlands, Norway, and the United States. Markets for canola meal are opening up in other countries.

Rapeseed offers a very important alternative crop to grain production in Western Canada since the land, tillage, harvesting, storage, and transportation equipment and facilities are virtually identical. A disadvantage is that the crushing industry in Western Canada is almost wholly dependent on rapeseed. Hence, large fluctuations in crop production as a result of changes in world grain prices or policies of marketing agencies can have serious consequences for this segment of the industry and for stability of export markets. The large decrease of over one-third in the area planted to rapeseed in 1981 as compared with 1980 vividly illustrates this point. It is estimated that 2.64 million hectares of rapeseed are required to meet Canada's existing markets.

Canadian plant breeders are currently working toward developing winter type canola varieties for production in Central Canada, reducing the fiber content in canola seed, and producing low glucosinolate HEAR varieties for industrial use. Currently some 80,000–120,000 hectares are devoted to the production of HEAR oil for industrial use but unfortunately the meal is of high glucosinolate type. Research is in progress to develop new low glucosinolate HEAR varieties which will produce an oil containing 50% or more erucic acid for industrial use and provide a meal acceptable for the domestic market.

B. China

China is the second largest producer of rapeseed in the world (Table IV). Most of their rapeseed is a winter form of *Brassica napus*, high in erucic acid and high in glucosinolates. There is less interest in reducing the erucic acid content of the oil than there is in developing low glucosinolate varieties.

TABLE XIII

Canadian Crushings of Rapeseed and Production of Oil and Meal (1975–1981)[a] (in tonnes)

Component	1975–1976	1976–1977	1977–1978	1978–1979	1979–1980	1980–1981
Seed	347,161	549,174	630,300	725,100	897,300	1,330,000
Oil	141,698	225,805	259,000	296,300	364,900	418,200
Meal	197,376	314,903	357,500	416,700	520,800	573,600

[a] From Statistics Canada (1977–1982).

This latter interest stems from their 10-year (1978–1988) plan to expand and modernize their dairy cattle, swine and poultry industries, for which they will need large quantities of high quality meal. Increased animal production is said to have a high priority (Bell, 1980).

In some areas of China crop rotation is practiced. Two crops of rice are planted during the summer and a crop of rapeseed during the "winter." Rapeseed is sown in mid-September, transplanted in mid-November, and harvested in June. In general, rapeseed is grown as a "winter crop" at low altitudes and at latitudes ranging from 25° to 42°N (Stefansson, 1980). This situation is somewhat unusual because most countries grow rapeseed in northern (Canada and Europe) or high altitude regions (India).

The rapeseed crushing industry in China is labor intensive and is carried out mainly at the commune level using small expellers and hydraulic presses, although there is some solvent extraction. The plants are small and do not extract the oil efficiently, i.e., about 12% of the oil is left in the meal. Of the rapeseed meal currently produced, more than 90% is used as fertilizer, with limited amounts being fed to ruminants (Bell, 1980).

Most of the rapeseed oil is said to be utilized at the commune level mainly in the crude oil form. Some of the rapeseed oil is refined (i.e., degummed, alkali refined, bleached, and deodorized) but little is hydrogenated for production of butter substitutes (Boulter, 1980).

China is not self-sufficient in oilseeds and has had a total net import of nearly 700,000 tonnes over the past four years (Earl, 1980). The Chinese are striving to achieve greater self-sufficiency in oil and meal using rapeseed as a means to that end. Production of rapeseed in China has almost doubled since 1978 (Table IV). Effective research and development programs are under way to develop low glucosinolate varieties with technical and scientific advice from Canada. It is noteworthy that the Chinese consumption of vegetable oil is currently estimated at 6 to 9 kg/capita/year compared to the Canadian consumption of over 14 kg/capita/year (Sarsons, 1980).

C. India

India is the world's third largest producer of rapeseed and mustard (Table IV). The Indian crop is generally a mixture of mustard and rapeseed. Both rapeseed and mustard are important edible oil crops and together account for about 20% of all oilseed production in India. Most of the production is in the northern regions at high altitude; the largest production is in the state of Uttar Pradesh (Garg and Bhan, 1978).

The area planted to rapeseed (Table II) and oil production (Table V) has shown little change over the 1971–1981 period, and thus the gap between supply and demand has widened. In recent years India has had to import large quantities of rapeseed oil to meet its rising demand (Table X). The per capita availability of all edible vegetable oils was estimated at 3 kg/year in 1973–1974 (Garg and Bhan, 1978). According to these authors the very low yields of rapeseed (~400 kg/ha) are due to lack of good quality seed, high susceptibility of the crop to diseases and pests, lack of resources to extend available technology, and instability of prices. These authors believe there is great scope for improvement in production technology. However, based on performance over the 1971–1981 period little change can be expected in the near future.

As in China, crushing of the seed is carried out mostly at the village level in small expeller and hydraulic press equipment. The efficiency of extraction is low leaving about 12% oil in the meal. The oil is utilized mainly in the crude form at the local level. There are some modern crushing plants in the Bombay area where the oil is extracted by modern techniques. Substantial refining facilities are also available to produce fully refined and hydrogenated products. A major product is "Vanaspati," a hydrogenated shortening used for cooking and as a substitute for ghee.

Essentially all the rapeseed and mustard oil produced in India is consumed domestically. In addition, since 1977 India has been the largest importer of rapeseed oil in the world taking fully 50% of total imports in 1978. In 1980, imports of rapeseed oil were 150,000 tonnes with 117,500 tonnes coming from Canada (Statistics Canada, 1982). The domestically produced rapeseed and mustard meal are used both as fertilizer and for feeding cattle and buffalo.

It appears that there are good opportunities for increasing rapeseed and mustard oil production in India if agricultural conditions can be improved.

D. European Countries

Europe is the major rapeseed producing area in the world. Rapeseed production increased three- to fourfold after World War II to alleviate the severe shortage of edible oils and to lessen their dependence on imported soybeans. This development has been highly beneficial in terms of a better balanced agriculture especially in Northern Europe. Most European production is of winter rape varieties which give higher yields (Table III), and modern oil extraction techniques ensure that oil yields from the seed are also high. Yield of oil from one hectare in Europe equals that from about 4 hectares in India.

In recent years expanded rapeseed production has been chiefly in France, West and East Germany, United Kingdom, Denmark, and Czechoslovakia. Production by the two other large rapeseed producers, Poland and Sweden, seems to have plateaued (Table IV). French production has doubled in the last year, a factor attributed mainly to a very favorable market condition (subsidies are paid to both the producers and processors of EEC grown rapeseed). The lower cost of domestically produced rapeseed encouraged crushers to use this oilseed preferentially.

From the data presented in Table XIV, which are averages for the 5 year period from 1977 to 1981, some facts can be derived as to the production and trade of rapeseed. The Eastern European countries are major producers of rapeseed and rapeseed oil but export relatively little (<10%). On the other hand, Sweden and Denmark also produce large amounts of rapeseed, but half of the rapeseed and 60% of the oil is exported. The third group of countries which includes France, West Germany, United Kingdom, Netherlands, Italy, and Finland, characteristically has both a high production and consumption of rapeseed oil. From the rapeseed and rapeseed oil export–import figures, it is clear that considerable trade occurs as well.

Low erucic, low glucosinolate varieties are now under development or have been developed in France, West Germany, Denmark, and Sweden but with minor exceptions are not in commercial production. Over the next few years these new varieties can be expected to replace most of the single low rapeseed varieties and ensure a supply of high quality, low glucosinolate meals for their livestock industries. The development of double low rapeseed varieties is progressing much more slowly in the Eastern European countries.

TABLE XIV

Approximate Annual Production and Trade of Rapeseed and Rapeseed Oil of Selected European Countries Based on the Years 1977–1981[a] (in 1000 tonnes)

Country	Rapeseed				Rapeseed oil		
	Production	Import–export[b]	Net	Total crushed	Production	Import–export[b]	Net
Poland	640	−40	600	510	210	−20	190
Germany, East	290	−20	270	250	110	−7	100
Czechoslovakia	150	10	160	160	60	2	60
Hungary	80	−10	70	70	30	−4	20
Sweden	270	−90	180	180	70	−40	30
Denmark	120	−110	10	10	5	−3	2
France	620	−90	530	520	220	−160	60
Germany, West	310	400	710	620	240	−110	130
United Kingdom	180	140	320	300	130	−10	120
Netherlands	30	80	110	100	40	−10	30
Italy	2	70	70	70	30	40	70
Finland	50	—	50	50	20	−3	15

[a] Calculated from data given in U.S. Department of Agriculture (1981).
[b] Export values were subtracted from import values.

In the past the European Common Market countries have imported and crushed oilseeds, mainly soybean, in order to produce large quantities of high protein meals for animal production. In recent years these countries have become partly self-sufficient in oil and meal due largely to their rapeseed production. France and West Germany are the principal exporters of rapeseed oil (Table XIV).

E. Other Countries

Japan is the largest importer of rapeseed in the world (Table VIII), largely because they prefer rapeseed oil for cooking. Japan itself produces relatively little rapeseed (Table IV), but their imports of rapeseed have increased steadily since 1971 to a total of 1,200,000 tonnes in 1981. Most of this comes from Canada. As would be expected, the production of rapeseed meal in Japan has also increased during the past decade to 684,000 tonnes in 1981. As the Canadian production of rapeseed has shifted to the canola types, so has the Japanese crushing since it is derived largely from the import of Canadian seed. The canola meal in Japan is now used extensively in feeding livestock.

In South America, Brazil is actively exploring the production of rapeseed with first production recorded in 1980 and 1981 (Table IV). Rapeseed grows well there and is said to be reasonably competitive (R. K. Downey, personal communication). Chile made good progress in developing their rapeseed industry in the 1970s backed by an excellent research and development program. Peru has a pilot rapeseed development project but no commercial production.

African production of rapeseed is currently mainly limited to Ethiopia (Table IV). Algeria, Morocco, and Nigeria have imported considerable quantities of rapeseed oil in recent years (Table X).

To date, the United States has not produced sizable quantities of rapeseed, probably because the agronomic conditions there are highly suitable for growing other oilseed crops, such as soybean, peanut, and cottonseed. Nevertheless, there are relatively large areas in the Midwestern United States and Alaska which are suitable for rapeseed production, and one may expect to find this oilseed crop to gain broader acceptance in the future.

Common mustard seed (*B. juncea*) including the oil is used in the United States in substantial quantities as a condiment on hamburgers, hot dogs, ham, etc. Most of the mustard is imported as seed from Canada and processed in the United States for subsequent use as table mustard. Importations of mustard seed to the United States averaged 34,000 tonnes annually from 1973 to 1977 (Anonymous, 1977). Some of the seed is extracted to produce a mustard meal for use in processed meat; the oil is used for industrial purposes. In addition, the United States imports about 6000–8000 tonnes of high erucic acid rapeseed oil per year chiefly from Europe, but about 50% comes from Canada (Anonymous, 1976).

Interest in producing canola has been increasing in areas suitable for its production, i.e., in Alaska (Wooding et al., 1978) and in some of the Midwestern States. In Alaska, some canola is being grown in commercial quantities on recently developed land. In the Midwest, the estimated area in 1979 was 40,000 hectares (equivalent to about 36,000 tonnes of rapeseed). Canola was produced on "set aside" acreage and the seed was exported to Canada and Japan (Clancy, 1979). Present information indicates that canola is not being crushed in the United States.

IV. CANOLA MEAL—A MAJOR BREAKTHROUGH

A. Importance

Apart from price and availability considerations, the composition of an oil is only one of the factors that enter into decisions to choose one oilseed over another in world markets. Two other important considerations are the proportion of meal in the seed and the quality of the meal. For example, soybeans contain only 18% oil hence a much higher proportion of their total value lies in the high protein meal than in the oil. Conversely, rapeseed peanuts, sunflower, and cottonseed contain 30–40% oil, therefore oil is the more valuable component. Then the quality of the meal becomes a very critical factor in determining the commercial value of one oilseed versus another. Sunflower seed, for example, produces an inferior meal, which is high in fiber and low in protein (23%); meals from cottonseed and peanut are of medium quality.

In the past, Canadian and European rapeseed meals high in glucosinolates have been fed to livestock but there were many restrictions on their use. In the Orient, the high glucosinolate rapeseed meals have traditionally been returned to the land for fertilizer along with the night soils and animal manures. But the newly genetically selected canola meal, with its improved palatability, low glucosinolate, high protein (38%) content, and excellent amino acid profile is a high quality meal and very competitive with soybean meal for essentially all types of animal and poultry production. A decided preference for canola is now clearly evidenced in world markets. Markets for canola seed now depend to a considerable extent on the demand for canola meal. The following brief summary of the major nutritional features of canola meal is presented to update the reader on these very recent developments. For greater details see an excellent publication by Clandinin (1981).

B. Major Nutritional Features

Canola meal is compared with a well recognized high quality standard meal, namely soybean meal in Tables XV and XVI. Recommended levels of feeding are compared with the high glucosinolate rapeseed meal in Table XVII.

The amino acid profiles of soybean and canola meal are very similar. Canola meal is somewhat lower in lysine but substantially higher in methionine than soybean meal. Thus, the two proteins are complementary as has been confirmed by feeding trials.

The crude fiber and ether extract content of canola meal are both higher

TABLE XV

Proximate and Amino Acid Composition of Canola Meal and Soybean Meal[a]

	Canola meal		Soybean meal	
	As fed (%)	In protein (%)	As fed (%)	In protein (%)
Proximate composition				
Moisture	7.49		11.00	
Crude fiber	11.09		7.3	
Ether extract	3.78		0.8	
Protein (Nx6.25)	37.96		45.01	
Amino acid composition				
Alanine	1.73	4.56	1.89	4.20
Arginine	2.32	6.11	2.90	6.44
Aspartic acid	3.05	8.03	5.04	11.20
Cystine	0.47	1.23	0.29	0.65
Glutamic acid	6.34	16.69	8.10	18.00
Glycine	1.88	4.96	2.07	4.60
Histidine	1.07	2.81	1.08	2.40
Isoleucine	1.51	3.98	2.11	4.69
Leucine	2.65	6.97	3.37	7.49
Lysine	2.27	5.98	2.80	6.22
Methionine	0.68	1.78	0.63	1.40
Phenylalanine	1.52	4.01	2.16	4.80
Proline	2.66	7.00	2.20	4.89
Serine	1.67	4.39	2.25	5.00
Threonine	1.71	4.50	1.71	3.80
Tryptophan	0.44	1.16	0.54	1.20
Tyrosine	0.93	2.46	1.26	2.80
Valine	1.94	5.11	2.25	5.00

[a] From Clandinin (1981).

TABLE XVI

Mineral, Vitamin, and Energy Content of Canola and Soybean Meals[a]

Description	Canola meal (as fed)	Soybean meal (as fed)
Minerals		
Calcium, %	0.68	0.29
Copper, mg/kg	10.4	21.5
Iron, mg/kg	159.0	120.0
Magnesium, %	0.64	0.27
Manganese, mg/kg	53.9	29.3
Phosphorus, %	1.17	0.65
Potassium, %	1.29	2.0
Selenium, mg/kg	1.0	0.1
Zinc, mg/kg	71.4	27.0
Vitamins		
Choline, %	0.67	0.28
Biotin, mg/kg	0.90	0.32
Folic acid, mg/kg	2.3	1.3
Niacin, mg/kg	159.5	29.0
Pantothenic acid, mg/kg	9.5	16.0
Riboflavin, mg/kg	3.7	2.9
Thiamine, mg/kg	5.2	4.5
Energy		
DE kcal/kg (cattle)	2830	3178
DE kcal/kg (swine)	2900	3300
ME kcal/kg (cattle)	2400	2606
ME kcal/kg (growing chickens)	1900	2249
ME kcal/kg (adult chickens)	2000	2249
ME kcal/kg (swine)	2700	2825
TDN % (cattle)	64	72
TDN % (swine)	66	75

[a] From Clandinin (1981).

than for soybean meal. The high ether extract values result from including in the canola meal the gums obtained from refining the oil. This increases the energy level of canola meal.

In general, canola meal is a better source of available Ca, Fe, Mn, P, Se, and Mg than soybean meal, whereas the latter is a richer source of Cu, Zn, and K. In vitamin content canola meal is superior to soybean meal in its content of thiamine, riboflavin, niacin, biotin, folic acid, and chloline, whereas soybean meal contains more pantothenic acid (Table XVI).

TABLE XVII

Levels of Canola and Rapeseed Meals Recommended in Rations for Various Species and Types of Animal Production

Species	Production phase	Canola meal[a] (%)	Rapeseed meal[b] (%)
Chickens	Starter, grower	20	15
	Layer, breeder	10	5
Turkeys	Starter, grower	20	10
	Breeder	15	10
Ducks	Starter, grower	20	—
Geese	Starter, grower	20	—
Pigs	Weaning to 20 kg	12	5
	Growing, finishing	10–15	5
	Breeding	12	3
Cattle	Calves	NR[c]	20
	Lactating cows	NR	5

[a] Clandinin (1981).

[b] Hussar (1977).

[c] NR, no restriction.

Soybean meal is higher than canola meal in available energy content (Table XVI). This is due in part to the higher crude fiber content of canola meal. Thus, when formulating rations for very young animals or birds, e.g. broiler chickens, it may be necessary to add a high energy ingredient such as fat to increase the energy content of canola supplemented rations.

Canola meal is much more palatable than the high glucosinolate rapeseed meal and can be fed at much higher levels to high producing dairy cows and growing finishing pigs (Table XVII).

The glucosinolate levels in canola meal are only one-eighth to one-tenth of those found in the high glucosinolate rapeseed meals. Hence, canola meal can be fed at much higher levels to animals and birds that are sensitive to glucosinolates, e.g. some strains of laying hens and breeding gilts or sows (Table XVII).

V. CONCLUSIONS

The ability of rapeseed to grow in harsh environments under a wide range of temperature and moisture conditions will enable rapeseed production to expand in many areas unsuited to other oilseed crops. This will include a number of developing countries. The easy interchangeability of rapeseed and grain production using the same equipment for cultivating, harvesting, storage, and handling favors rapeseed as an alternative crop for grain growing areas and will encourage expansion in some situations.

Between 1971 and 1981 total world production of rapeseed oil increased by 93% mainly because of very large increases in China, Canada, and Europe. Over the same period production of total fats and oils increased only 37%.

Canada, the world's largest producer and exporter of rapeseed, has achieved almost a complete changeover to the low erucic, low glucosinolate (canola) type.

China, the second largest producer of rapeseed, is striving to achieve greater self-sufficiency in oil and meal and is steadily expanding rapeseed production to achieve this end. Low glucosinolate varieties are under development but reducing the erucic acid in the oil has little or no priority.

India, the third largest producer appears to be making little progress in increasing production. Large quantities of rapeseed oil are being imported. Nevertheless, the potential exists to greatly improve yields through better quality seeds, improved pest control, and extension of existing knowledge.

Europe, the largest producing area of rapeseed in the world, is like Canada, changing over to the low erucic, low glucosinolate type rapeseed. The production and use of rapeseed and rapeseed products has increased markedly in the past decade with the availability of the new rapeseed varieties and favorable market conditions.

Low glucosinolate canola meal with an amino acid profile equal in quality to soybean meal is a suitable protein source for livestock. Protein concentrates developed from canola are also suitable for human foods.

ACKNOWLEDGMENTS

The author expresses thanks and appreciation to Mr. Henri Pellicer and Mr. David Durksen, Marketing and Economics Branch, Agriculture Canada, Ottawa, and Mr. Wayne House, Grain Marketing Office, Department of Industry, Trade and Commerce, Ottawa, for helpful information and critically reviewing the manuscript; to Dr. R. K. Downey, Agriculture Canada, Saskatoon, for helpful suggestions.

REFERENCES

Anonymous (1976). "Report of High Erucic Oil Mission to the US." Canada Department of Industry, Trade and Commerce, Ottawa.

Anonymous (1977). "Annual Review of Oilseeds, Oilcakes and Other Commodities". Frank Fehr & Co. Ltd., Prince Rupert House, London.

Bell, J. M. (1980). In "Report on Canola Technical Seminars, Peoples' Republic of China, Nov. 9–22, 1980," pp. 18–21. Canada Department of Industry, Trade and Commerce, Ottawa.

Boulter, G. S. (1980). In "Report on Canola Technical Seminars, Peoples' Republic of China, Nov. 9–22, 1980," pp. 47–48. Department of Industry, Trade and Commerce, Ottawa.

Canada Grains Council (1981). Canada Grains Industry Statistical Handbook. Canada Grains Council, Winnipeg, Manitoba.

Clancy, B. (1979). *Feedstuffs* **51** (25), 36.

Clandinin, D. R. (1981). "Canola Meal for Livestock and Poultry," Pub. No. 59. Canola Council of Canada, Winnipeg, Manitoba.

Duncker, H. W. (1981). "Market Commentary—Grains and Oilseeds," pp. 40–41. Agriculture Canada, Ottawa.

Earl, A. E. (1980). *In* "Report on Canola Technical Seminars, Peoples' Republic of China, Nov. 9–22, 1980," pp. 7–13. Canada Department of Industry, Trade and Commerce, Ottawa.

Editorial (1981). *J. Am. Oil Chem. Soc.* **58**, 723A–729A.

Garg, J. S., and Bhan, S. (1978). *Indian Farming* **27**, 13–17.

Hussar, N. (1977). *Proc. Symp. Rapeseed Oil, Meal By-Prod. Util., 1977,* (J. M. Bell, ed.), Publ. No. 45, pp. 137–154. Rapeseed Association of Canada, Winnipeg.

Ohlson, R. (1972). *In* "Rapeseed" (L.-A. Appelqvist and R. Ohlson, eds.), pp. 9–39. Elsevier, Amsterdam.

Sarsons, K. D. (1980). *In* "Report on Canola Technical Seminars, Peoples' Republic of China, Nov. 9–22, 1980," pp. 49–58. Canada Department of Industry, Trade and Commerce, Ottawa.

Statistics Canada 1977–1982. Grain and Oilseeds Review, Cat. No. 22-007, Statistics Canada, Agr. Stat. Div., Ottawa.

Statistics Canada 1978–1982. Field Crop Reporting Series, Cat. No. 22-002, Statistics Canada, Agr. Stat. Div., Ottawa.

Statistics Canada 1982. Exports by Commodities, Cat. No. 65-004, Statistics Canada, Export Trade Div., Ottawa.

Stefansson, B. R. (1980). *In* "Report on Canola Technical Seminars, Peoples' Republic of China, Nov. 9–22,1980," pp. 14–17. Canada Department of Industry, Trade and Commerce, Ottawa.

U.S. Department of Agriculture (1981). *In* "Oilseeds and Products" (W.L. Brant, ed.), FOP 21-81. U.S. Dept. of Agriculture, Foreign Agricultural Service, Washington, D.C.

Wooding, F. J., Lewis, C. E., and Sparrow, S. D. (1978). *Agroborealis* **10**, 35–38.

3

The History and Marketing of Rapeseed Oil in Canada

G. S. BOULTER

High and Low Erucic Acid Rapeseed Oils
Copyright © 1983 by Academic Press Canada
All rights of reproduction in any form reserved,
ISBN 0-12-425080-7

I. BACKGROUND

The earliest beginnings of rapeseed cultivation are shrouded in the mists of antiquity; its early history is not well documented. Plants of the species grown in Canada undoubtedly made their first impression on early agriculture as weeds. Domestication occurred whenever the economic value of locally grown weeds was recognized.

Sanskrit records indicate that rapeseed was used as a source of cooking and illumination oil as early as 2000–1500 BC. The crop was introduced to Japan from China (possibly via Korea) about 2000 years ago in 35 BC. Ancient civilizations in Asia and along the Mediterranean used rapeseed for lighting because it produced a smokeless flame. In Europe, rapeseed has been cultivated for a very long time in those countries which did not have poppyseeds or olives; records of cultivation are available from the thirteenth century. Rapeseed was the most important lamp oil until it was replaced by petroleum based oils.

Rapeseed oil had rather limited industrial application until the development of steam power when it was found that rapeseed oil would cling to water and steam washed metal surfaces better than any other lubricant. This was due to its composition, providing the desired viscosity, adhesiveness, and solubility in mineral oil when processed for lubricants. For this reason rapeseed oil became an essential component of marine engine lubricants in naval and merchant ships.

II. INTRODUCTION OF RAPESEED TO CANADA

A. Origin of Seedstocks

Prior to World War II the Forage Crop Division of the Experimental Farm Services, Canada Department of Agriculture, had obtained seed of both forage and oil rapeseeds and had distributed them for small plot tests at experimental farms and research stations across the country. These research plots indicated that rapeseed could be successfully grown in both Eastern and Western Canada.

In 1936, Mr. Fred Salvoniuk, a farmer at Shellbrook Saskatchewan, obtained seed from a friend or relative in Poland from which country he had emigrated in 1927. Mr. Salvoniuk planted this seed in his garden and found it adapted well. He continued growing a small plot for a few years. Later it was established that the rapeseed he grew was the *Brassica campestris* species known as turnip rape in Europe. However, because of its origin it became known as Polish rapeseed. Because there were no established markets for rapeseed in Canada at that time, field scale production did not occur.

B. First Commercial Production

The stimulus for the commercial production of rapeseed in Canada arose as the result of a crisis early in World War II. When supplies of rapeseed oil from Europe and Asia were cut off by blockade, the Canadian, British, and American navies were faced with a serious problem. The necessity of finding an alternative source of supply of this essential oil prompted the Wartime Agricultural Supply Board to contact Dr. T. M. Stevenson about the possibility of Canadian production of rapeseed.

From his knowledge of small scale tests, Dr. Stevenson, who was the Head of the Forage Crop Division of Canada Department of Agriculture, Ottawa, was able to assure the Board that Canadian production was possible. He was requested to initiate production as rapidly as possible.

In the spring of 1942, Dr. Stevenson supplied a few federal experimental farms and stations with the small quantity of seed he had on hand. The harvest of that year amounted to 2600 pounds (52 bushels) of seed of *Brassica napus* species. A considerably larger quantity of seed than this was required for planting in 1943 to relieve the serious shortage of rapeseed oil. Dr. Stevenson located and purchased a total of 41,000 pounds from U.S. seed companies. This seed had originally been secured from Argentina and the name Argentine for the *Brassica napus* was widely used in the early years of production, and still is applied to the varieties of the *Brassica napus* species.

With a supply of seed on hand, the next task was to find growers who would undertake production, in spite of their lack of knowledge of the crop and its potential. Provincial Departments of Agriculture on the prairies arranged contracts with farmers at a guaranteed price of six cents per pound, with marketing to be done through the Canadian Wheat Board. With a wheat acreage reduction program in effect because of the large stock on hand, and with the 1942 Saskatchewan farm price of wheat at around 75 cents per bushel, some farmers were prepared to gamble with this unknown crop. A relatively small number of farmers in Manitoba and Saskatchewan entered into production contracts.

In Saskatchewan many of the contracts were arranged with registered seed growers. When the crop in 1943 reached the flowering stage, growers discovered to their consternation that it contained a mixture of plant and flower types, some of which were suspiciously like the weed species of mustard.

The registered seed growers were highly sensitive to the hazard of the introduction of weeds on their farms and in a few cases legal action was threatened against the government. Ultimately the growers were pacified.

The performance of this first crop in 1943 was gratifying. From the 43,600 pounds of seed grown on 3200 acres, a harvest of 2,200,000 pounds was realized. At an average yield of 700 pounds per acre netting six cents per pound, the per acre returns were substantially better than from cereals. This result stimulated the desired expansion of production in the following year. Production of the Argentine species *Brassica napus* was launched.

As information about rapeseed and the fact that there was now a market for it in Canada became general knowledge, the Shellbrook farmer increased production of the *Brassica campestris* species and sold seed to his neighbors. Because of the Polish origin of both Mr. Salvoniuk and the seed he had initially obtained, this rapeseed species came to be called Polish.

Because seed of the *Brassica napus* species was distributed widely at the outset of production, it dominated the acreage for a few years. Yield tests showed that *Brassica napus* outyielded *Brassica campestris*. However, earlier maturity and more shattering resistance of *Brassica campestris* were more agronomically desirable features particularly in the northern areas of production, and *Brassica campestris* soon occupied more acres than *Brassica napus* species.

In the period 1943–1948, acreage and production of rapeseed in Western Canada increased progressively from year to year by a factor of 2 to 3 times. Production occurred in both Manitoba and Saskatchewan for the first 5 years but was confined to Saskatchewan in the latter part of this period.

III. EARLY USE AND DEVELOPMENT OF RAPESEED

A. As a Marine Engine Lubricant

As has already been observed, vegetable oils are superior to mineral oils in their ability to cling to metal surfaces, particularly where the surface is being washed by steam or water. When ships were powered by steam, vegetable oils were an essential item in formulating lubricants for their engines. Rapeseed oil was the much preferred oil.

The first commercial production of rapeseed in Canada provided the raw material for the first processing of Canadian produced rapeseed oil. The 1943 production of Canadian rapeseed was crushed, refined and blown at the W. R. Carpenter (Canada) plant in Hamilton, Ontario. The production of subsequent years was handled at a plant built and operated by J. Gordon Ross in Moose Jaw, Saskatchewan, under the name of Prairie Vegetable Oils Ltd.

The bulk of the oil was used for lubricants. The crude oil was alkali refined, bleached, and blown. Blown or oxidized oils are those in which polymerization is produced by passing a stream of air through the oil at a temperature of 95°–120°C for several hours. As oxidation progresses the viscosity of the oil rises and the specific gravity increases. It is this thickening effect which in turn modifies the adhesive properties of rapeseed oil when it is blended with petroleum products in the manufacture of lubricants. The most critical quality factor of the oil for this purpose was its degree of unsaturation or "iodine value." An iodine value of 103–105 was preferred and oils with a value of over 108 were considered unsuitable for blowing. Oils from *Brassica napus* or Argentine seed had the desired iodine value of about 105, while that from *Brassica campestris*, or Polish seed, had an undesirably high iodine value of about 115.

B. As an Edible Oil

Western nations did not utilize rapeseed oil for edible purposes to any extent until the mid-1940s. During the war there was also a serious shortage of off-shore edible oils and fats in Canada so that small quantities of rapeseed oil were diverted to edible use. The quality of the early oil produced was extremely variable so that refiners such as Canada Packers Limited had great difficulty in processing the crude oil for inclusion in blended shortenings. The original seed stocks were not uniform in maturity. Resulting admixtures of immature seed gave a very green oil which required rigorous bleaching and made good deodorization difficult.

C. Quality Improvement through Plant Breeding

Simultaneously with the sowing of the first crop, Dr. W. J. White started a breeding program at the Federal Department of Agriculture Dominion Forage Crops Laboratory in Saskatoon. Dr. H. K. Sallans and Mr. G. D. Sinclair at the Prairie Regional Laboratory of the National Research Council at Saskatoon did the early work of quality evaluation on the oil in terms of iodine value, acid value, refractive index, and oil color.

Under the incentive of a fixed government price and a severe oils and fat shortage, production and crushing of rapeseed increased steadily through the war years, but with the return to adequate supplies after the war, controls were phased out and producers and crushers were left to their own devices.

In the immediate postwar period, the ready availability of the traditional edible oils such as cottonseed, peanut, and soybean caused the use of rapeseed oil in edible products in Canada to be dropped. However, the demand for industrial grade oil continued and in 1949 Mr. Ross was able to negotiate a substantial sale of oil but because of world market conditions, the 1949 crop returned producers only 3.65 cents per pound, and in 1950 the crop nearly disappeared. It might well have disappeared completely at this point if it had not been for the persistence of a relatively small group of people including Mr. Ross who continued to contract production and to find markets for oil, meal, and seed. Mr. Ross continued to supply Canadian rapeseed oil until 1959.

Another advocate of rapeseed and rapeseed oil was Dr. Sallans, who pointed out on many occasions that the crop was well adapted to the Prairies, had the potential of replacing much of our imported vegetable oils, and could be essential in the case of national emergency. Dr. Sallans and his colleagues at the Prairie Regional Laboratory, Dr. B. M. Craig, Dr. L. R. Wetter and Dr. C. G. Youngs, did much to maintain interest in the crop and provided basic information on both oil and meal.

At the Division of Applied Biology of the National Research Council in Ottawa, Dr. N. H. Grace, Dr. H. J. Lips, and Dr. A. Zuckerman were investigating the edible uses of rapeseed oil. They concluded that with careful refining, bleaching, hydrogenating, and deodorizing, rapeseed oil could be substituted for soybean oil in edible products without loss in quality. However, edible oil processors remained reluctant to use the oil. They felt that flavor stability was inferior to other oils available to them and because there was no firm market, seed production and oil supplies were erratic.

It was not until 1956–1957 that rapeseed oil was again extracted and used for edible purposes. Since then, food use of rapeseed oil in Canada and Western Europe has surpassed the industrial usage. This era also marked the beginning of a Western Canadian based industry now consisting of nine crushing plants with a rated daily crushing capacity of approximately 5.0 thousand tonnes. The market has expanded rapidly as growers, processors, and refiners have learned how to handle the crop and produce quality products.

Cooperative research on many aspects of rapeseed processing and utilization has resulted in rapeseed oil (now termed canola oil) becoming the most widely used edible oil in the domestic market in Canada. Canola oil is the oil derived from the rapeseed varieties grown in Canada which are low in erucic acid and low in glucosinolates.

IV. THE STORY OF PARTNERSHIP AND TEAMWORK

The wartime impetus for rapeseed production and utilization in Canada was kept alive by a small group of men during the 1950s and early 1960s. Their hard work and determination launched the rapeseed improvement program which has brought us to the low erucic, low glucosinolate varieties we have today. With the internationally registered trademark of the term "Canola" designating the source of the products derived from these double low varieties, Canada has established a new image worldwide. This is based on the recognition of Canada as a world leader in plant breeding and research, and on the improved quality of the canola crop.

The survival of the crop came about through the faith stimulated by Mr. J. Gordon Ross and Dr. H. K. Sallans that rapeseed should find a place in

Canadian agriculture. The research and development involving plant breed-ing, analytical methods, processing conditions, formulation of products, and nutritional quality of oil and meal were carried out by a small number of individuals in universities, federal agencies, and Canadian industries work-ing together without any formal organization or structure. It seems appropri-ate to quote from a statement made by the late Director of the Prairie Re-gional Laboratory, Dr. H. K. Sallans, "without the co-operation of scientists in government laboratories, universities and industries, it would be impos-sible to carry a project of this diversity and magnitude through to a success-ful completion. All workers contributed ideas, time and financial resources to the development and monitoring of the program despite the fact that these were employed in separate and discrete administrative units. The program moved solely on its merits, and the fact that we have scientists in Canada who are anxious and willing to tackle problems that will contribute to im-proved economic conditions in Canada."

The nucleus for the rapeseed improvement program was developed in Saskatoon. The Saskatchewan Wheat Pool operated an extraction plant for flax and undertook the crushing of rapeseed for the J. Gordon Ross Syndi-cate after the plant of Prairie Vegetable Oils, Moose Jaw, was dismantled. The equipment was sold in 1951 to Canadian Vegetable Oil Processing, successor to W. R. Carpenter Canada, located in Hamilton. The Saskatche-wan Wheat Pool provided the industrial input into the program for primary processing of the seed to produce oil and meal. The Prairie Regional Labora-tory of the National Research Council was opened in 1948 to carry out research on agricultural products. It contained a pilot plant laboratory head-ed by Dr. Sallans, capable of assessing the effect of processing conditions for extracting rapeseed oil. Staff included Dr. C. G. Youngs, a chemical engi-neer with the responsibility for engineering research, Dr. L. R. Wetter, a biochemist with a background of experience in enzymology and proteins, and Dr. B. M. Craig, a biochemist with experience in the chemistry of fats and oils. Plant breeding work at Saskatoon was under the direction of Dr. W. J. White, who was in charge of the Federal Department of Agriculture's rapeseed breeding and development research program at Saskatoon. The University of Saskatchewan maintained an active interest in the feeding po-tential of rapeseed through Dr. J. M. Bell of the Animal Science Department and Professor J. B. O'Neil of the Poultry Science Department. The existence of these facilities at a single location contributed greatly to the development of a team approach.

This small group of individuals and organizations had the capacity to car-ry out research on the chemistry of the oil protein and glucosinolates, to develop better rapeseed varieties, and to study processing conditions to im-prove the quality of oil and meal. The knowledge acquired from the labora-tory research was readily evaluated in the industrial extraction plant, both in terms of processing and the acceptance of oil and meal by secondary industries. Research on the utilization of rapeseed oil in margarines, shortenings, and salad oils or secondary processing was contributed by Canada Packers Limited in Toronto and Montreal. Valuable input was provided by personnel such as Mr. R. A. Burt and Mr. B. F. Teasdale. When the nutritional quality of rapeseed oil became a concern to the development of the industry, the Food and Drug Directorate, National Health and Welfare, undertook to carry out research work in this area. As a consequence these groups who were associated with different organizations had the capacity to advance the development of the total industry.

In the early 1950s experimental lots of margarine were produced by the Dairy Pool in Saskatoon and Canada Packers, St. Boniface, Manitoba. This, combined with the Wheat Pool crushing plant in Saskatoon, provided the necessary facilities for testing rapeseed oil in commercial margarine production. Unfortunately, the margarine produced in these early trials did not spread satisfactorily and production at the Dairy Pool was short-lived. Experimental lots of margarine using rapeseed oil blends were also prepared in 1953 and 1954 by Canada Packers at St. Boniface with oil from the Co-op plant at Altona, but this did not reach any real commercial production at that time.

Oil from rapeseed crushed in Canada continued to be exported for industrial use by Mr. Ross. The Saskatchewan Wheat Pool plant in Saskatoon, where most of the seed was crushed, had installed alkali refining and bleaching equipment in 1949. From 1949 to 1956, all oil from this plant was shipped as refined, bleached oil. The Co-op Vegetable Oil plant in Altona, in addition to refining and bleaching, also had a deodorizer for the production of salad and cooking oil from sunflower seed. From 1951–1956 relatively small quantities of rapeseed oil were processed in this plant to provide an edible liquid oil which was marketed by Gattuso Corporation in Montreal. This represented the first significant use of rapeseed oil for the domestic market in Canada since the war years.

Experimental work to increase the edible use of rapeseed oil continued in Canada Packers, St. Boniface, Montreal, and Toronto plants and in their Research and Development Laboratories.

The use of rapeseed oil for edible purposes was also encouraged by two of the Associate Committees of the National Research Council. The Associate Committee on Grain Research recommended "That in view of the productivity of rape as an oil crop in Western Canada, the Canada Committee on Fats and Oils be asked to foster and promote commercial tests by industrial producers of edible products." As a result of this recommendation, Canada Packers Limited intensified their research program with cooperation from the Saskatchewan Wheat Pool through Mr. J. R. Reynolds, their chief chemist, and Drs. Craig and Youngs of the Prairie Regional Laboratory of the

National Research Council and to whom the author is deeply indebted for their gracious agreement in permitting the use of much of the material recorded in this chapter.

By 1955 Canada Packers, Montreal, was producing salad oil from rapeseed oil on a continuous basis. During 1956–1957 the use of rapeseed oil in edible products was temporarily interrupted while the Nutrition Issue (see the following section) was being investigated. Work was resumed when this issue was clarified.

The need was urgent to develop edible uses for the oil at that time. Markets in lubricating formulas were decreasing as diesel engines replaced steam and as new additives were developed for petroleum oils. Seed production was swinging to the *Brassica campestris* or Polish species because of its shorter time to maturity, but the iodine value of oil from this species was higher than desired for lubricants. Surplus wheat was building up and production of rapeseed could be greatly expanded as a "cash crop."

V. THE NUTRITION ISSUE

A. Early Concerns

In 1956 small but increasing quantities of rapeseed were finding their way into the edible liquid oil market. The potential looked good. The Saskatchewan Wheat Pool decided to install deodorizing equipment for expanded edible liquid oil production. Crushing and production was still being contracted by Mr. Ross and the deodorized oil marketed by Gattuso Corporation, Montreal, was being favorably received. On July 23 of that year, the Food and Drug Directorate of the Department of National Health and Welfare ruled that rapeseed oil was not an approved edible oil in Canada and that, in view of literature reports of abnormalities arising from feeding rapeseed oil to experimental animals, sales of this oil for edible purposes was to cease immediately and all stocks currently on retail shelves were to be withdrawn.

The literature reports referred to were primarily those of Dr. K. K. Carroll, Department of Medical Research, University of Western Ontario. Dr. Carroll had been studying the nutritional effects of rapeseed oil and erucic acid in the diet of rats since 1949 and had found that both resulted in reduced weight gains, increased cholesterol content, and increased weight of the adrenal glands.

Mr. Ross was able to persuade the Department of Health and Welfare to withdraw its objection to the use of rapeseed oil as an edible oil, pending a submission showing the oil to be a safe ingredient in the diet of man. At the request of Mr. Ross, Dr. Sallans and Dr. Craig prepared a review of the available information on the nutritional properties of rapeseed oil, and the subject received a thorough review at the meeting of the Canadian Committee on Fats and Oils in October 1956. The conclusion was that there was nothing to indicate that the rapeseed oil in its current limited use represented a hazard to health, but further information on its nutritional properties was urgently required. Three objectives were set out: to obtain more nutritional data, to determine the feasibility of reducing the erucic acid content through plant breeding, and to review the health records and autopsy reports in Germany during the war when rapeseed oil was virtually the only edible oil available. The last was given a low priority because it was felt stress and possible nutritional deficiencies during the war might have a bearing on the results.

Nutritional studies were initiated immediately by Dr. J. A. Campbell and Dr. J. L. Beare-Rogers of the Food and Drug Directorate, in addition to continued investigations by Dr. Carroll. Samples of oil from known seed sources, fractions of the oils, synthetic oils, and analytical data on these oils and on fat deposits in test animals were supplied by the Prairie Regional Laboratory. Work on plant breeding to reduce erucic acid content was begun in 1958 by Dr. B. R. Stefansson, Plant Science Department, University of Manitoba and Dr. R. K. Downey, Dominion Forage Crop Laboratory, Saskatoon. In 1960 these two gentlemen isolated the first seed variety with low erucic acid content. Special mention should be made in this connection of the very significant accomplishment of Dr. Craig in perfecting the gas liquid chromotographic technique as an analytical tool for half seed analyses, thereby speeding up the development of new varieties.

B. Nutritional Improvement through Processing and Breeding

1. PROCESSING

The use of rapeseed oil as an edible liquid oil had been growing slowly. Retail quantities of oil from the Saskatchewan Wheat Pool plant in Saskatoon were marketed in 1957. During 1958, rapeseed oil had completely replaced soybean oil in Canada Packers' St. Boniface products. In Montreal and Toronto, Canada Packers were using significant quantities of rapeseed oil in shortening and salad oils. By mid-1958 Canada Packers, Montreal, was producing shortening made entirely from rapeseed oil. The question of nutritional problems still hung over the oil and at the urging of Dr. Craig in 1958, the Edible Oil Institute requested clarification of the status of rapeseed oil from the Food and Drug Directorate. Following a comprehensive investigation with rats extending over some 18 months, the Food and Drug Directorate reported its experiment indicated no harmful effects were observed from feeding rapeseed oil at levels that would ordinarily be consumed by humans. As a result the Food and Drug Directorate announced it had no objection to the use of rapeseed oil in moderate amounts in foods in Canada.

Although not exactly a hearty endorsement of the oil this did open the door to expanded edible use, and the next 5 years saw rapid development of the rapeseed industry. Canada Packers led the way in the use of rapeseed oil in liquid oils, shortenings, and margarines, and various other food processors followed. Introduction and expanded use of the oil was gradual as all of these companies had to do a great deal of testing to develop suitable formulations, and to maintain the required quality in their products. Utilization was assisted by a marked improvement in the quality of rapeseed oil as plant breeders developed improved varieties of rape.

Improved crushing techniques also contributed to the expanded use as several of the western oil seed processing plants changed from straight expeller or screw press crushing to the prepress solvent extraction process.

Dr. Sallans viewed this time as critical in the development of the industry. Arrangement was made for Dr. Craig and Dr. Wetter from the Prairie Regional Laboratory and Mr. Reynolds from the Wheat Pool to spend 6 weeks in Europe studying processing and utilization of rapeseed, its oil, and its meal. This was the beginning of many exchange missions between Canada and other countries on the subject of rapeseed.

In 1960 the Wheat Pool commenced operation of a solvent extraction plant. This process, which dispensed with prepressing was the first of its type to be applied to rapeseed in the world. Hydrogenation problems, which had been occurring occasionally became serious with the start-up of this plant. Mr. J. R. Reynolds, Mr. A. B. Cameron, plant superintendent at the Wheat Pool Vegetable Oil Plant, and Dr. Youngs at the Prairie Regional Laboratory traced this problem to the seed preparation prior to oil extraction and established procedures that subsequently became widely adopted around the world. It is worth noting here that Mr. Cameron possessed a great deal of crushing experience, having been directly involved with the start-up and operation of the expeller plants at Moose Jaw, Altona, and Saskatoon prior to the new all solvent plant at Saskatoon.

Production and edible use of the oil continued to expand with the opening of processing plants by Western Canadian Seed Processors (now Canbra Foods) at Lethbridge in 1960, and Agra Vegetable Oil (now CSP Foods) at Nipawin in 1963. Rapeseed was also crushed in Canlin's expeller plant in Montreal since 1967.

The indispensable role played by Canada's rapeseed crushing industry in its multifaceted approach has added immeasurably to the development of rapeseed as Canada's number one oilseed. Throughout the inchoate stage, it bore a large share of the cost burden not only financially but also in terms of supplying expertise, manpower, plant and equipment, production time, and products for testing and quality evaluation; all this to the end that the growth and development of this fledgling industry be firmly and soundly established.

2. BREEDING LOW ERUCIC ACID RAPESEED VARIETIES

The results of the nutritional studies removed the objections to erucic acid in foods, but in the meantime the other approach to reduce the erucic acid content through plant breeding was more successful than hoped for. In 1960 Dr. Stefansson and Dr. Downey isolated seed from the forage rape "Liho" which contained little or no erucic acid. By 1964 this characteristic had been transferred to the oilseed rape varieties. Small quantities of seed were supplied by the Research Station, Canada Agriculture at Saskatoon, extracted at Prairie Regional Laboratory and the oil sent to Canada Packers, Toronto laboratory, for evaluation. Although low erucic acid rapeseed oils did not appear to have any particular advantage in margarine or shortening manufacture, they gave excellent yields of stabilized salad oil after treatment with light hydrogenation and winterizing, a process that is applied to soybean oil. A patent on the production of such salad oil from the low erucic acid rapeseed (LEAR) was obtained by Canada Packers.

From 1965–1969 the Saskatchewan Wheat Pool contracted production of low erucic acid seed obtained from Agriculture Canada, Saskatoon, and crushed commercial quantities for evaluation of the oil by Canada Packers and other edible oil processors. Higher costs were involved in handling the low erucic seed and processors were unwilling to pay a premium; therefore, production remained at a low level. With an abundance of European sunflower oil available for salad oil production, interest in low erucic and stabilized salad oil was also dampened.

The LEAR varieties came into their own, however, when questions arose again in 1970 over the nutritional effects of erucic acid. Papers presented at the International Conference on the Science, Technology and Marketing of Rapeseed and Rapeseed Products in Ste. Adèle, Quebec, reporting on investigations by Dutch, French, and Canadian nutritionists, linked erucic acid to fat accumulation in the hearts of young test animals and to heart-related problems in older animals. Although there was no evidence of harmful effects in humans, the Minister of National Health and Welfare stated it was considered prudent as a sound public health measure to change over to erucic acid free rapeseed as soon as practicable.

The continued development of the LEAR varieties carried on by plant breeders and the experience gained by continued production made it possible by December 1, 1973, to limit the erucic acid content in processed edible food products to less than 5%.

VI. QUALITY STANDARDS OF RAPESEED AND CANOLA OIL

Another landmark in the development and marketing of rapeseed oil was the setting of standards for the oil in 1965 under the auspices of the Edible Oils Institute. Samples of the oil from the four Western Crushers were examined in six refiner's laboratories. Specifications for free fatty acids, moisture and impurities, flash point, refined bleached color, green color in crude oil, refining loss and phosphatide content were approved and published by the Canadian Government Specifications Board. These standards have been revised periodically, the latest being in 1976.

A. Canadian Government Specifications Board Standards

Below are excerpts from this revision publication CAN2-32300M-76. Because of improvements in the quality of the oil produced in recent years the

TABLE I

Canadian Government Specifications Board Standards for Low Erucic Acid Rapeseed Oils

	Requirement	
Characteristic	Type 1	Type 2
Free fatty acid (as oleic acid) max. % by mass	1.0	1.0
Moisture and impurities combined max. % by mass	0.5	0.3
Flash point, min. °C	150	150
Refined bleached color, max.	1.5 Red[a]	1.5 Red[a]
Green color, crude oil stipulation lighter than	Std A[a]	Std A[a]
Neutral oil, min. % by mass	98.0	98.5
Loss,[b] max. % by mass	2.0	1.5
Phosphorus content, max. % by mass	—	0.02[c]
Erucic acid, max. % by mass	5.0	5.0

[a] In some years it may not be possible to meet these color requirements because of adverse climatic conditions. Under these circumstances other acceptable maximum color limits may be the subject of negotiation between the purchaser and the supplier.

[b] Loss includes free fatty acids, color bodies, phospholipids, gums, etc.

[c] This value is approximately equivalent to a phosphatide content of 0.60% by mass when determined as Acetone Insoluble Matter by the method given in the National Standard. Phosphorus determined by AOCS Method Ca 12-55 shall apply in the event of a dispute.

National Standard referred to in Table I is regarded today as minimal for both domestic and export sales.

Scope: This standard applies to the oil of rapeseed, intended for use in the manufacture of food products and containing less than 5% erucic acid.

Classification: Types—Low erucic acid rapeseed oil shall be of the following types.

> Type 1—Crude
> Type 2—Crude, degummed

Requirements: The low erucic acid rapeseed oil shall comply with the requirements specified in Table I.

B. Trading Specifications

Domestically Canadian canola oil is sold as crude, degummed (Type 2). The quality standards used by the trade at present are individual specifications mutually agreed on by the buyer and seller. While these standards are in accordance with the specification established by the Canadian Government Specifications Board previously discussed, as a general rule the current standards are more restrictive and cover a broader range of quality factors. For example, most crushers now offer canola oil containing a maximum of 2.0% erucic acid and having a maximum sulfur level of 5–7 parts per million. Refiners have suggested that as a guideline for the chlorophyll content, that it be not more than 10 parts per million and the free fatty acid be reduced to 0.7% (as oleic acid) maximum.

One of the crushers in Western Canada is prepared to sell a semirefined canola oil which can be physically refined. The specifications for this grade of oil are the same as for crude degummed with the following exceptions: moisture and impurities, 0.2% maximum; neutral oil, 99.0% minimum; phosphorus, 50 parts per million maximum; and sulfur, 5 parts per million, maximum.

For export, Canadian canola oil is sold primarily as crude, degummed (Type 2), the occasional exception is the trading of regular crude canola oil (Type 1). The quality standard used for export purposes are those established by the Canadian Government Specifications Board. These specifications are monitored by arrangement between the buyer and seller. The general practice is for export shipments of oil to be surveyed and analyzed by an independent laboratory such as Warnock Hersey Services or General Testing Laboratories, division of SGS Supervision Services Inc.

C. Standards for Canola

Canola is genetically improved rapeseed. There are two distinguishing characteristics, one relating to the oil component, the other to the meal,

TABLE II

Canola Oil or Low Erucic Acid Rapeseed Oil[a]

(a)	Shall be the oil produced from the low erucic acid oil-bearing seeds of varieties derived from the *Brassica napus* L. and *Brassica campestris* L. species;
(b)	Shall be refined, bleached, and deodorized;
(c)	Shall have
	(1) A relative density (20°C/water at 20°) of not less than 0.914 and not more than 0.920;
	(2) A refractive index (n_D 40°C) of not less than 1.465 and not more than 1.467;
	(3) A saponification value (milligrams potassium hydroxide per gram of oil) of not less than 182 and not more than 193;
	(4) An iodine value (Wijs) of not less than 110 and not more than 126;
	(5) An unsaponifiable matter content of not more than 20 g/per kilogram;
	(6) An erucic acid content of not more than 5% (w/w) of the component fatty acids;
	(7) An acid value of not more than 0.6 mg potassium hydroxide per gram of oil, and
	(8) A peroxide value of not more than 10 mEq peroxide oxygen per kilogram of oil; and
(d)	May contain oxystearin.

[a] This standard will be promulgated under the Canada Agricultural Products Standards Act.

covered by the term canola. These differentiate it from the other rapeseed varieties and are defined as follows under the Canadian Agricultural Products Standards (CAPS) Act.

Canola oil is the oil extracted from whole seeds of varieties from *Brassica campestris* and *Brassica napus* species with low levels of both erucic acid and glucosinolates, which are commonly processed for edible purposes. It shall not contain more than 5% erucic acid. For complete details see Table II.

Canola meal is the meal obtained after the removal of most of the oil from whole seeds of varieties with low levels of both erucic acid and glucosinolates, by a prepress solvent extraction or direct solvent extraction process. It shall contain not more than three milligram equivalents of 3-butenyl isothiocyanate per gram of oil-free dried meal.

Rapeseed meal contains small amounts of sulfur compounds that have growth inhibiting effects when fed to certain types of livestock. These sulfur compounds are called glucosinolates.

The modern rapeseed varieties known as canola are also termed "double low" because of the two above-mentioned characteristics. Presently the Canadian production of canola or "double low" varieties is testing about 2% erucic acid with the expectation of soon reaching 1% or less. Virtually all (90%) of the rapeseed now grown in Canada is of the canola type.

VII.   MARKETING

Rapeseed ranks fifth in world production of oilseed crops, surpassed only by soybean, sunflower seed, cottonseed, and peanut. It is a major source of dietary fat in many countries. Fifty countries are known to import rapeseed oil and 16 countries are major producers. Rapeseed oil has been used by humans for many centuries as a safe and acceptable dietary source of vegetable oil.

In Canada concurrent with the spectacular developments in plant breeding, agronomics, production, and processing to produce a high quality nutritious oil, emphasis was also given to implementing an aggressive marketing program. As quality improvements became evident, Canadian processors and consumers recognized the versatility of this home-grown product. In addition to the normal commercial promotion carried on by the firms directly concerned, the Federal Government through the Departments of Agriculture and Industry Trade and Commerce gave unstinting encouragement and support to this aspect of the enterprise.

A.   Domestic Acceptance and Marketing

The term canola, accentuating the excellent quality factors the plant breeders have introduced into canola seed, is a significant step forward in marketing. The nutritional content of double low canola seed is superior to the old varieties of rapeseed in respect to both the oil and meal. Since Canada is leading the world at the moment in the development of the double low varieties this gives her a distinct advantage in expanding market share for these products.

In 1978, Dr. A. B. Morrison, the Assistant Deputy Minister of the Health Protection Branch, Health and Welfare, Canada, issued the following statement regarding the safety of Canadian rapeseed oil: "On the basis of research conducted by the Health Protection Branch and studies undertaken by Agriculture Canada and elsewhere, the Health Protection Branch has given the product a clean bill of health as a source of fat in Canadian diets. Canadian rapeseed oil is a safe and nutritious ingredient of margarine, salad, cooking and frying oils and shortenings."

When properly processed, canola oil has an extremely good "chill test," which makes it an ideal oil for the manufacture of salad dressings and mayonnaise. Deodorized canola oil has a "bland" flavor that makes it desirable for processes that cannot tolerate any flavor carrying through to the finished product. Partially hydrogenated canola oil is a good base oil for the manufacture of margarines and is also a stable oil that can be used in deep frying. Canola oil is a good oil for an "all-purpose" shortening.

The extent to which canola oil has become accepted is well illustrated by

TABLE III

Canadian Exports of Rapeseed Oil[a]

Country of destination	1978 Jan.–Dec.	1979 Jan.–Dec.	1980 Jan.–Dec.
Algeria	—	6,030	—
Australia	3,314	3,348	4,280
Bangladesh	9,014	2,698	—
Brazil	—	—	707
Chile	500	12,178	3,344
Dominican Republic	—	—	878
Ethiopia	—	—	799
Hong Kong	5,592	5,987	13,358
India	45,994	70,069	117,524
Japan	12,516	8,665	9,769
Morocco	2,818	3,528	3,148
Netherlands	—	—	6,000
New Zealand	—	—	631
Peoples Republic of China	—	—	696
Spain	—	—	5,999
Singapore	—	—	752
Somalia	—	—	742
United States	1,650	2,607	2,831
Other	950	4,366	1,228
Total	82,348	119,476	172,686
Valued at	$53,414,000	$85,073,000	$118,783,000

[a] From Statistics Canada 65004.

growth in its consumption relative to soybean oil which has been the major oil source for many years. In the five years, 1975–1980, production of canola oil nearly doubled; it moved from slightly under 95,000 tonnes or 33% of the total vegetable oil production in Canada in 1975 to 183,000 tonnes or 47% of the total production in Canada in 1980. In 1980, canola accounted for 67% of the salad oil, 32% of the vegetable oil going into margarine, and 42% of the vegetable oil used in the production of shortening in Canada (see Chapter 10). The only other edible oil that contributed appreciably to the Canadian production is soybean, which accounted for 32% of the total in 1980. Together canola and soybean supply nearly 80% of the refined vegetable oil used annually in Canada. Canola oil is used in almost all of the foods that are designed for fat or to which fat may be added as an ingredient or as a cooking medium. The rapid and widespread acceptance of canola oil by industry and by the consumer attests to its fine quality.

In the publication "Canola Oil, Properties Processing and Food Quality,"

Professor Marion Vaisey-Genser (1979) states, "The high total fat intake of Canadians has alerted public health authorities to encourage moderation in fat consumption. Accordingly, the potential for further development of the home market for canola oil must be considered from the standpoint of decreasing per capita fat intake." Opportunities remain to increase the domestic market by further substitution of canola oil for other oils. During the past 30 years vegetable oil sources have been steadily gaining a larger portion of the domestic fats and oils market while the contribution from animal and marine sources have steadily declined. In 1950 vegetable oils made up 80% of all domestically deodorized oils. This proportion has increased to almost 90% in 1980. Within the vegetable oil group canola oil has become established as Canada's major vegetable oil.

Even though Canada is canola's best customer consuming between 182,000–183,000 tonnes of canola oil in 1980, Canada cannot possibly consume all the domestic production of oil. In 1980, 173,000 tonnes were exported, India being the largest customer taking two-thirds of the total. See Table III for Canadian exports of rapeseed oil in the past 3 years.

B. Export Market Development

Reference has already been made to the first technical mission sent to Europe in May 1961. This group consisted of representatives of the Prairie Regional Laboratory and Industry. It went to Europe to look at the utilization of rapeseed. In September 1961, the first trade delegation concerned primarily with rapeseed was headed by J. Gordon Ross and traveled to Britain and Europe. It included representatives of grain handling companies, research scientists, and trade officials of the federal government.

In the mid-1960s the interest in rapeseed became worldwide and Canada's exports began to increase. The trips to Europe in 1961 were the beginning of a number of trade and scientific exchanges between interested countries which has continued until the present. In the early promotional thrust, teamwork was particularly important because no one individual had sufficient detailed knowledge to deal adequately with all the varied aspects related to production, processing, and the utilization of rapeseed and its products. In order to overcome this difficulty Canadian trade missions usually included experts in rapeseed processing, oil utilization, oil chemistry, feeding of rapeseed meal, as well as experts in production and plant breeding. Similarly, when overseas trade missions visited Saskatoon, they had ready access to specialists in all these fields at a single location.

In April 1964 the first Canadian rapeseed trade delegation traveled to Japan, it being Canada's largest customer for rapeseed. The Japanese have been growing rapeseed for many generations but the pressure of a larger population on a small land base has made them look abroad for farm commodities. The Japanese imported 20,000 tonnes of Canadian rapeseed in 1960–1961, 700,000 tonnes in 1972–1973, and 1,100,000 tonnes in 1980–1981.

The favorable climate for development of the Japanese market traces back to much hard promotional work by people within the industry who went to Japan and who hosted Japanese delegations in Canada. A mutual exchange of rapeseed technology between Japan and Canada has benefited both countries. It is interesting to note in passing that since the mid-1970s Japan has entered into at least three joint ventures with Canadians in the crushing and processing of rapeseed in Western Canada.

But all the promotional work has not been aimed at Japan. During the years since 1961 numerous trade and technical missions from Canada have visited countries around the world, under the sponsorship of the Department of Industry, Trade and Commerce, Rapeseed Association of Canada (now the Canola Council of Canada), Canada Grains Council, Agriculture Canada, and others. Seminars have been held in Europe, and South America, Japan, Korea, Taiwan, India, China, Algeria, Morocco, the United States, Cuba, and Mexico; numerous samples of various rapeseed products have been supplied for testing purposes. Some of the missions were designed to discuss technical aspects of processing and the use of rapeseed oil and meal, while others explained the development program in Canada. A very significant contribution to market development was made at the time of Canada's changeover to low erucic varieties, by technical missions which visited all Canada's major customers for rapeseed.

In recent years the emphasis has been on sales of processed rapeseed products and some countries, including Japan, have purchased these "value added" products. Canada now is the world's largest exporter of rapeseed and has a domestic crushing capacity of around 1.5 million tonnes per year.

VIII. INDUSTRY ASSOCIATIONS

A. Rapeseed Association of Canada

In the space of 30 years since the crop was first produced on the Canadian Prairies, Canada achieved the distinction of becoming both the world's largest producer and exporter of rapeseed. For the first 20 years, rapeseed held a very uncertain place in the field crop plans of western producers. Producers pursued an in-and-out pattern of production, but since 1965 this new oilseed crop of the prairie regions has secured a permanent place in Canadian agriculture.

The surge in rapeseed production in 1965 gave rise to a need for market promotion. At this time the concept of a national association to promote the use of rapeseed and its products began to develop.

In March 1967 acting on a proposal put forward to the Canadian Barley and Oilseeds Conference a year earlier by Mr. James McAnsh, a meeting was held chaired by Mr. A. M. Runciman at which time a decision was made to form the Rapeseed Council of Canada (later changed to the Rapeseed Association of Canada). Mr. Runciman was elected its first President and James McAnsh became Executive Director.

This association encompassing in its membership the entire spectrum of parties interested in any way in rapeseed including researchers, growers, handlers, exporters, marketers, crushers, refiners, and feed manufacturers has done a great deal to advance and consolidate the rapeseed industry in Canada.

In its first printed publication it set out its main objectives as follows:

(a) Organize trade development programs, worldwide and on a long range basis, to ensure that Canadian rapeseed growers will have an expanding market for their crop.

(b) Encourage research work in the field of plant breeding and animal feeding, and become involved in programs related to continued improvement in quality of rapeseed, rapeseed oil and rapeseed meal.

(c) Disseminate information on production, marketing and all aspects of the rapeseed industry, and keep abreast of changes in regulations, trading practices and all matters pertinent to world markets.

(d) Maintain close contact with growers to keep them informed on market conditions and make them aware of the role they must play in market development.

(e) Conduct market surveys and studies, and strive to keep consumers and potential buyers of Canadian rapeseed and rapeseed products fully informed on the results of research and quality control programs conducted in Canada.

In the same year a noticeable shift in Federal Government policy occurred and the first official public support for the industry began with the setting up of a Rapeseed Utilization Assistance Program. This was funded by Industry, Trade and Commerce, and operated under the able direction of the late Dr. B. Weinberg, thus providing mechanisms to foster the team approach to solutions of problems in the rapeseed industry. This program is under the direct administrative control of the Research Committee of the Rapeseed Association. Much of the attention of the Research Committee has been directed toward improving the quality of rapeseed products in relation to marketing. This is especially the case with meal utilization because of problems in this area, but work has also been done to provide additional information on oil quality and utilization. This latter was particularly true between 1970 and 1975 along with the problems of the physiological effects of erucic acid. Currently the priorities for oil research are (I) improving refining techniques, (2) alternatives to present solvent extraction methods, (3) reduction of phospholipids, sulfur and chlorophyll in canola oil, (4) production of low *trans*-fatty acid canola products, and (5) new uses for canola oil.

In the late 1960s the Rapeseed Association first advanced the idea of a rapeseed pilot plant to support the expanding industry. This concept was picked up and acted on by the Federal Government authority responsible for the Canadian Wheat Board, who in turn commissioned studies to examine the opportunities for increasing Canadian participation in the manufacture and marketing of food proteins, starches, and component products from cereal grains and oilseeds. As a result of these investigations the POS Pilot Plant Corporation (POS represents Protein, Oil, Starch) was established on the campus of the University of Saskatchewan at Saskatoon and has been in operation there since 1977. This facility operates a versatile pilot plant for the development of value added food and feed technologies related to Canadian cereal, oilseed, and legume crops and represents a cooperative blend of many interests including federal and provincial governments, universities, industrial companies, and trade associations, which have combined to create a truly unique concept. Thus, this organization, while expanding on it, has incorporated in it the original pilot plant concept of the Rapeseed Association of Canada.

B. Canola Council of Canada

In July 1980 the Rapeseed Association of Canada relinquished its charter and letters patent when new letters patent were issued to the Canola Council of Canada by the Government of Canada.

The Canola Council of Canada is carrying on in the pattern and tradition established by the founders of the Rapeseed Association of Canada and is solidly dedicated to the original aims and objectives. With renewed devotion it is directing its time, energy, and resources to the development of foreign markets, while at the same time aiming for maximum expansion of the home market. A nation of only 25 million people has definite limitations on what it can consume. The per capita intake of fats and oils in Canada is already high; therefore, the main marketing thrust must be in the area of substitution of canola oil for competing oils. Particular attention is now being given to market development. There is a vital need for this industry to enlarge additional and alternate markets in order to add as much stability as possible to prices and provide continued demand for a volume of product. An expanded customer base is essential to sustain the present production and processing capacity. In addition to developing new markets diligent care must be given to serving and maintaining existing markets and to assure them of a regular supply of quality products.

The Council enjoys a close working relationship with the Canadian International Grains Institute and annually nominates a number of invitees selected from overseas customer countries to the five week "international" course at which rapeseed and canola are given special emphasis.

Renewed efforts are being given to multiplying the number of customers beyond the few being served at present, Japan and India being the largest. The Market Development Committee and the Executive of the Canola Council are embarked on a vigorous program for the expansion of off-shore outlets not only for Canadian rapeseed but more especially for canola products. To achieve these objectives a market development coordinator has been appointed and the Committee has been augmented to reflect a broader industry and government (provincial and federal) involvement.

The Canola Council does not intend to become directly involved in buying and selling products but rather to concentrate on developing markets and creating potentials so that the merchandisers can move in to make the sales. There is an important market development and market coordinating function to perform in order to complete the picture and thus assure the continuing prosperity of Western Canada's most exciting "value-added" industry in the agriculture sector.

ACKNOWLEDGMENTS

The author wishes to thank Mr. A. D. McLeod, the editor, and Drs. B. M. Craig and C. G. Youngs, contributors, for their kind permission to use portions of the publication "The Story of Rapeseed in Western Canada" in preparing this chapter.

REFERENCES

Appelqvist, L. A., and Ohlson, R. (1972). "Rapeseed, Cultivation, Composition, Processing and Utilization." Elsevier, Amsterdam.

Bell, J. M. (1977). "Rapeseed Oil, Meal and By-Product Utilization," *Proc. Symp.*, Publ. No. 45. Rapeseed Association of Canada, Winnipeg, Manitoba.

Clandinin, D. R. (1981). "Canola Meal for Livestock and Poultry." Publ. No. 59. Canola Council of Canada, Winnipeg, Manitoba.

Downey, R. K., Pawlowski, S. H., and McAnsh, J. (1970). "Rapeseed Canada's 'Cinderella' Crop." Publ. No. 8. Rapeseed Association of Canada, Winnipeg, Manitoba.

Harapiak, J. T., ed. (1975). "Oilseeds and Pulse Crops in Western Canada—A Symposium." Western Cooperative Fertilizers Limited, Calgary, Alberta.

McLeod, A. D., ed. (1974). "The Story of Rapeseed in Western Canada." Saskatchewan Wheat Pool, Regina, Saskatchewan.

Perkins, P. R. (1976). "An Economic Review of Western Canada's Rapeseed Processing Industry." A study for and in cooperation with the provincial governments of Alberta, Saskatchewan and Manitoba and the management of CSP Foods Ltd., Canbra Foods Ltd., United Oilseed Products Ltd., N.A.R.P. Co-op Ltd. and Alberta Food Products.

"Proceedings of the Fourth International Rapeseed Conference Giessen, West Germany" (1974). Foto-Druck Lenz D-63, Giessen, West Germany.

"Proceedings of the Fifth International Rapeseed Conference Malmö, Sweden" (1978). Organizing Committee of the Fifth Internatonal Rapeseed Conference, Lagerblads Trycker: AB, Karlshamn, Sweden.

"Proceedings of the Fourteenth Annual Convention—Canola Council of Canada." (1981). Canola Council of Canada, Winnipeg, Manitoba.

Rapeseed Association of Canada (1979). "Canola Canada's Rapeseed Crop." Publ. No. 56. Rapeseed Association of Canada, Winnipeg, Manitoba.

Rapeseed Association of Canada (1970). "Proceedings of the International Conference on the Science, Technology and Marketing of Rapeseed and Rapeseed Products, St. Adèle, Quebec." Rapeseed Association of Canada, Winnipeg, Manitoba.

Salmon, R. E., and Biely, J. (1978). "Canadian Rapeseed Meal." Publ. No. 51. Rapeseed Association of Canada, Winnipeg, Manitoba.

Stefansson, B. R., and McAnsh, J. (1977). "Grains and Oilseeds, Handling, Marketing, Processing," 2nd ed. Canadian International Grains Institute, Winnipeg, Manitoba.

Vaisey Genser, M., and Eskin, N. A. M. (1979). "Canola Oil, Properties, Processes and Food Quality," Publ. No. 55. Rapeseed Association of Canada, Winnipeg, Manitoba.

4

Chemical Composition
of Rapeseed Oil

R. G. ACKMAN

I. INTRODUCTION

If rapeseed (oil) did not exist, it would have to be invented. This para-
phrase of a statement originally applied to rubber has literally become true
for the oil currently produced in Canada and elsewhere which is widely
known as LEAR (low erucic acid rapeseed) oil. If produced from seed of
Brassica napus or *Brassica campestris* low in both erucic acid (<5% of total
fatty acids) and glucosinolates (<3 mg/g glucosinolates measured as 3-bute-
nyl isothiocyanate) the oil can be called canola oil, since "canola" is a

High and Low Erucic Acid Rapeseed Oils
Copyright © 1983 by Academic Press Canada
ISBN 0-12-425080-7

trademark of the Canola Council of Canada. As such it has been registered in Canada and by mid-1981 eleven other countries that cultivate or utilize rapeseed have either also accepted registration or are considering this step.

The need for a domestic edible vegetable oil has existed in most countries and the choice of the indigenous or introduced crop which is grown, is primarily a matter of climate (Downey, 1976). For this discussion, protein, government policies, and many other peripherally related factors (Craig et al., 1973; Slinger, 1977; Ohlson, 1979; Ziemlanski and Budzynska-Topolowska, 1978; Flanzy, 1979; Morice, 1979) are not relevant and are omitted. It may be observed that although the cultivation of rapeseed was anciently known in India, the main thrust of development, reviewed by Appelqvist (1972) and Ohlson (1972), has been in latitudes of 40–45°N or S. Selected growing areas favored by climate, such as the Peace River District of Canada, or Skåne in Sweden, are in fact at or about latitude 55°N. These growing conditions eliminate nearly all other oilseed crops except sunflower, although recently interest has been shown in white lupin (*Lupinus albus*, not the domestic garden variety) as an oilseed crop perhaps suited for the same areas (Hansen, 1976).

II. FATTY ACIDS

The principal fatty acids of these oils from temperate latitudes are compared in Table I. The unique characteristics of rapeseed oil are primarily due to the erucic (*cis*-13-docosenoic; 22:1 *n*-9) acid. This fatty acid is associated primarily with plants of the genus *Brassica* (rapeseed and mustard are the best known oil-yielding members of the family), but is also found in other plants such as several lupin varieties (Hansen, 1976; Favini et al., 1980), and meadowfoam (*Limnanthes alba*) (Pollard and Stumpf, 1980). Interestingly enough, a common Canadian weed seed contaminating harvested rapeseed, stinkweed (*Thlaspi arvenese*), contains 42% erucic acid (Rose et al., 1981), and wild mustard (*Sinapis kabar* D.C.) grows in the same areas and may also contribute erucic acid to oil from inadequately screened LEAR seed (Ackman and Sebedio, 1981). Rapeseed grown in other countries is similarly affected (Wilson, 1981). Mustard seed oil (Appendix I) does not properly belong in this review, although some comparative data are included in discussions of triglycerides since it is presumed that the occurrence and influence of erucic acid will follow the same rules in mustard and rapeseed oils. Erucic acid is distinguished by its melting point of 33°C, very high for a *cis* monoethylenic fatty acid. The major monoethylenic component of most vegetable oils, oleic (*cis*-9-octadecenoic; 18:1 *n*-9) acid, melts at 16°C. In an oil the fatty acids do not occur in free acid form, but in various combinations with other fatty acids in triglycerides and these melting points

TABLE I

Comparison of Major Fatty Acids, or Those of Special Interest, in Some Oils from Seeds of Plants Grown in Temperate Latitudes

Fatty acids	Plant seed				
	HEAR[a]	LEAR[b]	Sunflower[c]	Soybean[c]	White lupin[d]
16:0	4	3	6	10	7
18:0	1	1	4	4	3
18:1	15	60	18	24	55
18:2 *n*-6	14	20	68	51	15
18:3 *n*-3	9	13	<1	10	9
20:0	1	1	1	1	1
20:1	10	1	—	—	5
22:0	<1	<1	1	<1	3
22:1	45	<1	—	—	2

[a] Hadorn and Zürcher (1967); Swiss seed extract; 18:3 and 20:1 arbitrarily divided from a total of 19%.
[b] Sebedio and Ackman (1981); *B. campestris* cv. Candle seed extract.
[c] Hadorn and Zürcher (1967); authentic seed extract.
[d] Favini *et al.* (1980).

can only be taken as guides to the resulting physical properties of the oil in bulk. Since humans prefer an ambient temperature of about 20°C the difference generated by reducing the content of erucic acid and increasing that of oleic acid is quite critical for salad oil use. High erucic acid rapeseed (HEAR) oil is reported to melt at 4°–6°C (Ohlson, 1970) and would be cloudy in domestic refrigerators at or near such temperatures, whereas LEAR oils remain clear indefinitely at 5°C (Persmark and Bengtsson, 1976).

Of the other fatty acids of rapeseed oil, the 16:0 and 18:0 saturated acids are relatively low compared to the other oils of Table I, totaling about 5–6% in older LEAR oils (Ackman, 1977) and are not much different in newer LEAR/canola varieties (Tables II and III). The recently developed HEAR variety R-500 has oil with about the same saturated acid total as other *B. campestris* varieties (based on two analyses of one lot of seed), but these seem to have less saturated acid than *B. napus* varieties.

The polyunsaturated acids linoleic (*cis,cis*-9,12-octadecadienoic; 18:2 *n*-6)

TABLE II

Major Fatty Acids, and Those of Special Interest, in the Oils from Selected Seed Samples[a] of Current Canadian LEAR (Canola) and HEAR oils (w/w%)

Fatty acid	B. campestris				B. napus				
	Torch[b] 1973[c]	Candle 1977	Tobin 1981	R-500[d] 1976	Midas[b] 1973	Tower 1975	Altex 1978	Regent 1977	Andor 1980
14:0	0.04	0.05	0.05	0.08	0.05	0.08	0.06	0.07	0.04
16:0	3.48	3.55	3.12	1.72	3.84	5.31	4.27	4.29	3.87
18:0	1.50	1.38	1.30	0.93	1.69	1.40	1.74	1.85	1.34
20:0	0.42	0.43	0.37	0.86	0.55	0.53	0.60	0.60	0.50
22:0	0.27	0.20	0.16	1.16	0.33	0.25	0.26	0.30	0.39
Total saturated	5.71	5.61	5.00	4.75	6.46	7.57	6.93	7.11	6.14
16:1	0.22	0.28	0.23	0.25	0.29	0.22	0.28	0.27	0.18
18:1	61.65	55.58	56.64	12.18	64.33	59.52	60.95	61.52	58.21
20:1	1.38	1.78	0.97	4.55	1.16	1.33	1.27	1.29	1.55
22:1	0.44	1.63	0.11	54.07	0.02	0.02	0.04	0.01	0.01
Total monoethylenic	63.69	59.27	57.95	71.05	65.80	61.09	62.54	63.09	59.95
18:2	19.69	21.87	22.42	12.80	18.93	20.45	20.78	19.86	21.61
18:3	10.65	12.99	14.02	8.72	8.11	10.50	9.25	9.49	12.05
Total polyethylenic	30.34	34.86	36.44	21.52	27.09	30.95	30.03	29.39	33.66

[a] Courtesy of J. Daun, Canadian Grain Commission Laboratories, Winnipeg, Manitoba, except for Tobin which was courtesy of D. I. McGregor, Agriculture Canada, Saskatoon, Saskatchewan. Except for R-500 (total tabulated 97.3%) the components listed are 99% or more of total fatty acids.

[b] LEAR, but not double zero (i.e., contains glucosinolate >3 mg/g), hence technically not canola.

[c] Year licensed.

[d] HEAR (cv. Sarson); temporary license, 1976; not listed are 24:0 at 0.39% and 24:1 at 0.96%.

TABLE III

Oil Recovery, Percentages of Important Fatty Acids in Oil, and Percentage of Glucosinolates in Seed for European Varieties of HEAR and LEAR[a]

Variety	% oil recovery from seed	Weight percentages of fatty acids						% gluco-sinolates in seed
		16:0	18:0	18:1 n-9	18:2 n-6	18:3 n-3 20:1	22:1	
Gorzcanski	47.98	3.30	1.15	14.03	13.52	16.16	51.12	2.99
Major	46.04	3.03	1.09	18.09	13.32	18.69	45.00	2.95
Coriander	46.08	3.63	0.77	21.71	12.99	16.06	44.84	3.40
Brink	45.17	5.38	1.44	51.25	18.26	14.63	8.99	2.15
SW 7419	44.43	4.42	1.53	60.90	12.27	11.37	3.71	3.10
Lesira	44.33	4.46	1.53	59.03	20.22	11.79	2.78	1.87
Kara	45.23	4.50	1.51	60.68	19.04	11.62	2.65	2.30
Erra	42.15	5.03	1.45	58.88	21.01	11.65	2.27	2.50
Rapora	45.26	3.93	1.49	61.78	18.29	12.11	2.16	1.63
WW 748	43.76	4.22	1.67	60.27	20.09	12.06	1.69	3.40
Blanka	44.29	4.49	1.69	63.30	19.28	9.63	1.61	2.40
Primor	43.64	4.61	1.48	59.47	21.13	11.81	1.53	2.17
Girita	44.56	4.02	1.47	61.57	18.29	12.67	1.29	1.83
WW 766	44.53	4.45	1.44	62.45	17.95	11.88	0.92	3.00
Expander	41.67	4.14	1.61	60.04	20.82	12.24	0.79	2.35
Quinta	44.43	4.46	1.44	60.45	20.33	12.75	0.53	1.43

[a] From Sehovic et al. (1980).

89

and α-linolenic (*cis,cis,cis*-9,12,15-octadecatrienoic; 18:3 *n*-3) are both liquid acids for all practical temperatures and confer liquidity on the oils shown in Tables I to III. An adequate content of linoleic acid is a matter of concern to nutritionists (Alfin-Slater and Aftergood, 1976; Anonymous, 1980a; Food and Agriculture Organization, 1977; Houtsmuller, 1975; Wollbeck *et al.*, 1981). It is usually referred to as the "essential" fatty acid, although the structurally related γ-linolenic (*cis,cis,cis*-6,9,12-octadecatrienoic; 18:3 *n*-6) acid and arachidonic (*cis,cis,cis,cis,*-5,8,11,14-eicosatetraenoic; 20:4 *n*-6) acid are also ranked as essential (Bereziat, 1978; Halushka *et al.*, 1979; Lands *et al.*, 1977). The nutritional role of α-linolenic acid is much debated (Tinoco, 1982) and although it was in fact once regarded as an unwelcome competitor of linoleic acid (Holman, 1977; Houtsmuller, 1975), more recent views indicate a dietary need of 18:3 *n*-3 of 0.5% of calories (Holman, 1981). Although the related longer chain *cis,cis, cis,cis,cis*-5,8,11,14,17- eicosapentaenoic (20:5 *n*-3) acid has apparently desirable biochemical properties in the circulatory system (Bang and Dyerberg, 1980, 1981; Hirai *et al.*, 1980), it was thought not be derived from 18:3 *n*-3 in man (Dyerberg *et al.*, 1980; Sanders and Naismith, 1980). More recent work indicates that 18:3 *n*-3 is the precursor of 20:5 *n*-3 in man (Sanders and Younger, 1981). An objective of current breeding programs is to decrease 18:3 *n*-3 to less than 4–5% of fatty acids in LEAR oils, preferably with a concurrent increase in 18:2 *n*-6 to more than 30%, as well as to reduce erucic acid to less than 0.1% (Downey, 1979.) However, the balance between 18:2 *n*-6 and 18:3 *n*-3 probably should not be too radically altered in view of new evidence for the role of 18:3 *n*-3 in relation to 20:5 *n*-3, and presumably to 22:6 *n*-3 as well. Other countries have had similar objectives (Röbbelen, 1976; Hiltunen *et al.*, 1979) but a comparative study (Sehovic *et al.*, 1980) of sixteen European varieties shows (Table III) that insofar as oil fatty acids are concerned the reduction of erucic acid to 1% or less has, as in canola, resulted in a high-oleate oil, with proportions of other fatty acids (especially 18:2 *n*-6 and 18:3 *n*-3) being relatively unchanged. The total for 18:1 and 22:1 (gondoic acid, 20:1 *n*-9, was recorded together with 18:3 *n*-3 by Sehovic *et al.* 1980) is in fact remarkably constant at about 60–65% of oil, or in the same range as the total monoethylenic acids of the Canadian varieties listed in Table II. This inverse relationship between 18:1 and 22:1 had been forecast as early as 1965 and formed the basis for plant breeding to lower 22:1 (Craig, 1970) in rapeseed oils. The same relationship is observed in rapeseed phospholipids (Alter and Gutfinger, 1982).

In the HEAR and LEAR varieties available about 1970–75, for example Canadian LEAR cv. Span (Kramer *et al.*, 1975a) or French LEAR cv. Primor (Guillaumin *et al.*, 1980), the proportion of total C_{20} fatty acids was found to be correlatable with total C_{22} fatty acids (J. Daun, private communication). Figure 1 shows the regression line for the relationship, which has been extensively used for estimating 20:1 in analyses by gas–liquid chromatogra-

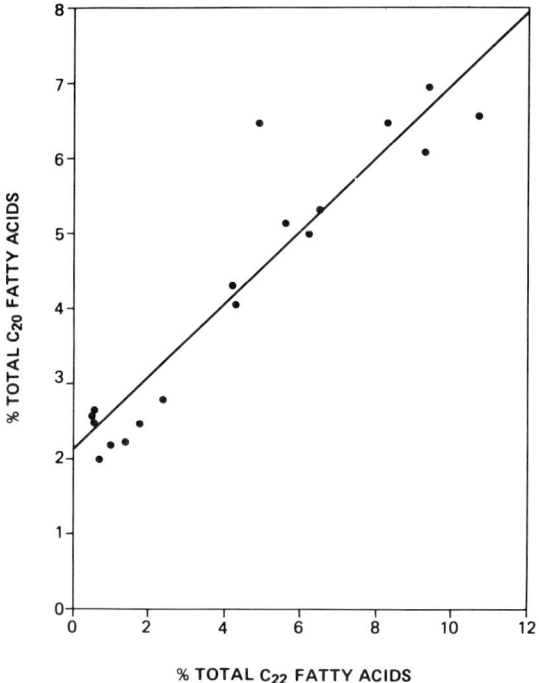

Fig. 1.   Regression line for determing total C_{20} fatty acids of rapeseed oils from total C_{22} fatty acids (J. Daun, private communication). The formula is $\%C_{20} = 2.13 + 0.49(\%C_{22})$.

phy (GLC) where 20:1 coincides with 18:3 *n*-3. The proportion of 20:0 in LEAR oils is nominally 0.4–0.5%, and of 20:2 about 0.1% or less, whereas that of 22:0 is usually 0.2–0.4%, even though 22:1 is as low as 0.01%. The 24:0 and 24:1 both remained at about 0.1% of LEAR oils even when 22:1 was low (Ackman, 1977).

The minor fatty acids of two LEAR oils (*B. napus* cv. Tower and *B. campestris* cv. Candle) are listed in Table IV with the exception of some of the monoethylenic isomer details which will be found in Tables V and VI. The minor fatty acid details were compared for a HEAR (cv. Target) and a LEAR (cv. Tower) and it was concluded that the odd-chain, *iso*, *anteiso*, and *cis* and *trans* 15:1 *n*-10 fatty acids had not been modified by the genetic manipulation necessary to change HEAR varieties to LEAR varieties (Sebedio and Ackman, 1979). This finding applied to both *B. napus* (cv. Tower) and *B. campestris* (cv. Candle) oils (Ackman and Sebedio, 1979).

Most vegetable oils such as those listed in Table V contain low proportions of *cis*-vaccenic (*cis*-11-octadecenoic; 18:1 *n*-7) acid relative to the ubiquitous 18:1 *n*-9. The exact origin of the edible oils examined by Kuemmel (1964) is not known, but Hougen and Wasowicz (1979) crushed their own

TABLE IV

Weight Percentages of Fatty Acids in Two Contemporary Canadian LEAR (Canola) Oils

Fatty acid	cv. Candle		cv. Tower refined
	Crude	Refined	
14:0	0.05	0.05	0.05
15:0	0.01	0.02	0.01
16:0	4.51	3.82	3.88
17:0	0.05	0.04	0.04
18:0	1.39	1.23	1.56
20:0	0.42	0.35	0.50
22:0	0.21	0.20	0.28
24:0	0.06	0.04	0.14
Total saturated	6.70	5.75	6.46
14:1	0.01	0.01	0.01
cis 15:1 n-10	0.01	0.02	0.02
trans 15:1 n-10	0.01	0.01	0.01
15:1 n-8	Trace	Trace	Trace
16:1	0.25	0.24	0.29
17:1 n-8	0.05	0.03	0.06
18:1	51.61	53.50	64.02
19:1	0.02	0.03	0.02
20:1	1.42	1.37	1.24
22:1	1.16	1.00	0.08
24:1	0.21	0.25	0.09
Total monoethylenic	55.69	56.46	65.80
16:2 n-6	0.03	0.02	0.09
16:2 n-4	Trace	Trace	Trace
16:3 n-3	0.13	0.15	0.08
18:2 n-6	24.47	23.52	18.79
18:3 n-3	13.58	13.82	8.59
20:2 n-6	0.14	0.11	0.05
20:3 n-3	ND[a]	ND	0.01
Total polyethylenic	38.68	37.70	27.61
Other[b]	0.33	0.17	ND

[a] Not detected.

[b] Mostly geometrical isomers of 18:3 n-3 as discussed by Ackman et al. (1974); traces of 14:2 n-6, iso 14:0, anteiso 15:0, and 19:1, are not included, and also omitted are the conjugated diethylenic acids discussed by Ackman et al. (1981) and Ackman and Sebedio (1981).

TABLE V

Percentages of cis-9-Hexadecenoic, Total Octadecenoic, and Total Eicosenoic Acids in Some Vegetable Oils, and Proportions of 18:1 n-9 (Oleic) and 18:1 n-7 (cis-Vaccenic) Acid Isomers

Oil	Weight % 16:1 n-7	Weight % 18:1	Isomer %		Weight % 20:1
			18:1 n-9	18:1 n-7	
LEAR (cv. Candle)[a]	0.3	54	94.3	5.7	1.4
LEAR (cv. Tower)[b]	0.3	64	95.6	4.4	1.2
HEAR (cv. R-500)[c]	0.2	12	96.2	3.8	4.6
HEAR (cv. Target)[b]	0.2	35	95.1	4.9	11.0
HEAR[d]	0.2	16	93.2	6.8	10.0
Corn[d]	0.1	26	98.4	1.6	<0.1
Corn[c,e]	0.4	27	97.4	2.6	0.2
Cottonseed[d]	0.4	17	95.2	4.8	<0.1
Cottonseed[c,e]	0.5	16	95.5	4.5	0.3
Olive[d]	0.9	77	97.5	2.5	0.2
Olive[c,e]	0.8	74	96.1	3.9	0.5
Peanut[d]	0.1	51	99.1	0.9	1.4
Peanut[c,e]	<0.1	38	98.7	1.3	1.9
Safflower[d]	0.1	14	95.6	4.4	0.3
Safflower[c,e]	0.1	14	95.6	4.4	0.2
Soybean[d]	0.1	27	95.6	4.4	0.2
Soybean[c,e]	0.1	24	94.1	5.9	0.1
Sunflower[c,e]	<0.1	16	96.9	3.1	0.2
Sesame[c,e]	<0.1	40	97.5	2.5	0.2
Palm[c,e]	0.1	38	98.1	1.9	<0.1
Coconut[c,e]	0.1	10	98.0	2.0	0.1

[a] Sebedio and Ackman (1981).
[b] Sebedio and Ackman (1979).
[c] Isomer ratios from open-tubular GLC only.
[d] Kuemmel, 1964; see also Ackman (1966) for HEAR oil.
[e] Hougen and Wasowicz (1978).

seed. However, the former worker used isolation and oxidative fission for isomer proportionation, while the latter workers depended on isomer resolution by open tubular gas–liquid chromatography. Where edible oils (of different origin) were examined by the two techniques the proportions of 18:1 n-7 to 18:1 n-9 show reasonable agreement (Table V). Hougen and Wasowicz (1979) concluded that the 18:1 n-7 of edible oils was without influence as a dietary factor in the feeding of rats. As shown in Table V, LEAR and HEAR oils have 18:1 n-7 to 18:1 n-9 proportions similar to other well-established edible vegetable oils.

It is, however, of interest to explore the influence of the genetic changes in the new varieties of rapeseed and canola on the proportions of n-7 iso-

TABLE VI

Percentages of Monoethylenic Isomers in Oil and of Isomers within Each Chain Length for Three HEAR Oils and Two LEAR (Canola) Oils

		HEAR[a] cv. unspecified	B. napus		B. campestris	
			Target[b]	Tower[b]	Candle[c]	R-500[d]
% of 16:1		0.2	0.2	0.3	0.3	0.2
% isomers	n-9	33	13	22	15	18
	n-7	67	78	70	81	75
	n-5	—	9	8	4	7
% of 18:1		16	35	64	54	12
% isomers	n-9	93.6	95.1	95.6	94.3	96.2
	n-7	6.4	4.9	4.4	5.7	3.8
% of 20:1		10.0	11.0	1.2	1.4	4.6
% isomers	n-9	92.9	95.6	97.6	96.2	73.7
	n-7	7.1	4.4	2.4	3.8	26.3
% of 22:1		44	23	0.1	1.2	54
% isomers	n-9	97.7	99.1	97.7	98.9	98.1
	n-7	2.3	0.9	2.3	1.1	1.9

[a] Kuemmel (1964); see also Ackman (1966).
[b] Sebedio and Ackman (1979).
[c] Sebedio and Ackman (1981).
[d] Isomer ratios from open-tubular GLC only; R. G. Ackman (unpublished results).

mers in the longer chain lengths of monoethylenic fatty acids. The isomers of 16:1 listed in Table VI are more or less constant in proportion, and the total 16:1 does not vary much. Similarly the proportion of 18:1 n-7 to 18:1 n-9 is fairly constant at 4–6% of total 18:1. It is curious that there is a relation between total 18:1 (Tower>Candle>Target), and the proportion of 22:1 n-7 in total 22:1 (Tower>Candle>Target), which itself is inversely related (Target>Candle>Tower) to total 18:1. To add complexity to this interesting relationship, the proportion of 20:1 n-7 in total 20:1 (Target>Candle>Tower) is then also inversely related to the proportion of total 18:1 and related directly to total 20:1. This does not seem to be a matter of chance and more than one elongation mechanism may be operating, especially as 24:0 and 24:1 are still obvious minor fatty acids when 22:1 is greatly reduced. Perhaps one mechanism is the conventional stepwise two-carbon chain extension, and the other a four-carbon (double two-step) chain extension. Each could be susceptible to the proportion of n-7 in the primer, or to the actual chain length of the primer (cf. discussion of 22:1 formation in Pollard and Stumpf, 1980). The possibility of the four-carbon extension is supported indirectly by several pieces of work (Appleby *et al.*, 1974; Biacs, 1979; Cassagne and Lessire, 1979).

The R-500 has an atypically high proportion of 20:1 *n*-7, but some other *B. campestris* cultivars (cv. Polar and cv. Span) seem to have had proportions of 7 or 8% of 20:1 *n*-7 in total 20:1 (Hougen and Wasowicz, 1979). There may be other factors such as maturity of seed which also require investigation, as they affect fatty acid composition (Izzo *et al.*, 1979) or distribution (Harris and Norton, 1979), but it seems that the LEAR varieties wih a low erucic acid content have C_{18}, C_{20} and C_{22} monoethylenic fatty acids dominated by the "*n*-9" isomers, with 4–6% of monoethylenic "*n*-7" acid in the C_{18} chain length, and lesser proportions in the C_{20} and C_{22} chain lengths. These *n*-7 isomer details are similar to the scanty information available for other vegetable oils.

The few minor fatty acids of rapeseed oil not included in the above discussion are present in trace amounts only and fall into groups which are possibly partly natural, and possibly partly artifacts. The conjugated octadecadienoic fatty acids found in some edible seed oils (Ackman *et al.*, 1981) require further study as the proportions found in rapeseed oils were even lower than in other vegetable oils (<0.1%). Conjugated acids are known to arise during commercial bleaching operations (Eicke, 1971; Van den Bosch, 1973; Helme, 1980) but it is also possible to envision a peroxidase in a seed altering linoleic acid into a hydroperoxy conjugated octadecadienoic acid (Porter *et al.*, 1980) and later having the hydroperoxy group removed by another enzyme (a hydroperoxidase). Rapeseed appears to be virtually unique among common oilseeds in lacking an active lipoxygenase, although erucic acid itself is an inhibitor of soybean and peanut lipoxygenases (St. Angelo *et al.*, 1979). The several geometrical isomers of 18:3 *n*-3 (linolenic) acid in rapeseed and other oils were originally thought to be artifacts and were attributed to deodorization (Ackman *et al.*, 1974; Devinat *et al.*, 1980b) but some have since also been found in very low levels in laboratory extracts of seeds (Sebedio and Ackman, 1979, 1981). Nutritionally, these two types of minor fatty acids seem to be of little significance (Emken, 1979; Applewhite, 1981) and are probably found in a variety of edible plant oils.

III. TRIGLYCERIDES

The combination of fatty acids on the glycerol moiety of HEAR oil was originally found to be based on the saturated and very long chain fatty acids (C_{20}–C_{24}) occurring in the 1- and 3-positions, while the C_{18} acids, especially the 18:2 *n*-6 and 18:3 *n*-3, were concentrated in the 2-position (Appelqvist and Dowdell 1968; Appelqvist 1972). More specifically the stereospecific distribution of fatty acid is given in Table VII for HEAR and LEAR oils and a mustard oil, and compared with the 2-position data found by Litchfield (1971), Sergiel (1973), Zadernowski and Sosulski (1979), and Mordret and Helme (1974) for LEAR oils. The percentage of 18:2 n-6 among the fatty acids in the 2-position of the oils in Table VII is roughly half of the highest values reported for popular North American retail vegetable oils (Carpenter et al., 1976). Litchfield (1971, 1973) investigated the positions of 18:1 n-9, 18:2 n-6, and 18:3 n-3 in oils of Cruciferae generally and produced a graphical correlation, including that of an early LEAR (*B. napus*) variety. This approach supports views of Jáky and Kurnick (1981), who investigated the concentration of 18:2 n-6 in the 1,3- and 2-positions. They suggest that in HEAR oils with 14.3 and 12.7% of 18:2 n-6 in the oil, at least 95% was concentrated in the 2-position, whereas in a LEAR oil with 18.8% of 18:2 n-6 in the oil the percentage of 18:2 n-6 in the 2-position was only 54%. The increased level of 18:2 n-6 evidently distributed itself into the 1- and 3-positions. This distribution has been plotted for a number of rapeseed oils (including cv. Oro with 0.6% 22:1) by Ohlson et al. (1975), and it is indicated that as total 22:1 is reduced, the excess 18:2 n-6 prefers the 1-position, and the low level of 20:1 and 22:1 prefers the 3-position.

Hydrogenation of unsaturated fatty acids such as 18:1, 22:1, and 18:2 n-6 proceeds more rapidly in the 1- and 3-positions than in the 2-position (Paulose et al., 1978; Kaimal and Lakshminarayana, 1979). These distributions are thus a factor in selectivity of hydrogenation. Linolenic (18:3 n-3) acid has a similar distribution to 18:2 n-6 (Ohlson et al., 1975), but on hydrogenation this acid may behave differently from 18:2 n-6 (Ilsemann et al., 1979). Rapeseed oil free of excessive sulfur compounds (see below) hydrogenates satisfactorily compared to other commercial vegetable oils (Ahmad and Ali, 1981; El-Shattory et al., 1981; Koman et al., 1981).

This distribution of fatty acids is also of some interest to nutritionists as the effects of fatty acids in rats can be modified by redistributing the fats by interesterification (Mukherjee and Sengupta, 1981). As much as 40% of the dietary fat consumed in the Federal German Republic in recent years has been subjected to transesterification processes such as described by Perron and Broncy (1978) and Monseigny et al. (1979). This makes surveys of composition of edible fats rather difficult in terms of fatty acids (Heckers and Melcher, 1978), and enforces the examination of unsaponifiable materials to discover the original oil composition (Strocchi, 1981).

The analysis of the stereospecific structure of rapeseed oil triglycerides with pancreatic lipase is often accepted as quite straightforward (cf. Breckenridge, 1979), but in studies with lipases of plant (Rosnitschek and Theimer, 1980), microbial (Kroll et al., 1973) and animal (Myher et al., 1979; Vajreswari and Tulpule, 1980) origins, oils rich in erucic acid tended

TABLE VII

Proportions of Fatty Acids Found in Various Positions of Rapeseed Oil Triglycerides and a Mustard Oil Triglyceride

Oil and position	Fatty acid								
	16:0	16:1	18:0	18:1	18:2 n-6	18:3 n-3	20:1	22:0	22:1
HEAR (in mole %)[a,b]									
1-	4.1	0.3	2.2	23.1	11.1	6.4	16.4	1.4	34.9
2-	0.6	0.2	—	37.3	36.1	20.3	2.0	—	3.6
3-	4.3	0.3	3.0	16.6	4.0	2.6	17.3	1.2	51.0
HEAR (in mole%)									
1,3-[c]	5	—	1	10	5	5	15	—	59
2-[c]	1	—	1	32	41	22	—	—	3
2-[d]	<1	<1	—	34	42	22	1	—	1
HEAR, cv. Sinus (in weight %)[e]									
2-	0.2	0.1	0.2	44.6	37.1	17.5	0.2	—	0.1
LEAR, cv. Oro (in mole %)[f]									
1-	6.4	—	1.7	64.7	15.8	7.4	1.8	0.2	0.7
2-	0.2	—	0.1	52.7	30.6	15.7	0.3	—	0.1
3-	7.8	—	1.8	70.4	10.3	5.4	2.7	0.7	1.0
LEAR, cv. Janpol (in weight %)[e]									
2-	0.5	0.1	0.2	56.8	29.1	13.1	0.1	—	—
LEAR, canbra (in weight %)									
2-[g]	1.2	0.5	0.5	52.4	31.8	12.3	—	—	—
2-[h]	0.6	0.3	—	56.3	30.6	12.1	—	—	—
2-[d]	0.3	0.2	—	50.0	33.5	15.9	0.1	—	—
LEAR, cv. Primor (in weight %)[g]									
2-	1.6	0.1	0.8	49.4	32.8	14.5	0.2	—	—
Mustard (in mole %)[i]									
1-	4.3	0.1	1.9	25.9	10.7	4.4	14.3	—	36.1
2-	2.0	0.3	1.3	38.4	32.3	27.7	0.2	—	0.5
3-	2.4	—	0.3	5.3	2.8	7.0	18.3	—	59.0

[a] Brockerhoff and Yurkowski (1966).
[b] Brockerhoff (1971).
[c] Litchfield (1973).
[d] Litchfield (1971).
[e] Zadernowski and Sosulski (1979).
[f] Töregard and Podlaha (1974).
[g] Mordret and Helme (1974).
[h] Sergiel (1973).
[i] Myher et al. (1979).

to be poor substrates. Notwithstanding this, free fatty acids in rapeseed oil can arise from fungal lipolysis (El Azzabi et al. 1981). Additional information on the distribution of fatty acids in oils from seeds of Cruciferae can be found in Franzke et al. (1972), Kroll and Franzke (1974), Tiscornia and Bertini (1973), Töregard and Podlaha (1974), and Osman and Fiad (1975).

The direct analysis of triglycerides by GLC is currently a reasonably standard procedure, although differences in response factors for carbon numbers are very common (Karleskind et al., 1974; Hamilton and Ackman, 1975; Kuksis et al., 1975; Myher, 1978; Aneja et al., 1979; Grob, 1979, 1981; Monseigny et al., 1979; Grob et al., 1980; D'Alonzo et al., 1981).

This technique is very useful in comparing different oils. Table VIII shows that most vegetable oils, including LEAR oils (0.4–4% of 22:1), have a maximum at C_{54} (i.e., $3 \times C_{18}$ fatty acids). The gas chromatogram for triglycerides of this type of oil (Fig. 2, LEAR cv. Cresor) is quite simple with one major peak at C_{54}. The triglycerides of HEAR oils have a more complex type of

TABLE VIII

Distribution of Triglyceride Molecules for HEAR and LEAR Oils with Mustard, Sunflower, Soybean and Peanut Oils for Comparison

Oil	Mole % erucic acid	Triglyceride profile								
		C_{48}	C_{50}	C_{52}	C_{54}	C_{56}	C_{58}	C_{60}	C_{62}	C_{64}
HEAR (cv. Victor)[a]	48	—	—	1	2	6	8	18	61	5
HEAR (cv. Target)[b]	38	—	<1	2	4	10	20	25	39	<1
HEAR[c]	?	—	—	4	6	12	16	24	39	<1
HEAR[d]	45	1	2	6	13	17	20	20	20	2
HEAR[e]	40	<1	2	6	9	11	18	19	32	<1
HEAR[f]	25	—	—	11	21	46	11	5	2	3
Mustard[g]	32	—	<1	3	8	14	29	20	21	4
LEAR[d]	4	2	6	18	61	7	2	1	1	—
LEAR (cv. Primor)[c]	?	—	—	15	81	5	—	—	—	—
LEAR[d]	0.7	3	4	18	70	5	1	<1	—	.—
LEAR[e]	?	—	1	13	75	6	3	1	<1	--
LEAR[f]	0.4	4	5	24	63	3	—	—	—	—
Sunflower[c]	—	—	2	21	75	1	—	—	—	—
Sunflower[d]	—	—	4	29	67	—	—	—	—	—
Soybean[d]	—	—	5	42	52	—	—	—	—	—
Peanut[c]	—	—	4	26	60	7	4	—	—	—

[a] Appelqvist (1972).
[b] Harlow et al. (1966).
[c] Karleskind et al. (1974).
[d] Defrancesco (private communication); erucic acid in weight %.
[e] Eyres (1979); erucic acid in weight %.
[f] Adapted from Defrancesco et al. (1981); erucic acid in weight %.
[g] Myher et al. (1979).

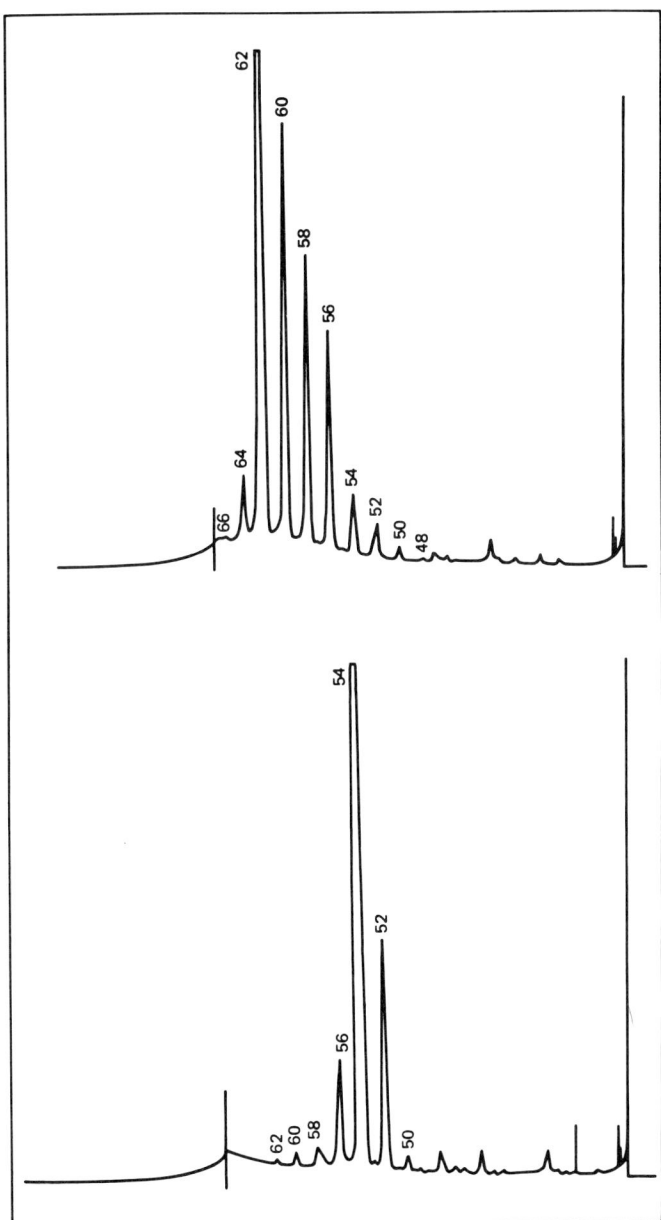

Fig. 2. Gas–liquid chromatograms of triglycerides of (above), rapeseed HEAR (cv. Cresus) and, (below) rapeseed LEAR (cv. Cresor). Column, Pyrex glass, 6 m in length and 0.4 mm i.d., methyl silicone coating by *in situ* polymerization. Oven programmed from 200°C to 365°C at 4°C/min, injector (glass) 400°C, detector (FID) 375°C, carrier gas helium at 6 ml/min. Reproduced from Monseigny *et al.* (1979) by permission of *Rev. Fr. Corps Gras* and authors.

TABLE IX

Molecular Species of Triacylglycerols (in mole %) of Mustard Seed Oil[a]

Molecular species[b]	Mustard seed oil		Molecular species[b]	Mustard seed oil	
	Calc.[c]	Reconst.[d]		Calc.[c]	Reconst.[d]
16:0 18:1 18:1	0.088	0.65	18:1 18:2 22:1	4.93	5.22
18:1 18:1 16:0	0.24		22:1 18:1 18:1	0.62	
			20:1 18:2 20:1	0.85	0.79
16:0 16:0 22:1	0.05	0.07	18:3 18:1 22:1	1.00	0.93
22:1 16:0 16:0	0.02		22:1 18:1 18:3	0.97	
16:0 18:1 20:1	0.30	0.73	18:2 18:2 22:1	2.03	1.36
20:1 18:1 16:0	0.13		22:1 18:2 18:2	0.32	
18:1 18:1 18:1	0.53	1.24	18:1 18:3 22:1	3.78	4.00
18:1 18:1 18:2	0.28	0.46	22:1 18:3 18:1	0.47	
18:2 18:1 18:1	0.22		20:1 18:3 20:1	0.65	0.61
18:1 18:2 18:1	0.44	1.04	18:3 18:2 22:1	0.84	0.78
18:1 18:1 18:3	0.69	0.46	22:1 18:2 18:3	0.81	
18:3 18:1 18:1	0.09		18:2 18:3 22:1	1.56	1.04
18:1 18:2 18:3	0.58	0.39	22:1 18:3 18:2	0.25	
18:3 18:2 18:1	0.08		18:3 18:3 22:1	0.65	0.60
18:1 18:3 18:3	0.45	0.30	22:1 18:3 18:3	0.62	
18:3 18:3 18:1	0.06		18:0 18:1 22:1	0.42	0.78
			22:1 18:1 18:0	0.04	
16:0 18:1 22:1	0.98	1.33	20:1 16:0 22:1	0.17	0.28
22:1 18:1 16:0	0.33		22:1 16:0 20:1	0.13	
18:1 16:0 22:1	0.30	0.32	18:1 18:1 22:1	5.87	6.22
22:1 16:0 18:1	0.40		22:1 18:1 18:1	0.74	
18:1 18:1 20:1	1.82	2.04	18:0 18:2 22:1	0.36	0.65
20:1 18:1 18:1	0.29		22:1 18:2 18:0	0.04	
16:0 18:2 22:1	0.83	1.12	20:1 18:1 20:1	1.01	0.95
22:1 18:2 16:0	0.28		18:2 18:1 22:1	2.42	1.62
18:2 18:1 20:1	0.75	0.66	22:1 18:1 18:2	0.39	
20:1 18:1 18:2	0.15				
18:1 18:2 20:1	1.53	1.71	16:0 22:1 22:1	0.01	0.02
20:1 18:2 18:1	0.25		22:1 22:1 16:0	0.04	
16:0 18:3 22:1	0.63	0.86	22:1 16:0 22:1	0.42	0.42
22:1 18:3 16:0	0.22		20:1 18:1 22:1	3.25	5.34
18:2 18:2 20:1	0.63	0.55	22:1 18:1 20:1	2.54	
20:1 18:2 18:2	0.13		18:1 20:1 22:1	0.03	0.03
18:1 18:3 20:1	1.17	1.31	22:1 20:1 18:1	0.0	
20:1 18:3 18:1	0.19		20:1 18:2 22:1	2.73	4.48
18:2 18:3 20:1	0.48	0.43	22:1 18:2 20:1	2.13	
20:1 18:3 18:2	0.10		20:1 18:3 22:1	2.09	3.44
			22:1 18:3 20:1	1.64	

TABLE IX (*CONTINUED*)

Molecular species[b]	Mustard seed oil		Molecular species[b]	Mustard seed oil	
	Calc.[c]	Reconst.[d]		Calc.[c]	Reconst.[d]
22:1 18:0 22:1	0.27	0.27	20:1 22:1 22:1	0.04	0.07
22:1 18:1 16:0	8.19	8.15	22:1 22:1 20:1	0.03	
18:1 22:1 22:1	0.08	0.08	22:1 18:1 24:1	0.46	0.83
22:1 22:1 18:1	0.01		24:1 18:1 22:1	0.42	
20:1 20:1 22:1	0.02	0.03	22:1 20:1 22:1	0.04	0.04
22:1 20:1 20:1	0.01				
22:1 18:2 22:1	6.88	6.84	22:1 22:1 22:1	0.11	0.11
22:1 18:3 22:1	5.27	5.24			

[a] Myher et al. (1979); reproduced by permission of *Canadian Journal of Biochemistry* and authors.
[b] The molecular species are represented (left to right) by fatty acids in the *sn*-1, *sn*-2 and *sn*-3 positions.
[c] Calculated values derived from 1-random 2-random 3-random calculation using the fatty acid compositions of the appropriate *sn* positions.
[d] Reconstituted values derived by random combination of experimental 1,3-diacylglycerols with the fatty acids in the *sn*-2 position.

chromatogram (Fig. 2, HEAR cv. Cresus) with a peak at C_{62} probably reflecting mostly the combination $2 \times C_{22} + 1 \times C_{18}$. As the proportions of 22:1 decline from 48% to 38% this shifts the emphasis from C_{62} to lesser carbon numbers as the probability of two 22:1 fatty acids being in one triglyceride molecule is low at 38% 22:1 by weight. A small amount of 24:0 and 24:1 helps account for the C_{64} carbon number ($C_{18} + C_{22} + C_{24}$) components, but the proportion of these declines as 22:1 is reduced.

The individual classes of triacylglycerols are chiefly studied by argentation chromatography (Persmark, 1972; Sergiel, 1973). Crambe (*Crambe abyssinica*) seed oil (Gurr et al., 1972) and mustard (*Brassica juncea*) seed oil (Myher et al., 1979) are probably very similar to rapeseed in types of acylglycerols (Harris and Norton, 1979). The molecular species for mustard oil are presented in Table IX. This particular sample had (in mole %) 16:0, 2.6%, 18:0, 0.9%; 18:1, 23.0%, 18:2 n-6, 15.3%; 18:3, 12.6%; 20:0, 0.6%; 20:1, 11.2%; 22:1, 31.6%; 24:1, 2.1% acids distributed on the glycerol molecule as shown in Table VII and combined by carbon numbers as shown in Table VIII. The analyses of tri- and diacylglycerols conducted on nonpolar GLC phases by Myher et al. (1979) will likely be extended by the use of novel polar phases (Takagi and Itabashi, 1977; Aneja et al., 1979).

TABLE X[a]

A. Calculated Carbon Numbers and Number of Double Bonds of Triglyceride Fractions of LEAR (cv. Lobra) Oil

Fraction[b]	PN	TG class	Area %	Calculated values					
				System 1[c]		System 2[c]		System 3[c]	
				CN	NDB	CN	NDB	CN	NDB
1	38	54:8	0.8	56.4	3.8	53.4	5.3	53.8	7.0
2	40	54:7	0.8	59.7	3.6	54.0	6.4	53.9	6.7
3		54:7	3.0	53.9	3.9	56.9	4.3	54.0	6.8
4		54:6	0.6	Not visible		53.4	5.1	53.8	5.6
5	42	54:6	7.7	59.0	4.3	53.9	5.9		
6		52:5	1.3	56.7	4.5	52.3	4.5	52.1	4.7
7		54:5	5.4	54.0	2.5	53.9	4.9		
8	44	54:5	14.7	55.0	3.5	53.9	5.0		
9		52:4	2.9	53.6	1.2	52.4	4.1		
10	46	54:4	19.7	57.2	3.5	54.0	4.1		
11		52:3	5.1	51.1	2.0	52.2	3.1		
12		54:3	25.2	54.0	3.0				
13	48	52:2	6.8	58.7	3.8	52.4	2.2		
14	50	56:3	0.8	55.8	4.1	55.7	3.0		
15		54:2	1.7	54.0	1.9				

B. Main TG Types and Main Fatty Acid Contents of Triglyceride Fractions of LEAR Oil

TG class	Area %	Main TG type	Main fatty acids (mole %)		
54:8	0.8	L-Ln-Ln	L-27.5	Ln-54.0	
54:7	0.8	L-L-Ln	L-62.7	Ln-31.6	
54:7	3.0	O-Ln-Ln	O-33.5	Ln-63.1	
54:6	0.6	L-L-L	L-88.1		
54:6	7.7	O-L-Ln	O-33.3	L-32.5	Ln-32.5
52:5	1.3	P-L-Ln	P-28.3	L-27.7	Ln-29.5
54:5	5.4	O-L-L	O-34.1	L-61.6	
54:5	14.7	O-O-Ln	O-62.8	Ln-31.3	
52:4	2.9	P-O-Ln	P-27.2	O-31.7	Ln-31.6
54:4	19.7	O-O-L	O-64.9	L-33.O	
52:3	5.1	P-O-L	P-28.8	O-33.2	L-32.8
54:3	25.2	O-O-O	O-96.6		
52:2	6.8	P-O-O	P-27.3	O-61.6	
56:3	0.8	O-O-C20:1	O-59.2	20:1-32.1	
54:2	1.7	St-O-O	St-23.9	O-52.1	

[a] Petersson et al. (1981); reproduced by permission of *J. Am. Oil Chem. Soc.* and authors. PN = partition number; CN = carbon number; NDB = number of double bonds; M = myristic acid, 14:0; P = palmitic acid, 16:0; ST = stearic acid, 18:0; O = oleic acid, 18:1; L = linoleic acid, 18:2; Ln = linolenic acid, 18:3.

[b] Numbered in order of elution from HPLC column.

[c] System 1: column 75 cm; acetonitrile/acetone, 50:50; system 2: column 100 cm; acetonitrile/acetone, 50:50; system 3: column 100 cm; acetonitrile/acetone, 60:40.

103

Combined high performance liquid chromatography (HPLC)/GLC technologies will probably be applied to LEAR oil with results similar to soybean oil (Wada *et al.*, 1977), although a few problem separations remain even with the latest HPLC technology (El-Hamdy and Perkins, 1981; Plattner, 1981).

The virtual elimination of 22:1 and 20:1 from LEAR oil means that the triacylglycerol analyses become much simpler. With only minor amounts of C_{16} and C_{20} acids present the application of newer techniques, especially of HPLC, for example as originally applied to rapeseed oil (HEAR, cv. Major) by Karleskind *et al.* (1978), becomes more practical as a means of examining oils for triglyceride details. Petersson *et al.* (1981) have examined a LEAR oil (0.5% 22:1 *n*-9) by HPLC and found, for example, that 25% of the total triglyceride is triolein (Table X).

Molecular distillation was of only limited use in preparing triacylglycerol fractions from the *B. campestris* cv. Span with 4.8% 22:1 (Kramer *et al.*, 1975a, b), but direct-inlet mass spectroscopy of triacylglycerols of rapeseed oil with chemical ionization yields extra information such as the proportion of different numbers of ethylenic bonds within one carbon number group (Murata and Takahashi, 1977). For example the C_{62} group is stated to have 13.4% of species with two ethylenic bonds, 44.2% of species with three ethylenic bonds, 27.9% of species with four ethylenic bonds, and 14.5% of species with five ethylenic bonds. Allowing for the effect of a probable difference in total 22:1, this is reasonably close to the C_{62} species calculated by Myher *et al.* (1979) for mustard seed oil (see Table IX). These can be summarized as having two (1.2%), three (48%), four (33%), and five (25%) ethylenic bonds.

IV. PHYSICAL PROPERTIES

The physical properties of HEAR oils have not changed since they were defined by Appelqvist and Ohlson (1972) as reproduced in Table XI. Only a few of these values have been redefined for LEAR, including canola oils, but selected properties such as molar polarization and dipole moment have been related to the 22:1 content of the triglyceride molecules (Rutkowski and Witwicka, 1975). Specific gravity has been altered slightly by elimination of the long 22:1 chain (Ackman and Eaton, 1977) in LEAR oils.

The solubility of rapeseed oil in various solvents, viewed as turbidity on cooling, has been used to distinguish HEAR oils from other common vegetable oils, chiefly as the Crismer value (AOCS method Cb 4-35). A regression line has been developed for LEAR oils and HEAR oils with up to 25% of 22:1 (Sahasrabudhe, 1977). An alternative and simpler solvent system has been proposed by McGregor (1977), and is probably more useful at low (0.5–5%)

TABLE XI

Some Physical Properties of HEAR and LEAR Oils

Description	HEAR[a]	LEAR
Compressibility, $\times 10^{-6}$	60.9	—
Dielectric constant, cgs units	3.06	—
Flash point, open cup, °C	282	—
Smoke point	218[b]	238[c]
Iodine value (Wijs)	97–108	112–131[d]
Pour point, °C	−12	—
Refractive index at 25°C	1.470–1.474	—
Refractive index at 40°C	1.465–1.469	1.465–1.467[e]
Saponification number	170–180	190[f]
Specific gravity at 25°/25°C	0.906–0.914	0.916–0.917[g]
Cold test (hr) of lightly hydrogenated and winterized oils	6[h](IV87)	12[h](IV93)
Specific heat at 20°C	0.488	—
Surface tension, dynes/cm² at 12°C	36.6	—
Thermal conductivity cals/cm²/sec for 1°C and 1 cm thick	41	—
Titer °C	11.5–15.0	—
Crismer value	76–82[i]	68[i]
Viscosity (Saybolt) 38°C	260	—
Viscosity (Saybolt) 100°C	54	—
Viscosity (centipoises) 50°C	—	17.0[c]

[a] Except as indicated values are taken from Appendix II in "Rapeseed" (Appelqvist and Ohlson, 1972).
[b] Costigliola (1970).
[c] Vaisey-Genser and Esken (1979).
[d] Anonymous (1981).
[e] Kovacs et al. (1978).
[f] Mordret and Helme (1974).
[g] Ackman and Eaton (1977).
[h] Mertens et al. (1970).
[i] Sahasrabudhe (1977).

percentages of 22:1. In most respects there is no obvious physical property distinguishing LEAR or canola oils from other common vegetable oils.

The increased use of LEAR (canola) oil in Canada, and modifications in processing conditions, have led to some alterations in physical properties of refined and processed oils. These are summarized by Vaisey-Genser and Eskin (1982). The high proportion of C_{18} fatty acids in LEAR oils does however lead the triglyceride crystal, which is β' in traditional HEAR oils (Kawamura, 1981), to rapidly adopt the β configuration in LEAR oils (Persmark and Bengtsson, 1976; Ohlson, 1979; Hernqvist, 1979). On cooling, the coexistance of liquid and solid phases gives LEAR oil some novel properties (Perron and Broncy, 1978). This does not affect salad oil utilization, but the wholly C_{18} triglycerides are very important in partially hydrogenated oils intended as stocks for margarines and shortenings (Ohlson, 1970; Teasdale et al., 1970; deMan, 1978, 1979; Eh-Shattory et al., 1981). This problem does not apply in the case of totally hydrogenated HEAR oil (iodine value 2–3) used as a stabilizer in peanut butter (Nolen, 1981). The polymorphic behaviour of LEAR oil crystals is not materially changed by random rearrangement of the fatty acids (Kawamura, 1981).

V. STEROLS

HEAR and LEAR oils contain brassicasterol, a C_{28} sterol (Fig. 3) characteristic of *Brassica* oils. It does not occur in other common edible vegetable oils (Table XII) except for mustard oil. Brassicasterol is, thus, a key factor in identifying *Brassica* oils either by themselves or after blending with other edible oils, since fatty acid compositions cover a range for each oil due to cultivar, climate, or maturity, and overlaps in physical properties are common (Spencer et al., 1976). Among recent authors examining this question

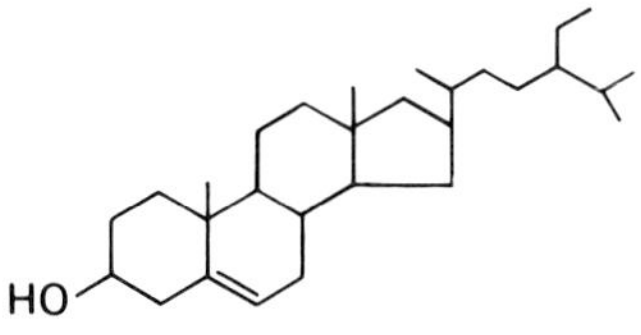

Stigmast-5-ene-3β-ol
β-SITOSTEROL

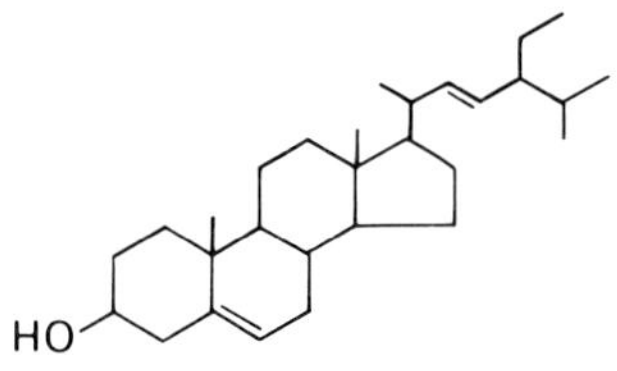

Stigmasta-5,22-diene-3β-ol
STIGMASTEROL

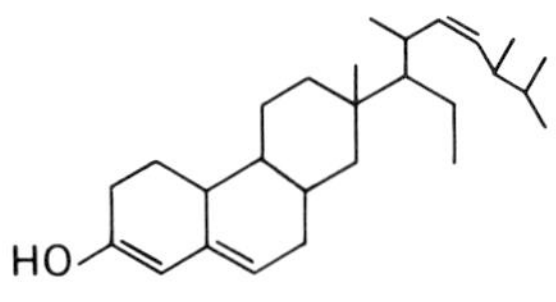

$\triangle^{5,22}$-Ergostadien-3β-ol
BRASSICASTEROL

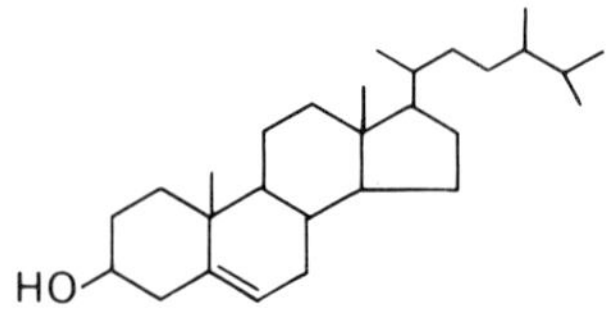

$\triangle^{5}$-24-Isoergosten-3β-ol
CAMPESTEROL

Fig. 3.   The four important sterols common to most commercial vegetable oils. Rapeseed oil is distinguished from the other oils by the presence of brassicasterol and virtual absence of stigmasterol. For further details see Tables XII–XV.

TABLE XII

Distribution of Important Sterols in Some Commercial Vegetable Oil[a]

Oil	% in oil	Composition in relative percent					
		Brassica-sterol	Chole-sterol	Campe-sterol	Stigma-sterol	Sito-sterol	Other
LEAR (cv. Oro)	0.53	7.4	—	35.5	—	57.1	—
Soybean	0.37	—	—	21.3	19.1	53.5	5.8
Corn	0.90	—	—	22.1	7.2	67.2	3.5
Sunflower	—	—	—	12.7	10.7	63.5	13.0
Peanut	0.25	—	—	12.8	10.8	74.6	1.8

[a] From Mordret and Helme (1974); Mannino and Amelotti (1975); Seher and Vogel (1975).

of oil identifications through unsaponifiables are Wolff (1980) and Strocchi (1981).

The unsaponifiables in rapeseed oils (LEAR, unrefined and refined) were found to range from 1.30 to 1.50% of the oil in a recent study (Maxwell and Schwartz, 1979; Maxwell et al., 1981). The total sterols in HEAR and LEAR oil were found to range from 0.528 to 0.975% of oil (Kovacs et al., 1978). This study in one laboratory by one method suggested that, as indicated by Niewiadomski (1975), conventional refining resulted in a loss of about 2 mg sterol/g of oil, accounting for part of this range. Kanematsu et al. (1973) showed that a HEAR oil with 1.39% unsaponifiables had total sterol reduced from 6.85 mg/g to 5.72 mg/g oil on refining. The change in each sterol accompanying the reduction in total sterol was found by both groups to be campesterol relatively unchanged, β-sitosterol slightly increased, and brassicasterol slightly decreased. Cortesi et al. (1973) observed a similar change

TABLE XIII

Content of Free Sterols, Esterified Sterols and Steryl Esters[a] in Oils from Brassica campestris and B. napus Seeds[b]

Species	Cultivar	Concentration (% of the oil)		
		Free sterols	Esterfied sterols	Calculated steryl esters
B. campestris				
Winter type	Sv 76-15069	0.27	0.36	0.6
Summer type	CDA Span	0.35	0.25	0.4
B. napus				
Winter type	Sv Brink	0.31	0.51	0.9
Summer type	WW Olga	0.36	0.71	1.2

[a] Assuming that the esters are all sitosteryl linoleate.
[b] Appelqvist et al. (1981); reproduced by permission of Phytochemistry and authors.

TABLE XIV

The Free Sterols and Esterified Sterols of Some Seeds of *Brassica* and *Sinapis* Species[a]

Species	Supplier and cultivar or line	Sterol composition (%)[b,c]											
		Cholesterol		Brassica-sterol		Campesterol		Sitosterol		Δ⁵-Avena-sterol		Δ⁷-Stigma-sterol	
		F	E	F	E	F	E	F	E	F	E	F	E
Brassica napus													
Summer rape	WW 77 2902	Tr	—	14.1	6.5	31.0	37.9	54.1	52.2	0.8	2.5	Tr	0.9
Summer rape	WW 77 3185	Tr	—	15.7	6.4	31.6	39.8	46.1	48.6	6.6	5.1	Tr	Tr
B. campestris													
Winter turnip rape	Sv 76 15069	Tr	—	12.7	5.5	28.2	36.3	58.0	52.6	1.1	5.6	Tr	—
Winter turnip rape	Sv 01030	Tr	—	13.8	6.1	28.8	37.1	55.8	52.0	1.6	4.7	Tr	Tr
Winter turnip rape	WW 78 2849	Tr	—	20.9	8.1	22.2	30.8	56.9	56.4	Tr	4.6	Tr	Tr
Summer turnip rape	CDA Span	Tr	—	12.8	4.8	26.7	35.2	60.5	57.3	Tr	2.7	Tr	—
Summer turnip rape	Sv 74 10105	Tr	—	13.8	4.2	26.6	34.0	59.6	57.8	Tr	3.1	Tr	0.7
Summer turnip rape	Sv 75 10223	Tr	—	10.4	4.8	29.0	38.8	60.6	53.4	Tr	2.9	Tr	Tr
Summer turnip rape	WW 77 5009	Tr	—	17.1	6.6	27.4	37.5	54.5	51.5	1.0	4.4	Tr	Tr
B. juncea	Sv	Tr	—	19.2	9.1	23.6	34.0	57.2	55.2	—	1.7	—	Tr
B. nigra	Sv	Tr	—	19.3	9.5	22.8	33.6	57.9	56.2	Tr	0.7	—	Tr
Sinapis alba	Sv Savor	2.0	2.3	10.2	2.6	24.5	32.7	52.0	41.7	11.3	20.7	—	—
Sinapis alba	Sv	2.4	1.2	11.4	3.4	25.3	34.2	48.4	44.4	12.5	16.8	—	—
S. arvensis (*Brassica kabar*)	SSTCI	—	Tr	7.8	5.9	25.8	32.5	66.4	56.1	—	4.4	—	1.1

[a] Appelqvist *et al.*, 1981; reproduced by permission of *Phytochemistry* and authors.

[b] Traces of stigmasterol were observed in several but not all samples. Also, traces of a component with the relative retention time of Δ⁷-avenasterol were noted in a few free sterol samples. A component with the same relative retention time as Δ⁷-avenasterol was observed in many esterified sterol samples at levels from traces up to a few percent.

[c] F = free sterols, E = esterified sterols.

TABLE XV

Content and Composition of 4-Demethylsterols in Some Vegetable Oils[a]

Oil	Demethyl- sterols (mg/100 g)	4-Demethylsterols (%)								
		Chole- sterol	Brassica- sterol	Campe- sterol	Stigma- sterol	Sito- sterol	Δ^5-Avena- sterol	Δ^7-Stigma- sterol	Δ^7-Avena- sterol	Others
Coconut	102	1	—	8	13	47	26	1	2	2
Cottonseed	510	Tr[b]	—	7	1	86	3	Tr	Tr	3
Grape seed	534	Tr	—	10	7	69	5	2	1	6
Linseed	471	1	1	27	8	42	13	—	1	6
Maize	1441	Tr	—	17	6	60	10	Tr	1	6
Olive	150	Tr	—	3	3	82	4	Tr	Tr	6
Palm kernel	140	1	—	10	13	69	7	—	—	8
Peanut	321	Tr	—	18	8	63	7	1	1	—
Rapeseed	954	Tr	10	33	Tr	48	3	1	1	2
Sesame	331	Tr	—	19	6	57	6	1	1	6
Soybean	394	Tr	—	20	18	51	4	2	1	10
Sunflower	494	Tr	—	7	7	59	8	6	5	3
Wheat germ	1425	Tr	—	19	4	60	7	2	2	5
RRT[c] of individual components on OV-1		1.88	2.07	2.35	2.52	2.87	2.92	3.21	3.29	

[a] Adapted from Kornfeldt and Croon (1981); by permission of *Lipids* and the authors.

[b] Tr = less than 0.5%.

[c] Cholestane is given the relative retention time of 1.00 and served as the internal standard for quantitation.

in brassicasterol relative to the two C_{29} sterols on laboratory hydrogenation over PtO_2. In soybean oil free sterols were found to be affected more than steryl esters when each step in the refining process was examined (Johansson and Hoffmann, 1979), and in rapeseed oil free brassicasterol exceeds the sterol in esterified form (Table XIV).

Thin-layer chromatography as a sole analytical method for study of sterols can lead to erroneous results for rapeseed oil, as an artifact from oxidation of campesterol may mimic cholesterol (Seher, 1976). This could account for persistent reports of cholesterol in allegedly all-vegetable margarines (e.g., Kanematsu et al., 1976), as well as purportedly pure vegetable oils (Itoh et al., 1973; Tiscornia and Bertini, 1973). Only the best GLC technology can be trusted in these analyses (Zürcher et al., 1976; Homberg, 1977; Wolff, 1980). On the other hand Appelqvist et al. (1981) report that *Sinapis alba* oil contains 2.0–2.4% cholesterol in the total sterols, but that the oil from several *Brassica* species contains only traces of this sterol. The work of this group in determining both free and esterified sterols is reproduced as Tables XIII and XIV. Independently, Itoh et al. (1973) also. identified 28-isofucosterol, stigmasterol, and 24-methylenecholesterol, with traces of others as minor components of HEAR oil. Itoh et al. (1981) later examined *Brassica napus* oil and identified two as natural *trans*-22-dehydrocholesterol and stigmasta-5,25-dienol, but fucosterol was thought to be an artifact.

The comparison of common (4-demethyl) sterols in rapeseed oils with other vegetable oils is shown in Table XV, taken from the recent work of Kornfeldt and Croon (1981). These authors also examined 4-monomethyl sterols and 4,4-dimethyl sterols and have identified the key components of these classes useful for identifying specific vegetable oils by mass spectral patterns. None suitable for this purpose was present in rapeseed oil. Two 3,5-dien-7-oxosteroids, campesta-3,5,22-trien-7-one and stigmasta-3,5-dien-7-one, respectively, are reported as minor components of *B. napus* rapeseed oil (Darmati et al., 1978). Unlike sunflower and poppy seeds, the sterols in rapeseeds do not change in composition during storage (Johansson and Appelqvist, 1979).

VI. TOCOPHEROLS

Since publication of a previous review on rapeseed oil (Ackman, 1977), little new data on tocopherols in rapeseed oil have appeared. However, analytical technology has changed/and been improved for both GLC (Mordret and Laurent, 1978; Slover and Thompson, 1981; Scott et al., 1982) and HPLC (Fujitani, 1979; Cortesi and Fedeli, 1980) of tocopherols. Some recent results for vegetable oils (Müller-Mulot et al., 1976, and others) are presented in Table XVI. The point of having 1 IU vitamin E/g polyunsaturat-

TABLE XVI

Tocopherol Contents of Some Selected Vegetable Oils

Oil[a]	Tocopherol isomer (mg/kg)			
	α	β	γ	δ
Rapeseed (refined)	70	16	178	7.4
HEAR[b]	268	—	426	—
HEAR[c]	160	—	431	—
LEAR (Canbra)[b]	192	—	431	40
LEAR (Primor)[b]	260	—	613	—
LEAR (Primor, rel.%)[d]	(32.5)	(1.6)	(64.3)	(1.5)
Soybean (refined)[a]	116	34	737	275
Soybean[e]	55	—	435	149
Soybean[f]	90	—	680	230
Safflower	223	7	33	3.9
Safflower[f]	480	—	—	—
Safflower[f]	600	—	—	—
Sunflower (refined)	608	17	11	—
Peanut	169	5.4	144	13
Peanut[f]	210	—	15	—
Corn	134	18	412	39
Corn[f]	180	—	750	—
Cottonseed	402	1.5	572	7.5
Olive	93	—	7.3	—
Olive[f]	140	—	—	—

[a] Results from Müller-Mulot et al. (1976) except as indicated.
[b] Mordret and Helme (1974).
[c] Persmark (1972).
[d] Mordret and Laurent (1978).
[e] Kato et al. (1981); free tocopherols made up 99% of the sample, with 1% in esterified form.
[f] Carpenter et al. (1976).

ed fatty acid (PUFA) was elaborated by Beringer and Dompert (1976), with special reference to the biological value of different isomers. The formula for calculating the biological value (x) from tocopherols (T) and tocotrienols (T-3) used was $1.49 (\alpha\text{-}T + \alpha\text{-}T\text{-}3) + 0.4(\beta\text{-}T + \beta\text{-}T\text{-}3) + 0.2 (\gamma\text{-}T + \gamma\text{-}T\text{-}3) + 0.02 (\delta\text{-}T + \delta\text{-}T\text{-}3) = x$ IU vitamin E. This emphasizes the merits of the α-isomer but rapeseed oils have a comfortable excess of the somewhat less valuable γ-isomer. Müller-Mulot et al. (1976) did not find any tocotrienols in rapeseed oil. Bieri and McKenna (1981) have reviewed the terms and units in current use and IUPAC–IUB descriptors and nomenclature are recommended (Anonymous 1980b). The amount of tocopherol in the Western diet appears to be satisfactory (Thompson et al., 1973; Beringer and Dompert, 1976) even if linoleic acid intake increases slightly.

Tocopherols are destroyed or removed from vegetable oils at various stages of refining of oils, especially deodorization (Juillet, 1975; Sleeter, 1981). Persmark (1972; see also Ackman, 1977) found no major change in rapeseed oils relative to the natural oil variability. This point of variability is also discussed by Beringer and Dompert (1976), whereas Hunter (1981) and Carpenter et al. (1976) respectively review the effect on tocopherols of current processing technology in the United States for soybean oil, and the tocopherols in various retail oils (Table XVI) in the same country. The ratio of vitamin E to PUFA was considered favorable even in "specially processed" oils. Part of the stability of LEAR oils toward autoxidation (Vaisey-Genser and Eskin, 1979) may be due to the reasonable content of tocopherols (Table XVI). For the same reason LEAR oil on heating, produces about the same proportion of cyclic monoenes as many other common edible oils (Guillaumin et al., 1980).

VII. POLAR LIPIDS (GUMS)

Crude HEAR and LEAR oils inevitably contain some nonglyceride impurities from the biosynthetically active parts of the seed (Carr, 1978). The polar lipids from a HEAR (cv. Sinus) and a LEAR (cv. Janpol) were recently examined by Sosulski et al. (1981). The results (Tables XVII and XVIII) are even more comparable in polar lipid class composition to soybean "lecithin" (Scholfield, 1981; Pardun, 1982) than previously reported by Zajac and Niewiadomski (1975) for HEAR lipids (Ackman, 1977). European HEAR (cv. Norde) phospholipids and fatty acids have been compared to those of LEAR (cv. Oro) by Alter and Gutfinger (1982). The proportion of neutral lipid is a variable that results from the type of processing and would normally be minimized. It is of interest that an alleged soy lecithin evaluated for mink feed closely resembled in fatty acid composition the HEAR lipids (Table XVIII), with 2.0–5.0% of 22:1 (Lund, 1980). If this was in fact rapeseed lecithin the beneficial results of this diet can be added to other nutritional studies (e.g., McCuaig and Bell, 1980, 1981) on HEAR and LEAR gums.

The phosphorus contents of gums and other polar and nonpolar lipid fractions are often chemically determined and converted to phospholipids. The most recent conversion factors are given by Pardun (1981), and methods have also recently been compared by Daun et al. (1981). Ashing of oil to determine phosphorus may be unnecessary (Totani et al., 1982). The amount of phosphorus in crude LEAR (or soy) oils does not necessarily predict the quality of the corresponding refined oil (Sambuc et al., 1982).

Phytic acid salts may be included with phospholipids under some circumstances (Ackman and Woyewoda, 1979). These are not strictly a lipid class and are more apt to be a problem in meals and protein products, not only from LEAR seed but also from other oilseeds (Jaffe, 1981).

TABLE XVII

Weight Percentage Composition of HEAR and LEAR Lipids and Phospholipid Components[a]

Lipid classes	Phospholipid components	HEAR (cv. Sinus)		LEAR (cv. Janpol)	
		Total lipid	Phospholipid	Total lipid	Phospholipid
Neutral lipids		95.8 ± 0.6		95.5 ± 0.4	
Polar lipids					
Phospholipids		3.3 ± 0.3		3.6 ± 0.3	
	Phosphatidylcholine		49.2 ± 0.6		48.1 ± 0.6
	Phosphatidylinositol		17.2 ± 0.4		19.5 ± 0.4
	Phosphatidylethanolamine		7.7 ± 0.5		8.9 ± 0.6
	Other phosphorus compounds				
	X_1		11.6 ± 0.7		11.0 ± 0.3
	X_2		6.5 ± 0.5		5.5 ± 0.4
	X_3		7.9 ± 0.6		7.0 ± 0.6
Glycolipids		0.9 ± 0.1		0.9 ± 0.1	

[a] Sosulski et al. (1981); reproduced by permission of J. Am. Oil Chem. Soc. and the authors.

TABLE XVIII

Fatty Acid Composition of the Principal Phospholipid and Glycolipid Components of HEAR and LEAR Lipid in Weight % Methyl Ester[a]

Components[b]	16:0	16:1	16:3	17:0	18:0	18:1	18:2	18:3	20:1	22:1
Phospholipids										
Phosphatidylcholine										
HEAR	10.1	—	—	—	0.8	47.7	33.1	5.4	1.5	1.4
LEAR	8.7	0.8	—	—	1.2	55.8	30.9	1.9	0.5	—
Phosphatidylinositol										
HEAR	25.6	1.4	—	—	3.0	30.4	35.9	3.4	—	—
LEAR	21.8	0.8	—	—	1.9	33.6	38.1	3.6	—	—
Phosphatidylethanolamine										
HEAR	19.2	0.6	—	—	2.6	41.4	30.6	2.9	1.4	1.3
LEAR	17.7	1.8	—	—	2.0	47.7	27.3	2.7	0.5	—
Glycolipids										
Digalactosyl diglyceride (DGDG)										
HEAR	10.9	1.8	0.9	2.2	2.2	14.0	48.4	10.3	2.5	7.2
LEAR	20.2	3.4	Tr	3.4	8.4	23.2	31.9	9.3	—	—
Monogalactosyl diglyceride (MGDG)										
HEAR	21.3	3.7	—	1.0	11.4	30.9	12.8	3.6	4.7	10.3
LEAR	19.1	5.5	—	—	6.7	43.3	15.9	9.5	—	—
Esterified sterol glucoside										
HEAR	22.6	0.9	0.6	0.8	5.9	43.0	11.4	3.9	3.7	7.4
LEAR	7.0	5.7	—	0.9	8.9	47.9	14.0	6.1	—	—

[a] Sosulski et al. (1981); reproduced by permission of J. Am. Oil Chem. Soc. and authors.

[b] Given as number of carbon atoms: number of double bonds.

VIII.   SULFUR

The change from rapeseed cultivars high in glucosinolates (HEAR and LEAR) to rapeseed cultivars low in glucosinolates might be thought to also reduce the sulfur level. Indeed the reduction of the glucosinolate level in LEAR seed seems to have had a beneficial effect on the sulfur in oil (El-Shattory et al., 1981). Daun and Hougen (1976) showed that LEAR cultivars gave crude oils with 18–31 ppm of sulfur, which could be reduced to 4–9 ppm by degumming and washing, and to 3–5 ppm by bleaching. Devinat et al. (1980a) found that 1975 industrial "colza" oil (probably HEAR) had 25–40 ppm sulfur, and LEAR oil (cv. Primor) 20–108 ppm. Two industrial rapeseed oils, INRA-0-THIO and INRA-00 had 21 and 7 ppm of sulfur, respectively. A Canadian LEAR (B. napus, cv. Regent, canola type, see Table II) extracted in the laboratory gave an oil with only 17 ppm and INRA-00 treated similarly had only 12 ppm of sulphur. The 1975 "colza" and Primor could not be refined and deodorized to less than about 10 ppm, but 1.8–1.9 ppm sulfur has been reported for a refined and deodorized LEAR oil in Europe (Thomas, 1982a), and comparable values have been reported in Canadian LEAR oils (El-Shattory et al., 1981). In two papers (Devinat and Biasini, 1980; 1980a), the types of sulfur compounds were broken down into volatile, thermolabile and nonvolatile, and it was shown that deodorization removed sulfur compounds in the proportions given in Table XIX. It was also found that the rate of hydrogenation on Ni–Fe Ziegler catalyst was accelerated more than 10-fold by the deodorization.

Other oils from seeds which do not contain glucosinolates also have sulfur compounds with the characteristics of those in laboratory-extracted rapeseed oil, but industrially produced oils may have a greater variety of

TABLE XIX

Modification of the Distribution of Sulfur Compounds in a Rapeseed Oil Neutralized and Bleached, Compared to the Untreated Oil, or to the Oil with Prior Deodorization[a]

Oil and treatment	Sulfur (ppm)			
	Total	Volatile	Thermolabile	Nonvolatile
Crude	46	20	6	20
Neutralized, bleached	23	6	3	14
Acid washed, dried, bleached, deodorized[b]	29	0	2	27
Washed, dried, deodorized[b], then neutralized and bleached	25	2	1	22

[a]   Adapted from Devinat and Biasini (1980).
[b]   2 hours at 200°C

sulfur compounds than oils prepared in laboratories (Johansson, 1977). HEAR oil hydrogenated with copper–chromite catalyst behaved much like soybean oil (Johansson, 1979; Johansson and Lundin, 1979a,b).

IX. SELENIUM, HEAVY METALS, AND ORGANIC TOXINS

Selenium is allied with sulfur chemically, and is regarded as an essential micronutrient (Beare-Rogers et al., 1974; Levander, 1975). The levels of selenium in crude rapeseed oils (Elson, 1980) are given in Table XX. There is no evidence that the Se level varies between HEAR and LEAR oils, and the levels approach or fall below the limits of detectability on refining. There is certainly no public health concern.

The same applies to heavy metals in LEAR oils examined by Elson et al. (1979). Zinc, lead, cadmium, and copper values (Table XXI) are similar to literature values assembled by Ackman (1977). Mercury and arsenic in rapeseed oils included in that report are at levels similar to those for soybean oil (Thomas, 1982b). They meet or surpass Codex Alimentarius Proposed Standards (see below). Correlations of some metals pairs in oilseeds, including mustard, have been reported (Deosthale, 1981) but do not necessarily relate to oils. Trace metals in oils have to be considered as factors in stability (Flider and Orthoefer, 1981), as analytical challenges (Pickford, 1981), or as micronutrients (Chesters, 1981). Metals such as iron, and even phosphorus, in crude soy and LEAR oils are poor predictors of the quality of refined oils (Sambuc et al., 1982).

Toxic weed seed contaminants have been considered in soybean oils, but most are alkaloids and are not extracted with the oil during processing (List et al., 1979). Grain is accompanied by a variety of weed seeds (Daun and

TABLE XX

Selenium Content of Oils and Seed[a]

Sample	Se content (ng/g) = (ppb)
Crude rapeseed oil	8.9 ± 2.5
Fully refined rapeseed oil	2.8 ± 1.8
Crude LEAR oil (cv. Tower), canola type	7.9 ± 3.8
Fully refined LEAR oil (cv. Tower) canola type	1[b]
Desolvenated HEAR oil (cv. Target)	3.8 ± 1.3
Crude soybean oil	1[b]
LEAR seed (cv. Tower), canola type	70 ± 12

[a] Elson (1980); reproduced by permission of Canola Council of Canada and author.
[b] Se level well below detection limit.

TABLE XXI

Metal Content of Oils[a]

Sample and source	Metal level (ppm)			
	Zinc	Lead	Cadmium	Copper
Crude LEAR (cv. Tower)—C.S.P. Foods, Altona	2.4 ± 0.4	0.24 ± 0.02	Bdl	Bdl
Deodorized LEAR (cv. Tower)—C.S.P. Foods, Altona	Bdl[b]	0.07 ± 0.02	Bdl	Bdl
Crude LEAR (cv. Tower)—C.S.P. Foods, Nipawin	1.0 ± 0.2	0.06 ± 0.02	Bdl	Bdl
Crude LEAR (cv. Tower)—received spring, 1975 (source unknown)	3.6 ± 0.4	0.22 ± 0.02	Bdl	Bdl
Refined LEAR (cv. Tower)—received spring, 1975 (source unknown)	1.1 ± 0.2	Bdl	Bdl	Bdl
Degummed LEAR—C.S.P. Foods, Saskatoon	2.1 ± 0.3	Bdl	Bdl	Bdl
Crude LEAR (cv. Span)—Canbra Foods, Lethbridge	2.9 ± 0.5	Bdl	Bdl	Bdl
Corn oil—Mazola brand, lot No. 0586	Bdl	Bdl	Bdl	Bdl
Peanut oil—Planters brand, lot No. R3612	Bdl	Bdl	Bdl	Bdl

[a] Elson et al. (1979); reproduced by permission of J. Am. Oil Chem. Soc. and authors.

[b] Bdl = below detection limit (<0.05 ppm Cu, <0.005 ppm Cd, <0.02 ppm Pb, <0.8 ppm Zn).

Tkachuk, 1976), but a different assortment of weed seeds occur in rapeseed screenings (Ackman and Sebedio, 1981; Rose et al., 1981). These did not produce an oil with any adverse effect on the health of rats, but dockage oil is reported to affect the stability of LEAR oils (Ismail et al., 1980). Many common edible oils or seeds contain potentially toxic components (Dhopeshwarkar, 1981), but these are eliminated or reduced by oil refining (Scott, 1975; Thomas, 1976; Carr, 1978; Young, 1978; Helme, 1980).

X. SPECIFICATIONS

Various countries or companies have specifications for rapeseed crude oils (Thomas, 1982). That of Canada CAN-32-300M-76 for "rapeseed oil, low erucic acid, crude and crude, degummed" is given in Chapter 3.

On the international level the Codex Alimentarius Commission (Hlavacek, 1981; Wessels, 1981) has a draft standard for "Edible low erucic acid rapeseed oil" at step 8 of the Codex Procedure. This standard is reproduced in the version available in early 1982 (Appendix II) and is not the final standard.

APPENDIX I

Weight Percentages of Important Fatty Acids in Some Mustard Seed Oils of Diverse Origins[a]

Fatty acids	Yugoslavia	Turkey	Spain	Israel	England
14:0	0.03	0.03	0.03	0.04	0.02
16:0	2.98	3.56	3.83	3.00	3.26
18:0	1.26	1.04	1.07	1.09	0.83
20:0	0.62	0.56	0.68	0.71	0.54
22:0	1.09	1.15	1.09	1.41	1.20
24:0	0.34	0.38	0.35	0.54	0.55
Total saturated	6.32	6.72	7.05	6.79	6.40
16:1	0.14	0.28	0.44	0.15	0.18
18:1	14.76	10.96	8.84	9.88	9.11
20:1	17.03	16.08	13.67	14.36	11.14
22:1	32.99	35.09	38.17	40.20	44.55
24:1	1.01	0.80	0.97	0.99	1.19
Total monounsaturated	65.93	63.21	61.25	65.58	66.17
18:2	15.17	12.88	12.51	12.93	12.71
18:3	11.74	14.93	16.45	13.36	13.11
20:2	0.81	1.15	1.58	1.20	1.07
20:3	0.04	1.10	0.40	0.13	0.18
22:2	—	—	0.31	—	0.34
Total polyunsaturated	27.76	30.06	31.25	27.62	27.41

[a] Extracted from seed supplied by J. Daun, Canadian Grain Commissioners Laboratory, Winnipeg. All are *Sinapis arvensis* except for wild mustard from Spain.

APPENDIX II:

**Draft Standard for Edible Low Erucic Acid Rapeseed Oil
(at Step 8 of the Codex Procedure)**

1. SCOPE: This standard applies to edible low erucic acid rapeseed oil but does not apply to low erucic acid rapeseed oil which must be subjected to further processing in order to render it suitable for human consumption.

2. DESCRIPTION: Low erucic acid rapeseed oil (synonyms: low erucic acid turnip rape oil; low erucic acid colza oil) is produced from the low erucic acid oil-bearing seeds of varieties derived from the *Brassica napus* L., *Brassica campestris* L. species.

3. ESSENTIAL COMPOSITION AND QUALITY FACTORS:
3.1   Identity characteristics
3.1.1   Relative density (20°C/water at 20°C)   0.914–0.917
3.1.2   Refractive index (n_D 40°C)   1.465–1.467
3.1.3   Saponification value (mg KOH/g oil)   188–193
3.1.4   Iodine value (Wijs)   110–126
3.1.5   Crismer value   67–70
3.1.6   Unsaponifiable matter   Not more than 20g/kg
3.1.7   Brassicasterol   Not less than 5% of total sterols
3.1.8   Erucic acid   Not more than 5% (m/m) of the
                                                                      component fatty acids
3.1.9   GLC ranges of fatty acid composition (%)[a]

C14:0	<0.2
C16:0	2.5–6.0
C16:1	<0.6
C18:0	0.9–2.1
C18:1	50–66
C18:2	18–30
C18:3	6–14
C20:0	0.1–1.2
C20:1	0.1–4.3
C22:0	<0.5
C22:1	<5.0
C24:0	<0.2

3.2   Quality Characteristics
3.2.1   Color: Characteristic of the designated product.
3.2.2   Color and taste: Characteristic of the designated product and free from foreign and rancid odor and taste.
3.2.3   Acid value: Not more than 0.6 mg KOH/g oil.
3.2.4   Peroxide value: Not more than 10 mEq peroxide oxygen/kg oil.

4. FOOD ADDITIVES:
4.1   Colors: The following colors are permitted for the purpose of restoring natural color lost in processing or for the purpose of standardizing color, as long as the added color does not deceive or mislead the consumer by concealing damage or inferiority or by making the product appear to be of greater than actual value.

[a] See paragraphs 52 and 53 of ALINORM 79/17.

APPENDIX II (*CONTINUED*)

	Maximum level of use
4.1.1 Beta-carotene	
4.1.2 Annatto[b]	
4.1.3 Curcumin[b]	
4.1.4 Canthaxanthine	Not limited
4.1.5 Beta-apo-8'carotenal	
4.1.6 Methyl and ethyl esters of Beta-apo-8'-carotenoic acid	

4.2 Flavors: Natural flavors and their identical synthetic equivalents, except those which are known to represent a toxic hazard, and other synthetic flavors approved by the Codex Alimentarius commission are permitted for the purpose of restoring natural flavor lost in processing or for the purpose of standardizing flavor, as long as the added flavor does not deceive or mislead the consumer by concealing damage or inferiority or by making the product appear to be of greater than actual value.[b]

	Maximum level of use
4.3 Antioxidants	
4.3.1 Propyl, octyl, and dodecyl gallates[b]	100 mg/kg, individually or in combination
4.3.2 Butylated hydroxytoluene (BHT)[b]	200 mg/kg, individually or in combination
4.3.3 Butylated hydroxyanisole (BHA)[b]	
4.3.4 Tertiary butylhydroquinone (TBHQ)	
4.3.5 Any combination of gallates with BHA, BHT, and/or TBHQ[b]	200 mg/kg, but gallates not to exceed 100 mg/kg
4.3.6 Ascorbyl palmitate	500 mg/kg, individually or in combination
4.3.7 Ascorbyl stearate	
4.3.8 Natural and synthetic tocopherols	Not limited
4.3.9 Dilauryl thiodipropionate	200 mg/kg
4.4 Antioxidant synergists	
4.4.1 Citric acid	Not limited
4.4.2 Sodium citrate	Not limited
4.4.3 Isopropyl citrate mixture	100 mg/kg, individually or in combination
4.4.4 Monoglyceride citrate	
4.4.5 Phosphoric acid	
4.5 Antifoaming agent: Dimethyl polysiloxane (syn: Dimethyl silicone) singly or in combination with silicon dioxide	10 mg/kg
4.6 Crystallization inhibitor Oxystearin	1250 mg/kg
5. CONTAMINANTS:	
5.1 Matter volatile at 105°	0.2% m/m
5.2 Insoluble impurities	0.05% m/m
5.3 Soap content	0.005% m/m
5.4 Iron (Fe)	1.5 mg/kg
5.5 Copper (Cu)	0.1 mg/kg

[b] Temporarily endorsed.   *(Continued)*

APPENDIX II (CONTINUED)

5.6	Lead (Pb)[b]Fn.[b]	0.1 mg/kg
5.7	Arsenic (As)	0.1 mg/kg

6. HYGIENE: It is recommended that the product covered by the provisions of this standard be prepared in accordance with the appropriate sections of the General Principles of Food Hygiene recommended by the Codex Alimentarius Commission (Ref. No. CAC/RCP 1-1969).

7. LABELING: In addition to Sections 1, 2, 4, and 6 of the General Standard for Labeling of Prepackaged Foods (Ref. NO. CAC/RS 1-1969), the following specific provisions apply:
7.1   Name of the food
7.1.1   All food products designated as low erucic acid rapeseed oil, low erucic acid turnip rape oil, low erucic acid colza oil, must conform to this standard.
7.1.2   Where low erucic acid rapeseed oil has been subject to any process of esterification or to processing which alters its fatty acid composition or its consistency, the name low erucic acid rapeseed oil or any synonym shall not be used unless qualified to indicate the nature of the process.
7.2   List of ingredients
7.2.1   A complete list of ingredients shall be declared on the label in descending order of proportion.
7.2.2   A specific name shall be used for ingredients in the list of ingredients except that class titles may be used in accordance with Subsection 3.2(c)(ii) of the General Standard for the Labeling of Prepackaged Foods.
7.3   Net contents: The net contents shall be declared in accordance with Subsection 3.3(a) of the General Standard for the Labeling of Prepackaged Foods.
7.4   Name and Address: The name and address of the manufacturer, packer, distributor, importer, exporter, or vendor of the product shall be declared.
7.5   Country of origin
7.5.1   The country of origin of the product shall be declared if its omission would mislead or deceive the consumer.
7.5.2   When the product undergoes processing in the second country which changes its nature, the country in which the processing is performed shall be considered or to be the country of origin for the purpose of labeling.
7.6   Lot identification: Each container shall be embossed or otherwise permanently marked in code or in clear to identify the producing factory and the lot.
7.7   Date marking
7.7.1   The date of minimum durability of the food shall be declared in clear.
7.7.2   In addition to the date, any special conditions for the storage of the food should be indicated if the validity of the date depends thereon.
7.8   Bulk packs: (to be elaborated)

8. METHODS OF ANALYSIS AND SAMPLING: The methods of analysis and sampling referred to hereunder are international referee methods and are subject to endorsement by the Codex Committee on Methods of Analysis and Sampling.
8.1   Determination of relative density: According to the FAO/WHO Codex Alimentarius Method (FAO/WHO Methods of Analysis for Edible Fats and Oils, CAC/RM 9-1969, Determination of Relative Density at t/20°C). Results are expressed as relative density at 20°C/water at 20°C.
8.2   Determination of refractive index: According to the IUPAC (1964) method (IUPAC Standard Methods for the Analysis of Oils, Fats and Soaps, 5th edition, 1966 II.B.2 *Refractive*

APPENDIX II (*CONTINUED*)

Index). Results are given as the refractive index relative to the sodium D-line at 40°C (n_D 40°C).

8.3   Determination of saponification value (I_s): According to the IUPAC (1964) method (IUPAC Standard Methods for the Analysis of Oils, Fats and Soaps, 5th edition, 1966 II.D.2 *Saponification Value (I_s)*. Results are expressed as the number of milligrams KOH per 1 g oil.

8.4   Determination of iodine value (I_I): According to the (Wijs) IUPAC (1964) method (IUPAC Standard Methods for the Analysis of Oils, Fats and Soaps, 5th edition, 1966 II.D.7.1., II.D.7.2 and II.D.7.3 *The wijs Method*). Results are expressed as percentage of mass per mass absorbed iodine.

8.5   Determination of Crismer value (I_C): According to the AOCS method (Official and Tentative Methods of the American Oil Chemists' Society; AOCS Official Method Cb 4–35, Crismer Test, Fryer and Weston Modification, and Ca 5a-40, Free Fatty Acids, calculating the acidity as oleic acid). Results are expressed by a conventional value (I_C) as described in the method.

8.6   Determination of unsaponifiable matter: According to the IUPAC (1964) *diethyl ether method* (IUPAC) (Standard Methods for the Analysis of Oils, Fats and Soaps, 5th edition, 1966, II.D.5.1 and II.D.5.3). Results are expressed as grams unsaponifiable matter per kilogram oil.

8.7   Determination of fatty acid composition: According to IUPAC Methods II.D.19 and II.D.25.

8.8   Determination of sterols: according to IUPAC Method II.C.8.

8.9   Determination of acid value (I_A): According to the IUPAC (1964) method (IUPAC Standard Methods for the Analysis of Oils, Fats and Soaps, 5th edition, 1966 II.D.1.2 Acid Value (I_A). Results are expressed as the number of milligrams KOH required to neutralize 1 g oil.

8.10   Determination of peroxide value (I_P): According to the IUPAC (1964) method (IUPAC Standard Methods for the Analysis of Oils, Fats and Soaps, 5th edition, 1966 II.D.13 *Peroxide Value*). Results are expressed in milliequivalents active oxygen per kilogram oil.

8.11   Determination of matter volatile at 105°C: According to the IUPAC (1964) method (IUPAC Standard Methods for the Analysis of Oils, Fats and Soaps, 5th edition, 1966 II.C.1.1 *Moisture and Volatile Matter*). Results are expressed as percentage of mass per mass.

8.12   Determination of insoluble impurities: According to the IUPAC (1964) method (IUPAC Standard Methods for the Analysis of Oils, Fats and Soaps, 5th edition, 1966 II.C.2 *Impurities*). Results are expressed as percentage of mass per mass.

8.13   Determination of soap content: According to the FAO/WHO Codex Alimentarius Method (FAO/WHO Methods of Analysis for Edible Fats and Oils, CAC/RM 13-1969, *Determination of Soap Content*). Results are expressed as percentage of mass per mass sodium oleate.

8.14   Determination of iron[c]: According to the FAO/WHO Codex Alimentarius Method (FAO/WHO Methods of Analysis for Edible Fats and Oils, CAC/RM 14-1969, *Determination of Iron Content*). Results are expressed as milligrams iron per kilogram.

8.15   Determination of copper[c]: According to the AOAC (1965) method (Official Methods of Analysis of the AOAC, *International Union of Pure and Applied Chemistry Carbamate Method*, 24.023-24.028). Results are expressed as milligrams copper per kilogram.

8.16   Determination of lead[c]: According to the AOAC (1965) method, after complete digestion, by the colorimetric *dithizone determination procedure* [Official Methods of Analysis of the AOAC, 1965, 24.053 (and 24.008, 24.009, 24.043j, 24.046, 24.047 and 24.048)]. Results are expressed as milligrams lead per kilogram.

8.17   Determination of arsenic: According to the colorimetric *silver diethyldithiocarbamate method* of the AOAC (Official Methods of Analysis of the AOAC, 1965, 24.011-24.014, 24.016-24.017, 24.006-24.008). Results are expressed as milligrams arsenic per kilogram.

[c] Might be replaced by Atomic Absorption Spectrophotometry in the future.

REFERENCES

Ackman, R. G. (1966). *J. Am. Oil Chem. Soc.* **43**, 483–486.
Ackman, R. G. (1977). *Proc. Symp. Rapeseed Oil, Meal By-Prod Util Vancouver, 1977* Publ. No. 45, pp. 12–36.
Ackman, R. G., and Eaton, C. A. (1977). *J. Am. Oil Chem. Soc.* **54**, 435.
Ackman, R. G., and Sebedio, J.-L. (1979). *Proc. Int. Rapeseed Conf 5th Malmö 1978* Vol. 2, pp. 9–12.
Ackman, R. G., and Sebedio, J.-L. (1981). *J. Am. Oil Chem. Soc.* **58**, 594–598.
Ackman, R. G., and Woyewoda, A. D. (1979). *J. Chromatogr. Sci.* **17**, 514–517.
Ackman, R. G., Hooper, S. N., and Hooper, D. L. (1974) *J. Am. Oil Chem. Soc.* **51**, 42–49.
Ackman, R. G., Eaton, C. A., Sipos, J. C., and Crewe, N. F. (1981). *Can. Inst. Food Sci. Technol. J.* **14**, 103–107.
Ahmad, I., and Ali, A. (1981). *J. Am. Oil Chem. Soc.* **58**, 87–88.
Alfin-Slater, R. B., and Aftergood, L. (1976). *Med. Chem. (Academic)* **2**, 43–80.
Alter, M., and Gutfinger, T. (1982). *Riv. Ital. Sostanze Grasse* **59**, 14–18.
Aneja, R., Bhati, A., Hamilton, R. J., Padley, F. B., and Steven, D. A. (1979). *J. Chromatogr.* **173**, 392–397.
Anonymous (1980a). "Report of the Ad Hoc Committee on the Composition of Special Margarines." Health and Welfare Canada, Ottawa.
Anonymous (1980b). *J. Nutr.* **110**, 8–15.
Anonymous (1981). "Western Canadian Oilseeds 1980," *Crop Bull.* **149**. Can. Grain Comm., Grain Res. Lab., Winnipeg, Manitoba.
Appelqvist, L.-A. (1972). *In* "Rapeseed" (L.-A. Appelqvist and R. Ohlson, eds.), pp. 1–8. Elsevier, Amsterdam.
Appelqvist, L.-A., and Dowdell, R. J. (1968). *Ark. för Kemi* **28**, 539–549.
Appelqvist, L.-A., and Ohlson, R., eds. (1972). "Rapeseed." Elsevier, Amsterdam.
Appelqvist, L.-A. D., Kornfeldt, A. K., and Wennerholm, J. E. (1981). *Phytochemistry* **20**, 207–210.
Appleby, R. S., Gurr, M. I., and Nichols, B. W. (1974). *Eur J. Biochem.* **48**, 209–216
Applewhite, T. H. (1981). *J. Am. Oil Chem. Soc.* **58**, 260–269.
Bang, H. O., and Dyerberg, J. (1980). *Adv. Nutr. Res.* **3**, 1–22.
Bang, H. O., and Dyerberg, J. (1981). *Medd. Groenl., Man & Soc.,* **2**, 3–18.
Beare-Rogers, J. L., Nera, E. A., and Heggtveit, H. A. (1974). *Nutr. Metab.* **17**, 213–222.
Bereziat, G. (1978). *Rev. Fr. Corps Gras* **25**, 463–473.
Beringer, H., and Dompert, W. U. (1976). *Fette, Seifen, Anstrichm.* **78**, 228–231.
Biacs, P. A. (1979). *In* "Advances in the Biochemistry and Physiology of Plant Lipids" (L.-A. Appelqvist and C. Liljenberg, eds.), pp. 275–280. Elsevier, Amsterdam.
Bieri, J. G., and McKenna, M. C. (1981). *Am. J. Clin. Nutr.* **34**, 289–295.
Breckenridge, W. C. (1979). *Hand. Lipid Res.* **I**, 197–232.
Brockerhoff, H. (1971). *Lipids* **6**, 942–956.
Brockerhoff, H., and Yurkowski, M. (1966). *J. Lipid Res.* **7**, 62–64.
Carpenter, D. L., Lehmann, J., Mason, B. S., and Slover, H. T. (1976). *J. Am. Oil Chem. Soc.* **53**, 713–718.
Carr, R. A. (1978). *J. Am. Oil Chem. Soc.* **55**, 765–771.
Cassagne, C., and Lessire, R. (1979). *In* "Advances in the Biochemistry and Physiology of Plant Lipids" (L.-A. Appelqvist and C. Liljenberg, eds.), pp. 393–398. Elsevier, Amsterdam.
Chesters, J. K. (1981). *Chem. Soc. Rev.* **10**, 270–279.
Cortesi, N., and Fedeli, E. (1980). *Riv. Ital. Sostanze Grasse* **57**, 16–19.
Cortesi, N., Mariani, C., and Fedeli, E. (1973). *Riv. Ital. Sostanze Grasse* **50**, 411–412.

Costigliola, B. (1970). *Proc. Int. Conf. Sci., Technol. Market. Rapeseed Rapeseed Prod. St. Adèle, 1970*, pp. 203–212.

Craig, B. M. (1970). *Proc. Int. Conf. Sci., Technol. Market. Rapeseed Rapeseed Prod. St. Adèle, 1970*, pp. 158–170.

Craig, B. M., Mallard, T. M., Wight, R. E., Irvine, G. N., and Reynolds, J. R. (1973). *J. Am. Oil Chem. Soc.* **50**, 395–399.

D'Alonzo, R. P., Kozarek, W. J., and Wharton, H. W. (1981). *J. Am. Oil Chem. Soc.* **58**, 215–227.

Darmati, S., Bastic, M., Javanoric, J., Gasic, M. J., Spiteller, M., and Piatak, D. M. (1978). *Bull. Soc. Chim., Beograd* **43**, 567–572.

Daun, J. K., and Hougen, F. W. (1976). *J. Am. Oil Chem. Soc.* **53**, 169–171.

Daun, J. K., and Tkachuk, R. (1976). *J. Am. Oil Chem. Soc.* **53**, 661–662.

Daun, J. K., Davidson, L. D., Blake, J. A., and Yuen, W. (1981). *J. Am. Oil Chem. Soc.* **58**, 914–916.

Defrancesco, F., Casagrande, S., Defrancesco, C., Cescatti, G., and Boccardi, A. (1981). *Riv. Ital. Sostanze Grasse* **58**, 175–178.

deMan, J. M. (1978). *Can. Inst. Food Sci. Technol. J.* **11**, 194–203.

deMan, J. M. (1979). *Proc. Int. Rapeseed Conf., 5th Malmö 1978* Vol. 2, pp. 169–171.

Deosthale, Y. G. (1981). *J. Am. Oil Chem. Soc.* **58**, 988–990.

Devinat, G., and Biasini, S. (1980). *Rev. Fr. Corps Gras* **27**, 563–565.

Devinat, G., Biasini, S., and Naudet, M. (1980a). *Rev. Fr. Corps Gras* **27**, 229–236.

Devinat, G., Scamaroni, L., and Naudet, M. (1980b). *Rev. Fr. Corps Gras* **27**, 283–287.

Dhopeshwarkar, G. A. (1981). *Prog. Lipid Res.* **19**, 107–118.

Downey, R. K. (1976). *Chem. Ind. (London)* May 1, pp. 401–406.

Downey, R. K. (1979). *Proc. Int. Rapeseed Conf., 5th Malmö 1978* Vol. 1, pp. 106–112.

Dyerberg, J., Bang, H. O., and Aagaard, O. (1980). *Lancet,* 199.

Eicke, H. (1971). *Seifen, Oele, Fette, Wachse* **97**, 712–715.

El Azzabi, T. S., Clarke, J. H., and Hill, S. T. (1981). *J. Sci. Food Agric.* **32**, 493–497.

El-Hamdy, A. H., and Perkins, E. G. (1981). *J. Am. Oil Chem. Soc.* **58**, 867–872.

El-Shattory, T., deMan, L., and deMan, J. (1981). *Can. Inst. Food. Sci. Technol. J.* **14**, 53–58.

Elson, C. M. (1980). *In* "6th Progress Report: Research on Canola Seed, Oil, Meal and Meal Fractions" (E. E. McGregor, ed.), Publ. No. 57, pp. 246–250. Canola Council of Canada, Winnipeg, Manitoba.

Elson, C. M., Hynes, D. L., and MacNeil, P. A. (1979). *J. Am. Oil Chem. Soc.* **56**, 998–999.

Emken, E. A. (1979). *In* "Geometrical and Positional Fatty Acid Isomers" (E. A. Emken and H. J. Dutton, eds.), pp. 99–129. Am. Oil Chem. Soc., Champaign, Illinois

Eyres, L. (1979). *Chem. N. Z.* **43**, 237–239.

Favini, G., Domenichini, M., and Fedeli, E. (1980). *Riv. Ital. Sostanze Grasse* **57**, 27–30.

Flanzy, J. (1979). *Ann. Biol. Anim., Biochim. Biophys.* **19**, 467–470.

Flider, F. J., and Orthoefer, F. T. (1981). *J. Am. Oil Chem. Soc.* **58**. 270–272.

Food and Agriculture Organization (1977). "Dietary Fats and Oils in Human Nutrition." FAO Food Nutri. Pap. 3. FAO United Nations, Rome.

Franzke, C., Kroll, J., and Seedorf, C. (1972). *Zentralbl. Pharm.* **111**, 1025–1034.

Fujitani, T. (1979). *Yukagaku* **28**, 468–473.

Grob, K., Jr. (1979). *J. Chromatogr.* **178**, 387–392.

Grob, K., Jr. (1981). *J. Chromatogr.* **205**, 289–296.

Grob, K., Jr., Neukom, H. P., and Battaglia, R. (1980). *J. Am. Oil Chem. Soc.* **57**, 282–286.

Guillaumin, R., Rondot, G., and Coquet, B. (1980). *Rev. Fr. Corps Gras* **27**, 189–196.

Gurr, M. I., Blades, J., and Appleby, R. S. (1972). *Eur. J. Biochem.* **29**, 362–368

Hadorn, H., and Zürcher, K. (1967). *Mitt. Geb. Lebensmittelunters. Hyg.* **58**, 351–384.

Halushka, P. V., Knapp, D. R., and Grimm, L. (1979). *Curr. Top. Hematology* **2**, 75–143.

Hamilton, R. J., and Ackman, R. G. (1975). *J. Chromatogr. Sci.* **13**, 474–478.

Hansen, R. P. (1976). *N. Z. J. Agric. Res.* **19**, 343–345.

Harlow, R. D., Litchfield, C., and Reiser, R. (1966). *Lipids* **1**, 216–220.

Harris, J. F., and Norton, G. (1979). *Proc. Int. Rapeseed Conf., 5th Malmö 1978* Vol. 2, pp. 16–20.

Heckers, H., and Melcher, F. W. (1978). *Am. J. Clin. Nutr.* **31**, 1041–1049.

Helme, J. P. (1980). *Oleagineux* **35**, 93–103.

Hernqvist, L. (1979). *Scand. Symp. Lipids [Proc.], 10th, Nyborg, 1979*, F224–F229.

Hiltunen, R., Huhtikangas, A., and Hovinen, S. (1979). *Acta Pharm. Fenn.* **88**, 31–34.

Hirai, A., Hamazaki, T., Terano, T., Nishikawa, T., Tamura, Y., Kumagai, A., and Sajiki, J. (1980). *Lancet*, 1132–1133.

Hlavacek, R. J. (1981). *J. Am. Oil Chem. Soc.* **58**, 232–234.

Holman, R. T. (1977). *In* "Polyunsaturated Fatty Acids" (W.-H. Kunau and R. T. Holman, eds.), pp. 163–182. Am. Oil Chem. Soc., Champaign, Illinois.

Holman, R. T. (1981). *Chem. Ind. (London)* Oct. 17, pp. 704–709.

Homberg, E. (1977). *Fette, Seifen, Anstrichm.* **79**, 234–241.

Hougen, F. W., and Wasowicz, E. W. (1979). *Proc. Int. Rapeseed Conf., 5th Malmö 1978* Vol. 2, pp. 13–15.

Houtsmuller, U. M. T. (1975). *In* "The Role of Fats in Human Nutrition" (A. J. Vergroesen, ed.), pp. 331–351. Academic Press, New York.

Hunter, J. E. (1981). *J. Am. Oil Chem. Soc.* **58**, 283–287.

Ilsemann, K., Reichwald-Hacker, I., and Mukherjee, K. D. (1979). *Chem. Phys. Lipids* **23**, 1–5.

Ismail, F., Eskin, N. A. M., and Vaisey-Genser, M. (1980). *In* "6th Progress Report: Research on Canola Seed, Oil, Meal and Meal Fractions" (E. E. McGregor, ed.), Publ. No. 57, pp. 234–239. Canola Council of Canada, Winnipeg, Manitoba.

Itoh, T., Tamura, T., and Matsumoto, T. (1973). *J. Am. Oil Chem. Soc.* **50**, 122–125.

Itoh, T., Komagata, H., Tamura, T., and Matsumoto, T. (1981). *Fette, Seifen, Anstrichm.* **83**, 123–125.

Izzo, R., Lotti, G., and Pioli, L. (1979). *Riv. Soc. Ital. Sci. Aliment.* **8**, 191–198.

Jaffe, G. (1981). *J. Am. Oil Chem. Soc.* **58**, 493–495.

Jáky, M., and Kurnik, E. (1981). *Fette, Seifen, Anstrichm.* **83**, 267–270.

Johansson, A. (1977). *Scand. Symp. Lipids, [Proc.], 9th, Visby, 1977*, H225–H228.

Johansson, A., and Appelqvist, L.-A. (1979). *J. Am. Oil Chem. Soc.* **56**, 995–997.

Johansson, A., and Hoffmann, I. (1979). *J. Am. Oil Chem. Soc.* **56**, 886–889.

Johansson, L. E. (1979). *J. Am. Oil Chem. Soc.* **56**, 987–991.

Johansson, L. E., and Lundin, S. T. (1979a). *J. Am. Oil Chem. Soc.* **56**, 974–980.

Johansson, L. E., and Lundin, S. T. (1979b). *J. Am. Oil. Chem. Soc.* **56**, 981–986.

Juillet, M. T. (1975). *Fette, Seifen, Anstrichm.* **77**, 101–105.

Kaimal, T. N. B., and Lakshminarayana, G. (1979). *J. Am. Oil Chem. Soc.* **56**, 578–580.

Kanematsu, H., Maruyama, T., Niiya, I., Imamura, M., and Matsumoto, T. (1973). *Yukagaku* **22**, 814–817.

Kanematsu, H., Maruyama, T., Kinoshita, Y., Niiya, I., and Imamura, M. (1976). *Eiyo to Shokuryo* **29**, 39–43.

Karleskind, A., Valmalle, G., and Wolff, J.-P. (1974). *Rev. Fr. Corps Gras* **21**, 617–630.

Karleskind, A., Valmalle, G., Midler, O., and Blanc, M.(1978). *Rev. Fr. Corps Gras* **25**, 551–556.

Kato, A., Tanabe, K., and Yamaoka, M. (1981). *Yukagaku* **30**, 515–516.

Kawamura, K. (1981). *J. Am. Oil Chem. Soc.* **58**, 826–829.

Koman, V., Sahajova-Hojerova, J., and Csicsayova, M. (1981). *J. Am. Oil Chem. Soc.* **58**, 102–105.

Kornfeldt, A., and Croon, L.-B. (1981). *Lipids* **16**, 306–314.

Kovacs, M. I. P., Ackman, R. G., Anderson, W. E., and Ward, J. (1978). *Can. Inst. Food Sci. Technol. J.* **11**, 219–221.

Kramer, J. K. G., Hulan, H. W., Mahadevan, S., and Sauer, F. D. (1975a). *Lipids* **10**, 505–510.

Kramer, J. K. G., Hulan, H. W., Mahadevan, S., Sauer, F. D., and Corner, A. H. (1975b). *Lipids* **10**, 511–516.

Kroll, J., and Franzke, C. (1974). *Fette, Seifen, Anstrichm.* **76**, 385–390.

Kroll, J., Franzke, C., and Genz, S. (1973). *Pharmazie* **28**, 263–269.

Kuemmel, D. F. (1964). *J. Am. Oil Chem. Soc.* **41**, 667–670.

Kuksis, A., Myher, J. J., Marai, L., and Geher, K. (1975). *J. Chromatogr. Sci.* **13**, 423–430.

Lands, W. E. M., Hemler, M. E., and Crawford, C. G. (1977). *In* "Polyunsaturated Fatty Acids" (W.-H. Kunau and R. T. Holman, eds.), pp. 193–228. Am. Oil Chem. Soc., Champaign, Illinois.

Levander, O. A. (1975). *J. Am. Diet. Assoc.* **66**, 338–344.

List, G. R., Spencer, G. F., and Hunt, W. H. (1979). *J. Am. Oil Chem. Soc.* **56**, 706–710.

Litchfield, C. (1971). *J. Am. Oil Chem. Soc.* **48**, 467–472.

Litchfield, C. (1973). *Fette, Seifen, Anstrichm.* **75**, 223–231.

Lund, R. S. (1980). *Kraftfutter* **5**, 232, 234, 236.

McCuaig, L. W., and Bell, J. M. (1980). *In* "6th Progress Report: Research on Canola Seed, Oil, Meal and Meal Fractions" (E. E. McGregor, ed.), Publ. No. 57, pp. 190–194. Canola Council of Canada, Winnipeg, Manitoba.

McCuaig, L. W., and Bell, J. M. (1981). *Can. J. Anim. Sci.* **61**, 463–467.

McGregor, D. I. (1977). *Can. J. Plant Sci.* **57**, 133–142.

Mannino, S., and Amelotti, G. (1975). *Riv. Ital. Sostanze Grasse* **52**, 79–83.

Maxwell, R. J., and Schwartz, D. P. (1979). *J. Am. Oil Chem. Soc.* **56**, 634–636.

Maxwell, R. J., Reimann, K. A., and Percell, K. (1981). *J. Am. Oil Chem. Soc.* **58**, 1002–1004.

Mertens, W. G., Mag, T. K., and Teasdale, B. F. (1970). *Proc. Int. Conf. Sci., Technol. Market. Rapeseed Prod. St. Adèle, 1970,* pp. 213–222.

Monseigny, A., Vigneron, P.-Y., Levacq, M., and Zwoboda, F. (1979). *Rev. Fr. Corps Gras* **26**, 107–120.

Mordret, F., and Helme, J. P. (1974). *Proc. Int. Rapskongr., 4th Giessen, 1974,* pp. 283–289.

Mordret, F., and Laurent, A. M. (1978). *Rev. Fr. Corps Gras* **25**, 245–250.

Morice, J. (1979). *Ann. Biol. Anim. Biochim. Biophys.* **19**, 471–477.

Mukherjee, S., and Sengupta, S. (1981). *J. Am. Oil Chem. Soc.* **58**, 287–291.

Müller-Mulot, W., Rohrer, G., and Medweth, R. (1976). *Fette, Seifen, Anstrichm.* **78**, 257–262.

Murata, T., and Takahashi, S. (1977). *Anal. Chem.* **49**, 728–731.

Myher, J. J. (1978). *Handb. Lipid Res.* **1**, 123–196.

Myher, J. J., Kuksis, A., Vasdev, S. C., and Kako, K. J. (1979). *Can. J. Biochem.* **57**, 1315–1327.

Niewiadomski, H. (1975). *Nahrung* **19**, 525–536.

Nolen, G. A. (1981). *J. Am. Oil Chem. Soc.* **58**, 31–37.

Ohlson, R. (1970). *Proc. Int. Conf. Sci., Technol. Market. Rapeseed Prod. St. Adèle 1970,* pp. 171–189.

Ohlson, R. (1972). *In* "Rapeseed" (L.-A. Appelqvist and R. Ohlson, eds.), pp. 9–35. Elsevier, Amsterdam.

Ohlson, R. (1979). *Proc. Int. Rapeseed Conf., 5th Malmö 1978,* Vol. 2, pp. 152–167.

Ohlson, R., Podlaha, O., and Töregard, B. (1975). *Lipids* **10**, 732–735.

Osman, F., and Fiad, S. (1975). *Nahrung* **19**, 641–647.

Pardun, H. (1981). *Fette, Seifen, Anstrichm.* **83**, 240–242.

Pardun, H. (1982). *Fette, Seifen, Anstrichm.* **84**, 1–11.

Paulose, M. M., Mukherjee, K. D., and Richter, I. (1978). *Chem. Phys. Lipids* **21**, 187–194.

Perron, R., and Broncy, M. (1978). *Rev. Fr. Corps Gras* **25**, 525–531.

Persmark, U. (1972). *In* "Rapeseed" (L.-A. Appelqvist and R. Ohlson, eds.), pp. 174–197. Elsevier, Amsterdam.

Persmark, U., and Bengtsson, L. (1976). *Riv. Ital. Sostanze Grasse* **53**, 307–311.

Petersson, B., Podlaha, O., and Töregard, B. (1981). *J. Am. Oil Chem. Soc.* **58**, 1005–1009

Pickford, C. J. (1981). *Chemical Soc. Rev.* **10**, 245–254.

Plattner, R. D. (1981). *In* "Methods in Enzymology" (J. M. Lowenstein, ed.), Vol. 72, pp. 21–34. Academic Press, New York.

Pollard, M. R., and Stumpf, P. K. (1980). *Plant Physiol.* **66**, 649–655.

Porter, N. A., Weber, B. A., Weenen, H., and Khan, J. A. (1980). *J. Am. Chem. Soc.* **102**, 5597–5601.

Röbbelen, G. (1976). *Fette, Seifen, Anstrichm.* **78**, 1–8.

Rose, S. P., Bell, J. M., Wilkie, I. W., and Schiefer, H. B. (1981). *J. Nutr.* **111**, 355–364.

Rosnitschek, I., and Theimer, R. R. (1980). *Planta* **148**, 193–198.

Rutkowski, A., and Witwicka, J. (1975). *Acta Aliment. Pol.* **25**, 333–338.

Sahasrabudhe, M. (1977). *J. Am. Oil Chem. Soc.* **54**, 323–324.

St. Angelo, A. J., Kuck, J. C., and Ory, R. L. (1979). *J. Agric. Food Chem.* **27**, 229–234.

Sambuc, E., Devinat, G., and Naudet, M. (1982). *Rev. Fr. Corps Gras* **29**, 117–123.

Sanders, T. A. B., and Naismith, D. J. (1980). *Lancet*, 654–655.

Sanders, T. A. B., and Younger, K. M. (1981). *Br. J. Nutr.* **45**, 613–616.

Scholfield, C. R. (1981). *J. Am. Oil Chem. Soc.* **58**, 889–892.

Scott, A. D. (1975). *Chem. Ind.*, Jan. 18, pp. 74–79.

Scott, C. G., Cohen, N., Riggio, P. P., and Weber, G. (1982). *Lipids* **17**, 97–101.

Sebedio, J.-L., and Ackman, R. G. (1979). *J. Am. Oil Chem. Soc.* **56**, 15–21.

Sebedio, J.-L., and Ackman, R. G. (1981). *J. Am. Oil Chem. Soc.* **58**, 972–973.

Seher, A. (1976). In "Lipids" (R. Paoletti, G. Porcellati, and G. Jacini eds.), Vol. 2, pp. 293–313. Raven Press, New York.

Seher, A., and Vogel, H. (1975). "Annual Report of the State Institute for Fat Research, E10-E11 Münster, West Germany.

Sehovic, D., Javonovic, K., Basic, V., Stojak, L., Koludrovic, B., Jesic, L., Balzer, I., and Hrust, V. (1980). *Hrana Ishrana* **21**, 47–49.

Sergiel, J.-P. (1973). *Rev. Fr. Corps Gras* **20**, 137–141.

Sleeter, R. T. (1981). *J. Am. Oil Chem. Soc.* **58**, 239–247.

Slinger, S. J. (1977). *J. Am. Oil Chem. Soc.* **54**, 94A–99A.

Slover, H. T., and Thompson, R. H. Jr. (1981). *Lipids* **16**, 268–275.

Sosulski, F., Zadernowski, R., and Babuchowski, K. (1981). *J. Am. Oil Chem. Soc.* **58**, 561–564.

Spencer, G. F., Herb, S. F., and Gormisky, P. J. (1976). *J. Am. Oil Chem. Soc.* **53**, 94–96.

Strocchi, A. (1981). *Riv. Ital. Sostanze Grasse* **58**, 271–279.

Takagi, T., and Itabashi, Y. (1977). *Lipids* **12**, 1062–1068.

Teasdale, B. F., Helmel, G. A., and Swindells, C. E. (1970). *Proc. Int. Conf. Sci., Technol. Market. Rapeseed Prod. St. Adèle 1970*, pp. 190–202.

Thomas, A. (1976). *Fette, Seifen, Anstrichm.* 78, 141–144.

Thomas, A. (1982a). *J. Am. Oil Chem. Soc.* **59**, 1–6.

Thomas, A. (1982b). *Fette, Seifen, Anstrichm.* **84**, 133–136.

Thompson, J. N., Beare-Rogers, J. E., Erdody, P., and Smith, D. C. (1973). *Am. J. Clin. Nutr.* **26**, 1349–1354.

Tinoco, J. (1982). *Prog. Lipid. Res.* **21**, 1–45.

Tiscornia, E., and Bertini, G. C. (1973). *Riv. Ital. Sostanze Grasse* **50**, 251–268.

Töregard, B., and Podlaha, O. (1974). *Proc. Int. Rapskongr., 4th Giessen 1974*, pp. 291–300.

Totani, Y., Pretorius, H. E., and du Plessis, L. M. (1982). *J. Am. Oil Chem. Soc.* **59**, 162–163.

Vaisey-Genser, M., and Eskin, N. A. M. (1979). "Canola Oil: Properties, Processing and Food Quality," Publ. No. 55. Rapeseed Association of Canada, Winnipeg, Manitoba.

Vaisey-Genser, M., and Eskin, N. A. M. (1982). "Canola Oil: Properties and Performance," Publ. No. 60. Canola Council of Canada, Winnipeg, Manitoba.

Vajreswari, A., and Tulpule, P. G. (1980). *Lipids* **15**, 962–964.

Van den Bosch, G. (1973). *J. Am. Oil Chem. Soc.* **50**, 487–493.

Wada, S., Koizumi, C., and Nonaka, J. (1977). *Yukagaku* **26**, 95–99.

Wessels, H. (1981). *Fette, Seifen, Anstrichm.* **83**, 82–84.

Wilson, N. L. (1981), *J. Sci. Food Agric.* **32**, 1103–1108.

Wolff, J. P. (1980). *Riv. Ital. Sostanze Grasse* **57**, 173–178.

Wollbeck, D., von Kleist, E., and Elmadfa, I. (1981). *Fette, Seifen, Anstrichm.* **83**, 317–323.

Young, V. (1978). *Chem. Ind.* Sept. 16, pp. 692–703.

Zadernowski, R., and Sosulski, F. (1979). *J. Am. Oil Chem. Soc.* **56**, 1004–1007.

Zajac, M., and Niewiadomski, H. (1975). *Acta Aliment Pol.* **25**, 63–70.

Ziemlanski, S., and Budzynska-Topolowska, J. (1978). *Ann. Nutr. Aliment* **32**, 781–800.

Zürcher, K., Hadorn, H., and Strack, C. (1976). *Deutsche Lebensmittel Rundsch.* **10**, 345–352.

5

Pathways of Fatty Acid Biosynthesis in Higher Plants with Particular Reference to Developing Rapeseed

P. K. STUMPF AND M. R. POLLARD

I. INTRODUCTION

In 1964, R. K. Downey and B. M. Craig published the results of an investigation in the genetic control of fatty acid biosynthesis in rapeseed and in their introduction stated:

High and Low Erucic Acid Rapeseed Oils

One explanation is that the C_{16} and C_{18} fatty acids are synthesized in a manner similar to other plant systems, whereas the monounsaturated C_{20} and C_{22} acids result from a chain elongation system similar to that shown by Wakil and Mead in animal tissues. The proposed system would involve oleic acid as the precursor and the addition of one acetate to the carboxylic end of the fatty acid molecule to form eicosenoic acid and two acetates to form erucic acid.

Seventeen years later, the basic concepts of fatty biosynthesis have been elucidated (Stumpf, 1980). In this chapter we shall apply these principles to the biochemical pathways occurring in the developing seeds of the rapeseed and will show that the suggestions proposed by Downey and Craig were entirely appropriate.

II. THE BASIC PATHWAY

The enzymatic reactions responsible for the synthesis of fatty acids in higher plants are very similar to those found in bacteria and animal systems (Shimakata and Stumpf, 1982). These are summarized in Table I.

However, the molecular organization of these enzymes differs drastically when prokaryotic and eukaryotic systems are examined. With the exception of the more advanced prokaryotic organisms, such as some species found in the Actinomycete Order or in the Corynebacter family, prokaryotes synthesize fatty acids by employing a nonassociated series of enzymes that catalyze the reactions listed in Table I. Each of these enzymes can be readily isolated, purified, crystallized and their kinetic and molecular properties studied in great detail (Vagelos, 1974). Most of these organisms employ an anaerobic mechanism for the introduction of a single cis double bond, which is inserted while the hydrocarbon chain is being formed. The enzyme responsible for this important reaction is a β-OH-decanoyl-ACP dehydratase which removes an element of water by a β,γ-elimination to form β,γ-cis-decenoyl-ACP. This product does not undergo reduction by the enoyl-ACP reductase but is further elongated to form the final cis-monoenoic fatty acid:

$$\beta\text{-OH-decanoyl-ACP} \xrightarrow[\text{dehydratase}]{-H_2O} \textit{cis}\text{-3-decenoyl-ACP} \xrightarrow[\text{additions}]{4C_2} \textit{cis}\text{-11-vaccenoyl-ACP}$$

In all animal and yeast systems, fatty acid synthesis is catalyzed by a polyfunctional hetero- (yeast) or homodimer (animal) which is localized in the cytosol of the cell (Stoops and Wakil, 1980). The introduction of the single $\Delta 9$ double bond with a cis configuration occurs on the preformed hydrocarbon chain, and is catalyzed by a membrane-bound stearoyl-CoA desaturase that requires in addition a reductant (two electrons) and molecular oxygen:

TABLE I

General Reactions in Fatty Acid Synthesis

1. $CH_3COSCoA + ACP\cdot SH \xrightarrow{\text{acetyl transferase}} CH_3CO\cdot S\cdot ACP + CoASH$

2. $HO_2C\cdot CH_2COSCoA + ACP\cdot SH \xrightarrow{\text{malonyl transferase}} HO_2C\cdot CH_2CO\cdot S\cdot ACP + CoASH$

3. $CH_3CO\cdot S\cdot ACP + HO_2C\cdot CH_2CO\cdot S\cdot ACP \xrightarrow{\text{β-ketoacyl-ACP synthase}} CH_3COCH_2CO\cdot S\cdot ACP + ACP\cdot SH + CO_2$

4. $CH_3COCH_2CO\cdot S\cdot ACP + NADPH + H^+ \xrightarrow{\text{β-ketoacyl-ACP reductase}} \text{D-}CH_3CH(OH)CH_2CO\cdot S\cdot ACP + NADP^+$

5. $CH_3CH(OH)\cdot CH_2CO\cdot S\cdot ACP \xrightarrow{\text{β-hydroxyacyl ACP dehydratase}} CH_3CH'=CHCO\cdot S\cdot ACP + H_2O$

6. $CH_3CH=CHCO\cdot S\cdot ACP + NADPH + H^+ \xrightarrow{\text{enoyl-ACP reductase}} CH_3CH_2CH_2CO\cdot S\cdot ACP + NADP^+$
   (butyryl-ACP)

7. Butyroyl·S·ACP now reacts with a second molecule of malonyl·S·ACP and proceeds through reactions 3–6 to form hexanoyl·S·ACP, etc., until palmitoyl·S·ACP is formed.

$$\text{Stearoyl-CoA} + \text{NADPH} + \text{H}^+ + \text{O}_2 \xrightarrow[\text{desaturase}]{\text{Cyt } b_5} \text{oleoyl-CoA} + \text{NADP}^+ + 2\text{H}_2\text{O}$$

In higher plants, however, a different picture has recently emerged, but one which embodies features from both prokaryotic and yeast/animal systems. In essence, plants employ a nonassociated series of enzymes to synthesize the fatty acid chain as do prokaryotes, while employing aerobic desaturation on the preformed chain to produce unsaturated fatty acids as do animals and yeast.

There is now good evidence that in leaf cells, the only site for fatty acid synthesis (i.e., palmitic and oleic acids) is the chloroplast (Ohlrogge et al., 1979) and in developing seed cells (cotyledonous tissue), the only site is the proplastid (Simcox et al., 1977). Further modification of oleic acid occurs in modifying compartments where the following reactions can occur: (a) hydroxylation, in the endosperm of developing castor bean seeds where oleate is converted to ricinoleate, (b) desaturation, presumably on the endoplasmic reticuli of the leaf cell where oleate (as β-oleoyl phosphatidylcholine) is further desaturated to linoleic which is then transferred to another carrier substrate to be desaturated to linoleneate, and finally (c) elongation, which has been demonstrated in homogenates of developing rape, meadowfoam, and jojoba seeds, where addition of C_2 units to oleoyl-CoA results in the synthesis of 20:1(11) or 22:1(13) fatty acids (Pollard et al., 1979; Pollard and Stumpf, 1980a,b). These ideas are summarized in Fig. 1.

The molecular structure(s) of the enzymes responsible for the synthesis of palmitic acid, the principal saturated fatty acid in higher plants and oleic acid, the precursor of a number of commercially important fatty acids can now be described quite precisely. There is good evidence that the enzymes responsible for the synthesis of palmitic acid are very similar to the enzymes in prokaryotic organisms, namely they are nonassociated and easily separable, that is, they are not components of a polyfunctional protein (Shimakata and Stumpf, 1982). The acyl carrier protein, ACP, is a separable protein that must be present for de novo synthesis. It is solely localized in leaf cells in the chloroplast and presumably also in the proplastids of seed cells. Thus, by definition all substrates associated with ACP and all enzymes that utilize acyl-ACPs as substrates must be localized in these organelles. Equally interesting, acetyl-CoA synthetase responsible for the conversion of free acetate to acetyl-CoA is also localized in the chloroplast of the leaf cell (Kuhn et al., 1981). Its distribution in the seed cell has as yet not been determined. Although palmitoyl-ACP is the prime end product of fatty acid biosynthesis in plants, an elongation system rapidly converts palmitoyl-ACP, with malonyl-ACP as the C_2 component, to stearoyl-ACP. Either free stearic acid or stearic

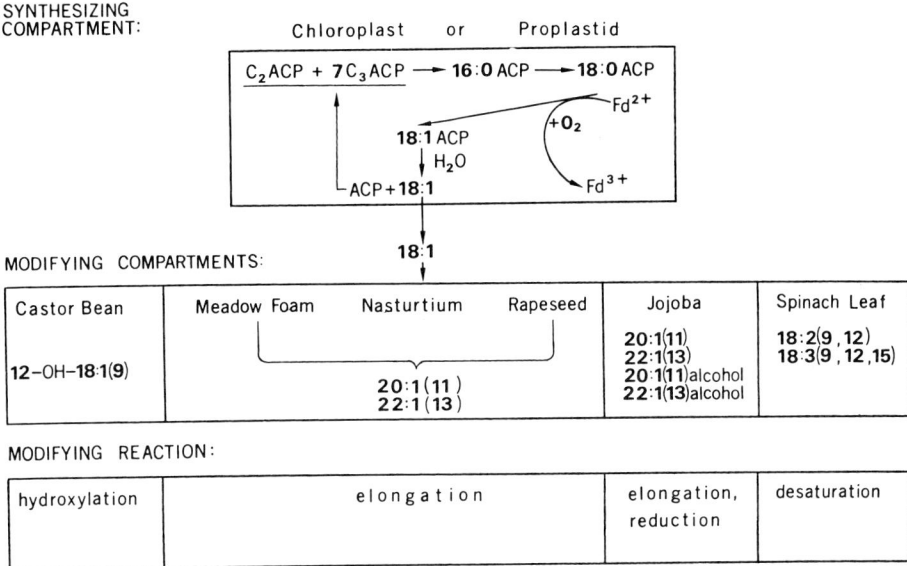

CENTRAL ROLE OF OLEIC ACID

Fig. 1. The central role of oleic acid in the synthesis of fatty acids in different plant tissues. C_2ACP, acetyl ACP; C_3ACP, malonyl ACP; Fd, ferredoxin.

acid associated with neutral or complex lipids is only found in trace quantities in higher plants. The reason for this probably relates to the observation that a soluble highly active stearoyl-ACP desaturase is solely localized in the chloroplasts of leaf cells (Ohlrogge et al., 1979). Thus, as soon as stearoyl-ACP is synthesized, it is rapidly desaturated to oleoyl-ACP. In addition to the presence of the desaturase in the chloroplast, another important enzyme, namely oleoyl-ACP hydrolase, rapidly converts oleoyl-ACP to free oleic acid and ACP (Ohlrogge et al., 1978). When isolated chloroplasts utilize [^{14}C]acetate as a substrate under aerobic conditions, the major end product of ^{14}C-incorporation is free [^{14}C]oleic acid. This end product can readily move out of the chloroplast into the cytosolic compartment to undergo further changes as indicated in Fig. 1 In certain cases it is apparent that appreciable amounts of palmitoyl-ACP can be hydrolyzed to palmitic acid and exported from the chloroplast or plastid (Pollard and Stumpf, 1980b). This is due to the wide specificity of the acyl-ACP hydrolase (Ohlrogge et al., 1978).

Having described briefly the present status of fatty acid biosynthesis in higher plants, it would be worthwhile now to apply this information to the biosynthesis of the fatty acids found in rapeseed.

III. BIOSYNTHESIS OF ERUCIC ACID IN DEVELOPING RAPESEED

A. Early Experiments

In common with a few other oilseeds (listed in Fig. 1), rapeseed can contain appreciable quantities of fatty acids with chain lengths greater than the usual 18 carbon atoms. Significant amounts of polyunsaturated acids are also present. Typical analyses of the fatty acids, present in high and zero erucate strains of rapeseed, are given in Table II, these strains being the ones used in the biochemical experiments reported in Section III,B of this chapter.

In 1961, Craig had shown a clear linear relationship between oleic and erucic acids in the seed oils of low and high erucic acid strains of *Brassica napus* L., that is, a low oleic acid and a high erucic acid level were found in high erucic acid strains and the converse was observed with low erucic acid strains (Craig, 1961; see also Table II). Obviously, the elongation of oleic acid to erucic acid was the pathway of biosynthesis of erucic acid. Of interest, no direct effect on total fatty acid production was observed, i.e., the seed oil content of the different strains remained essentially constant. Downey and Craig in 1964 published data on the incorporation of [^{14}C]acetate in developing seeds of *Brassica napus*. As would be expected, oleic, eicosenoic and erucic acids were labeled. When the [^{14}C]monoenoic acids were cleaved by the permanganate–periodate reagent to mono- and dicarboxylic acids, the C_9 mono- and the C_9 dicarboxylic acids derived from [^{14}C]oleic were equally labeled. The specific activities of the C_9 monocarboxylic acids

TABLE II

Fatty Acid Composition of Total Lipids from Developing Rapeseed of Different Varieties

Rapeseed variety	Percent fatty acid composition of total lipids[a]					
	16:0	18:1	18:2	18:3	20:1	22:1
High erucate strain of *Brassica juncea*	2.5	9	17	14.5	6	51
Zero erucate strain of *Brassica campestris*	2.5	68	19	9.5	1	1
Zero erucate strain of *Brassica napus*	3.5	64.5	22	10	<1	<1

[a] Trace amounts of long chain saturated fatty acids, namely, stearic (18:0), arachidic (20:0), behenic (22:0) and lignoceric (24:0) acids, present in the seed oil, have not been quantitated.

obtained from the oxidation of [^{14}C]eicosenoic and [^{14}C]erucic acids remained quite constant whereas the specific activity of the [^{14}C]C_{11} dicarboxylic acid derived from eicosenoic acid had five times the radioactivity of the C_9 dicarboxylic acid derived from the oxidation oleic acid. The C_{13} dicarboxylic acid obtained from [^{14}C]erucic acid oxidation had four times the radioactivity found in the C_{11} dicarboxylic acid derived from [^{14}C]eicosenoic acid. Downey and Craig (1964) therefore correctly concluded that the following scheme would explain their results:

$$9C_2 \xrightarrow{} \underset{18:1\Delta^9}{\text{oleic acid}} \xrightarrow{C_2} \underset{20:1\Delta^{11}}{\text{eicosenoic acid}} \xrightarrow{C_2} \underset{22:1\Delta^{13}}{\text{erucic acid}}$$

In Chapters 1 and 6 of this volume, a genetic analysis of erucic production is discussed. In brief, however, there appear to be two genes, acting in an additive manner, that control the level of erucic acid. The assignment of the precise gene products associated with these genes would clarify the understanding of the mechanisms regulating erucic acid biosynthesis.

B.   Recent Experiments

A number of experiments were designed to confirm and extend the earlier results of Downey and Craig (1964). Varieties of rape representing compositional extremes were employed, namely, a high erucate strain of *B. juncea*, and zero erucate strains of *B. napus* or *B. campestris* (Table II).

When [^{14}C]acetate was incubated with developing seed tissue at 26°C, a number of fatty acids became labeled. The percentage of added [1-^{14}C]acetate incorporated into lipid-extractable material is given in Table III, along with the [^{14}C]fatty acid distribution in total lipids, triacylglycerols, and polar lipids. The high erucate strain of *B. juncea* gave, relatively, a high incorporation of [^{14}C]acetate into lipids, predominantly into C_{20} and C_{22} acids esterified into triacylglycerols. Labeled C_{16} and C_{18} acids were found largely in the polar lipid fraction. This labeling pattern did not change during the course of the incubation, or with pulse-chasing with cold acetate, reminiscent of experiments performed with developing nasturtium seeds (Pollard and Stumpf, 1980a). The data for reductive ozonolysis of [^{14}C]monoenoic fatty acids isolated from *B. juncea* lipids by Ag$^+$ TLC followed by preparative GLC are shown in Table IV. Most of the ^{14}C label is incorporated during the chain elongation of 18:1. This explains the low incorporation of [^{14}C]acetate into lipids by intact seeds of zero erucate strains of *B. campestris* during an identical incubation (Table III), confirming the absence of an effective chain-elongation step.

Low incorporations of [U-^{14}C]glucose and [1-^{14}C]oleic acid into acyl

TABLE III

Incorporation of Label into Fatty Acids from the Incubation of [1-^{14}C] Acetate with Developing Rapeseed Slices

Incubations	Percent of total ^{14}C incorporated	Percent ^{14}C distribution in fatty acids[a]								
		12:0	14:0	16:0	18:0	18:1	20:0	20:1	22:0	22:1
Brassica juncea seeds										
Total lipids, 9 hr[b]	43	—	—	5	2	17	3.5	6	13.5	53
Triacylglycerols, 9 hr	27.5	—	—	—	—	—	—	2	19	79
Polar lipids, 9 hr	9	—	—	17.5	11	57	1	1.5	6	6
Brassica campestris seeds										
Total lipids, 9 hr[b]	5	4	12	17	—	54	—	12	—	—

[a] It is typical of these types of incubations using intact seed tissue that significant amounts of labeled saturated acids can be formed, though present in only trace amounts in endogenous lipids (Pollard and Stumpf, 1980a,b).

[b] The distribution of incorporated label between lipid classes was *B. juncea*: triacylglycerols, 64%; diacylglycerols, 8%; all polar lipids, 21%; phosphatidylcholine, 10%; *B. campestris*: triacylglycerols, 31%; diacylglycerols, 13%; all polar lipids, 40%; phosphatidylcholine, 13%.

TABLE IV

Ozonolysis of [^{14}IC] Monoenoic Fatty Esters Produced by Incubation of [1-^{14}C] Acetate with *Brassica juncea* Seeds

[^{14}C] Fatty ester	Percent distribution of ^{14}C between fragments from ozonolysis		Type of biosynthesis, ratio of ^{14}C in preformed carbon atom to ^{14}C in elongation carbon atom
	Aldehyde	Aldehyde ester	
18:1(9c)	42(C_9)	58(C_9)	*de novo*, —
20:1(11c)	14(C_9)	86(C_{11})	Chain elongation, 0.05:1
22:1(13c)	6(C_9)	94(C_{13})	Chain elongation, 0.035:1

groups by developing rapeseed meant that no other meaningful comparisons could be drawn from *in vivo* experiments with zero and high erucate strains.

In vitro experiments were carried out. In a typical experiment, 1 weight of fresh, developing rapeseed was homogenized to a paste in a pestle and mortar with 2 volumes of buffer (0.3 M sucrose, 0.08 M HEPES, 5 mM ascorbate, 5 mM β-mercaptoethanol, 5 mg ml^{-1} BSA, at pH 7.2). Filtration of the paste through Miracloth resulted in a cell-free homogenate. Each elongation assay contained ATP, 1 μmole; NADH, 250 nmoles; NADPH, 250 nmoles; malonyl-CoA, 100 nmoles; $MnCl_2$, 2.5 μmoles; $MgCl_2$, 2.5 μmoles; enzyme preparation, 0.25 ml; and [1-^{14}C]oleoyl-CoA, 25 nmoles, in a total volume of 0.5 ml. Incubations were in open tubes, at 27°C, and were terminated after 45 min by addition of 5% methanolic KOH (3 ml). Methyl esters were prepared and were analyzed on a radio-GLC with a 10% DEGS column. In two experiments elongation activity was observed for *B. juncea* (conversions of 8% 20:1 plus 1% 22:1, and 13% 20:1 plus 9% 22:1, respectively), but not for *B. campestris* or for *B. napus* (<1% 20:1 in each case). Thus, the elongation system present in the high erucate strain of *B. juncea* is absent in *B. campestris* and *B. napus* zero erucate strains.

IV. CONCLUSION

The data presented here clearly showed that oleoyl-CoA is elongated by the addition of malonyl-CoA via a CoA-elongation pathway. Since both eicosenoic and erucic acids are intensely labeled in the 1-^{14} C and 1,3-^{14}C positions respectively, and since the parent oleic acid has a distribution of radioactivity distinctly different from the elongation products, it is quite clear that the *de novo* and the elongation pathways are separated. We would assume that the *de novo* pathway occurs in a proplastid in which the synthesis of oleic acid takes place via the ACP pathway as outlined in Table I. Then, just as with chloroplasts, the free oleic moves into the second or modifying compartment, there to be elongated by a presumably membrane-bound series of enzymes via a CoA pathway to the final product, namely erucic acid.

Earlier data suggest that two genes are responsible, in an additive manner, for controlling the level of erucic acid in rapeseed. The gene products under the control of these two genes are as yet not identified. Our work indicates that at least one of these genes must code for the elongation system, which is defective in the zero erucate strains, but further elaboration is not yet possible. Elongation will involve a large number of enzyme activities, of the general type as found in fatty acid synthesis (Table I, nos. 2–6), so several genes may code for the entire system. The most likely point of genetic control would be the first committed step in the elongation sequence, namely the β-ketoacyl-CoA synthetase. Furthermore, the biochemical evidence to date does not show whether a single elongation system is involved, or whether two separate systems are required, one for 18:1 → 20:1 and another for 20:1 → 22:1. In the latter case, a defective gene would be required for each elongation system. Another possible point for genetic control is the supply of malonyl-CoA. The most likely enzyme candidate here is the cytosolic acetyl-CoA carboxylase, which is responsible for the generation of malonyl-CoA from acetyl-CoA. However, as the source of cytosolic acetyl-CoA is not known, a preceding step could be involved instead. Concluding this topic, we feel that it will be possible to relate the genetic controls on erucic acid production to the enzymology, but only when the system has been more fully characterized biochemically.

There are now several examples of seed oils that contain monoenoic fatty acids derived from the elongation of oleoyl-CoA by a malonyl-CoA elongation pathway. These include the developing jojoba seed, which elongates oleoyl-CoA to cis-11-eicosenoyl-CoA and cis-13-docosenoyl-CoA (erucoyl-CoA) (Pollard et al., 1979), and meadowfoam, which accumulates cis-13-docosenoic acid as well as cis-5-eicosenoate among other acids (Pollard and Stumpf, 1980b; Moreau et al., 1981). Cis-13-docosenoic acid is formed by the elongation of oleoyl-CoA with malonyl-CoA but the cis-5-eicosenoate is formed by the direct Δ^5 desaturation of eicosanoyl-CoA. Finally, Crambe abyssinica seeds, which have high levels of erucic acid in their seed oil, have been examined and the pathway of erucic acid biosynthesis appears to be consistent with those proposed earlier (Gurr et al., 1974).

Discussion so far has centered on long chain fatty acid biosynthesis. However, significant amount of linoleic and α-linolenic acids are also present in rapeseed oil. Their biosynthesis has not been studied in rapeseed, but has been studied in seed oils rich in polyunsaturated acids, namely, safflower, soybean, and linseed (Slack et al., 1978; Stymne and Appleqvist, 1978). By analogy, it is reasonable to assume that in rapeseed these acids are synthesized by a pathway involving oleoyl-CoA incorporation into phosphatidylcholine, with subsequent desaturations occurring on this phospholipid, producing first linoleate, then α-linolenate.

ACKNOWLEDGMENTS

We wish to thank Dr. A. Benzioni, Research and Development Authority, Ben-Gurion University of the Negev, Israel, for technical assistance; P. F. Knowles, University of California at Davis, Agricultural Extension Service, for the supply of developing rapeseed; and Ms. Billie Gabriel for her assistance in the preparation of this manuscript. We are indebted for support to NSF Grant FCM 79-03976.

REFERENCES

Craig, B. M. (1961). Can. J. Plant Sci. **41**, 204-210.

Downey, R. K., and Craig, B. M. (1964). J. Am. Oil Chem. Soc. **41**, 475–478.

Gurr, M. I., Blades, J., Appleby, R. S., Smith, C. G., Robinson, M. P., and Nichols, B. W. (1974). Eur. J. Biochem. **43**, 281–290.

Kuhn, D. N., Knauf, M., and Stumpf, P. K. (1981). Arch. Biochem. Biophys. **209**, 441–450.

Moreau, R. A., Pollard, M. R., and Stumpf, P. K. (1981). Arch. Biochem. Biophys. **209**, 376–391.

Ohlrogge, J. B., Shine, W. E., and Stumpf, P. K. (1978). Arch. Biochem. Biophys. **189**, 382–391.

Ohlrogge, J. B., Kuhn, D. N., and Stumpf, P. K. (1979). Proc. Natl. Acad. Sci. U.S.A. **76**, 1194–1198.

Pollard, M. R., and Stumpf, P. K. (1980a). Plant Physiol. **66**, 641–648.

Pollard, M. R., and Stumpf, P. K. (1980b). Plant Physiol. **66**, 649–655.

Pollard, M. R., McKeon, T., Gupta, L. M., and Stumpf, P. K. (1979). Lipids **14**, 651–662.

Shimakata, T., and Stumpf, P. K. (1982). Arch. Biochem. Biophys. **217**, 144–154.

Simcox, P. D., Reid, E. E., Canvin, D. T., and Dennis, D. T. (1977). Plant Physiol. **59**, 1128–1132.

Slack, C. R., Roughan, P. G., and Balasingham, N. (1978). Biochem. J. **170**, 421–433.

Stoops, J. K., and Wakil, S. J. (1980). Proc. Natl. Acad. Sci. U.S.A **77**, 4544–4548.

Stumpf, P. K. (1980). In "The Biochemistry of Plants" (P.K. Stumpf and E.E. Conn, eds.) Vol. 4, Chapter 7, pp. 177–204. Academic Press, New York.

Stymne, S., and Appelqvist, L.-A. (1978). Eur. J. Biochem. **90**, 223–229.

Vagelos, P. R. (1974). MTP Int. Rev. Sci.: Biochem. Lipids **4**, 100–140.

The Development of Improved Rapeseed Cultivars

B. R. STEFANSSON

High and Low Erucic Acid Rapeseed Oils
Copyright © 1983 by Academic Press Canada
All rights of reproduction in any form reserved.
ISBN 0-12-425080-7

I. INTRODUCTION

A. Historical Background

Rapeseed breeding in Canada began soon after the crop was introduced during World War II, when H. G. Neufeld at Nipawin, Saskatchewan, made several selections from seed stocks introduced from Argentina. He lacked facilities for plant breeding, and gave his selections for evaluation to W. J. White at the Canada Agriculture Research Station, Saskatoon (formerly known as Dominion Forage Laboratory of the Federal Department of Agriculture). Dr. White selected for uniformity, lodging resistance, high oil content in the seed, and low iodine value in the oil. The chemical analyses were performed by the Prairie Regional Laboratory of the National Research Council, Saskatoon. This work led to the release in 1954 of the first licensed Canadian cultivar of summer rape, Golden.

The rapeseed breeding program at Saskatoon was continued on a small scale until 1962 when the rapeseed breeder began to devote full time to rapeseed breeding. Expansion of this program continued, and now the staff involved in rapeseed breeding includes two plant breeders, two plant pathologists, a chemist, and a cytologist.

The rapeseed breeding program at the University of Manitoba, Winnipeg, was initiated in 1953. Soybeans, sunflowers and safflowers were evaluated as potential edible oilseed crops for several years. Work on these crops was discontinued when it became evident that rapeseed was the best adapted edible oilseed for large-scale production in the Prairie Provinces. Since 1966, the plant breeder at Winnipeg has devoted most of his time to rapeseed. In 1958, a chemist joined the staff. The staff now includes one breeder, a chemist, and a pathologist.

The rapeseed breeding program at the University of Alberta, Edmonton, began in 1969, primarily to develop cultivars for regions of Alberta where climatic conditions differ from those in most parts of Manitoba and Saskatchewan.

In the initial stages, rapeseed breeding in Canada was largely restricted to agronomic characteristics and oil content. In 1956, following published reports from the Department of Medical Research, University of Western Ontario, concerning the role of erucic acid in nutrition, the Food and Drug Directorate of the Department of National Health and Welfare issued orders prohibiting the use of rapeseed oil as a human food in Canada. While these orders were rescinded within months to permit review of the data, they provided a stimulus for attempts to reduce the level of this component in the oil. At this time the plant breeders at Saskatoon and Winnipeg began to acquire the facilities and personnel to perform analyses for the chemical components of their rapeseed samples. Further details of these early developments are available in "The Story of Rapeseed in Western Canada" (McLeod, 1974).

B. Major Changes in Rapeseed Oils and Meals

Until recently the fatty acid composition of rapeseed oil was quite different from that of other edible vegetable oils; from 40 to 60% of the fatty acid components of rapeseed oil consisted of the long chain fatty acids, erucic and eicosenoic. This unusual fatty acid composition has been the subject of numerous nutritional studies. Detrimental effects attributed to the long chain fatty acid components of rapeseed oil stimulated plant breeders to search for genetically controlled variation in these components. Rape plants which produce seed oil essentially without erucic acid were isolated (Stefansson et al., 1961) and this characteristic was incorporated into cultivars suitable for commercial production. The new "low erucic acid" rapeseed oils contain only the fatty acid components found in other edible vegetable oils traditionally used as food in the Western World.

The fatty acid compositions of the high and low erucic acid rapeseed oils (Brassica napus) in Table I were selected to illustrate the differences between the two kinds of oil which have generally occurred in a large number of analyses performed for cultivar development during a period of several years. The absence of erucic acid has always been accompanied by a reduction in eicosenoic acid from levels ranging around 10% to approximately 1.5%, and by a major increase in oleic acid from around 15% to approximately 60%. Some tendency toward higher levels of the polyunsaturated 18-carbon fatty acids, linoleic and linolenic, usually occurs in the low erucic acid oils. The content of the short chain saturated fatty acids (16:0 and 18:0) tends to be slightly higher and the long chain fatty acid behenic (22:0) tends to be lower in the low erucic acid oils. Thus, the elongases that are involved in the biosynthesis of erucic acid from oleic also appear to influence the

TABLE I

Fatty Acid Composition of Canadian Rapeseed Oils with High and Low Erucic Acid Content

Cultivar	Fatty acids (% of total fatty acids)									
	16:0	8:0	18:1	18:2	18:3	20:1	20:2	22:0	22:1[a]	22:2
Regent	3.8	1.7	62.7	19.9	19.0	1.4	Trace	0.2	Trace	
Reston	2.9	1.0	16.3	14.0	9.9	9.3	0.5	0.5	44.2	0.6

[a] Erucic acid.

quantitative distribution of chain lengths in the saturated fatty acids. The accumulation of oleic acid substrate (18:1) in the low erucic acid oils apparently enables the desaturases to produce somewhat higher levels of polyunsaturated 18-carbon fatty acids.

The amino acid composition of rapeseed protein is known to be favorable for use as a protein supplement in animal rations. However, effective utilization of rapeseed meal in animal rations has been limited by the presence of minor constituents known as glucosinolates. Glucosinolates occur in many species of the mustard family, and their breakdown products impart the hot or sharp taste to mustard, radishes, and turnips. The detrimental effects of feeding large amounts of products containing glucosinolates to farm animals (Bowland et al., 1965) stimulated plant breeders to search for genetically controlled variation in the glucosinolate content of rapeseed. Reduced levels were found in the rape cultivar, Bronowski (personal communication, J. Krzymanski, formerly of Agriculture Canada Research Station, Saskatoon, Canada) and this characteristic was incorporated into cultivars suitable for commercial production.

C. Commodity Name Not Adequate to Describe Rapeseed Oil and Meal

The magnitude of the variation in the composition of rapeseed oil and meal now commercially available has created a need for new terms to describe the products derived from rapeseed. The fatty acid composition of most edible vegetable oils such as soybean, sunflower, or cottonseed oils, varies within narrow limits. Thus, the species or commodity name (e.g., soybean oil) provides a reasonable description of the fatty acid composition of soybean oil. In contrast, the erucic acid content of commercially available rapeseed oil may vary from near zero to 55%, and the oleic acid from 10 to more than 60%. A number of terms have been proposed or utilized to describe the new rapeseed oil whose fatty acid composition has been altered by the elimination of erucic acid; these include low erucic acid rapeseed oil (LEAR), canbra, and canola. Similar terms such as high erucic acid rapeseed oil (HEAR) and common or traditional rapeseed oil have been used to describe rapeseed oil whose fatty acid composition includes substantial amounts of erucic acid.

D. "Double Low" (Canola) Rapeseed

After the rape cultivar, Tower, which produces seed oil very low in erucic acid content and seed low in glucosinolate content was released in 1974, the so-called "double low" rapeseed became commercially available in Canada. Large quantities of seed (approximately one million pounds) were distributed in 1974 and Canadian rapeseed crushers began crushing the new rapeseed. Conversion of the crushing industry was rapid, and now practically all Canadian rapeseed crushers crush the "double low" rapeseed almost exclusively. The need for a simple term to describe the new improved oil and meal derived from the "double low" cultivars became critical when these products became commercially available. The rapeseed crushers in Western Canada adopted the name canola, and developed specifications for the erucic acid content of the oil and the glucosinolate content for seed and meal eligible to qualify for use of the name canola. The term canola is being generally accepted in Canada to distinguish the new improved rapeseed from the old, less desirable seed and products.

II. RAPESEED BREEDING

A. Environmental Effects

Many of the most difficult problems in genetics and plant breeding involve distinguishing between effects due to genetics and environment. Several statistical methods are used to separate genetic and environmental effects. However, the efficiency of selection or plant breeding can also be improved by knowledge of the manner in which environmental factors influence particular species and cultivars.

Each species tends to be adapted to a particular range of environmental conditions and when an individual factor, such as temperature, deviates from this range, the plant is under stress. Stresses from different factors sometimes produce similar effects. In addition to influencing oil and protein content, different environmental stresses may produce a similar syndrome of effects (Stefansson, 1970). For example, high temperature, drought, and long days all hasten the maturity of rapeseed, reduce the time from flowering to maturity, and probably reduce the photosynthate available for seed production. Canvin (1965) grew rapeseed at controlled temperatures of 10°, 16°, 21°, and 26.5°C for the period of seed development. The highest oil content (52%) was observed at the lowest temperature and oil content continually decreased with increasing temperature, down to 32% at 26.5°C. Days from flowering to maturity also decreased with increasing temperature. While other environmental factors produce less dramatic effects, high levels of nitrogen often tend to reduce oil and increase protein content (Röbbelen and Thies, 1980a; Stefansson, 1970). Other nutrient deficiencies or excesses may also influence specific aspects of biosynthesis.

B. Analytical Methods

Rapid and accurate methods of analysis are required to investigate the range of variation of the particular chemical components that may have an important bearing on rapeseed quality. Only a few of the developments from the rapidly expanding field of analytical chemistry which have been used in plant breeding will be mentioned.

The methods used by Canadian plant breeders to determine the oil content of rapeseed may serve as an example of the evolution and adoption of more effective analytical procedures. The methods used for this purpose evolved from the Goldfisch method through the Comstock press method (Comstock and Culbertson, 1958) and the Swedish method (Troeng, 1955) to use of the extremely efficient low resolution nuclear magnetic resonance (NMR) instruments designed to determine oil content. The NMR analyzer permits one technician to perform many times the number of analyses possible with earlier methods.

Canadian plant breeders began using gas chromatographs equipped with silicone columns in 1967 to investigate the variability in the chain length of fatty acids in oils from a large number of small samples of rapeseed. Investigations of this kind have shown that the erucic acid content of rapeseed oil may range from less than 0.1 to more than 60%. Gas chromatographs equipped with improved columns, automatic sampling, and computed output can provide the quick and accurate analyses required for investigation of the variability in the polyunsaturated fatty acids which occur in rapeseed oil.

The procedures for analysis of glucosinolate content in rapeseed were too cumbersome and time consuming for screening large numbers of seed samples until Youngs and Wetter (1967) developed a rapid method of analysis for small samples of rapeseed or rapeseed meal. Use of this method led to the discovery of low glucosinolate content in the seed from the rape cultivar Bronowski. The tes-tape test (Lein, 1970) which is extremely rapid, is very useful in isolating plants that produce seeds with low glucosinolate content from the progeny of crosses used in breeding programs. Gas chromatographic analyses of the trimethylsilyl derivatives of the glucosinolates (Röbbelen and Thies, 1980b) appears to provide the most precise and reliable quantification of particular glucosinolates. This method has been used to demonstrate the presence of indolyl glucosinolates in rapeseed (Olsen and Sørensen, 1980).

C. Oil and Protein Content

1. INTRODUCTION

Oilseeds are grown to produce oil; however, the residue that remains after the oil is extracted usually is high in protein content, and is used as a protein supplement in animal rations. In rapeseed breeding considerable emphasis has been placed on breeding for higher oil content but little emphasis has been placed on maintaining the protein content of the seed or meal. In contrast, soybean breeders have emphasized selection for both oil and protein content for several decades. The low quality image that has been associated with rapeseed meal, due to its glucosinolate content and sometimes due to poor processing, probably discouraged breeding for higher protein content. Improved processing techniques, especially involving temperature control, and cultivars with reduced levels of glucosinolates have improved the quality of rapeseed meal. These changes have increased the value of canola meal and probably will lead to greater emphasis on protein content in rapeseed breeding programs.

2. SELECTION FOR OIL AND PROTEIN

In spite of environmental influences, progress has been made in improving the oil and protein content of rapeseed. Although the number of genetic studies is limited, a number of genes condition the level of oil content (Grami et al., 1977; Olsson, 1960). Gene action appears to be largely additive. Oil content is determined by the genetic constitution of the plant which produces the seed rather than by the genotype of the embryo (Grami and Stefansson, 1977; Olsson, 1960).

Grami et al. (1977) estimated the broad sense heritability for percentage oil and percentage protein in progeny from two rapeseed cultivars as approximately 0.26 for each single trait, but as 0.33 for the sum of oil and protein as a percentage of the seed. Phenotypic and genotypic correlation between oil and protein averaged −0.81 and −0.71, respectively. Similar values have been obtained by Röbbelen (1978). These high negative correlations indicate that it is necessary to select for both oil and protein to maximize the intrinsic value of the seed. Further increases in oil and protein as a percentage of the seed may be achieved using the pleotropic effects associated with yellow-seeded forms of rapeseed (Pawlowski and Youngs, 1969; Jonsson and Bengtsson, 1970; Stringam et al., 1974).

Selection for both oil and protein has resulted in significant progress. For example, the sum for oil and protein for Tower selected for both oil and protein is approximately 2.9% higher than the sum for Midas selected primarily for oil content (Stefansson and Kondra, 1975). The improved protein content has permitted some Canadian rapeseed crushers to merchandize canola meal with 36 or 38% instead of the previously guaranteed 34% protein content (at 8% moisture). However, much work remains to be done to bring the sum of oil and protein in the cultivars of turnip rape up to the standards now available in rape and to maximize the production of these two components of the seed in rape and turnip rape.

D. Erucic Acid

Until recently the seed oils from all cultivars of rape and turnip rape contained substantial amounts of the long chain fatty acid, erucic acid. The
amount usually varied from 24 to 50% of the total fatty acids in the oil.
During the 1960s and 1970s, a large number of nutritional experiments (see
Chapter 11) indicated that consumption of large amounts of rapeseed oil
with high levels of erucic acid could be detrimental to a number of animal
species. These experiments stimulated plant breeders to search for genetically controlled low levels of erucic acid in rapeseed oil. Selection of individual plants from two strains of rape (B. napus), Liho and a strain from
Budapest, resulted in the isolation of strains of rape with seed oil essentially
free from erucic acid (Stefansson et al., 1961; Stefansson and Hougen,
1964). In turnip rape (B. campestris) a similar procedure resulted in the isolation of plants with seed oil free from erucic acid (Downey, 1964). These
sources of genes for the absence of erucic acid have been used widely to
develop cultivars with seed oil low in erucic acid content and appear to be
the only genes thus far used for this purpose throughout the world.

The frequency of genes for the absence of erucic acid in rape, turnip rape,
and other closely related species appears to be very low. If seed oils low in
erucic acid content were available within leaf mustard (Brassica juncea) this
species, now grown for edible oil in India and China, could become an
edible oil crop in other countries. Efforts to find genes for the absence of
erucic acid in this species have been under way for a number of years in
several countries, and individual plants from B. juncea which produce seed
oils essentially free from erucic acid have been isolated recently (Kirk and
Oram, 1981).

The erucic acid content is largely controlled by the genotype of the developing seed rather than by the genotype of the maternal plant (Downey and
Harvey, 1963; Harvey and Downey, 1964; Stefansson and Hougen, 1964;
Kondra and Stefansson, 1965). For this reason, a technique could be developed whereby one cotyledon or a part of one cotyledon could be used as a
sample for fatty acid analysis, while the remainder of the seed could be used
to produce a plant (Downey and Harvey, 1963). This technique is useful in
breeding and can also be used to facilitate genetic studies, since the seeds
on an F_1 plant represent the F_2 population.

The absence of erucic acid is conditioned by one locus in turnip rape and
two loci in rape. Multiple alleles occur at each locus (Jönsson, 1977a; Krzymanski and Downey, 1969; Stefansson and Hougen, 1964). Homozygous
genotypes with different alleles produce levels of erucic acid ranging from
less than 0.1 to 60%. While one locus in turnip rape and two loci in rape are
adequate to account for the absence of erucic acid, additional loci seem to
be needed to account for the levels of erucic acid that occur in the progeny

of certain crosses. The low frequencies of erucic acid levels that reach the levels of the high parent in crosses between plants with high and low erucic acid content (Dorrel and Downey, 1964) support this view. This is not surprising since, as fatty acids are measured as a percentage of total fatty acids (or of oil), a change in the percentage of any one fatty acid must be accompanied by an equal change in other fatty acids (Stefansson and Storgaard, 1969). Thus, genes conditioning the levels of other fatty acids such as linoleic and linolenic probably will have some effect on the level of erucic acid, especially if erucic acid is the major constituent of the oil.

The gene action of alleles conditioning erucic acid is largely additive while that for eicosenoic acid is largely dominant (Stefansson and Hougen, 1964; Kondra and Stefansson, 1965). Thus, an allele that produces 2% erucic and 8% eicosenoic acid can easily be distinguished from one that produces almost no erucic acid (less than 0.1) and 2–3% eicosenoic acid (Jönsson, 1977a). The eicosenoic acid levels for rapeseed oils from strains of rape and turnip rape with erucic acid contents from 10 to 40% usually are above 10% while at higher levels of erucic (50–60%) eicosenoic acid levels are lower, sometimes less than 5%. So far no *Brassica* genotype has been found that will produce seed oil with more than 65% erucic acid (Röbbelen and Thies, 1980a; Calhaun et al., 1975).

The polyunsaturated fatty acids, linoleic and linolenic, remain largely unchanged in the very high erucic acid oils; therefore, reduction in the levels of these acids appears to be needed to achieve higher levels of erucic acid in *Brassica* species.

Since only one or two gene loci are involved in turnip rape and rape, respectively, the development of low erucic acid cultivars appeared to be relatively easy. However, the original gene source for absence of erucic acid was a forage crop cultivar, Liho. Thus many characteristics, undesirable in oilseed rape, were associated with the gene source. For this reason, most plant breeders used some form of backcrossing to develop low erucic acid oil seed cultivars. This procedure facilitated recovery of characteristics such as plant type, earliness, and oil content in the new cultivars. The first Canadian low erucic acid summer rape cultivars Oro in 1968 and Zephyr in 1971 and the turnip rape cultivar Span in 1971, which had a reduced seed and oil yield potential, were quickly replaced by superior varieties Midas and Torch in 1973.

It was more difficult to develop low erucic acid cultivars for northern Europe because most of the rapeseed in this area is derived from winter rape. European cultivars were highly developed for seed and oil yield; more time was required to grow a plant from seed to maturity because a cold treatment (vernalization) was required so that the plant could proceed from vegetative to reproductive development. Furthermore, a characteristic had to be transferred from summer rape to winter rape, a cross that produces a

large number of undesirable segregates. In spite of these difficulties, satisfactory cultivars were soon developed. It took less than 15 years to develop
low erucic acid cultivars (Brink in Sweden, Quinta in Germany, and Jet Neuf
in France) equal or superior to the old high erucic acid cultivars. These
cultivars were essentially developed at the same time (Röbbelen and Thies,
1980a).

E. Polyunsaturated Fatty Acids

1. LINOLENIC ACID

Successful elimination of erucic acid from rapeseed stimulated interest in
genetic manipulation involving the polyunsaturated fatty acids. Soybean
and rapeseed oils are the only major edible vegetable oils that contain substantial quantities of linolenic acid (approximately 10%). A reduction in the
linolenic acid content would improve the flavor and oxidative stability of
these oils and reduce the need for partial hydrogenation of edible oils used
in the liquid form. A substantial body of opinion suggests higher levels of
linoleic (up to 50%) might be desirable for nutritional reasons. However, the
highest possible levels of oleic and low levels of both linoleic and linolenic
might be desirable for maximum stability in countries where refrigeration is
not generally available. A rapeseed oil of this type might provide an economical substitute for olive oil.

Screening procedures, the analyses and selection of large numbers of cultivars, strains, individual plants and half-seeds carried out in several countries, have been relatively ineffective in establishing genetically controlled
low levels of linolenic acid in rape and turnip rape. For this reason, large-
scale mutation experiments were initiated in Germany (Röbbelen and Rakow, 1970) and in France (Morice, 1975). Levels of 3.5% linolenic acid in
low erucic acid summer rape selections from the mutation experiments have
been reported (Röbbelen and Thies, 1980a).

Due to the large variation in erucic acid content (0.1–60%) in the seed
oils from rape and turnip rape, the separation of genetically controlled variation of erucic acid from environmental variation was relatively easy. The
invariable association of substantial amounts of eicosenoic acid (e.g. 6%)
with an allele for the presence of erucic acid which might condition the
production of a low level of erucic acid (e.g., less than 2%) ensured selection of the genetically lowest level of erucic acid.

The limited range of variation for linolenic acid content in rape and turnip
rape and the large environmental effects on this fatty acid makes it difficult
to separate genetic and environmental effects. Furthermore, the effect of
each gene conditioning linolenic acid content may be quite small. Since the

seed oils from both the species (*B. campestris* and *B. oleracea*) whose genomes make up the amphidiploid rape (*B. napus*) contain linolenic acid, at least two loci conditioning the levels of linolenic acid must be present in rape. Two biosynthetic pathways (one via 18:1 → 18:2 → 18:3 and the other via 12:3 → 14:3 → 16:3 → 18:3) for the synthesis of linolenic have been established, and both pathways probably are present in rapeseed (Brar and Thies, 1978). If this assumption is correct, then at least two loci condition linolenic acid content in *B. campestris* and *B. oleracea* and at least four in *B. napus*. In this case there would be eight alleles conditioning the 10% (approximately) level of linolenic acid in rapeseed and each allele possibly would contribute about 1.25% linolenic acid. Furthermore, if, as in the alleles for erucic acid content, a variation in effectiveness exists, analytical procedures must be precise enough to detect differences of 1% or less of linolenic acid. Furthermore, environmental effects may influence linolenic acid content by several percentage points (Canvin, 1965). Thus, very precise analyses and several generations of progeny testing may be needed to establish that reduced levels of linolenic acid are genetically controlled.

In spite of these difficulties, breeding programs aimed at reducing the linolenic content of rapeseed oil are in progress in Canada, France, Germany, and Sweden (Jönsson, 1977b). The low linolenic mutants reported by Röbbelen and Thies (1980a) apparently carry deleterious effects from exposure to mutagens; therefore, the genes conditioning low linolenic acid content will have to be transferred to healthy, vigorous plants by crossing and backcrossing. While difficult, such a transfer appears to be possible. Thus, as Jönsson (1977b) suggested, the development of rape cultivars with less than 5% linolenic acid in the seed oil appears to be feasible.

2. LINOLEIC ACID

The elimination of erucic acid from rapeseed oil resulted in an increase in the linoleic acid content of the oil from approximately 13 to 21%. Further increases appear to be possible since the values as high as 50% have been reported. However, such values do not appear to be stable and were not recovered in progeny tests (Jönsson, 1977a). Values up to at least 30% linoleic are stable, that is, are under genetic control. Some variation in linoleic acid content seems to occur in many populations of low erucic acid rapeseed. Since a reduction in component fatty acids other than oleic and linoleic provide greater opportunity for variation in these components, it may be desirable to select for increased linoleic acid content in populations low in both erucic and linolenic acids and thus to add this characteristic after reduced levels of linolenic have been achieved.

F. Glucosinolates

After the oil has been extracted from rapeseed, the meal that remains contains 34–38% protein (at 8% moisture level). The amino acid balance of rapeseed protein is quite favorable; however, the use of rapeseed meal in animal rations has been limited by its glucosinolate content (Bowland et al., 1965). The antinutritional and goitrogenic cleavage products from these glucosinolates have a pungent taste that decreases the palatability of the feed.

Under some conditions cleavage products from the glucosinolates containing sulfur became dissolved in the oil. These sulfur-containing compounds act as catalyst poisons and make hydrogenation more difficult. Reynolds and Youngs (1964) developed a technique that involved heating of the seed at an early stage in processing to inactivate the enzyme, myrosinase, to minimize the breakdown of the glucosinolates during crushing and oil extraction. This technique, which has been adopted by all rapeseed crushers in Canada, greatly facilitated utilization of rapeseed oil and meal in Canada.

Numerous methods for removing the glucosinolates from rapeseed meal have been considered. However, the methods so far tried have not been practical for use in large rapeseed crushing plants. For this reason, plant breeders initiated a search for genetically controlled low levels of glucosinolates in rapeseed. Known genetic variation in glucosinolate levels within the species of rapeseed was relatively small until Krzymansky in 1967 discovered that the glucosinolate content of the Polish *B. napus* cultivar Bronowski was only approximately 10% of the usual average (personal communication, J. Krzymanski, formerly of Agriculture Canada Research Station, Saskatoon, Canada). Bronowski has since been used extensively as a source for low glucosinolate content in breeding programs in North America and in Europe. The low glucosinolate content of Bronowski has been transferred to *B. campestris*. Bronowski appears to be the main and possibly sole source of low glucosinolate content in rapeseed breeding programs in the Western World.

Genetic studies (Kondra and Stefansson, 1970; Krzymansky, 1970; Lein, 1970) indicate that three to five gene loci are involved in the inheritance of glucosinolate content. Analyses of seeds from reciprocal crosses indicate that the glucosinolate content is determined by the maternal genotype rather than the genotype of the embryo. The gene systems controlling the three major glucosinolates do not segregate completely independently of each other; some of the genes probably control early stages in the biosynthetic pathways.

Double low cultivars (low in erucic acid and glucosinolates) of summer rape, i.e., Tower (Stefansson and Kondra, 1975) and Erglu (Röbbelen, 1976), were developed by backcross methods and released for production in 1974 in Canada and Germany. In Canada the release of other double low cultivars soon followed, namely, the summer rape cultivars Regent in 1977, Altex in 1979, and Andor in 1981, and the summer turnip rape cultivars Candle in 1977 and Tobin in 1981. Production of double low rapeseed increased rapidly in Canada, and in 1980, 80% of the rapeseed acreage was occupied by double low cultivars.

G. Triazine Resistance in Rapeseed

Triazine resistance could be of considerable importance to rapeseed production throughout the world. While trifluralin is used as a preplant herbicide for rape to control annual grassy weeds, there is a need for more effective herbicides such as the triazines to control broad-leaved weeds (some of which are closely related to rapeseed) in the seedling stages of the crop. A triazine resistant form of *B. campestris*, sometimes called bird rape, has been found in corn fields in Ontario and Quebec (Maltais and Bouchard, 1978; Souza-Machado and Bandeen, 1979). The resistance is cytoplasmically inherited. In other words, when used as the female in crosses, all susceptible plants produced susceptible progeny and all resistant plants produced only resistant progeny (Souza-Machado *et al.*, 1978). Triazine resistant cultivars of rapeseed are being developed using backcross methods. After five backcrosses, seed yields of the triazine resistant form of Candle were substantially lower than those of Candle. The cytoplasm does not appear to be as detrimental to vigor and yield in backcrosses involving rape. Thus if seed yields can be recovered in this cytoplasm, triazine resistant analogues of current cultivars may soon become available. Field trials conducted in 1980 indicate that a high level of weed control can be achieved using the triazine herbicides on a triazine resistant form of rapeseed.

H. Hybrid Cultivars

Interest in developing hybrid cultivars was stimulated a decade ago when Shiga and Baba (1971) and Thompson (1972) reported cytoplasmic male sterility in rape. Two other cytoplasmic male sterile cytoplasms, one derived from a Japanese male sterile radish (Bannerot *et al.*, 1977) and one derived from *Diplotaxis muralis* (Shiga, 1980), have been discovered. Shiga (1980) reviewed the cytoplasmic male sterility in rape and related species and developed terminology for describing the systems. The sterility discovered in rape by Shiga and Baba (1971) and Thompson (1972) appeared to be identical, and the cytoplasm was designated "nap," the restorers "Rfn" and the genes for sterility "rfn." In a similar manner the cytoplasm derived from radish was designated "ogu" and the one from *D. muralis* "mur," and the maintainer genes "rfog" and "rfm," respectively.

The heterosis that occurs in the F_1 progeny from some crosses probably is sufficient to justify the development of hybrid cultivars. However, much work remains to be done to evaluate the cytoplasmic male sterility systems and to develop suitable inbred lines. The male sterility in many of the lines with "nap" cytoplasm is incomplete. While this probably can be overcome by suitable selection methods, it is difficult to work with this system because most Canadian and European varieties possess many restorers for this cytoplasm. The sterility mediated by the "ogu" cytoplasm is complete and all known rape cultivars are maintainers; however, a satisfactory restorer does not appear to be available. Restorers probably can be found in the genus *Raphanus* (Heyn, 1978; Shiga, 1980); however, most of the rape lines in the "ogu" cytoplasm exhibit a degree of yellowing when grown at 10°C or lower temperatures. While the initial reports (Shiga, 1980) are promising, much work will be needed to evaluate the potential of the *Diplotaxis* cytoplasm for use in hybrid varieties.

The need for canola type quality imposes additional difficulties on the development of hybrid rape cultivars. Concentration on quality in Canada and in Europe has resulted in a temporary decrease in the genetic diversity in the new canola cultivars. For this reason the probability of obtaining a satisfactory level of heterosis in the F_1 of crosses between these varieties is decreased. Thus, it will probably be necessary to find lines that provide high levels of heterosis in certain cross combinations and introduce the quality characteristics into them by backcrossing. The task of producing hybrids with canola quality is formidable and is not likely to be accomplished in less than 10 years.

Hybrid varieties could be developed more quickly without canola quality. This approach appears to have been taken in China. In that country, hybrid cultivars, some using incompatibility and others utilizing cytoplasmic male sterility, are under development and appear to be close to commercial utilization (Stefansson, 1980).

III. CANOLA IN OTHER COUNTRIES

A. Northern Europe

The changeover to rapeseed cultivars that produce seed oil low in erucic acid is essentially complete in most European countries. Cultivars low in both erucic acid and glucosinolates are being developed in several European countries. The development of canola cultivars has been somewhat slower in Europe than in Canada due to the necessity of transferring the canola characteristics to highly developed cultivars of winter forms of the crop. Some commercial production of winter canola is expected in Europe in

1983 (private communication, R. K. Downey, Agriculture Canada Research Station, Saskatoon, Canada), and production of this kind of rapeseed can be expected to increase rapidly thereafter.

B.   The Orient

Most of the rapeseed produced in the Orient is of the traditional kind, high in both erucic acid and glucosinolates. Little attention appears to have been given to the development of canola cultivars. Perhaps the need to increase food supplies discourages the deployment of resources for work on quality. Prakash (1980) suggests that alteration of the composition of the oil and meal is one of a number of possible future trends in India. The recent increase in international contacts probably will stimulate the development of canola cultivars in China.

IV.   FUTURE TRENDS

The two major changes in the composition of rapeseed, which have been made available commercially, have improved the nutritional value of the oil and meal. Further improvements probably will become commercially available during the present decade. The possibilities for changes in chemical composition of crops probably are not limited to rapeseed; the changes made in rapeseed may thus provide an indication of the manner in which other crops might be modified to provide better food and feed.

Nutritional studies, such as those summarized in this book, suggest or indicate ways in which human diets might be changed in an effort to improve human health. However, essentially all edible vegetable oils now produced are consumed. Therefore, suggestions or recommendations that consumers might improve their health by consuming fats rich in linoleic acid might help the more fortunate individual, but would reduce the availability of the desired component to the average consumer. The possibilities of providing more of the desired kinds of fats and oils by changing the crop kinds grown are limited because crops already are generally grown in their area of adaptation where yields are the highest. Therefore, the best hope for improving the nutritional quality of foods such as the edible oils for the average consumer is, as has been done with the erucic acid component of rapeseed oil, to modify the chemical composition of the oil through plant breeding.

The degree to which the morphological characteristics of plants have been modified to provide food and feed is truly remarkable. Crop cultivars within the genus *Brassica* are used to provide edible roots, stems, leaves, buds, flowers, and seeds (Gomez-Campo, 1980). Within a species cultivars have been developed for different purposes; for example, members of the species *Brassica campestris* with an enlarged root provide the turnip, the headed form, Chinese cabbage, and the same species developed for oilseed production, rapeseed.

It has taken centuries of cultivar development to achieve these morphological modifications. By comparison, attempts to modify the chemical composition of crop plants are quite recent. A number of developments suggest that this area may become increasingly important. The nutritional quality of several crops is being defined in terms of chemical composition. Unambiguous identification of antinutritional compounds and recognition of the desirable components provides motivation for efforts to modify chemical composition through plant breeding. Increasing knowledge of the variability in the chemical composition of crop plants and their wild relatives provides an indication of changes that can be achieved. Increasing knowledge of the relationships between crop plants and their wild relatives increases the possibilities of transferring characteristics from species to species and even from genus to genus. Knowledge of biosynthetic pathways provides a theoretical background for understanding variation, selection, and the effects of genes controlling chemical composition. Recent advances in genetics, especially the genetics of microorganisms, may eventually be used to modify the composition of crop plants. Higher plants do not require carbohydrates or other complex compounds as a source of energy and are much easier to control than microorganisms that must be produced under aseptic conditions to prevent contamination. For these reasons, it seems probable that in addition to providing food and feed, higher plants and not microorganisms will eventually become the organisms of choice for the biosynthesis of exotic compounds.

REFERENCES

Bannerot, H., Boulidard, L., and Chupeau, Y. (1977). *Eucarpia Cruciferae Newsl.* **2**, 16.
Bowland, J. P., Clandinin, D. R., and Wetter, L. R. (1965). *Can. Dep. Agric. Publ.* **1257**, 69–80.
Brar, G. S., and Thies, W. (1978). *Proc. Int. Rapeseed Conf., 5th, 1978* Vol. 2, pp. 27–30.
Calhaun, W., Crane, J. M., and Stamp, D. L. (1975). *J. Am. Oil Chem. Soc.* **52**, 363–365.
Canvin, D. T. (1965). *Can. J. Bot.* **43**, 63–69.
Comstock, V. E., and Culbertson, J. O. (1958). *Agron. J.* **50**, 113–114.
Dorrell, D. G., and Downey, R. K. (1964). *Can. J. Plant Sci.* **44**, 499–504.
Downey, R. K. (1964). *Can. J. Plant Sci.* **44**, 295.
Downey, R. K., and Harvey, B. L. (1963). *Can. J. Plant Sci.* **43**, 271–275.
Gomez-Campo, C. (1980). *In* "Brassica Crops and Wild Allies" (S. Tsunoda, K. Hinata, and C. Gomez-Campo, eds.), pp. 3–31. Jpn. Sci. Soc. Press, Tokyo.
Grami, B., and Stefansson, B. R. (1977). *Can J. Plant Sci.* **57**, 625–631.
Grami, B., Baker, R. J., and Stefansson, B. R. (1977). *Can. J. Plant Sci.* **57**, 937–943.
Harvey, B. L., and Downey, R. K. (1964). *Can. J. Plant Sci.* **44**, 104–111.

Heyn, F. W. (1979). *Proc. Int. Rapeseed Conf., 5th, 1978* Vol. 1, pp. 82–86.

Jönsson, R. (1977a). *Hereditas* **86**, 159–170.

Jönsson, R. (1977b). *Hereditas* **87**, 205–218.

Jönsson, R., and Bengtsson, L. (1970). *Sveriges Utsaedesfoeren. Tidskr.* **80**, 149–155.

Kirk, J. T. O., and Oram, R. N. (1981). *J. Aust. Inst. Agric. Sci.* **47**, 51–52.

Kondra, Z. P., and Stefansson, B. R. (1965). *Can. J. Genet. Cytol.* **7**, 505–510.

Kondra, Z. P., and Stefansson, B. R. (1970). *Can. J. Plant Sci.* **50**, 643–647.

Krzymanski, J. (1970). *Hodowla Rosl., Aklim. Nassien.* **14**, 95–133.

Krzymanski, J., and Downey, R. K. (1969). *Can. J. Plant Sci.* **49**, 313–319.

Lein, K.A. (1970). *Z. Pflanzenzuecht.* **63**, 137–154.

McLeod, A.D. (1974). "The Story of Rapeseed in Western Canada." Saskatchewan Wheat Pool, Regina, Canada.

Maltais, B., and Bouchard, C. J. (1978). *Phytoprotection* **59**, 117–119.

Morice, J. (1975). *C.R. Seances Acad. Agric. Fr.* **61**, 335–345.

Olsson, G. (1960). *Hereditas* **46**, 29–70.

Olsen, O., and Sørensen, H. (1980). *J. Agric. Food Chem.* **28**, 43–48.

Pawlowski, S. H., and Youngs, C. G. (1969). *Proc. 2nd Annu. Meet., Rapeseed Assoc., Can.* pp. 47–51.

Prakash, S. (1980). *In* "Brassica Crops and Wild Allies" (S. Tsunoda, K. Hinata and C. Gomez-Campo, eds.), pp. 151–163. Jpn. Sci. Soc. Press, Tokyo.

Reynolds, J. R., and Youngs, C. G. (1964). *J. Am. Oil Chem. Soc.* **41**, 63–65.

Röbbelen, G. (1976). *Fette, Seifen, Anstrichm.* **78**, 10–17.

Röbbelen, G. (1978). *Fette, Seifen, Anstrichm.* **80**, 99–103.

Röbbelen, G., and Rakow, G. (1970). *Proc. Int. Rapeseed Conf. 3rd, 1970*, pp. 476–490.

Röbbelen, G., and Thies, W. (1980a). *In* "Brassica Crops and Wild Allies" (S. Tsunoda, K. Hinata, and C. Gomez-Campo, eds.), pp. 253–283. Jpn. Sci. Soc. Press, Tokyo.

Röbbelen, G., and Thies, W. (1980b). *In* "Brassica Crops and Wild Allies" (S. Tsunoda, K. Hinata, and C. Gomez-Campo, eds.), pp. 285–299. Jpn. Sci. Soc. Press, Tokyo.

Shiga, T. (1980). *In* "Brassica Crops and Wild Allies" (S. Tsunoda, K. Hinata, and C. Gomez-Campo, eds.), pp. 205–221. Jpn. Sci. Soc. Press, Tokyo.

Shiga, T., and Baba, S. (1971). *Jpn. J. Breed.* **21**, Suppl.2, 16–17.

Souza-Machado, V., and Bandeen, J. D. (1979). *In* "Notes on Agriculture" (W. S. Young, ed.), Vol. XV, No. l, pp. 19–21. Ontario Agricultural College, University of Guelph, Ontario, Canada.

Souza-Machado, V., Bandeen, J. D., Stephenson, G. R., and Lavigne, P. (1978). *Can. J. Plant Sci.* **58**, 977–981.

Stefansson, B. R. (1970). *Proc. Int. Rapeseed Conf., 3rd, 1970*, 86–91.

Stefansson, B. R. (1980). *In* "Report on Canola Technical Seminars, Peoples' Republic of China, Nov. 9–22, 1980," pp. 14–17. Canada Department of Industry, Trade and Commerce, Ottawa.

Stefansson, B. R., and Hougen, F. W. (1964). *Can. J. Plant Sci.* **44**, 359–364.

Stefansson, B. R., and Kondra, Z. P. (1975). *Can. J. Plant Sci.* **55**, 343–344.

Stefansson, B. R., and Storgaard, A. K. (1969). *Can. J. Plant Sci.* **49**, 573–580.

Stefansson, B. R., Hougen, F. W., and Downey, R. K. (1961). *Can. J. Plant Sci.* **41**, 218–219.

Stringam, G. R., McGregor, D. I., and Pawlowski, S. H. (1974). *Proc. Int. Rapskongr., 4th, 1974*, pp. 99–108.

Thompson, K. F. (1972). *Heredity* **29**, 253–257.

Troeng, S. (1955). *J. Am. Oil Chem. Soc.* **32**, 124–126.

Youngs, C. G., and Wetter, L. R. (1967). *J. Am. Oil Chem. Soc.* **44**, 551–554.

7

The Introduction of Low Erucic Acid Rapeseed Varieties into Canadian Production

J. K. DAUN

I. INTRODUCTION

One of the most significant achievements in the field of plant breeding has been the changes in the chemical composition of rapeseed brought about during the last 20 years. Accomplishing these changes with little, if any, loss of agronomic quality made it possible for Canadian producers to rapidly accept the new varieties. Within the period 1970–75 the entire Canadian industry successfully converted from production of rapeseed with erucic acid levels in the oil ranging from 20–40% to less than 5% erucic acid levels in the oil. At the present time, the Canadian rapeseed industry is in the process of converting again, this time from low erucic acid rapeseed (LEAR) to canola (canola is a trademark of the Canola Council of Canada and refers to seed and seed products from varieties of *Brassica campestris* L. and *Brassica napus* L. with low levels of both erucic acid and glucosinolates).

This chapter will review the development of rapeseed varieties in Canada and will pay special attention to the introduction of low erucic acid rapeseed into the Canadian grain handling and processing systems.

Much of the historical information used in this chapter may be found in "The Story of Rapeseed in Western Canada" (McLeod, 1974). Most of the data used is unpublished data although it may have appeared in abbreviated form in crop reports or in the "Prairie Wide Co-operative Rapeseed Test" from which data is used with permission of the test coordinator. Useful background on the conversion to low erucic acid varieties may also be found in minutes of the "Annual Meetings of the Rapeseed Association of Canada" (Rapeseed Association of Canada, 1971–1975). For information on quality and composition see Chapter 4 or the publication "Canola Oil, Properties, Processes and Food Quality" by Vaisey-Genser and Eskin, 1979.

II. DEVELOPMENT AND QUALITY OF RAPESEED VARIETIES IN CANADA PRIOR TO 1970

A. Origins of Canadian Rapeseed Varieties

During the 1930s, the Forage Crop Division of the Experimental Farms Service of the Federal Department of Agriculture experimented with seed of both oil and forage crop rapeseed. Their tests established that summer types of rapeseed could be grown successfully in both Eastern and Western Canada. In 1936, a farmer in Saskatchewan received a sample of rapeseed from a relative in Poland. This sample of *Brassica campestris* seed was increased and eventually became the source of all *Brassica campestris* varieties in

Canada. Because of its origin this type of rapeseed became known as Polish rape in Canada (turnip rape in Europe).

In 1942, the Wartime Agricultural Supply Board instructed the Forage Crop Division at Saskatoon to initiate production of rapeseed. Because only a small quantity of seed was available in Canada, 41,000 pounds of seed were purchased from the United States. This seed, which had originated in Argentina, became the source of Canadian varieties of *Brassica napus* and rapeseed of the *Brassica napus* species came to be known as Argentine rape in Canada (rape in Europe).

B. Plant Breeding prior to 1970

The names Argentine and Polish became unofficial varietal names for the mixed seed stocks described above. Selection work to develop improved varieties was initiated at the Agriculture Canada Research Station in Saskatoon in 1944 *(Brassica napus)* and at the Agriculture Canada Research Station in Indian Head, Saskatchewan in 1948 *(Brassica campestris)*. An oilseed breeding program was initiated in 1952 at the University of Manitoba and in 1969 at the University of Alberta. Between 1950 and 1970 selection procedures concentrated on agronomic properties and increased oil content. Selections with low amounts of erucic acid were isolated as early as 1960 in Winnipeg *(Brassica napus)* and in 1963 in Saskatoon *(Brassica campestris)*.

TABLE I

Chemical Characteristics of Rapeseed[a] Released in Canada prior to 1970

Variety	Year released	Oil content (% dry basis)	Protein content (% in oil-free meal dry basis)	Glucosinolate content (μmoles/g oil-free air dry meal)
B. napus				
"Argentine"[b]	—	40	47	—
Golden	1954	41	47	—
Nugget	1961	43	46	—
Tanka	1963	42	47	—
Target	1966	43	44	98
Oro	1968	40	42	88
Turret	1970	44	44	106
B. campestris				
"Polish"[b]	—	41	42	—
Arlo	1958	42	42	53
Echo	1964	40	42	44
Polar	1969	42	42	62

[a] Based on data from the "Prairie Wide Co-operative Variety Tests" (Anonymous, 1972–1981).

[b] Rapeseed introduced into Canada.

TABLE II

Chemical Characteristics of Rapeseed Oils Released in Canada prior to 1970

Variety	16:0	18:0	18:1	18:2	18:3	20:0	20:1	22:0	22:1	24:0	Iodine value (g/100 g)	Saponification value (mg KOH/g)
B. napus												
"Argentine"[b]												
Golden	3.4	1.5	20.2	14.3	8.8	0.8	11.6	0.8	38.6	0.3	102	176
Nugget	3.1	1.8	22.2	13.7	7.9	1.0	12.6	0.8	36.5	0.5	99	176
Tanka	3.2	1.3	19.6	15.5	7.2	0.8	11.0	0.8	40.0	0.5	100	175
Target	3.2	1.3	23.0	14.7	7.8	0.8	11.5	0.8	36.5	0.5	101	176
Oro	4.1	1.3	59.8	19.8	9.1	0.7	2.5	0.4	2.1	0.3	114	190
Turret	3.1	1.2	19.0	14.2	9.3	0.8	10.5	0.7	40.9	0.5	103	175
B. campestris												
"Polish"[b]												
Arlo	2.7	1.1	28.1	17.1	7.4	0.6	9.8	0.7	32.1	0.3	104	178
Echo	2.3	1.1	33.7	18.7	11.7	0.6	9.3	0.6	21.5	0.5	114	181
Polar	2.5	1.3	26.4	16.4	9.5	0.7	10.7	0.8	31.3	0.6	107	178

Note: The first variety in each group ("Argentine" = 3.5, 1.1, 17.8, 14.9, 9.9, 0.7, 9.9, 0.8, 41.0, 0.3, 105, 175; "Polish" = 2.8, 1.3, 36.4, 17.0, 6.5, 0.7, 10.2, 0.7, 23.9, 0.5, 108, 180).

Fatty acid composition[a] (% of total fatty acids)

[a] Based on gas–liquid chromatographic analysis of authentic samples. Values vary depending on growing location and climatic conditions.
[b] Rapeseed introduced into Canada.

C.　Quality of Rapeseed prior to 1970

1.　AGRONOMIC CHARACTERISTICS

During the period 1943–1970, the farm yield of rapeseed increased from 700 kg/ha to greater than 900 kg/ha. This increase was due to a combination of factors, including improved understanding of production problems such as moisture requirements and herbicide use, and the introduction of new varieties which were earlier maturing, more resistant to lodging and had about a 12% yield advantage over the original rapeseed. By 1970, rapeseed yields in "Prairie Wide Co-operative Variety Tests" were 2350 kg/ha for *B. napus* types and 1750 kg/ha for *B. campestris* types compared with 2100 kg/ha for the original Argentine *B. napus* and 1560 kg/ha for Polish *B. campestris* types.

2.　CHEMICAL CHARACTERISTICS

Tables I and II outline some of the chemical characteristics of seed and oil from varieties of rapeseed released in Canada by 1970. Varieties released prior to 1970 all showed significant increases in oil content over the common types. They all had high levels of glucosinolates, averaging 100 μmoles/g (of oil-free air dry meal) for *B. napus* and 70 μmoles/g for *B. campestris*. Since protein content was not considered a selection criterion, there was no significant change in protein except for a possible slight decrease in high oil selections of *B. campestris*. Oil from early varieties of rapeseed was characterized (Table II) by high levels of erucic acid (35–40% in *B. napus* and 20–30% in *B. campestris*). Iodine values were generally about 99–103 for *B. napus* and 104–114 for *B. campestris*. Saponification values were 175–176 for *B. napus* and 178–181 for *B. campestris*. One variety of rapeseed (cv. Oro) with low erucic acid was not grown to any extent. Its lower agronomic and chemical qualities (compared with other contemporary varieties) along with large imports of low priced European sunflower oil prevented its successful competition for a share of the market.

III.　MECHANISM OF RAPESEED VARIETAL DEVELOPMENT

In order to understand the timeframe for the Canadian conversion to lower erucic acid rapeseed varieties, it is necessary to understand the system used in Canada to select new varieties for licensing. According to the Canada Grain Act nonlicensed varieties of grain can only receive the lowest grade possible in its class. Since it was not possible to distinguish between high and low erucic rapeseed, in the grain handling system, this ruling could not effectively be applied to discourage high erucic acid rapeseed. Hence the pedigreed seed system was used to ensure eventual complete conversion to low erucic acid rapeseed. By licensing only low erucic acid varieties it was planned that within a few years the only certified seed available to farmers would be of the low erucic acid type.

As previously noted, rapeseed breeding in Canada has taken place primarily at three locations, the Agriculture Canada Research Station at Saskatoon (and in early years Indian Head), the University of Manitoba, and also more recently at the University of Alberta. Plant breeding techniques that were used to develop new cultivars of rapeseed have been described in Chapter 6.

A.   Development of a New Variety for Licensing

Usually selection of promising strains takes 5–7 years with the last two to three years spent in cooperative yield trials. Selection of summer crops can be hastened somewhat by growing 2 to 3 generations per year, utilizing both greenhouses and outdoor plots in Canada in the summer, and California or Mexico in the winter. Since changes in chemical characteristics such as erucic acid content or glucosinolates are more easily determined than improvements in agronomic factors, some yield testing in the final stages was often omitted in the development of LEAR and canola strains.

After preliminary yield tests of 2–5 years, selections were entered into strain tests, normally for 1 or 2 years. One or two kilograms each of promising varieties were then collected and entered into the Canada-wide rapeseed co-operative tests. These selections were planted at about 20 different locations, mainly in Western Canada, in plots ranging from 6×1.25 m to 6×2 m in size. After harvesting and testing, results were evaluated by Advisory Committees on Grain Breeding, Quality and Plant Diseases. Varieties with good agronomic and quality records received support from these committees when the breeder applied to Plant Products division of Agriculture Canada for licensing.

B.   The Pedigreed Seed System

Varieties licensed by Plant Products Division of Agriculture Canada are increased for release to farmers under the pedigreed seed system. The seed under this system is classified into three types.

1. *Breeder seed.* This seed is controlled by the breeder who is responsible for maintaining varietal purity as well as supplying stocks to foundation seed growers. On occasion it has been possible for breeders to make minor improvements in varieties by reselection of breeders seed.

2. *Foundation seed.* This is seed stock one generation removed from breeder seed. Growers wishing to become foundation seed growers must comply with strict regulations regarding cultivation practice and must un-

TABLE III

Erucic Acid Levels in Certified Rapeseed[a]

Variety	1971/1972	1972/1973	1973/1974	1974/1975	1975/1976	1976/1977	1977/1978	1978/1979	1979/1980	1980/1981
Brassica napus										
Oro										
Erucic acid std. (%)	1.2	1.2	1.2	1.2	1.2	1.2	1.2	—	—	—
Average erucic acid (%)	0.7	—	—	0.5	1.0	—	—	—	—	—
% within std.	96	—	—	100	100	—	—	—	—	—
Zephyr										
Erucic acid std.	1.2	1.2	1.2	1.2	1.2	1.2	—	—	—	—
Average erucic acid	0.4	0.4	0.3	—	—	—	—	—	—	—
% within std.	99	99	100	—	—	—	—	—	—	—
Midas										
Erucic acid std.	—	—	1.2	1.0	1.0	1.0	1.0	1.0	1.0	1.0
Average eruci acid.	—	2.7	0.3	0.3	0.6	0.3	0.4	0.4	—	—
% within std.	—	—	100	99	97	100	100	100	—	—
Tower										
Erucic acid std.	—	—	1.2	1.0	1.0	1.0	1.0	1.0	1.0	1.0
Average erucic acid	—	—	1.3	0.3	0.3	0.3	0.4	0.4	0.4	0.5
% within std.	—	—	50	100	100	100	100	100	95	100
Regent										
Erucic acid std.	—	—	—	—	—	—	1.0	1.0	1.0	1.0
Average erucic acid	—	—	—	—	—	—	0.4	0.4	0.4	0.5
% within std.	—	—	—	—	—	—	96	100	99	99

(Continued)

TABLE III (CONTINUED)

Variety	1971/ 1972	1972/ 1973	1973/ 1974	1974/ 1975	1975/ 1976	1976/ 1977	1977/ 1978	1978/ 1979	1979/ 1980	1980/ 1981
Altex										
Erucic acid std.	—	—	—	—	—	—	—	1.0	1.0	1.0
Average erucic acid	—	—	—	—	—	—	—	0.4	0.4	0.5
% within std.	—	—	—	—	—	—	100	99	100	
Brassica campestris										
Span										
Erucic acid std.	3.8	3.8	3.8	1.2	1.2	1.2	1.2	1.2	1.2	1.2
Average erucic acid	3.3	3.6	3.4	2.8	2.1	1.9	—	—	—	—
% within std.	90	71	84	27	50	50	—	—	—	—
Torch										
Erucic acid std.	—	—	3.8	3.8	1.8	1.8	1.8	1.8	1.8	1.8
Average erucic acid	—	2.5	2.9	1.3	1.1	0.9	1.0	0.9	1.1	0.8
% within std.	—	—	93	95	89	97	97	97	86	100
Candle										
Erucic acid std.	—	—	—	—	—	—	2.0	2.0	2.0	2.0
Average erucic acid	—	—	—	—	—	—	1.4	1.5	1.7	2.0
% within std.	—	—	—	—	—	—	87	93	82	74

[a] Unpublished data supplied by Plant Products and Quarantine Directorate, Food Production and Inspection Branch of Agriculture Canada.

dergo a 3-year trial period supervised by the Canadian Seed Growers' Association.

3. *Certified seed.* This is seed one generation removed from foundation seed or two generations removed from breeders seed. Certified seed is usually sold for increase of seed for commercial production.

Crops of foundation and certified seed must be grown under strict standards of land purity requirements, area requirements and isolation from possible cross-pollination. Fields must be inspected by government inspectors when the crop is in the early flowering stage. After harvest, a 500 g sample of seed must be submitted for inspection and grading. Inspection includes germination tests, impurity content (i.e. freedom from weed seeds and other species), and tests for erucic acid content and possibly glucosinolate content. Seed from this system is certified according to grade and sold to growers for planting.

Because of the possibility of reversion or the presence of volunteer high erucic acid material, the standards for erucic acid level in certified seed have been much lower than the level accepted for food use. Table III outlines the erucic acid levels in certified rapeseed as well as the average value found and the percentage of seed tested meeting the standard.

IV. THE DECISION TO CONVERT TO LEAR VARIETIES

A. The Nutritional Considerations

Questions about the nutritional suitability of rapeseed oil were first raised in 1956 when the Food and Drug Directorate ruled that based on reports of animal feeding trials in the literature, rapeseed oil would not be approved as an edible oil. The objections of the Food and Drug Directorate to the use of rapeseed oil as an edible oil were withdrawn after a further review of the nutritional properties of rapeseed oil revealed that in its current limited use there was nothing to indicate that it was a hazard to health.

In 1958, following a comprehensive nutritional study with rats, the Food and Drug Directorate further reported that no harmful effects had been observed in rats fed on rapeseed oil at levels that would ordinarily be consumed by the public. The Food and Drug Directorate had no objection to the use of rapeseed oil in moderate amounts in food in Canada.

Questions about the nutritional effects of erucic acid were again brought up in August 1970 at the International Conference on the Science, Technology and Marketing of Rapeseed and Rapeseed Products in St. Adèle, Quebec. Nutritionists from Canada, France, and Holland reported investigations linking erucic acid to fat accumulation in young animals and heart problems in older animals. Although no evidence was presented which suggested health problems for humans, on August 12, 1970, the Hon. Mr. Munro, the Minister of National Health and Welfare stated that the Federal Government was concerned and felt that it was "prudent to accelerate Canada's change-over to erucic acid free rapeseed."

By 1973, the majority of Canadian crushing plants were able to produce oils with levels of erucic acid lower than 5% without blending with other oils.

In May 1975, amendment B.09.022 was made to Canadian Food and Drug Regulations which limited the amount of C_{22} monoenoic fatty acids in cooking oils, margarines, shortenings, or other oil products to not more than 5% of the total fatty acids (Statute Revision Commission, 1978).

B. Availability of Varieties

In 1970, when the decision was made to convert to LEAR, only one variety of rapeseed was available which had low levels of erucic acid. Oro rapeseed, a *B. napus* variety had been released in 1968 by the Agriculture Canada Research Station at Saskatoon but was not grown widely because of its relatively low yielding ability. Fortunately, the breeding programs at both Saskatoon and the University of Manitoba had lines available with low erucic acid contents and in 1971 the varieties Zephyr and Span were introduced. These two varieties were replaced by Midas and Torch in 1973.

TABLE IV

Chemical Characteristics of Rapeseed Released in Canada between 1971 and 1981

Variety	Year released	Oil content (% dry basis)	Protein content (% in oil-free meal dry basis)	Glucosinolate content (μmoles/g oil-free air dry meal)
B. napus				
Zephyr	1971	40	42	97
Midas	1973	43	42	115
Tower	1975	42	46	18
Regent	1977	42	45	18
Altex	1978	42	45	18
Andor	1980	42	44	18
Reston	1981	45	48	18
B. campestris				
Span	1971	39	41	62
Torch	1973	39	41	62
Candle	1977	42	42	18
Tobin	1981	43	41	18
R-500	—[a]	42	41	115

[a]Temporary license granted in 1976.

TABLE V

Chemical Characteristics of Rapeseed Oils Released in Canada between 1971 and 1981

Variety	Fatty acid composition[a] (% of total fatty acids)										Iodine value (g/100 g)	Saponification value (mg KOH/g)
	16:0	18:0	18:1	18:2	18:3	20:0	20:1	22:0	22:1	24:0		
B. napus												
Zephyr[b]	3.9	1.7	59.5	20.5	11.6	0.6	1.3	0.4	0.4	0.1	118	191
Midas[b]	4.0	1.8	59.3	21.5	11.1	0.6	1.1	0.4	0.1	0.1	118	191
Tower[c]	4.0	1.6	60.2	21.1	10.8	0.6	1.3	0.3	0.1	0.1	120	191
Regent[c]	4.0	1.7	61.9	19.7	10.1	0.7	1.5	0.4	0.1	0.1	117	191
Altex[c]	3.6	1.6	62.0	20.4	9.8	0.6	1.4	0.4	0.1	0.2	118	191
Andor[c]	3.7	1.5	57.5	21.2	13.3	0.6	1.5	0.4	0.1	0.2	122	191
Reston	3.5	1.1	18.8	14.5	9.7	0.7	8.5	0.5	42.3	0.4	108	175
B. campestris												
Span[b]	3.1	1.6	55.8	21.7	11.1	0.5	2.4	0.4	3.0	0.4	118	189
Torch[b]	2.8	1.4	62.8	19.4	11.2	0.5	1.2	0.2	0.5	0.1	117	190
Candle[c]	3.2	1.3	56.5	22.4	12.9	0.5	1.5	0.2	1.3	0.2	124	190
Tobin[c]	3.0	1.4	58.4	21.9	11.9	0.5	1.5	0.2	1.0	0.1	122	190
R-500	1.7	0.9	10.2	13.7	11.6	0.7	3.4	1.0	56.2	0.5	106	171

[a] Based on gas–liquid chromatographic analysis of authentic samples. Values vary depending on growing location and climatic conditions.
[b] LEAR.
[c] LEAR (canola type).

C. Quality of the Initial LEAR Varieties

The seed and oil characteristics of rapeseed varieties released between 1971 and 1981 are summarized in Tables IV and V.

The early LEAR varieties, especially Zephyr and Span were not well liked by farmers or processors. These varieties were low yielding compared to HEAR varieties and Zephyr had a major lodging problem. Both Span and Zephyr had lower oil contents than HEAR varieties, making them less profitable to process. Indeed, the quality of these varieties was such that they would not have been licensed had it not been extremely urgent to make LEAR types available to farmers. These varieties quickly disappeared from the Prairies when Midas and Torch became available.

The change in fatty acid composition of rapeseed oil brought about by the introduction of LEAR presented some difficulties to processors. Margarines made from HEAR varieties had extremely good plastic properties and a very stable microcrystalline (β') structure. With rapeseed oil containing less than 10% erucic acid, the microcrystal structure became unstable and margarines made from LEAR rapeseed oil tended to become grainy. This problem is also encountered in sunflower oil, which also has a large proportion of one fatty acid. The problem can be overcome by blending or by using emulsifiers (Vaisey-Genser and Eskin, 1979).

Removal of erucic acid from rapeseed oil also resulted in increased levels of linolenic acid in the oil. This has been considered undesirable because of the tendency of linolenic acid to oxidize. The high smoke point and heat stability of the traditional oil have been lost. LEAR oil, however, has been shown to be equal to soybean oil in oxidative stability (Vaisey-Genser and Eskin, 1979), and plant breeders are hoping to reduce the linolenic acid levels and increase the linoleic acid levels in future varieties.

V. MONITORING THE CONVERSION TO LEAR

Records that enable the introduction of new rapeseed varieties to be recorded include surveys of varieties planted and surveys of harvested grain. The "Prairie Grain Varietal Survey" is carried out annually with the cooperation of the three provincial elevator pools (Anonymous, 1972–1981). This annual report, since 1972, has combined data on area planted obtained from Statistics Canada and Provincial Departments of Agriculture with estimates on varietal distribution obtained from pool elevator agents. Information on the varietal distribution in rapeseed prior to 1972 appeared in the publication "Seedtime and Harvest" (Durksen, 1968–1971) published by the now disbanded Federal Grain Limited.

The monitoring program for harvested grain was carried out by the Canadian Grain Commission's Grain Research Laboratory. In 1971 the Commission established an oilseeds laboratory primarily to carry out analysis of erucic acid in Canadian rapeseed samples representing the annual new crop, rail carlot shipments, and export cargoes. More recently, samples of oil produced at Western Canadian crushing plants have also been analyzed. Results from new crop and carlot surveys have been published in annual New Crop Bulletins (Western Canadian Flax and Rapeseed) and in the annual report of the Grain Research Laboratory. Results from cargo surveys have been published monthly in the Rapeseed Digest or the Canola Digest, currently published by the Canola Council of Canada, Winnipeg, Manitoba.

Crushing plants, oil refineries, and Health and Welfare Canada have also been monitoring the level of erucic acid in rapeseed and rapeseed oil processed in Canada. Results from these programs are not available for publication.

A. The Prairie Grain Variety Survey

A summary of the conversion to LEAR as observed through the variety survey is shown in Table VI. Betweeen 1970 and 1972, the rapeseed area planted to low erucic acid types increased from essentially zero (some specially contracted LEAR cv. Oro had been grown on a very limited area) to 80%, with most of that change occurring in 1972. By 1977 almost all the rapeseed grown in Western Canada was of the LEAR type, and by 1980 about 80% of the rapeseed sown in Western Canada was low in both erucic acid and glucosinolates (canola). Small amounts of high erucic acid rapeseed are still grown in Western Canada under contract with crushing plants for industrial use.

Although the Prairie Grain Varietal Survey gives a fairly accurate estimate of the proportion of each variety sown in any given year, it is not possible to use this survey for accurately predicting the erucic acid content of rapeseed harvested or delivered into the grain handling system using the information since some increases in erucic acid occur while the crop is grown.

Craig et al. (1973) found that the erucic acid content of rapeseed increased by 0.5% as the seed was grown from isolation in California to commercial production in Western Canada. Most of this increase was thought to be due to the presence of volunteer high erucic acid plants contaminating the low erucic acid fields although some increases in erucic acid due to cross-pollination and environment were noted. Wild mustard [Brassica kaber (DC) L. C. Wheeler cv. pinnatifida (Stokes) L. C. Wheeler] has also been implicated in adding to the erucic acid content of rapeseed. Studies on the erucic acid content of wild mustard found in Canadian rapeseed have

TABLE VI

Percentage of Total Rapeseed Crop Planted to Each Variety in the Prairie Provinces (1968–1981)

Variety	1968	1969	1970	1971	1972	1973	1974	1975	1976	1977	1978	1979	1980	1981
B. napus														
Target	11.0	11.9	17.5	19.2	—	—	—	—	—	—	—	—	—	—
Tanka	4.9	2.6	1.6	—	—	—	—	—	—	—	—	—	—	—
Oro[a]	—	—	—	—	3.7	4.0	2.4	1.2	0.5	0.1	0.3	—	—	—
Turret	—	—	—	—	—	—	—	—	—	—	—	—	—	—
Zephyr[a]	—	—	—	—	27.9	19.8	5.7	1.7	0.8	0.5	0.2	—	—	—
Midas[a]	—	—	—	—	—	1.3	20.0	30.2	23.2	20.5	16.3	6.2	3.0	1.6
Tower[b]	—	—	—	—	—	—	1.9	11.9	24.1	31.0	31.8	20.7	13.9	8.8
Regent[b]	—	—	—	—	—	—	—	—	—	—	1.1	20.3	25.2	27.0
Altex[b]	—	—	—	—	—	—	—	—	—	—	—	0.8	14.3	21.5
B. campestris														
Polish	16.3	13.9	9.2	9.5	—	—	—	—	—	—	—	—	—	—
Arlo	22.0	17.2	15.9	10.5	—	—	—	—	—	—	—	—	—	—
Echo	42.6	52.8	51.8	46.6	—	—	—	—	—	—	—	—	—	—
Polar	—	—	—	—	—	—	—	—	—	—	—	—	—	—
Span[a]	—	—	—	3.2	48.7	60.2	32.2	10.1	6.7	2.3	1.1	0.4	0.5	—
Torch[a]	—	—	—	—	—	0.6	31.6	42.0	43.1	43.7	36.1	24.7	15.2	9.2
Candle[b]	—	—	—	—	—	—	—	—	—	0.9	11.6	25.3	26.3	29.8
Total LEAR	0	0	0	6.5	80.3	85.9	93.8	97.1	98.8	99.0	98.5	98.4	98.4	99.1

[a] LEAR.
[b] LEAR (canola type).

shown that at the maximum level allowed in rapeseed (5%), this contaminant might add 0.3% erucic acid to the oil.

The first *B. campestris* LEAR varieties had erucic acid contents on the order of 3.5% as breeder's seed. This level was sufficiently close to 5% to provide serious problems when the seed was grown commercially. Both LEAR cv. Span and cv. Torch were reselected after release to provide seed with significantly lower levels of erucic acid.

Although increases in the erucic acid content of pedigreed seed may be predicted and documented, the major cause of increases in erucic acid in rapeseed crops has been the use of nonpedigreed seed. Nonpedigreed seed stock often contained large amounts of volunteer material which in the early 1970s was of high erucic acid seed.

B. Surveys of the Erucic Acid Content of Rapeseed Grown in Western Canada for Commercial Use

1. METHODS FOR ESTIMATING ERUCIC ACID

The establishment of guidelines and regulations for levels of erucic acid in rapeseed and rapeseed oil required the establishment of uniform methods for determining these components. The Canadian Government Standards Board specification for low erucic acid rapeseed oil (CGSB, 1976) identifies two procedures for determining the erucic acid content of rapeseed oil, one for routine use and one for use in cases where the result is in dispute. Plant Products Division of Agriculture Canada recommended a procedure for determining the erucic acid content in foundation and certified rapeseed. They circulated a reference mixture of fatty acids and established an erucic acid check series through which they "licensed" laboratories to perform analyses (Barrette, 1976). Many laboratories found that they could modify the official procedure and still obtain acceptable results. For example, at the Grain Research Laboratory, accurate results were obtained for many years employing a one-step extraction and methylation technique described by Hougen and Bodo (1973).

Many early results for erucic acid actually included total C_{22} fatty acids since the analysis was performed on nonpolar columns in order to save time and increase sample throughput. Present day analyses are usually carried out using polar phases, including a mixed phase which allows separation of all major fatty acids according to chain length and degree of unsaturation. Erucic acid values reported from these analyses still contain very small amounts of vaccenic and brassidic acids. It is necessary to use open tubular gas chromatography to determine these isomers.

2. ERUCIC ACID CONTENT OF CANADIAN RAPESEED (1971–1981)

Surveys carried out by the Canadian Grain Commission on the erucic acid content of Canadian rapeseed are summarized in Fig. 1. By 1974, the average level of erucic acid in the New Crop Survey had decreased to below 5%. Because of carryover of stocks from previous years, it took until 1976 for the average level of erucic acid in exported seed (carlot and cargo surveys) to fall below 5%.

Table VII shows the percentage of rapeseed which met LEAR specifications over the years 1971–1980. By 1977, virtually all shipments of rapeseed clearing Canadian ports had erucic acid levels less than 5%. Data were collected between 1971 and 1975 on individual rapeseed rail car samples both on a completely random (20%) sampling basis and on every rapeseed carlot designated as LEAR. In 1971, since most of the LEAR was grown under contract to grain companies, most of the rail shipments designated LEAR were actually LEAR. In the following years, although it was made mandatory for farmers to declare whether a delivery was HEAR or LEAR, the export trade made little effort to segregate by erucic acid level, especially in terminal elevators. No urgency was felt concerning the segregation of LEAR seed since buyers of Canadian rapeseed were not prepared to pay a premium for it. At the end of 1977, analysis of individual carlots was discontinued since data from New Crop and Cargo surveys were providing sufficient indication of the erucic acid levels.

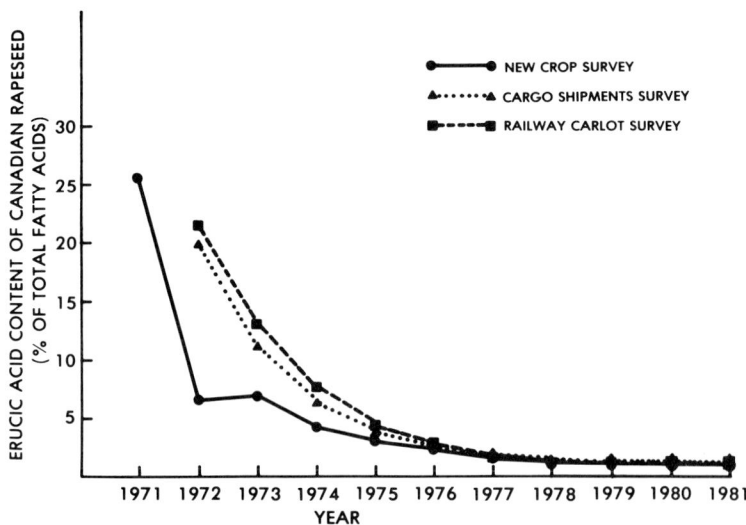

Fig. 1.   Yearly average of erucic acid in crop monitoring programs from the Canadian Grain Commission.

TABLE VII

Percentage of Samples of Rapeseed Containing Less than 5% Erucic Acid (1971–1980)

Crop Year	% of Samples with less than 5% erucic acid		
	New Crop Survey	Rail Carlot Survey[a]	Cargo Survey
1971	—	94	—
1972	74	74	4
1973	71	74	1
1974	81	77	26
1975	87	86	62
1976	92	—	81
1977	95	—	99[b]
1978	93	—	99[b]
1979	98	—	99[b]
1980	99	—	100
1981	97[c]	—	100[d]

[a] Carlots designated LEAR.

[b] Virtually 100% of the shipments were LEAR; sampling for these years included subsamples of which one or two tested slightly higher than 5%.

[c] Only 77 samples tested. Included two samples of HEAR grown for High Erucic Acid Market.

[d] To July 30, 1982.

3.   VARIATION IN ERUCIC ACID CONTENT OF CANADIAN RAPESEED IN
      RECENT YEARS

Although the average level of erucic acid in Canadian rapeseed has been reduced to about 1% in recent years, there is still some variation in these levels, especially in farm samples. *Brassica campestris varieties of LEAR have significantly higher levels of erucic acid than B. napus varieties.* Since *B. campestris* varieties have been grown in the north and west areas of the rapeseed growing region, there has been some localization of erucic acid values according to geographic growing area. Higher levels of erucic acid in *B. campestris* growing areas may also result from volunteer high erucic acid rapeseed since seed from this species may lie dormant for as long as 10 years.

Table VIII shows the average level of erucic acid in rapeseed from different areas of Western Canada. The trend toward higher values of erucic acid in rapeseed from northern and western areas is evident. Also, rapeseed from the Peace River region of Northern Alberta and British Columbia consistently has had erucic acid levels significantly higher than levels in seed from the rest of Western Canada. Farmers in this region of Canada continued to grow HEAR varieties for some time after farmers in the rest of Canada had converted to LEAR varieties.

TABLE VIII

Average Erucic Acid Content of Rapeseed Grown in Different Areas of Western Canada (1977–1981)

Regions[a]	1977	1978	1979	1980	1981[b]
Manitoba					
South	1.8	0.5	0.6	0.3	0.5
North	1.0	0.7	1.0	0.3	0.4
Saskatchewan					
Central	1.1	0.5	0.5	0.4	1.0
North	1.0	1.0	1.4	0.8	0.7
Alberta					
South	1.3	1.3	1.1	1.1	1.3
Central	1.5	1.5	2.1	1.5	1.3
North	4.1	4.4	3.3	2.7	2.5
All Areas	1.6	1.3	1.3	1.1	1.0

[a] Areas include provincial crop districts as follows:

Manitoba South	1,2,3,7,8,9,10,11
Manitoba North	4,5,6,12
Saskatchewan Central	5A,5B,6A,6B,7A,7B
Saskatchewan North	8A,8B,9A,9B
Alberta South	1,2,3
Alberta Central	4,5,6
Alberta North	7 including British Columbia

[b] Based on composite from 1032 samples.

TABLE IX

Relative Frequency Distributions of Erucic Acid in Western Canadian Rapeseed Grown and Exported from Western Canada 1977–1980

Crop year	New Crop Survey				Export Cargo Survey			
		% of samples with erucic acid content less than						
	n	5%	3%	2%	n	5%	3%	2%
1977	443	96	81	54	230	99	94	86
1978	490	93	89	83	315	99	97	87
1979	393	98	91	78	153	100	100	95
1980	453	99	94	84	79	100	99	93
1981	77	97[a]	95[a]	94[a]	75	100[b]	100[b]	99[b]

[a] Only 77 selected samples texted. Included two samples of HEAR grown for high erucic acid market.
[b] To July 30, 1982.

TABLE X

Erucic Acid Content of Oils Processed at Western Canadian Crushing Plants

	Erucic acid (% of total fatty acids)		
Plant	1979	1980	1981
A	1.4	0.8	0.7
B	0.7	0.9	0.8
C	0.3	0.3	0.7
D	1.7	1.4	1.3
E	1.9	1.9	1.4
F	—	—	1.7
G	3.7	3.5	3.1
H	1.5	1.3	0.9

There has been some consideration given to the possibility of lowering the maximum level of erucic acid allowed in rapeseed from 5% to a lower value, possibly 3%. Table IX shows the percentage of samples with erucic acid levels greater than 5, 3, and 2% for the crop surveys and export cargo surveys from 1977 to 1980. Data from this table suggest that it might be possible to lower the maximum level of erucic acid to 3.0%. The level of erucic acid in oils processed at Western Canadian Crushing plants (Table X) shows that even in 1981 at least one crushing plant in Western Canada would have difficulty meeting a 3% limit.

VI. CONVERSION TO CANOLA

In 1975, Tower, the first variety of rapeseed with low levels of glucosinolate and erucic acid, was licensed in Canada (Stefansson and Kondra, 1975). By 1981, six varieties of rapeseed with the above characteristics were licensed. In 1979, these varieties were given the trademark canola by the Canola Council of Canada. Canola types of rapeseed contain lower levels of glucosinolate in the meal than ordinary rapeseed (Table IV). The level of erucic acid in canola oil is equal to or lower than the level of erucic acid in LEAR oil (Table V). The only difference between canola and LEAR oils is that most canola oils contain substantially lower amounts of sulfur (from glucosinolates) than LEAR oils.

VII. HIGH ERUCIC ACID RAPESEED IN WESTERN CANADA

Although the Western Canadian rapeseed crops have completly converted to low erucic acid varieties, there is still a market for high erucic acid rapeseed oil to be used industrially. This market is for about 15,000 tonnes of oil, most of which is processed to provide erucic acid chemicals. Three high erucic acid rapeseed varieties have been grown to provide this oil, cultivars Turret, R-500, and Reston. Turret and Reston each contain about 40–45% erucic acid whereas R-500 contains 55–60%. Since minimum specifications for HEAR oil are 50% erucic acid, it is necessary to mix oils to produce a blend meeting specifications. The major problem with HEAR production has been the high glucosinolate level in the meal produced. This problem have been relieved somewhat by the introduction of Reston, which has a low level of glucosinolates.

The production of HEAR has been controlled by the Canadian Crushing Industry through contracts to meet expected demands for oil. Up to the present time only one company has been involved in this market and there has been no evidence of high erucic acid seed appearing in the edible oil export market. The Canadian Crushing Industry will continue to control this market as it is in their own best interests for the erucic acid content of Canadian rapeseed to remain low.

REFERENCES

Anonymous (1972–1981). "Prairie Grain Variety Survey." Canadian Co-operative Wheat Producers Limited, Regina.
Anonymous (1976). "National Standard of Canada. Rapeseed Oil, Low Erucic Acid, Crude and Crude, Degummed." Canadian Government Specifications Board CAN2-32-300M-76, Ottawa, Ontario.
Barrette, J. P. (1976). *J. Assoc. Off. Anal. Chem.* **59**, 855–858.
Craig, B. M., Mallard, T. M., Wight, R. E., Irvine, G. N., and Reynolds, J. R. (1973). *J. Am. Oil Chem. Soc.* **50**, 395–399.
Durksen, D. (1968–1971). "Seedtime and Harvest." Federal Grain Limited, Winnipeg, Manitoba.
Hougen, F. W., and Bodo, V. (1973). *J. Am. Oil Chem. Soc.* **50**, 230–234.
McLeod, A. D., ed. (1974). "The Story of Rapeseed in Western Canada." Saskatchewan Wheat Pool, Regina.
Rapeseed Association of Canada (1971–1975). "Annual Meetings of the Rapeseed Association of Canada." Canola Council of Canada, Winnipeg, Manitoba.
Statute Revision Commission (1978). "Consolidated Regulations of Canada," Vol. VIII, B.09.022, p. 6036. Government of Canada.
Stefansson, B. R., and Kondra, Z. P. (1975). *Can. J. Plant Sci.* **55**, 343–344.
Vaisey-Genser, M., and Eskin, N. A. M. (1979). "Canola Oil, Properties, Processes, and Food Quality," Publ. No. 55. Rapeseed Association of Canada, Winnipeg, Manitoba.

8

Rapeseed Crushing and Extraction

D. H. C. BEACH

I. A HISTORICAL REVIEW OF THE CANADIAN RAPESEED CRUSHING INDUSTRY

Rapeseed was first crushed in Canada by a small pressing plant in Moose Jaw, Saskatchewan during World War II. It ceased operation at or near the

High and Low Erucic Acid Rapeseed Oils
Copyright © 1983 by Academic Press Canada
ISBN 0-12-425080-7

end of the war. A post war pressing plant in Saskatoon was superseded by a small extraction plant in the late 1950s. A plant in Altona, Manitoba commenced the pressing of sunflower seed and rapeseed in modest quantities in the 1950s.

The full-press (pressing only) and solvent-only process were each known to have shortcomings. Published data favored a dual process of prepressing followed by solvent extraction for oilseeds with more than 20 or 25% oil. The management of both the Lethbridge, Alberta, and the Nipawin, Saskatchewan crushing plants were aware of these facts, and although there had been no Canadian experience of prepressing/solvent extraction of rapeseed, they both proceeded to adapt equipment that was then current in the milling of soybean and other oilseeds to the milling of rapeseed. This chapter will attempt to review some of the problems encountered by the industry to date and to outline the current Western Canadian rapeseed crushing technology.

II. PREPROCESSING

A. Cleaning

Conventional grain cleaning equipment has been adapted to the cleaning of rapeseed. The first stage of cleaning consists of scalping off any coarse material and then removing any cereal grains that may be present. The second and final operation selectively removes and separates the undersized as well as any oversized particles that escaped separation in the first stage. Air aspiration is employed at each stage. Rapeseed fragments usually form part of the undersized material. They are rich in oil and may justify a small re-cleaner for their recovery.

A major rapeseed cleaning innovation has recently been proved in practice and is now available to the industry. It consists of a single machine with air aspiration and only one moving element. It performs the functions of the former multiple units, requires less space, maintenance and surveillance. The quality of the product is excellent.

A good cleaning operation will provide rapeseed with not more than 1–2% foreign material. Cleaning of flax closely parallels that of rapeseed. Sunflower seed and soybeans are cleaned more easily and thoroughly than either flax or rapeseed.

B. Dehulling

The fibrous seed coats of both sunflowers and soybeans are easily stripped mechanically, leaving the seed core with essentially all of the oil and protein intact. The productive capacity of the process is thus enhanced, and both solvent loss and freight costs per unit of protein are minimized.

Since swine and poultry require low fiber for optimum growth and feed conversion efficiency the protein enhanced meals are preferred. However, since no satisfactory commercial method of dehulling rapeseed has been devised to date the rapeseed fiber accompanies the other seed constituents through the process emerging with the protein meal. Grinding of the meal followed by air classification has been only moderately successful in reducing the fiber content of rapeseed meal. The industry has not as yet installed any significant capacity for this purpose.

III. RAPESEED CONSTITUENTS AND THEIR POSSIBLE INTERACTION

The major constituents of rapeseed are oil, protein, fiber, and water. Some of the important minor constituents are free fatty acids, phosphatides (gums), enzymes (particularly myrosinase), and glucosinolates. The abundance of the major constituents remain relatively constant throughout processing with the exception of water, which is reduced.

There are two types of rapeseed grown in Canada, *Brassica campestris* and *Brassica napus*, both of which yield oil and meal with generally similar characteristics. Earlier varieties of each type contained significant quantities of erucic acid (in the oil) and glucosinolates (in the meal) both of which have been shown to be of dietary concern; erucic acid for humans and glucosinolates for animals. Canadian plant breeders have developed varieties of both the campestris and napus types which yield reduced levels of these substances.

Crushing and extraction leave the oil essentially unaltered. Some of the proteins may be denatured to some degree during processing. In the case of soybean meal, it is in fact toasted to denature a specific protein. In the case of rapeseed, toasting improves the palatability by altering some of the otherwise bitter tasting elements of the meal. Glucosinolates may also be transformed into other forms, either benign or detrimental, depending on the process conditions.

Glucosinolates in the presence of water and a suitable catalyst and at an appropriate temperature may undergo transformation. The indigenous enzyme myrosinase unfortunately serves the catalytic function only too well once the seed is crushed. The architectural destruction of the seed required to produce good oil release inevitably results in intimate contact of the glucosinolates, myrosinase and water, and the reaction then needs only an appropriate temperature and period of time to occur. The reaction rate starts to become significant at temperatures higher than 50°C and appears to terminate when the temperature reaches about 85°C. This termination is attributed to thermal inactivation/destruction of the enzyme.

Myrosinase resists deactivation when seed moisture levels are low and has been known to survive processing at temperatures of 88°C–93°C when seed moisture levels were 6% or less. For this reason the industrial practice has been to quickly raise the temperature of the crushed seed containing adequate moisture (7% minimum) beyond the deactivation level (85°C). Thereafter, the temperature may be increased or decreased and moisture lowered as required.

The glucosinolate hydrolysis products consist of many complex compounds, two of which are of particular importance: isothiocyanates and oxozolodinethiones, both of which are oil soluble. Since the oil comes into contact with the hydrogenation catalyst prior to the admission of hydrogen and since in any event these culprits are adsorbed preferentially by the catalyst, excess catalyst is required solely for their immobilization. Reactor turn around time may be lengthened as well. Some of the glucosinolate derivatives may also remain in the meal and produce unwanted dietary effects in animals and poultry. No method has been found to date that is both practical and economical for the removal of these compounds from the meal.

The advent of the "double low" seed varieties (varieties low in both erucic acid and glucosinolates) would be expected to have allowed some relaxation of the stringent deactivation procedure. However, since small amounts of glucosinolates still remain and since some hitherto obscure glucosinolate forms are now recognized, complete and total enzyme deactivation appears to still be advisable. The cottonseed processor has a similar problem with gossypol which must be detoxified by heating prior to expressing the oil. The deactivation of myrosinase and detoxification of gossypol by heating are variants of the cooking process, cooking being common to the processing of most oil seeds.

IV. PROCESSING

A. Flaking

The objective is to massively deform the seed structure by crushing and shearing and to leave the crushed seed with a large surface/volume ratio (a thin flake). Rapeseed flakes thinner than 0.2 mm are very fragile, whereas flakes thicker than 0.3 mm process less satisfactorily. Therefore, the general practice has been to flake the seed to within these limits. (Early experience with the "double low" varieties was frustrating; the seed "shattered" instead of "flaking," notwithstanding the seed temperature being moderate.) However, this problem has been solved by modifying the moisture content and altering the temperature prior to flaking.

The evaporative cooling that accompanied attempts to heat the seed with hot air not only suppressed the temperature rise, but the moisture loss further embrittled the seed. Indirect heating has provided the essential independent control of seed temperature and moisture and has been retrofitted by the author to one plant and included in the design of a recent plant with gratifying results. The warmed (32°–40°C) seed is sufficiently plastic to permit rolling into a thin flake without shattering.

Some "oiling out" on the flaking rolls has been observed when the seed was preheated to 40°C or higher, usually causing severe machine vibration. One processor has reported this condition to have occurred when the seed was only 32°C. Another processor is reported to be heating the seed to 82°C prior to flaking but is using substantial pressure on the roll scrapers. The enzyme survival or destruction after this process is as yet unknown. Some degree of preheating the rapeseed prior to flaking seems likely to become a general practice.

B. Cooking

Cooked oilseeds release their oil more readily than uncooked seed. The precise reasons for this fact are obscure but probably relate to the properties of the coexistent proteins as well as the form, size, and viscosity of the microscopic or submicroscopic oil units. Water also has an effect on oil release, perhaps by promoting, in conjunction with heating, the denaturing of proteins to types with lower adsorbtion properties, or by altering surface tension.

For reasons previously discussed, rapeseed flakes with 7–9% moisture should be rapidly heated through the 50°–80°C temperature range. The high erucic acid varieties have responded to 15–30 min of cooking at 100°–107°C. The current "double low" varieties usually process better at about 85°C.

On completion of cooking, moisture levels of 2.75–3.5% usually provide for satisfactory pressing and contribute to press cake of good quality. Cottonseed has been found to require rapid heating to 88°C with up to 15% moisture or 104°C with 12% or more moisture, followed by cooking and drying at temperatures as high as 132°C. Gossypol is obviously more resistant to denaturing by heating than is myrosinase. Figure 1 shows a cutaway view of a typical stack cooker.

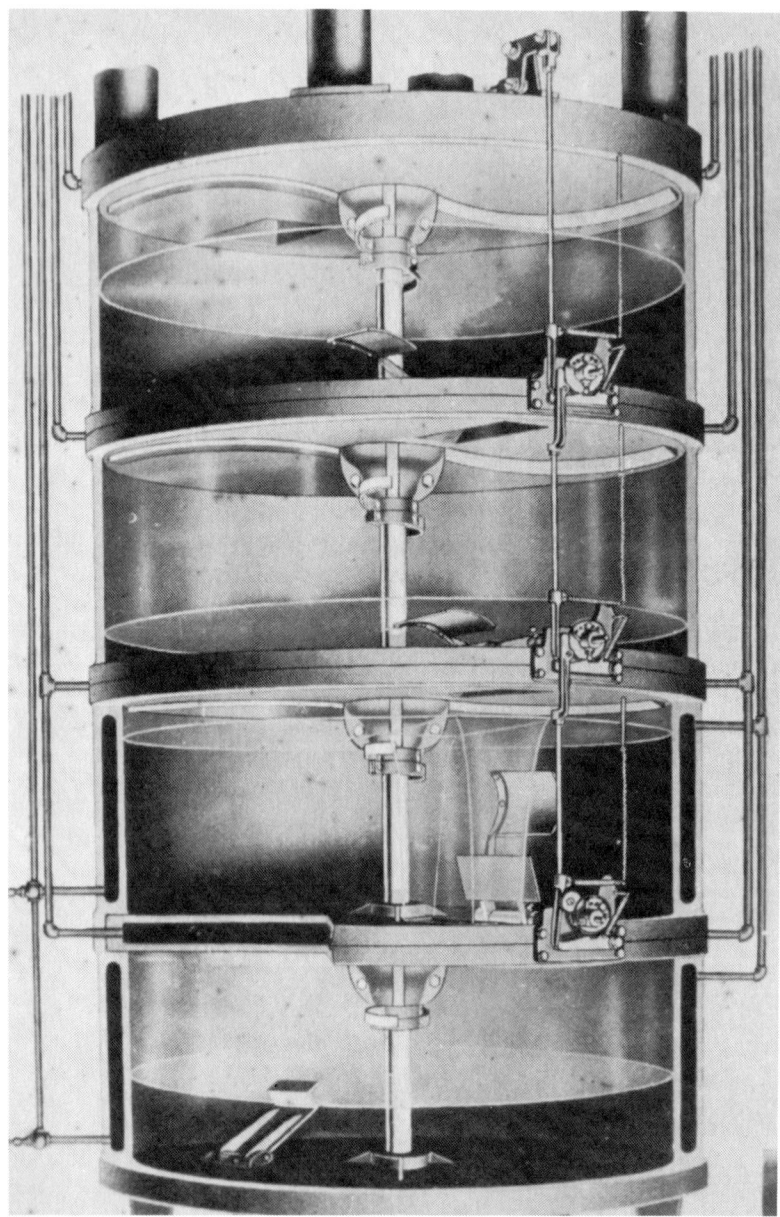

Fig. 1.   Phantom view of four-high stack cooker. Courtesy of the French Oil Mill Machinery Co.

Fig. 2. Screw press with one half of the barrel cage removed to show the worm sections. Courtesy of Krupp Canada Inc.

C. Pressing

Cooked rapeseed flakes with about 3 to 4% moisture and 44 to 45% oil are passed through a continuous screwpress which readily expels 75% or more of the oil, reducing the cake oil to about 16%. This mild pressing causes only a modest rise in temperature which, unlike full pressing (3–5% residual cake oil), does not threaten the stability and quality of the proteins. Good cake is spongy and permeable but does not disintegrate unduly with conveying. Figure 2 shows a screw press with one-half of the cage barrel removed to show the screw section.

D. New Developments in Presses

A screwpress has been introduced recently combining the functions of flaking, cooking, and pressing (prepressing). This machine repeatedly crushes and shears, compresses, and then relaxes the flow of seed. The mechanical energy input is supposed to provide for adequate heating. It is designed only for prepressing. The general observation seems to be critical of the high proportion of "foots" expelled with the oil and which must be dealt with either by recycling (thereby reducing net throughout) or by the addition of subsequent specialized "foots" presses. The crushing industry is watching with interest and caution.

Apprehension focuses on: (a) further developments that will hopefully decrease the yield of "foots" and (b) confirm moderate and acceptable service life of press components, particularily the worms and cage bars.

E. Oil Settling and Filtering

The expelled press oil with some entrained solid matter is gravity settled in a screening tank. The settlings are continuously dredged off, drained and (a) recycled back through the cooker for repressing or (b) repressed in a "foots" screw press, a specialized version of the main screw press. The repressed ("foots" press) cake when produced is directed to extraction along with the main cake stream. The expelled oil from a foots press is recycled back to the screenings tank for resettlement of the suspended fines or (c) the dredges may go directly to extraction along with the main cake stream provided their inclusion does not adversely affect the overall extraction operation.

The settled oil is continuously drawn off from the screenings tank. The remaining suspended fines in the oil are removed by either filtering or centrifuging. Filtering is common. The totally enclosed multiple screen type of filter is favored. The plates commonly consist of double sided No. 80 stainless steel screens precoated with the fines themselves. Many units are powered open and closed for cleaning. The plate and frame filter press is no longer specified for this service since it is highly labor intensive and replacement filter cloths or papers represents a continuous and significant operating cost. The filtered press oil is blended with the extracted oil. Since its phosphatide (gums) content is lower than that of the extracted oil, minimum degummed oil standards can often be met by degumming only the extracted oil and blending it with the undegummed press oil.

F. Extraction

The basis of solvent extraction is the interfacing of the oil or a rich oil/ solvent solution in the flake or cake crumble with a rich solvent/oil solution.

If the planar thickness of each phase were of monomolecular order, equilibrium would be prompt, resulting in a single uniform phase. Since the cake particles are finite, diffusion of solvent into the oil in the cake particle and diffusion of oil from within the particle out into the solvent/oil solution requires time. After a limited time period the somewhat strengthened (oil rich) miscella at the particle surface is replaced by miscella richer in solvent which reelevates the diffusion rate. This is accomplished by staged countercurrent movement of the cells of press cake and the solvent. Figure 3 is a

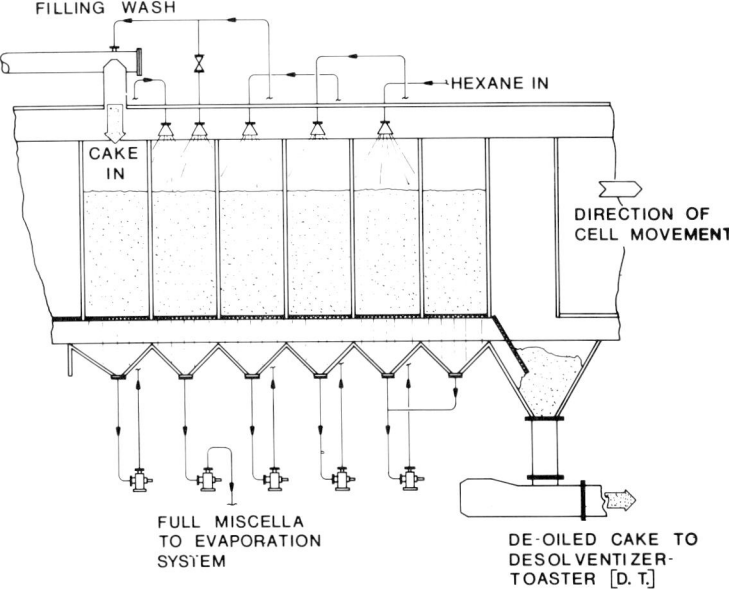

Fig. 3. Schematic diagram of a "staged" countercurrent extraction process. Courtesy of Beach-Doodchenko & Associates.

schematic diagram of a staged counter current extraction system. Most commercial extractors are layout variations of this concept.

The ideal cake particle would be one with maximum surface/volume ratio. The practical approximation to this ideal is the thinnest flake that is still durable enough to survive the mechanical and hydraulic abuse that the commercial process imposes on it. Soybean flakes retain their integrity quite well throughout the extraction process, due in part to the fact that they have not been previously severely crushed, compressed and sheared in a screw press. The author's experience has been that rapeseed cake flakes (flakes that are produced by passing broken-up press cake through a set of flaking rolls) are fragile, disintegrate severely, and seriously restrict solvent and miscella percolation through the bed of furnish in the extractor cell. Soybean flakes provide a higher rate of percolation and for these as well as other reasons, yield their oil to extraction more readily and completely than does the severely fragmented rapeseed cake.

Whereas continuous (rather than "staged") countercurrent flow systems of both furnish and solvent have been devised, none has come into general use. Modified batch type extraction is dominant wherein the furnish to be extracted is charged into discrete cavities or cells and repeatedly and in turn saturated and then flushed with increasingly solvent-rich solutions of solvent and oil (miscella).

Miscella percolation rate in itself and/or the number of miscella stages do not adequately predict extractor performance. Time is also a major factor. As many as seven or eight successive wash stages have been used; however, it has been found that thoroughly saturating the bed and then allowing it to "soak" (diffusion time) followed by flushing the enriched miscella from the surface of the particles and replacing it with a leaner miscella, repeated two or three times, results in extraction efficiencies as good as the former practice of using more stages with their attendant shorter soaking periods. The reduced number of pumps saves energy, capital and maintenance costs.

One school of thought favors a shallow (20–30 cm) bed of cake crumbles (or flakes in the case of soybeans or sunflowers) whereas another school believes that deep beds are not an obstacle to good extraction, are more compact and provide more capacity per unit of building volume. Provided the bed permits an adequate percolation rate the deep bed extractor seems to be equally as successful as the shallow bed type.

Deep bed extractors appear to exceed the shallow bed type in aggregate installed capacity in America. Several configurations of the deep bed extractor are manufactured including rotating cells with stationary or moving cell bottoms (Fig. 4 shows a cutaway view of an extractor of this type), or stationary cells with rotating filling chute, wash heads, miscella collectors, and discharge chute.

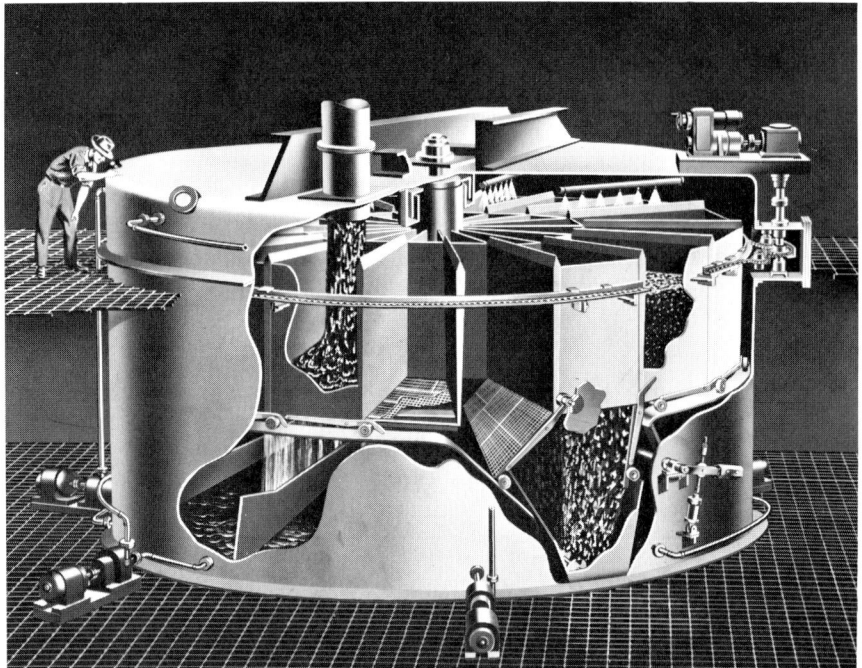

Fig. 4. Phantom view of "Rotocel" extractor. Courtesy of Dravo Engineers and Constructors, Solvent Extraction and Biochemicals.

The "planar loop" extractor provides for shallow to intermediate bed depths and is offered by one American and several European companies. In this type the extractor compartments move along a path in a vertical plane. Chains driven by sprockets are used to move the compartments along the path (around the loop). Vertical tower types of these extractors have also been built and operated successfully; however, building and maintenance requirements appear to have prevented their continued use. The horizontally elongated type of loop or belt extractor is offered by two European and one American supplier. The vertically elongated configuration appears to have been abandoned.

The common solvent is normal hexane which readily dissolves vegetable oils in all proportions. It has an appropriate boiling temperature and its relatively low latent heat of vaporization helps conserve energy in the distillation stages. Hexane is noncorrosive to metals, does not react chemically with the oil it dissolves, and is chemically stable under the process conditions. It is available in quantity at moderate cost. Since it is highly flammable the process equipment must be designed, maintained and operated outside the range of flammable air/hexane mixtures. The air/hexane mixture throughout the process at normal operating temperatures and conditions is well beyond the flammable limit. Hexane leaks may and will occur from time to time and will inevitably result in an explosive zone existing somewhere between the source of the leak and the zone of hexane-free atmosphere. This zone of high flammability is usually within the building enclosure itself which obviously dictates extreme design as well as operating safety measures. All electrical equipment must be explosion proof and no surface within the boundaries of the extraction operation should ever approach the ignition temperature of hexane/air mixtures. Good design and a high level of operator discipline are necessary for the safe operation of a hexane solvent extraction plant.

Although elevated temperatures reduce oil viscosity and enhance diffusion, hexane vapor pressure limits the practical operating temperature of the extractor and its contents to about 55°–60°C. Higher temperatures and the consequent higher vapor pressures unduly increase the volume of vapor which the recovery systems must capture and recycle. Futhermore, if the cake temperature is at or near the boiling temperature of the solvent, a vapor phase may occur at the interface between cake fragment and solvent (miscella), effectively thwarting liquid diffusion.

Except on start-up, the solvent does not normally require heating since the hot cake from the prepress process often provides more than sufficient heat to maintain the required extraction temperature. It has, in fact, been found necessary to allow the cake to dissipate as much heat as possible while in transit from the preparation process to the extraction plant. In addition, subcooling the solvent below the ambient temperature of the solvent work tank has been used successfully to control extractor temperatures. Notwithstanding the reduced cake temperatures resulting from the generally lower temperatures used to cook the current "double low" rapeseed varieties, some cooling of cake and/or subcooling of solvent is still found to be necessary in some plants.

G. Extraction Operation

The extraction operator can, within limits, control bed depth, solvent input and to some degree its (solvent/miscella) distribution within the extractor as well as extractor temperatures. However, the operator must accept the quality and quantity of press cake produced in the preparation plant. The quality of extracted product is highly dependent on a good press cake. Poor quality cake commits the extraction plant to poor oil recovery with high solvent loss and even high levels of solvent in the meal, both of which are beyond the extraction operator's power to effectively control.

Accidental entrainment of water with the solvent entering the extractor is an occurrence to be rigorously avoided. This can happen due to carelessness and/or inadequate design of the solvent/water separator. The results are restricted percolation, high residual oil and solvent in the extracted meal and in severe cases "blinding off" of the perforated cell floors. This latter condition may necessitate a complete plant shutdown and "*in situ*" floor cleaning, with substantial lost time.

Another accident to be avoided is an overflow of miscella into the extracted meal. The results are overloading of the desolventizer and higher than normal residual solvent in the meal. In addition the solvent recovery system is subjected to an overload with attendant higher than normal losses.

H. Distillation

Separation of the oil and solvent is accomplished by conventional distillation methods. Most of the solvent is recovered for repeated use. Distillation and recovery of the solvent requires energy and utilities; consequently, the operating objective must be to concentrate as much oil as possible in the least practical amount of solvent.

The full miscella leaving the extractor is usually directed to a surge tank from which it is pumped at a steady rate to the distillation equipment. Solvent evaporation is normally effected in three stages. The first and second use vertical shell and tube rising-film evaporators. The first stage, often termed "first effect evaporator" receives the moderate temperature (80°C) hexane and water vapors from the meal desolventizer on the shell side as a source of heat. The full miscella, consisting of about 20%–25% oil by weight, enters the tubes at about 50°C. As part of the hexane in the miscella is progressively evaporated, and as it rises in the heated tube, the volume/unit mass continuously increases sweeping the remaining liquid up the wall of the tube. The vapor and liquid phases are then separated, the vapors proceeding directly to water cooled condensers. The oil enriched liquid (miscella) is reheated in a steam heated exchanger and passes on to a second rising film evaporator heated with steam on the shell side. The hexane remaining in the twice enriched miscella now accounts for only about 5% of the miscella mass and is removed by countercurrent steam stripping in a disk and doughnut vacuum tower under a pressure of about 100 torr. The oil temperature at withdrawal should be well above 100°C, 110°C being common to ensure that the oil is thoroughly devoid of moisture and essentially solvent free.

I. Degumming

The hot oil may be cooled to 82°C and proceed directly to degumming by first being thoroughly mixed with 1% or 2% of hot (82°C) condensate water. Some processors also add a small (variable) amount of phosphoric acid to further promote the precipitation and coagulation of the gums. This practice is common where the degumming and alkali refining steps are combined. The oil-soluble gums hydrate and precipitate, are centrifuged off and pumped to the desolventizer to be mixed with the de-oiled meal in which they act as a dust suppressant in the meal and as a binder in the meal pellets. The "gums" also contribute some nutritional value to the product. The degummed oil being "wet" is heated to about 105°C and is sprayed into the top of a vacuum tower; the water "flashes" off as vapor, and the dry oil is drawn off, cooled, and pumped to storage. Since press oil usually contains substantially less phosphatide material than does solvent extracted oil, the industry requirements of not more than 0.5% acetone insolubles (gums) can often be met by degumming only the extracted oil and blending it with the undegummed press oil.

The residual oil in the extracted cake is usually targeted to be 1% or less. This objective is consistently met in the soybean industry but is difficult to attain with rapeseed, perhaps due to the fact that supplying the extractor with cake in flake form seems unattainable. Analysis of the extracted cake prior to desolventizing and of the desolventized meal for residual oil will show a consistent discrepancy, with the "apparent" residual oil in the desolventized meal being more abundant. The additional increment of petroleum ether-soluble matter extracted has been shown to be largely nonglyceride in nature. The reasons for this material to be extracted subsequent to desolventizing are not clear. In any event a true assessment of extractor effectiveness must be done on extracted flakes prior to desolventizing notwithstanding the practical difficulties of sampling at this point in the process.

J. Meal Desolventizing

The residual solvent in the meal as it leaves the extractor is usually about 25%–35% by weight. The desolventizer–toaster heats the meal and solvent and vaporizes and steam strips the solvent from the meal at atmospheric pressure.

The heating is accomplished by agitating the solvent-wet meal as it passes over a series of superimposed steam-heated trays. Live steam is blown upward through the bed of meal on one or two of the uppermost trays. Since the boiling temperature of hexane at standard pressure is some 69°C, most of it vaporizes readily. Most of the heat required is supplied by the condensing steam which saturates the meal with moisture in place of the displaced hexane. Steam, in addition to that which condenses, provides a partial pressure effect on the evaporating hexane, lowering its effective boiling temperature and enhancing the removal of hexane from the meal. The stripping steam is essential in addition to indirect heating since the system must function at atmospheric pressure; there is no practical way of introducing this solid material into and out of a vacuum environment.

Total direct contact steam averages about 1 kg of steam for each 12 to 14 kg of raw material (rapeseed) processed. Rapeseed meal is difficult to thoroughly desolventize; residual levels of hexane are typically 300 to 1200 ppm. This problem merits research for both safety and economic reasons. Moisture laden (25% or more) desolventized meal may be dried in the lower stages of the desolventizer in small plants. Larger plants use a separate rotating kiln-type steam tube drier to reduce the water content to 12–14%, which is then followed by an air cooler in which the moisture is further reduced to less than 12%, the evaporating moisture absorbing some of the heat.

Finished meal is ground to reduce any oversized particles before shipping to domestic users. Export meal is always pelletized.

The vapor stream from the meal and oil distillation systems consists of solvent vapor, water vapor (from the stripping steam) and air that has entered the system entrained in the voids of the press cake. Some additional air unavoidably gains entry into the vacuum systems. The vapor stream from the final condenser consists of air saturated with solvent and water vapors. The solvent vapors are selectively scrubbed in a packed tower by a counterflow of special mineral oil. The mineral oil is recycled after being heated, stripped of solvent, and cooled. The hexane vapor remaining in the vent stream should be below the lower explosive limit (1.3% by volume). Nevertheless, overall rapeseed plant losses per tonne flow through the extractor are typically much higher than they are for soybeans. This problem should be researched and, if possible, overcome.

ACKNOWLEDGMENTS

The author gratefully acknowledges the contributions of Mr. J. Reynolds, Alberta Food Products, for historical contributions; Mr. F. Olfert, CSP Foods, for manuscript suggestions; and Dr. C. G. Youngs, National Research Council, for reviewing the manuscript and providing helpful suggestions.

9

The Commercial Processing of Low and High Erucic Acid Rapeseed Oils

B. F. TEASDALE AND T. K. MAG

High and Low Erucic Acid Rapeseed Oils
Copyright © 1983 by Academic Press Canada
All rights of reproduction in any form reserved.
ISBN 0-12-425080-7

I. INTRODUCTION

A. Objectives of Processing

Rapeseed oil constitutes a substantial proportion of the supply of edible oil to consumers in many parts of the world. It includes the original HEAR (high erucic acid rapeseed) oils, LEAR (low erucic acid rapeseed) oils and canola oils (canola is the name adopted by Canadians for the oil derived from the new rapeseed low in both erucic acid and glucosinolates). Since there are great differences in the kinds of edible oil products which these consumers prefer, the type and degree of processing that are applied to the oils vary markedly from country to country, and even within different regions of some countries.

Using Western Europe as an example, Gander (1976) points out that in the north, "solid" fat products are preferred while in the south far more liquid oil is used. North America, Australia, and parts of India are examples of other regions where "solid" fat products predominate. Lesieur (1976) lists Japan, Brazil, and, again, parts of India as examples of where liquid oil is preferred.

In addition, there are subdivisions within the "solid-fat" and "liquid-oil" categories. Vanaspati users demand a "grainy" product with the texture of ghee, whereas consumers of margarine and shortening insist on a smooth, grain-free product. Some of the liquid-oil users prefer that the oil be subjected to very minimal processing whereas others require a light-colored, bland oil. For example, consumers in India and Pakistan who relish their cold-pressed, raw oil find that the fully refined deodorized, canola oil made for the Canadian market is too bland.

Government regulations also influence consumer preferences. North American regulations have inhibited the manufacture and sale of margarine to a greater extent than has been the case in Europe. As a result even today shortenings are more popular in North America than they are in Europe.

Finally, it must be recognized that those products designed for household use often do not fit the requirements of the large, very diversified, commercial and industrial markets. As a result of the specialized requirements of the various segments of these markets, a vast array of "tailor-made" edible oil products are manufactured. This is particularly true in the case of shortenings in Canada and the United States.

B. Comparison of Low Erucic Acid Rapeseed Oil with Other Vegetable Oils

Table I gives the ranges for the major fatty acids of several common vegetable oils. Based on the values given in Table I the following comments can

TABLE I

The Range of Fatty Acid Composition (%) for Commonly Used Vegetable Oils

Fatty acids		Canola oil[a]	HEAR oil[b]	Soybean oil	Sunflower oil	Corn oil
Palmitic	(16:0)	2–5	3–5	10–11	5–7	10–12
Stearic	(18:0)	1–3	1–3	4	3–5	2–4
Oleic	(18:1)	53–58	18–27	23–26	20–23	26–28
Linoleic	(18:2)	19–23	14–18	50–54	64–68	55–58
Linolenic	(18:3)	8–12	8–9	7–9	Tr–1	Tr–1
Gondoic	(20:1)	1–2	12–14	—	—	—
Erucic	(22:1)	tr–4	25–45	—	—	—

[a] Canola oil is the oil extracted from the new varieties of rapeseed low in erucic acid and gluconsinolates.
[b] HEAR, high erucic acid rapeseed oil.

be made: (1) canola and HEAR oils are lowest in palmitic and stearic acids, (2) canola oil has a high level of oleic acid, (3) canola and HEAR oils contain about 20% linoleic acid, (4) canola, HEAR and soybean oils contain appreciable amounts of linolenic acid, (5) canola oil contains small amounts of gondoic and erucic acids whereas HEAR oil has much higher levels of these acids, and (6) the above comments for canola oil apply equally for LEAR (low erucic acid rapeseed) oil since the oil from these two rapeseed types is the same.

Wettström (1972) points out that the linoleic and linolenic acid moieties of rapeseed oil are located preferentially in the more protected 2-position of the triglycerides, whereas in soybean oil the linolenic acid is distributed nearly randomly and linoleic acid has only a slight preference for the 2-

TABLE II

Nontriglyceride Constituents in Selected Vegetable Oils

Constituents	Canola oil	HEAR oil	Soybean oil	Sunflower oil	Corn oil
Free fatty acids (%)	0.4–1.0	0.5–1.8	0.3–1.0	0.5–1.0	0.5–1.8
Phospholipids (%)					
Nondegummed	Up to 3.5	Up to 3.5	Up to 4.0	0.2–0.7	1–2
Degummed	Up to 0.75	Up to 0.6	Up to 0.3	—	—
Unsaponifiables (%)	0.5–1.2	0.5–1.2	0.5–1.6	0.3–1.5	0.5–2.0
Chlorophylls (ppm)	5–25	5–70	Nil	Tr	Nil
Sulfur[a] (ppm)	3–10	5–25	Nil	Nil	Nil

[a] Method according to Daun and Hougen (1976).

position. Several authors have reported that a similar preferential position-ing prevails for LEAR oil (Appelqvist, 1971; Litchfield, 1971; Rocquelin et al., 1971; Sergiel, 1973). As a result of their particular triglyceride composi-tion HEAR and LEAR oils can be expected to have even more resistance to oxidation than is indicated by their fatty acid composition alone. It is gener-ally agreed that any of the tests applied to assess the stability of an oil, such as the Active Oxygen Method or Schaal Oven Stability, are of only limited value. The true test of acceptability rests with the consumer. (For information concerning the practical experience in Canada relative to the acceptability of canola oil see Section II,D.)

Table II gives typical values for nontriglyceride constituents present in five crude and water-degummed vegetable oils.

II.   UNIT PROCESSES

A.   Refining

The crude, water-degummed oil, as received at the refinery from the ex-traction plant requires the removal of free fatty acids and phosphatides to very low levels before other processes such as bleaching, hydrogenation, and deodorization can be carried out efficiently. Generally, the process of refining used is to contact the oil with an alkali, usually sodium hydroxide, and to separate the resulting aqueous soap phase, together with precipitated phosphatides and other oil-insoluble materials, from the oil. An alternative to this approach is the use of "physical" refining, in which the free fatty acids are removed from the oil by steam distillation after a suitable pretreat-ment of the oil to remove phosphatides and other, heat-sensitive materials. It is expected that this approach to refining will eventually supersede alkali refining because of the lower capital cost, possibly lower processing costs, and because it avoids the processing of soapstock, which is a by-product of alkali refining.

1.   ALKALI REFINING

Alkali refining consists of five main steps: (1) contacting the oil with phos-phoric acid, (2) neutralizing free acidity in the oil with sodium hydroxide solution, (3) separating the soap phase from the oil, (4) water-washing of the oil, and (5) drying of the oil. The purpose of the phosphoric acid treatment of the oil before contacting with alkali is to help precipitate nonhydratable phosphatides and some colored material, notably chlorophyll and related compounds (Ohlson and Svensson, 1976). Also, traces of prooxidant metals such as iron and copper are more efficiently removed (List et al., 1977). Advantages are achieved in lower refining losses, lower bleaching clay usage, and better flavor and flavor stability of the oil after deodorization. Acids other than phosphoric may be used, such as citric acid or oxalic acid, but it has been shown by Ohlson and Svensson (1976) that these other acids are not effective in the removal of chlorophylloid substances.

The neutralization step is usually carried out with sodium hydroxide solution of 2–3 N strength. The amount of solution is based on the free fatty acid content of the oil and the amount of phosphoric acid used in the pretreatment (two equivalents per mole); also, an excess of about 10% to 20% of the amount required to saponify the free fatty acids is usually added to ensure adequate removal of impurities.

The soap phase, which contains the precipitated, nonhydratable phosphatides, and other oil-insoluble material, is then separated from the oil phase. The oil is water washed to reduce the soap concentration to less than 50 ppm. A small amount of citric acid or phosphoric acid may be added to the washed oil to "split" remaining traces of soap. This improves the efficiency of subsequent bleaching. The oil is then dried. Most of the industry uses continuous process equipment, but batch process installations are also still in use.

A typical analysis of an alkali-refined oil compared to the crude oil is shown in Table III. The data show that free fatty acids, phosphorus (phosphatides), and soap are reduced to very low levels. Sulfur compounds are not removed completely, but at levels of 2 to 3 ppm, they do not present a serious problem in subsequent processing. Chlorophyll is reduced to some extent in refining as pointed out in connection with the use of phosphoric acid.

a.   *Batch Alkali Refining.*   In small plants, batch alkali refining still has significant advantages, since the capital investment is low, there are few maintenance problems, and changes in oil stock are very easy. Disadvantages are that a properly refined oil is difficult to achieve and losses are generally higher than in continuous processes. This is because there are

TABLE III

Analysis of Canola Oil

Constituents	Crude water-degummed	Alkali-refined
Free fatty acids	0.4–1.0%	0.05%
Phosphorus	150–250 ppm	0–5 ppm
Sulfur	3–10 ppm	2–7 ppm
Chlorophyll	5–25 ppm	0–25 ppm
Soap	—	0–50 ppm
Moisture	0.05%	0.05%

limitations in the process temperatures that can be used, intensity of agitation, and particularly in soapstock separation, which is by gravity only, rather than by centrifugal force. Also, about 24 hr are required to complete a refining cycle.

A typical batch refining installation is described by Norris (1964). Briefly, it consists of an open tank, or kettle, holding about 30 tons of oil (one tank car). The kettle is equipped with a two-speed paddle agitator, steam coils, and a cone bottom. Sparging pipes are arranged in a grid in the head space of the tank for the addition of the alkali solution and wash-water. Agitator speeds are usually 40 and 8 rpm.

A typical batch-refining cycle is as follows: (i) *Loading and deaeration:* 30 tons of oil are loaded into the kettle and the temperature adjusted to about 30°C while the oil is agitated. The agitator is then stopped and the charge left to deaerate for several hours, often overnight. (ii) *Phosphoric acid pretreatment:* 0.2–0.5% of 85% concentrated phosphoric acid is added to the oil while agitating at high speed. Heating may be done up to 70°C. Contact time of 30 min is allowed at the maximum temperature chosen. (iii) *Neutralizing:* Sodium hydroxide solution of 2–3 N strength is added while agitating at high speed for 10–15 min. (iv) *Soapstock agglomeration:* After 10–15 min of fast agitation, the speed of agitation is reduced, and, if neutralization was carried out at a temperature below 70°C, the charge is now heated to 70°C to facilitate agglomeration of small soapstock particles to larger ones, and to partially melt the soap to reduce entrainment of neutral oil. (v) *Soapstock settling.* Agitation is stopped to allow the soapstock particles to settle. At least 1 hr and often several hours are required to achieve good separation. The soapstock layer is then withdrawn from the bottom of the tank. (vi) *Water washing:* Four washes are required using 5 to 10% of hot water (based on the oil charge) in each. After the last wash, about 200 ppm of citric acid or phosphoric acid may be added to "split" traces of soap. (vii) *Drying:* The oil is heated to 105°C with high agitation until dry. Several hours are required.

b. *Continuous Alkali Refining.* Most oil is refined by continuous alkali refining. Installations are relatively high in capital cost, mainly because of the centrifuges required. Figure 1 shows a typical installation.

In operation, the oil is heated to 90°–95°C and contacted with 0.05–0.3% of phosphoric acid in a high intensity mixer and small contacting vessel to provide about 2 min residence time. This is followed by neutralization with 2–3 N sodium hydroxide in a high intensity mixer, and separation of the soapstock phase from the oil in a centrifuge. Sodium hydroxide solution and oil are in contact for only about 20 sec before centrifuging (short-mix system). This is in contrast to the U.S. practice in refining soybean oil, which calls for contact times of up to 15 min (Wiedermann, 1981)

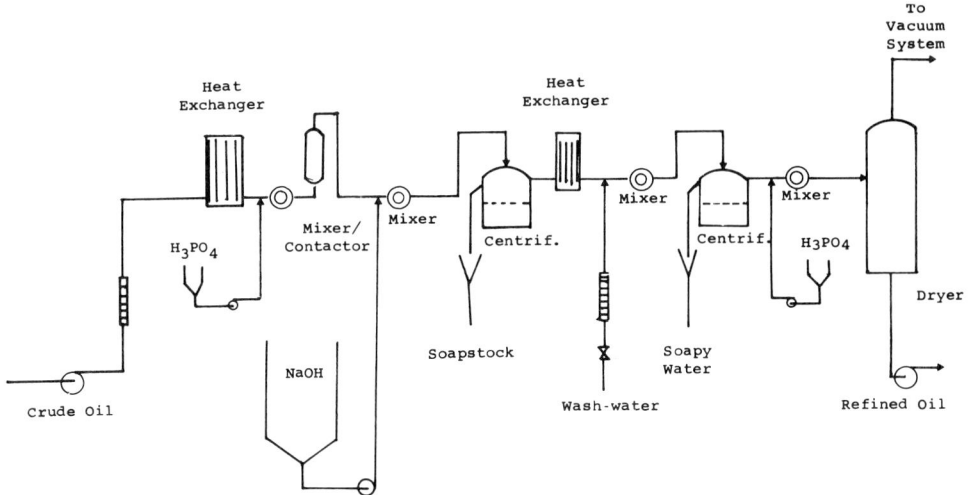

Fig. 1. Continous alkali-refining system. Adapted from Westfalia Separator AG.

and temperatures of 33°–75°C. The reason for this latter practice apparently derives from the refining of nondegummed soybean oil.

The centrifuged oil is brought into contact with about 10% of 90°C soft water in a high intensity mixer and centrifuged again to remove the wash-water. In some operations, two washing stages are used with about 5% of water in each. If a single wash is performed, the washed oil is contacted with a small amount of citric acid or phosphoric acid solution (200 ppm) to "split" remaining traces of soap. The oil is then continuously dried in a vacuum dryer.

As already pointed out, the centrifuges required in the process are quite expensive. They also can be a serious limitation if there is a need for frequent cleaning. Developments in centrifuge technology have been concerned mainly with this problem. In the past, hollow-bowl centrifuges were often used. They were relatively inexpensive, but required frequent shutdown for cleaning. Disc-bowl machines for larger capacities and equipped with pumps for discharge of both heavy and light phases, and the use of bowl-flush water at about 10% of the oil flow alleviated this problem somewhat. The present trend is to utilize split-bowl disc centrifuges, which allow periodic opening of the bowl during operation to discharge heavy sediment collecting at the bowl wall. No bowl-flush is required. Bowl opening for discharge is fully automated and the frequency of opening can be varied by a timing device. It is advantageous to use this type of machine for both soapstock and wash-water separation. Further, clean-in-place arrangements are offered by centrifuge manufacturers, which make it possible to reduce the frequency of centrifuge disassembly for cleaning from once, or even more often, per week to once every 2–3 months.

c. *Acidulation of Soapstock.* The aqueous soapstock phase containing the fatty acid soaps, precipitated phosphatidic material, and entrained neutral oil as the main fatty constituents, is usually acidulated with sulfuric acid to recover the fatty material from the water phase. Batch acidulation as well as continuous acidulation is practiced. The water phase arising in the process is highly acidic and often highly contaminated with fatty material and other organics contributing to a high biological oxygen demand. The fatty phase (acid oil) is utilized as raw material in the production of technical fatty acids.

Batch acidulation installations consist of a series of vats, often made of wood staves or lead-lined steel, and more recently also of fiberglass reinforced plastic (FRP), holding about 40–50 tons of soapstock. The vats are equipped with a steam line for heating with live steam. Soapstock and sulfuric acid are charged to the vats by overhead piping of mild steel; discharge from the vats is through a bottom outlet and copper piping. Acid water goes to the sewer after a neutralization stage, and the acid oil to a collection tank for water washing and storage.

In a typical batch operation, soapstock and wash-water from a refining run is charged to a vat and heated to about 70°C, if required. Concentrated sulfuric acid is then added to the soapstock until a "split" is obtained (pH of less than 3). Depending on the tendency of the acidulated soapstock to produce emulsions there may be a water layer with low fat concentration in the bottom of the vat, an emulsion layer above, and a layer of free acid oil at the top.

Stubborn emulsions must sometimes be heated for days to produce a "break." Considerable vat space can sometimes be required to handle emulsions. Vats, and building space to house them are expensive to maintain because of the corrosive nature of the process.

Continuous acidulation is increasingly practiced. The installation can be much more compact and it allows higher temperatures to be used, which is conducive to avoiding troublesome emulsions. Also, acid usage can be much better controlled. Descriptions of installations can be found in the literature, for example, by Crauer (1970), Braae (1976), and Duff (1976), the latter particularly on aspects of automating the process. The main disadvantage of the continuous processes described by Crauer and Braae is the use of centrifuges for separation of acid oil from acid water. It has been found that a decanter tank made of FRP, for example, and sized to give 2–3 hr residence time can give very good separation of acid oil in connection with a continuous, high temperature soapstock acidulating stage. This avoids the use of expensive centrifuges. As with batch acidulation, the acid water is neutralized to pH 5.5–6.0 before discharge to sewers or waste water treatment facilities. The acid oil is washed with 5–10% of hot water to reduce the mineral acid content sufficiently for storage and transport in mild steel tanks.

2. PHYSICAL REFINING

In this process, the free fatty acids are removed from the oil by steam distillation (steam refining) rather than by saponification, as in alkali refining. The equipment required is a deodorizer with certain modifications, which will be discussed in more detail in Section II. The free fatty acids are thus recovered directly; the process of acidulation and the attendant waste water treatment are therefore completely eliminated, together with the need for an alkali-refining operation. Steam refining and deodorizing of the oil can be carried out in one pass, if desired. These are very significant advantages.

In preparation for steam refining, the crude oil must be very thoroughly degummed to a phosphorus concentration of less than 5 ppm, and bleached. Degumming with an acid and water (instead of the conventional degumming with water alone), followed by bleaching, with an acid pretreatment stage, achieves the required removal of phosphatides and other, heat-sensitive materials.

The subject of steam-refining of oils has been discussed extensively in the recent literature on oil processing. The following may be cited for useful background reading: List *et al.* (1977) on pretreating of soybean oil for steam refining; Sullivan (1976) and Gavin (1978) on aspects of steam-refining deodorizers and conditions. No data specifically relating to canola or LEAR oil appear to have been published to date, but the findings available on soybean oil can be taken as applicable to canola and LEAR oil. It is interesting to note that steam refining of lard, tallow, and palm oil has been practiced for a considerable time, particularly in Europe.

B. Bleaching

Alkali-refined oil, or specially degummed crude oil, is bleached in preparation for hydrogenation or deodorization (including steam refining). Bleaching is an adsorption process in which surface-active clay is suspended in the oil under appropriate conditions to adsorb compounds that are sufficiently polar to be attracted to the active sites of the surfaces of clay particles. The compounds involved are soaps, oxidative breakdown products (Ney, 1964; Pardun *et al.*, 1968), and also colored compounds such as the chlorophylls, and to a minor extent, phosphatides and the carotenoids. The last are responsible for most of the reddish-yellow color that predominates in the oil, but which is far more efficiently removed by heat breakdown during hydrogenation or deodorization than by adsorption on bleaching clay. Removal of small concentrations of phosphatides and soap, left after refining or degumming, and of oxidation products and chlorophyll is essential for producing good quality deodorized oil.

Usually, acid-activated clays are used in the process. These clays are manufactured from inactive bentonite and montmorillonite clays by milling, activating with a mineral acid, usually sulfuric acid (Norris, 1964) and washing and drying to a moisture content of 10–15%. Particle sizes range from 10–15% above 80 μm to 30–40% below 20 μm according to Patterson (1976).

The use of acid-activated clay is essential in bleaching oils containing chlorophyll. Wettström (1972) indicates that acidic conditions destabilize the pigments. Carbon black is also used for removal of chlorophyll (Norris, 1964), but its use is much more costly.

1. BATCH BLEACHING

Batch bleaching is still practiced in smaller plants that require frequent stock changes. The tanks used are equipped with heating coils and an agitator, and may be open to the atmosphere, although it is advantageous to bleach under vacuum. The long contact times inherent in the process can lead to considerably reduced bleaching efficiency, especially in open kettles, because of chemical changes, including oxidation.

In operation, the oil is charged to the kettle and heated to about 70–80°C, the clay is added, and heating continued to about 95–105°C. About 0.5–1.5% clay is required depending on oil quality and product specifications. Contact time of 10–20 min is allowed at the maximum temperature before filtering is started. Filters may be of a variety of designs, but plate-and-frame type filters, rather than tank filters, are often used because they are somewhat easier to manipulate. In vacuum bleaching these operations are performed with the bleaching vessel under reduced pressure.

2. CONTINUOUS BLEACHING

Continuous bleaching is the most commonly practiced method. A typical system is shown in Fig. 2. The essential process steps are heating the oil to bleaching temperature, slurrying the clay with the heated oil, introducing it into the bleaching zone, which is usually under vacuum, and then filtering the oil–clay slurry. There are numerous process versions. The main differences between processes pertain to the manner of clay addition (slurried or dry), and to the design of the bleaching zone, which may be a continuous flow tank as in Fig. 2, or a tank divided into semicontinuously operating sections as in the Hobum design described by Liebing and Lau (1974), or simply a tubular section under pressure as described by Harris (1974) and Mag (1980). The advantages of these latter two types of systems are that they allow good control over the degree of drying of the clay in the oil which is

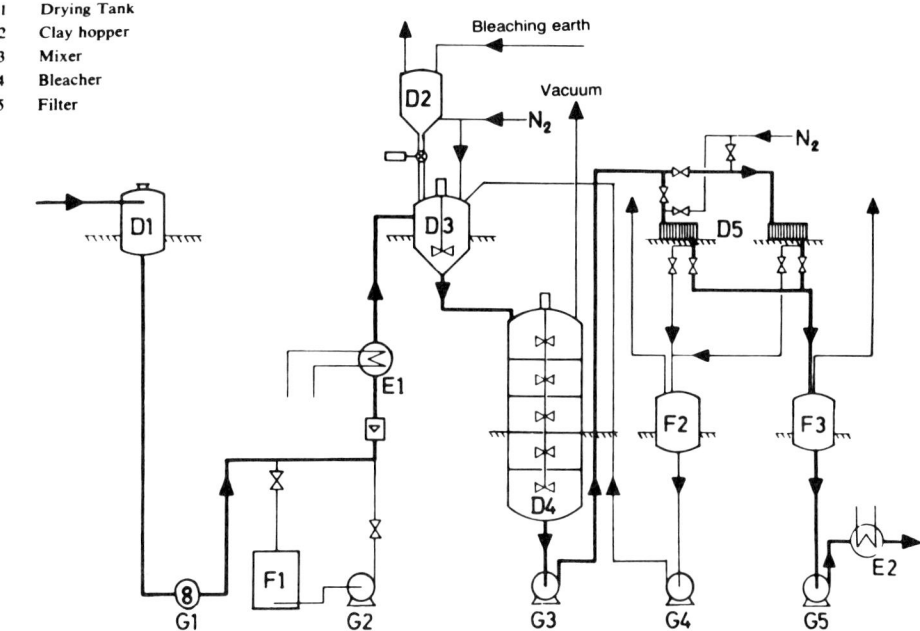

D1   Drying Tank
D2   Clay hopper
D3   Mixer
D4   Bleacher
D5   Filter

Fig. 2. Continuous vacuum bleaching system (LURGI). Adapted from Appelqvist and Ohlson (1972).

important to bleaching efficiency (Norris, 1964; Wettström, 1972), and that they involve only short contact time between oil and clay. The latter makes stock changes much easier.

Spent bleaching clay, which contains about 30% oil, is usually dumped in landfill sites. The two main methods for recovering the oil from the clay are extracting with hot water in the presence of a surface active agent (Svensson, 1976), which leaves about 2% oil, and extracting with hexane in specially adapted filters or in a separate process, which is relatively capital intensive. No efforts at regenerating the clay have been reported to date probably because clay is relatively inexpensive. Higher costs of disposal could change this in the future.

3. BLEACHING OF CRUDE OIL FOR PHYSICAL REFINING

Oil that has been degummed with acid and water to a phosphorus concentration of less than 50 ppm can be prepared for steam refining by bleaching. Usually, a small amount of phosphoric acid (0.05–0.1%) is used on the oil prior to the addition of bleaching clay to assist in the removal of the phosphatides. Also, it may be necessary to use somewhat more clay than is required in the bleaching of alkali-refined oil. Contact time and contact temperature are left unchanged from ordinary bleaching practice. Acid–water degummed oils after pretreating with acid and clay should have a phosphorus content of less than 5 ppm, and be well bleached (essentially free of chlorophyll) before steam refining.

C. Hydrogenation

The hydrogenation of canola oil serves two purposes: (1) to increase the oxidative stability, and (2) to change the melting behavior for use in margarines, shortenings, and other, more specialized products.

In the process of hydrogenation, the fatty acid moieties containing double bonds are progressively saturated with hydrogen. Reaction with oxygen at these sites is no longer possible; also, the melting point is raised as a result of the hydrogen addition to the fatty acid chain. Usually only a relatively small portion of the double bonds available is being saturated with hydrogen, that is, the oil is only partially hydrogenated. Iodine values in the range of 75–95 are typical for hydrogenated oil stocks.

I. HYDROGEN AND CATALYSTS

There are a number of processes for producing hydrogen gas, the two most important ones in Canada being the electrolytic process and the steam hydrocarbon process. Mattil (1964) gives detailed descriptions of these two and various other processes used. The main advantages of the electrolytic process are that it produces gas at very high purity directly without a purification stage, installations can be easily tailored for small capacities, and plants can be readily started up and shut down. By-product oxygen can be a contaminant of the hydrogen produced, but hydrogen purities of 99.8% are readily achievable. Research into improved electrode designs and electrolyte additives promise significantly improved process economics in the future.

The steam hydrocarbon process is usually intended for large capacities. It is said to be simple and reliable in operation and also produces hydrogen of better than 99.8% purity. The hydrocarbon used as raw material, usually natural gas, must be free of nitrogen and sulfur compounds. Carbon dioxide and carbon monoxide are by-products of the reactions taking place and must be removed by chemical conversion and scrubbing. One of the chief advantages has been the relatively low cost of natural gas and its availability.

Hydrogenation requires a catalyst to proceed at practical rates. Only nickel catalysts are of importance at present, although other metals, notably copper, were also employed to some extent in the past.

Two nickel catalyst manufacturing processes are in use: wet reduction and dry reduction, with the latter being the one of greater importance today.

These processes are described in some detail by Mattil (1964) together with theoretical background on catalyst functioning, and the poisoning and promotion of catalytic metals.

Dry-reduced catalysts have somewhat better filterability than wet-reduced catalysts. Usual nickel concentrations are in the range of 20–25%. The remainder is made up of the support (about 25%) and fully hardened oil (about 50%) to protect it from air and moisture and to put it into a dry, flaked form for handling and shipment.

2. HYDROGENATION EQUIPMENT

The process is usually operated in the batch mode. Continuous processes have been developed, notably by Lurgi in Germany, and some firms in the United States, but these have not found significant application in the industry. The main reason appears to be the difficulty in changing from one degree of hydrogenation to another without producing significant amounts of an intermediate, unwanted hydrogenated oil stock.

A typical batch arrangement is shown in Fig. 3 (Hastert, 1981). It consists of the hydrogenation vessel or converter, which is equipped for heating, cooling, and agitation, and means for evacuating, and for pressurizing with hydrogen gas. In addition, there is a catalyst filter, catalyst slurry tank, and a hydrogen gas meter. Often, equipment for posthydrogenation bleaching or other treatment of the oil is provided as an integral part of the process.

In operation, bleached oil is loaded into the vessel and heated under agitation and vacuum to ensure deaeration and drying. The catalyst is then added, either as a slurry or in dry, flaked form. Usually, indirect steam heating is used. Once the desired hydrogenation temperature is reached the evacuated vessel is pressurized with hydrogen gas. The temperature is controlled by cooling with water to remove the heat of reaction. At the end of the hydrogenation run, the hydrogen is evacuated from the vessel, and the oil cooled to filtration temperature and then filtered to remove the bulk of the catalyst. A separate cooling vessel is sometimes provided to achieve faster turnaround of the hydrogenation vessel. The filtered oil usually requires a posthydrogenation treatment with bleaching clay, or citric acid as a chelating agent, to remove nickel to concentrations of less than 0.5 ppm.

3. CONTROL OF THE PROCESS

Since the process of saturation of double bonds is not usually carried to completion, as already pointed out earlier, the control of the process to produce specific hydrogenated oil stocks is relatively complex. There are three main aspects which require consideration: (1) selection of process conditions, (2) hydrogenation end point control, and (3) hydrogenation selectivity and isomer formation during the process.

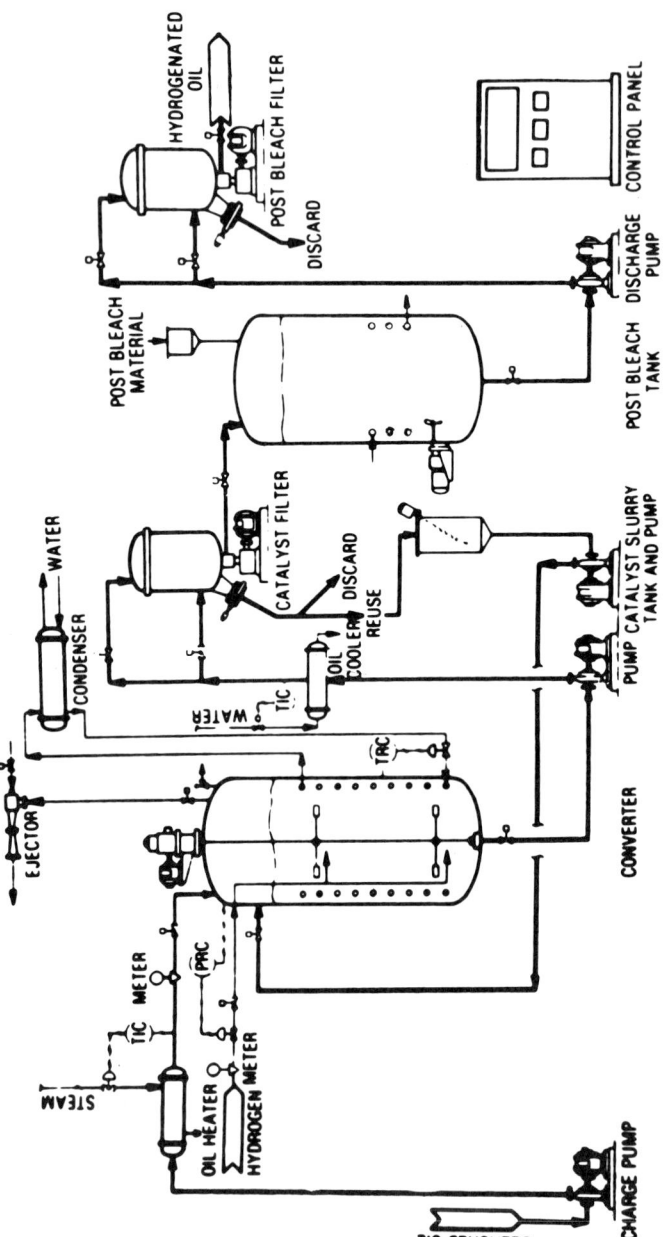

Fig. 3. Edible oil hardening plant (Hastert, 1981).

a. *Process Conditions.* There are four process variables that can be used to influence the reaction: temperature, pressure, catalyst concentration, and agitation. In practice, it is primarily temperature and, to a lesser extent, pressure, that are routinely manipulated to influence the reaction. Catalyst concentration is usually kept as low as consistent with the desired rate of hydrogenation. Agitation is usually fixed by equipment design. A typical set of conditions for making a margarine oil base stock is as follows: temperature (H_2 gas on), 165°C; temperature (control), 200°C; pressure, 30–70 kPa (gauge); catalyst concentration in oil, 0.1-0.2% (25% Ni); agitation (six-blade turbine, two sets), 80 revolution/min. In the case of a shortening base oil stock, the temperature at which the process is controlled would be chosen at 165°C, for example, and the pressure at 200 kPa.

These two sets of conditions are often referred to as selective and nonselective, respectively. The main practical effect is that under so-called selective conditions, the fat produced melts more rapidly with increasing temperature than under nonselective conditions, that is, it has a steeper solid fat index curve.

b. *End Point Control.* End point control uses a variety of methods, for example: (1) hydrogen gas metering and refractive index, (2) iodine value (I.V.), (3) solid fat index, (4) melting points (capillary, Wiley), or (5) differential thermal analysis. Of these five methods only (1), (2), and (3) are usually applied. Melting points and differential thermal analysis have some application in the production of certain specialty fats.

The usual practice for the process operator is to determine the refractive index of a sample of the oil during hydrogenation, since this can be done in a few minutes, on site. It correlates well with the iodine value of the oil and also the solid fat index. If a hydrogen gas meter is available, this will have been set at a precalculated amount of gas at which to stop the reaction for determination of the refractive index. Calculation of the amount of gas required is based on the stoichiometry of hydrogenation (Mattil 1964). When the desired refractive index is reached, the batch is cooled for filtration and the solid fat index is determined in the laboratory. This is by far the most important analysis method in hydrogenation control. Typical solid fat indices for selectively and nonselectively hydrogenated canola oil are given in Table IV (Teasdale, 1975).

c. *Selectivity and Isomerization.* Because of the presence of fatty acid moieties of different unsaturation and hence different reactivity, there is a tendency for hydrogenation to proceed with a degree of selectivity. The more reactive trienes tend to hydrogenate most readily followed by the dienes, and the monoenes. This is a desirable feature, but the relative amounts of trienes, dienes, and monoenes present and other factors also

TABLE IV

Typical Analysis of Selectively and Nonselectively Hydrogenated Canola Oil[a]

Sample[b]	Iodine value[c]	Solid fat index at				Principal fatty acids (%)					
		10.0°C	21.1°C	26.7°C	33.3°C	16:0	18:0	18:1	18:2	18:3	% trans
Original	118.5	—	—	—	—	4.9	1.9	57.0	24.0	10.4	—
S 1	86.2	10.8	1.4	0.1	—	4.8	4.3	78.1	10.6	Tr.	34.0
NS 1	86.0	6.2	1.8	1.2	0.4	4.9	10.0	67.5	14.4	0.4	24.6
S 2	72.8	41.3	22.5	15.9	5.3	4.8	12.9	76.7	3.5	—	51.9
NS 2	71.6	24.5	13.4	8.2	4.5	4.8	18.3	67.0	7.7	—	31.7

[a] Adapted from Teasdale (1975).

[b] The original oil before hydrogenation was compared to hydrogenated products prepared by selective (S) or nonselective (NS) hydrogenation procedures.

[c] Iodine value calculated from the fatty acid analysis.

influence the course of reaction with the result that all compounds are hydrogenated to some extent simultaneously. If it is desired to achieve maximum selectivity toward hydrogenation of the more unsaturated compounds, high temperatures and low pressures must be used as outlined earlier.

Since hydrogenation is not carried to completion there is considerable scope for isomerization to occur. The mechanism by which this takes place has been described by Allen and Kiess (1955). They indicate that selective conditions tend to favor isomer formation. Both positional and geometric isomers are formed. In positional isomers, double bonds have wandered from their original position along the fatty acid carbon chain; in geometric isomers the position of groups attached to carbon atoms have changed relative to each other in space from the natural cis to trans. Recently the nutritional properties of isomeric fatty acids have been questioned. Applewhite (1981) in a literature review has concluded that this concern is not substantiated by the available data.

4. POSTBLEACHING

Filtration of the oil directly out of the hydrogenation process is not reliable in removing nickel to the very low levels (less than 0.5 ppm) required. Concentrations above this level impair the color of the oil in subsequent deodorization. Often, several parts per million of nickel may still have to be removed due to the presence of nickel soaps and colloidal nickel which pass through the catalyst filter.

Commonly, the oil is bleached with 0.25–1% clay depending on the severity of the problem. The addition of citric acid or phosphoric acid to the oil to act as a sequestrant is helpful in stubborn cases and to achieve clay savings. Appropriate levels are 10–100 ppm added to the oil before bleaching. Nickel removal by bleaching is more efficient when moisture in the oil–clay mixture is maintained at about 0.1%.

D. Formulation

Before considering the utilization of canola oil in the formulation of margarine base oils, shortenings and specialty fats, it should be pointed out that this oil has proved to be an excellent salad and cooking oil both for general household use and also in the commercial manufacture of mayonnaise, sandwich spreads, and liquid and "spoonable" salad dressings.

Eskin and Frankel (1976) reported that canola oil performed well in their laboratory evaluations. When Dobbs (1975) surveyed consumers in Manitoba, he found that they were satisfied with the performance of canola oil for household frying. Despite the fact that its level of linolenic acid is somewhat higher than that of soybean oil, canola oil is giving excellent results without having to be partially hydrogenated and winterized. The 1981 Canadian consumption statistics show that canola oil represented 70.5% of the salad/cooking oil market (Anonymous, 1982). Canola's shares of the margarine oil and all-vegetable shortening segments in 1981 were 36.1 and 46.4% respectively.

In formulating margarine base oils, shortenings, and specialty fats, it is important to use a minimum number of base stocks. This system has several advantages as Latondress (1981) has pointed out: (1) the number of heels of hydrogenated batches that must be reworked is greatly reduced, (2) by blending two or more batches of the same hydrogenated base stock minor variations between individual batches tend to average out, and (3) scheduling of plant operations is greatly simplified. One important additional benefit is that inventory costs are reduced when the number of ingredient oils is closely controlled.

Table V gives the solid fat index (SFI) of typical hydrogenated canola oil, hydrogenated soybean oil, and palm oil at different temperatures.

Latondress (1981) and Wiedermann (1968, 1978) have stated that the preferred technique for formulation control is the SFI as determined either by dilatometry or nuclear magnetic resonance (NMR). The SFIs given in this chapter were determined by dilatometry using a modification of AOCS Method Cd 10–57. Control by SFI has two limitations which must be recognized: (1) Due to mutual solubility effects the SFI of a blend cannot be calculated directly from the SFIs of the ingredient oils. Factors must be applied which differ depending on the formula. (2) Different formulas having the same SFI do not always give identical finished products.

The simplest technique for formulating is to blend the various components together and by far the majority of margarine base oils and shortenings are formulated in this way. Other techniques such as fractionation and interesterification are sometimes used under special circumstances, for example, to produce high polyunsaturated and/or low trans margarines or to reduce recrystallization tendencies.

TABLE V

Solid Fat Index of Typical Hydrogenated Canola Oil, Hydrogenated Soybean Oil, and Palm Oil Used as Ingredients For Formulation

Temperature (°C)	Hydrogenated canola oil				Hydrogenated soybean oil		Palm oil
	C-1	C-2	C-3	C-4	SB-4	SB-5	
10	4	12	38	50	50	60	22–28
21.1	2	5	20	40	40	45	15–20
33.3	0	0	2	15	15	30	7–10

TABLE VI

Formulas for Print Margarine Oils

Ingredient[a]	Types of print margarines (PM)							
	PM-1	PM-2	PM-3	PM-4	PM-5	PM-6	PM-7	PM-8
Liquid canola oil	—	—	—	—	—	—	20	50
C-1	60[b]	60	51	51	—	55	45	—
C-2	—	—	—	—	65	—	—	—
C-3	—	—	—	—	—	20	—	—
C-4	40	—	—	34	35	—	—	—
SB-4	—	40	34	—	—	25	—	—
SB-5	—	—	—	—	—	—	35	50
Palm oil	—	—	15	15	—	—	—	—

[a] See Table V for ingredients from canola oil (C) and soybean oil (SB).
[b] Percent by weight.

1. FORMULATION OF MARGARINE BASE OILS

In Canada considerably more margarine is consumed directly in the home than is used in commercial products. As a result, formulation of margarine base oils is relatively simple, as compared to the shortening situation. Canola oil is used principally in two types of margarine base oils: (1) print or stick margarine, and (2) soft or tub margarine.

 a. Formulation of Print Margarine Oils. Table VI illustrates some typical formulas for print (stick) margarine. The SFI ranges for these formulas are 26–28 at 10°C, 13–15 at 21.1°C, and 2–3.5 at 33.3°C.

TABLE VII

Formulas for Soft Margarine Oils

Ingredient[a]	Types of soft margarine (SM)			
	SM-1	SM-2	SM-3	SM-4
Liquid canola oil	80[b]	—	—	68
C-1	—	85	75	—
C-4	—	15	—	—
SB-5	20	—	25	17
Palm oil	—	—	—	15

[a] See Table V for ingredients from canola oil (C) and soybean oil (SB).
[b] Percent by weight.

b. Formulation of Soft Margarine Oils.   Table VII shows some examples of possible formulas for soft (tub) margarine oils. The SFI ranges for these formulas are 10–14 at 10°C, 6–9 at 21.1°C, and 2–4 at 33.3°C.

2. FORMULATION OF SHORTENINGS

As was mentioned earlier the market for shortening is a very diverse one. Not only is there the household portion to consider but there is the larger, more specialized, commercial and industrial segment. The latter includes such customers as small bakers, and fish and chip friers, large bakeries, potato chip producers, and prepared cake mix manufacturers.

a. Plastic Shortenings.   Table VIII gives some examples of the wide array of "plastic" shortening formulas in which canola oil is used. Latondress (1981) gives some very useful guidelines for formulating shortening, as does Thomas (1978). The U.S. patent literature, as abstracted in two volumes (Gillies, 1974; Gutcho, 1979), is also of considerable value. Although much of the information deals with soybean oil, in most cases it can be adapted for canola oil use, providing proper care is taken to avoid the recrystallization problem that can occur with hydrogenated canola oil (see Section II,D,4).

b. Fluid Shortenings.   Although "plastic" shortenings predominate the market, there are two types of fluid shortenings that bear mention. One type is used for frying and the other for making baked goods such as bread and cakes. Ease of handling is the principal advantage for fluid shortening. For example, fast-food outlets find it useful to be able to pour the shortening from a container onto the griddle or into the fryer. Bakeries can meter the shortening, at ambient temperature, into their dough mixers.

Fluid shortenings consist of a liquid oil base, which may be either unhydrogenated or lightly hydrogenated, in which crystals of hard triglyceride and/or emulsifiers such as mono- or diglycerides are suspended. The type, size, and stability of the crystalline phase are critical. For fluid shortenings to be used for frying, the liquid oil base should be lightly hydrogenated, and possibly winterized. No emulsifiers are used, of course; antioxidants and methyl silicone are added. For baking applications, emulsifiers are required. Linteris and Thompson (1958) made fluid shortenings by dispersing hydrogenated "stearines," made from rapeseed oil and mustard seed oil, in cottonseed oil. Handschumaker and Hoyer (1964) have patented the manufacture of a fluid cake shortening using soybean oil, glyceryl monobehenate, and glyceryl monostearate.

3. SPECIALTY FATS

a. Frying Fat.   The principal requirement for a good frying fat is resistance to oxidative and thermal breakdown. Plasticity can be sacrificed to

TABLE VIII

Formulas for Plastic Shortenings

Ingredient[a]	Types of plastic shortenings (PS)							
	PS-1	PS-2	PS-3	PS-4	PS-5	PS-6	PS-7	PS-8
Liquid canola oil	—	—	45	—	—	—	—	20
C-1	60[b]	36	—	62	62	36	87	23
C-2	16	15	—	—	—	—	—	—
C-3	—	—	—	—	30	—	—	—
C-4	—	—	—	—	—	—	—	—
SB-4	16	15	—	30	—	31	—	—
SB-5	—	—	—	—	—	—	—	—
Palm oil	—	27	—	—	—	30	—	50
Hydrogenated palm oil[c]	8	7	7	8	8	3	13	7
Beef tallow	—	—	48	—	—	—	—	—
SFI[d] at 10°C	25–28	25–28	21–24	26–30	26–30	34–37	20–22	21–24
at 26.7°C	17–19	17–19	16–17	17–19	17–19	13–15	15–17	16–18
at 40°C	7–9	6–9	7–10	7–10	7–10	1–3	10–12	7–10

[a] See Table V for ingredients from canola oil (C) and soybean oil (SB).
[b] Percent by weight.
[c] Iodine value <5.
[d] SFI, solid fat index.

TABLE IX

Hydrogenated Canola Oil for Frying

Hydrogenation condition	Fatty acids (%)				SFI[a] at		
	16:0	18:0	18:1	18:2	10°C	26.6°C	40°C
Selective	4.8	12.9	76.7	3.5	41.3	15.9	Tr
Nonselective	4.8	23.8	64.6	4.6	36.3	16.3	4.1

[a] SFI, solid fat index.

achieve the necessary superior stability. Canola oil, hydrogenated to an I.V. of approximately 65–75, is used as a frying fat. Table IX gives the fatty acid compositions and SFIs of canola oil hydrogenated to an I.V. of 72 using "selective" conditions (205°C, 42 kPa, 0.05% of nickel as catalyst) and to an I.V. of 65 using "nonselective" conditions (135°C, 415 kPa, 0.05% of nickel as catalyst).

   b. Shortening for Roll-In Pastries. Shortenings for use in French and Danish pastries must be plastic but also rather "waxy" so that the finished baked goods will have a "flaky" texture. In U.S patent 3,985,911, Kriz and Oszlanyi (1976) describe the manufacture of such a shortening by interesterifying 70% lard and 30% hydrogenated soybean oil. Although not stated in the patent, it is probable that hydrogenated canola could be used in place of soybean oil, providing the proper SFIs are achieved and the chilling and plasticizing are done as described.

   c. Filling and Icing Fats. The special requirements for these products are that they have the ability to entrap air in a short mixing time and that they retain the air in the icing or filling. Kidger (1966) describes the manufacture of such a product in U. S. patent 3,244,536. Two components, A and B, are interesterified together. Component A, consisting of any vegetable oil which contains at least 50% C_{18} fatty acids, is hydrogenated to certain specifications. Component B is coconut or palm kernel oil. Component A comprises 50–90% of the blend and component B comprises 50–10%.

4. CRYSTALLIZATION

   In order to achieve and maintain the desirable body, texture, and performance characteristics, it is essential that the crystals in the solid fat portion of margarines and shortenings be very small (approximately 10 μ or less). The proper crystal structure is achieved by solidifying the product rapidly (see Section II,G).

   It is known that while most formulations retain the required fine crystal structure, others tend to recrystallize with time to form undesirable, large crystals. Although the true situation is probably more complex, it is generally agreed that fat crystals having the β-prime polymorphic form tend to remain small and needle-like, whereas β crystals even if small at first tend to grow larger. When a size of about 25 μ has been reached a "graininess" can be detected. If the crystals continue to grow the desirable qualities of the product will continue to deteriorate. In extreme cases the margarine or shortening becomes a mass of large crystals swimming in liquid oil.

Generally speaking the more diverse the triglyceride composition of the solid phase the greater is the likelihood that the β-prime form will be produced during chilling and that it will be maintained. Hydrogenated HEAR oil has excellent β-prime characteristics whereas hydrogenated canola oil with its predominance of C_{18} fatty acids in the triglycerides has a strong tendency to recrystallize.

Table X gives the average percentages of C_{18} fatty acids and non-C_{18} fatty acids for five vegetable oils based on the data from Table I, and also the ratio of the former to the latter. These data serve to illustrate why hydrogenated canola and sunflower oils are likely to give rise to recrystallization problems.

Recrystallization is, of course, more troublesome the greater the degree of hydrogenation since more solid phase is being produced as hydrogenation proceeds. At an SFI of 5 at 21.1°C there is no problem but at an SFI of 20 at 21.1°C recrystallization almost certainly will occur with canola oil. For this reason hydrogenated canola oil is often used as only the softer component of a margarine base oil or shortening.

Recrystallization can be inhibited in two ways. The addition of sorbitan tristearate at a level of about 0.3%, as suggested by Madsen and Als (1969) is usually very effective. Introduction of a non-C_{18} fatty acid into the triglycerides by interesterification can be helpful. For example, 10% of palm oil was interesterified with 90% of canola oil hydrogenated to an I.V. of 66. A blend consisting of 55% of the interesterified mixture and 45% of lightly hydrogenated canola oil was found to have good crystal stability.

The information available in the literature on the use of interesterification

TABLE X

C_{18} and Non-C_{18} Fatty Acids

Description	Canola oil	HEAR oil	Soybean oil	Sunflower oil	Corn oil
C_{18} (%)	93	48	89.5	94	89
Non-C_{18} (%)	7	52	10.5	6	11
Ratio	13.3	0.92	8.5	15.7	8.1

to inhibit recrystallization of products that contain hydrogenated sunflower oil could be expected to be usefully applied to hydrogenated canola oil. Two examples are British Patent 1,121,662 to Unilever Limited (1968) and Canadian Patent 830,938 to Gander et al. (1969).

E. Deodorizing

In this process, flavor and odor compounds, free fatty acids, and other degradation products still left in the oil, or newly formed in previous processes, are removed by steam distillation. Color is usually improved markedly due to heat breakdown of the reddish-yellow carotenoid compounds present. The odoriferous compounds, which are naturally present in canola oil and which give it the pungent mustard odor, are also removed. These compounds are derived from glucosinolates and their breakdown products. Andersen (1962) gives an overview of the range of compounds that must be removed in deodorizing. The effect of the process is a nearly bland tasting, light yellow oil, low in free fatty acid concentration, and of good flavor stability in storage.

The theoretical basis of steam distillation and the process conditions commonly used are summarized by Mattil (1964). Briefly, use is made of the volatility of the compounds to be removed from the essentially nonvolatile triglycerides at absolute pressures of 2–10 mm Hg and temperatures of 225°–265°C, with steam as the carrier gas for the volatiles. The use of very low pressures also protects the oil from oxidation during the process.

A deodorizing plant consists of three main components: (1) the deodorizing vessel or tower, (2) a high temperature heat source, usually utilizing a high boiling heat transfer medium, and (3) equipment for vacuum generation and vapor condensation. Details of equipment design and materials of construction have been reviewed recently by Gavin (1981). It is worth noting that because of the high temperatures involved all metal in contact with the hot oil is usually stainless steel to avoid iron contamination and iron catalyzed reactions.

Oil suitable for deodorization must be free of impurities that can undergo rapid heat degradation and which can impair the flavor and color stability. The following are notable: phosphatides, which should not exceed 5 ppm as phosphorus equivalent; chlorophyll and derivatives, which should be essentially undetectable, and nickel, which should be below 0.5 ppm. Further, the prooxidant metals iron and copper should not exceed 0.1 and 0.01 ppm, respectively (Evans et al., 1951).

1. BATCH DEODORIZING

Batch deodorizing is still operated in some plants, the main advantage being somewhat lower capital investment and simplicity of operation including start-up and shutdown. The main disadvantages are that the vacuum equipment for vapor handling must be relatively large which makes for higher operating costs, and that deodorized oil quality is not usually as good as with semicontinuous or continuous process installations. Also, heat recovery and vapor scrubbing, which are increasingly important aspects of the deodorizing process, are very difficult to accomplish, and oil losses are higher.

In a typical batch cycle, the oil is pumped to the deodorizer vessel, which is a cylindrical tank equipped with heating coils, a steam sparging line arranged in a grid near the bottom of the vessel, and a vacuum system. Sufficient head space is allowed in the loaded vessel to prevent oil from splashing into the vapor outlet to the vacuum train. During loading, the vessel is evacuated, and heating and steam sparging are started. The oil is heated to about 230°–250°C which takes about 2 hr. The maximum temperature is maintained for 2–3 hr. Sparging steam usage is about 5% per hour. The charge is then cooled to 50–60°C under vacuum and with sparging steam still being applied. In many installations a separate vessel is used for the cooling phase of the cycle to increase the throughput of the deodorizer. The cooled oil is usually pumped through a polishing filter and sparged with N_2. A complete cycle requires about 8 hr.

2. SEMICONTINUOUS AND CONTINUOUS DEODORIZING

The main advantage of these two process modes is the much more efficient exposure of the oil to the low absolute pressure and the stripping steam, resulting in deodorized oil of better quality. Numerous process designs are being used successfully. The semicontinuous processes have the advantage of allowing stock changes without any significant intermixing. This is usually an important factor when many different products must be made in a refinery, and a semicontinuous process is therefore often preferred. Figure 4 is an example of a typical installation.

The deodorizing vessel consists of a large cylindrical tank or shell in which four trays are positioned which essentially perform the various stages of the batch deodorizing cycle described previously, in succession, on a small quantity of oil (1–2.5 tons). The vessel is under vacuum. In operation, the oil is pumped to a measuring tank which batches the oil into tray-sized quantities. From there the oil enters the steam-heating tray of the deodorizer to be deaerated and heated to about 160°C in 15–20 min. Sparging steam at about 0.5% per hour is used for agitation. The subsequent stages are; heating to 250°–270°C in the high temperature heating tray, again with 0.5% sparging steam, deodorizing in the deodorizing section using about 1.5% of steam per hour, and finally cooling to about 50°–60°C using 0.5% steam for agitation in a cooling section. Total time in the deodorizer is therefore only 60–80 min and stripping steam usage about 3%. The cooled oil is pumped

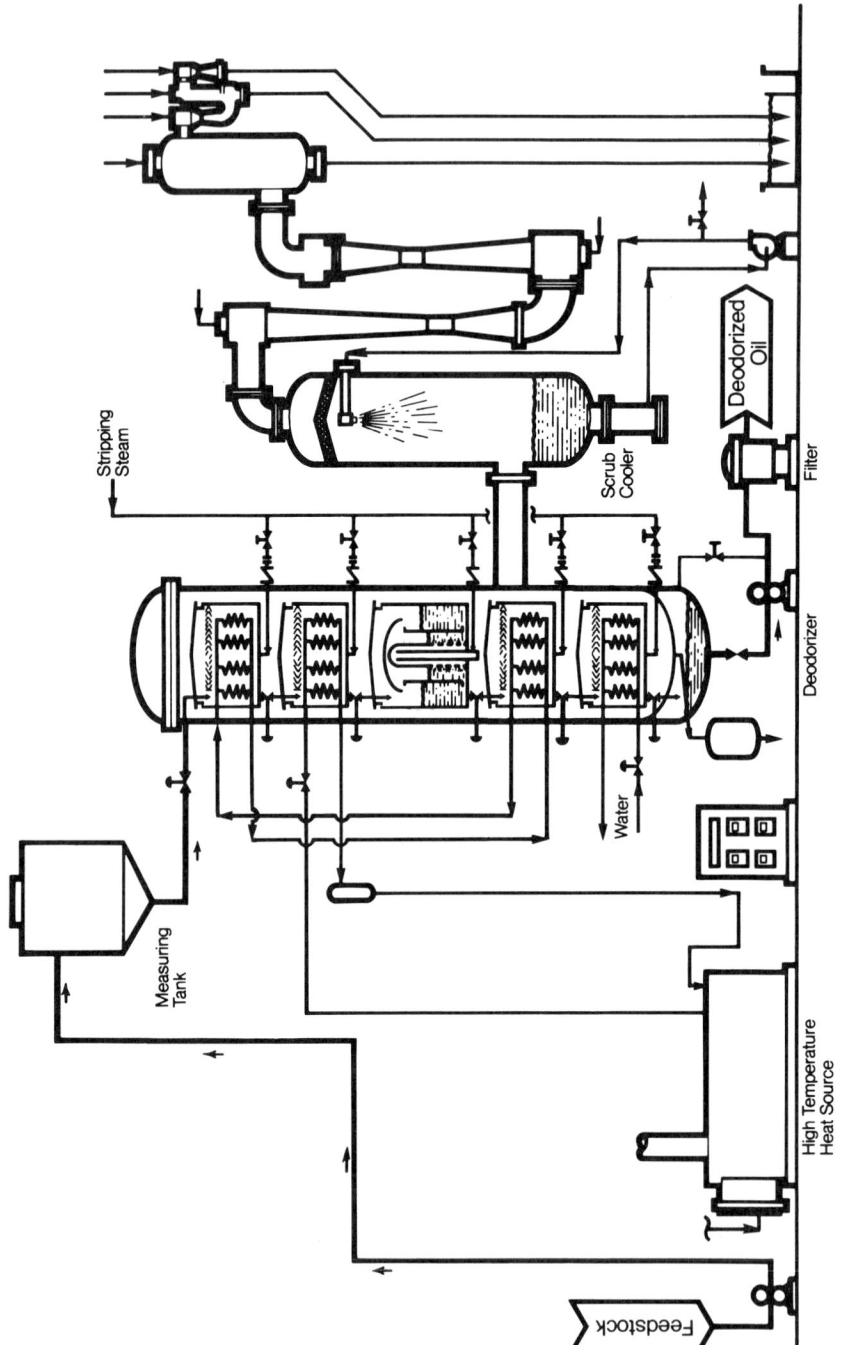

Fig. 4. Votator semicontinuous deodorizing plant (with heat recovery). Courtesy of Chemetron Process Equipment, Inc., Louisville, Kentucky.

through a polishing filter and sparged with nitrogen gas, as in batch operations. It is also good operating practice to add about 2 ppm of citric acid in aqueous solution to the oil in the cooling section to sequester traces of metallic pro-oxidants.

Most suppliers of semicontinuous deodorizing processes also offer process versions suitable for continuous operation. In addition, there are processes such as that by Wurster and Sanger (Mattil, 1964) designed for continuous operation only.

3. VACUUM AND HEAT

The most common method of vacuum generation is by a system of steam-jet pumps and water-spray condensers. Usually, a vacuum aggregate consists of one or two large steam-jet pumps, or boosters, in series, followed by a condenser, and then a two-stage steam-jet assembly with intercondenser as shown in Fig. 4. The booster assembly is sized to compress the total vapor load, which consists of stripping steam, fatty acids and other volatiles, and some air. The condenser removes essentially all vapor except air, which is further compressed and exhausted to the atmosphere in the two-stage steam-jet intercondenser assembly. Depending on local energy costs, it can be advantageous to replace the two-stage jet–intercondenser portion of the system with a mechanical vacuum pump–condenser assembly. It is usually not attractive to replace the complete system with vacuum pumps since the high vapor loads are more efficiently compressed by steam-jet boosters.

The high temperature heat source may be either high pressure steam, or a liquid–gas or a liquid–liquid heat transfer system based on a heat-stable, petroleum-derived oil. The advantage of the systems using heat transfer oils is that they operate at very moderate pressures of 200–300 kPa (gauge). The advantage of using steam is that it eliminates any question of contaminating the deodorized oil with a petroleum-derived product.

With the advent of high energy costs, it has become attractive to recover some of the heat contained in the deodorized oil by cooling with incoming oil. Semicontinuous and continuous deodorizers can be easily equipped for this, and most recent installations include this feature. About 40–50% of the process heat required can be recovered in semicontinuous operations.

4. DISTILLATE AND CONDENSER WATER

In the vacuum system described previously, the distilled fatty material is condensed together with the stripping steam in the barometric condenser. Condenser water and condensed vapors are discharged to a basin, which provides a seal against the atmosphere (see Fig. 4). Much of the fatty material floats to the surface in that basin as an emulsion and is skimmed off for recovery. Recovery is done by heating the emulsion in a tank and separating the oily layer by gravity.

In the past, the water from the basin was often discharged to sewers, lagoons, or streams. Water pollution abatement, the need to effect water savings, and odor problems associated with condenser water have forced considerable changes. Modern deodorizer installations usually include a scrub cooler, which condenses about 90–95% of the fatty material from the deodorizer vapors ahead of the barometric condenser (see Fig. 4). The condenser water is still skimmed after discharge to the seal basin, which is made quite large to achieve more complete separation of oily material from the water. The water is pumped over a cooling tower for reuse in the condenser. Where odors from the cooling tower are a problem, the cooling of the condenser water is done in heat exchangers using "clean" water, which in turn is cooled in the tower. Gavin (1981) has described a system incorporating these features.

5. STEAM REFINING

The steam-refining mode of the deodorizing process requires some changes in equipment and process conditions. With respect to process equipment, it is usual to make all parts, vapor-contacting as well as oil-contacting, of stainless steel, to supply somewhat larger capacity in the vacuum system for the section handling the condensible vapors, and, of course, to use a fatty acid scrub cooler. Changes in processing conditions may include longer deodorizing time and higher temperature or higher stripping steam usage rates. Steam-refining deodorizers and conditions were recently reviewed by Gavin (1981).

F. Final Formulation

1. MARGARINES

Traditional margarines contain 80% fat phase and 20% aqueous phase. Diet margarines consist of 40% fat and 60% aqueous phase. Canadian Food and Drug Regulations, which are similar to those of the Codex Alimentarius, control the composition of margarine in Canada. Different regulations will prevail in other jurisdictions.

Lecithin, as an antispattering agent, monodiglycerides as emulsifier, flavor, color, and vitamins are dissolved in the molten base oil. The aqueous phase usually contains milk or whey solids, salt, and flavor. Potassium sorbate or sodium benzoate as preservative and a metal sequestering agent such as citric acid may also be added. Diet margarines usually require additional emulsifiers and also stabilizers. The oil and water phases may be prepared separately and then mixed together either batchwise or continuously. In the larger, more sophisticated factories, the various ingredients are metered into the appropriate phase in a fully continuous mode. Since the mixture will separate if agitation is stopped, the fat phase must be solidified as quickly as possible.

2. SHORTENINGS

As is the case with margarine, shortening is a standardized food in Canada and the composition is controlled by Food and Drug Regulations.

The use of monoglycerides, or mono- and diglycerides and lactylated monoglycerides, are permitted at up to specified levels. Class IV preservatives (BHA, BHT, or propyl gallate as antioxidants, and citric acid or some citrates as metal chelating agents) are permitted up to 0.02%. Dimethylpolysiloxane is allowed up to 10 ppm as a foam inhibitor. In practice, household shortenings usually contain about 1.5–2.5% of monoglycerides. Commercial shortenings, especially for making prepared cake mixes, will contain up to the allowable maximum of 10%.

G. Chilling and Packaging

In order to achieve the smooth structure required for margarine and shortening it is necessary to chill very quickly to produce a large number of crystal nuclei. This is best done by using a continuous scraped-surface chiller. The Votator system (United States) is probably the most commonly used. Other units are the Kombinator (Germany), Gerstenberg (Denmark) and the recently introduced Groen system (United States) (Greenwell, 1981).

The incoming fat and water mixture (margarine), or fat only (shortening), is pumped through the annular space of the product chamber where it is chilled against the heat-transfer tube and immediately scraped away by the blades. In the case of print margarines the chilled material, which is partially solidified and partially supercooled, moves to a quiescent zone (B unit) where it solidifies before going to the forming and wrapping machines. With soft margarine and shortening an agitated B unit is used and the product is filled into packages as a semisolid. Approximately 12% of air or nitrogen is incorporated into shortening during the chilling operation to improve plasticity and "whiteness." After packaging, margarine is moved directly to cooler storage (10°–15°C). Shortening is stored at 25°–30°C for a few days to "temper," that is, to allow the crystals to stabilize.

H. Winterizing

The designation "winterizing" is a carryover from the days when cottonseed oil was left outside in the winter to allow the higher melting portion to solidify for subsequent removal by filtration. The liquid portion was used in the manufacture of salad dressing and mayonnaise. These products, which are oil-in-water emulsions, are usually stored in the refrigerator. If fat crystals should appear the emulsion will separate and the desirable properties will be destroyed.

As was pointed out earlier in this chapter, canola oil is a natural salad oil and normally does not require winterizing. However, when there is a requirement for a liquid oil of improved stability, Teasdale (1966) has shown that canola oil offers a better source than either HEAR oil or soybean oil. Low erucic acid rapeseed oil, hydrogenated to 0.8% linolenic acid, was winterized to give a yield of 94–95% of salad oil having a cold test of 12 hr. HEAR oil, also hydrogenated to the 0.8% linolenic level, gave a yield of 75% having a cold test of 6 hr.

Evans et al. (1964) gave the following data for hydrogenated, winterized soybean oil: 90, 75, and 60% yield for a linolenic acid content of 3, 1, and 0.5%, respectively. Current procedures for winterizing are described by Neumunz (1978) and Hastert (1981).

I. Interesterification

According to Going (1967) there are three types of interesterification: (1) interchange between a fat and free fatty acids ("acidolysis"), (2) interchange between a fat and an alcohol ("alcoholysis"), and (3) rearrangement of fatty acid radicals in triglycerides ("ester interchange"). The last type is the one commonly referred to when the term interesterification is used. It is sometimes called rearrangement or transesterification.

In this chemical process the triglyceride composition, but not the fatty acid composition, of a single fat or oil, or a mixture of two or more fats and/ or oils, is changed. As a result the physical properties of the fat, or mixture, are altered. In effect a new fat has been produced.

An alkaline catalyst, such as sodium methoxide, or metallic sodium is normally used. The fat must be dry and free of fatty acids to prevent destruction of the catalyst. When the reaction is conducted at a temperature that maintains the fat in a molten state the process leads to "random" interesterification. "Directed" interesterification takes place when a portion of the fat is in a solid state. In a typical case of random interesterification the fat is heated in a closed tank under vacuum to 135°–150°C, 0.1% of catalyst is added, and mixed thoroughly for about 0.5 hr. A sample is taken for analysis and if the reaction is not complete an additional amount of catalyst is added and mixed for another 0.5 hr. When the interesterification is complete approximately 0.05% of phosphoric acid is added with agitation to neutralize the catalyst. The batch is then washed or bleached.

There are many excellent papers on interesterification in the literature, for example those by Coenen (1974), Naudet (1974), Hustedt (1976), and Sreenivasan (1978). The following publications give examples of the application of interesterification of LEAR oil. Katzer et al.(1974) have reported laboratory tests in which 20–25% of LEAR oil, hydrogenated to an I.V. of 1.9 was interesterified with 75–80% of sunflower oil or soybean oil. The resultant fats were judged to be suitable for making margarine. They were low in trans fatty acids, they contained 39.5–53.5% linoleic acid and were capable of crystallizing in the β-prime polymorphic form.

AB Karlshamns Oljefabriker (1974) in a French patent application gives examples of margarine ingredients made by interesterifying LEAR oil with coconut oil. In one example, seven parts of LEAR oil, hydrogenated to I.V. 60, are interesterified with three parts of coconut oil. This product is blended (16–40%) with coconut oil, liquid vegetable oil and palm oil. A satisfactory margarine was produced which maintained its β-prime form for 6 weeks. The patent literature of Canada, Europe, and the United States contains many other examples of how interesterification can be used to produce specialty products.

III.   CURRENT APPLICATION OF HEAR OIL FOR EDIBLE USES

As has been stated in preceding sections many countries are following Canada's lead and are reducing, or eliminating, the production and use of high erucic acid rapeseed (HEAR) oil. The following are some examples of where HEAR oil continues to be used because of some desirable characteristic.

Fully hydrogenated HEAR oil has been used in the United States since 1961 as a stabilizer and thickener component in peanut butter, at a maximum concentration of 2% of the weight of the finished butter (Federal Register, 1977). A Canadian Patent was issued to Japikse (1969a) for use of about 1% of hydrogenated HEAR oil to produce a "flavor-improved stabilized peanut butter." Superglycerinated, fully hydrogenated HEAR oil has also been used since 1957 in cake mix formulations, as an emulsifier in shortening at a maximum concentration of 4% of the shortening, or at 0.5% of the total weight of cake mix (Federal Register, 1977). Seiden (1967) has patented the use of 0.6–1.4% of HEAR oil hydrogenated to an I.V. 10–30 to control the oil-off and slump of margarines containing palm oil and coconut oil. Going and Dobson (1966) have found that as little as 5% of hydrogenated HEAR oil is useful as a non-β forming hardstock in plastic shortening. Japikse (1969b) has patented the use of 2–8% of HEAR oil which has been hydrogenated to an I.V. no greater than 12 as a non-β tending hardstock in margarine. As was mentioned in Section II,D,2,b, Linteris and Thompson (1958) reported on the manufacture of fluid shortenings by dispersing HEAR oil stearines in cottonseed oil.

REFERENCES

AB Karlshamns Oljefabriker (1974). Demande de Brevet d'Invention, République Française 74/19568.

Allen, R. R., and Kiess, A. A. (1955). *J. Am. Oil Chem. Soc.* **32**, 400–405.

Andersen, A. J. C. (1962). *In* "Refining of Oils and Fats," (P. N. Williams ed.), 2nd rev. ed. pp. 155–158. Macmillan, New York.

Anonymous (1981). *Canola Dig.* **16** (2).

Appelqvist, L.-A. (1971). *J. Am. Oil Chem. Soc.* **48**, 851–859.

Appelqvist, L.-A., and Ohlson, R. (1972). "*In* Rapeseed." Elsevier, Amsterdam.

Applewhite, T. H. (1981). *J. Am. Oil Chem. Soc.* **58**, 260–269.

Braae, B. (1976). *J. Am. Oil Chem. Soc.* **53**, 353–357.

Coenen, J. W. E. (1974). *Rev. Fr. Corps Gras* **21**, 403–413.

Crauer, L. S. (1970). *J. Am. Oil Chem. Soc.* **47**, 210A–212A, 235A.

Daun, J. K., and Hougen, F. W. (1976). *J. Am. Oil Chem. Soc.* **53**, 169–171.

Dobbs, J. E. (1975). M.Sc. Thesis, Dept. of Food and Nutrition, University of Manitoba, Winnipeg, Canada.

Duff, A.J. (1976). *J. Am. Oil Chem. Soc.* **53**, 370–381.

Eskin, N. A. M., and Frankel, C. (1976). *Actes Congr. Mond.—Soc. Int. Etude Corps Gras, 13th, 1976* Section A, pp. 1–9.

Evans, C. D., Schwab, A. W., Moser, H. A., Hawley, J. E., and Melvin, E. H. (1951). *J. Am. Oil Chem. Soc.* **28**, 68–37.

Evans, C. D., Beal, R. E., McConnell, D. G., Black, L. T., and Cowan, J. C. (1964). *J. Am. Oil Chem. Soc.* **41**, 260–263

Federal Register (1977). **42**, 48335–48336.

Gander, K.-F. (1976). *J. Am. Oil Chem. Soc.* **53**, 417–420.

Gander, K.-F., Hannewijk, J., and Haighton, A. J. (1969). Canadian Patent 830,938.

Gavin, A. M. (1978). *J. Am. Oil Chem. Soc.* **55**, 783–791.

Gavin, A. M. (1981). *J. Am. Oil Chem. Soc.* **58**, 175–184.

Gillies, M. T. (1974). "Shortenings, Margarines and Food Oils," Food Technol. Rev. No. 10. Noyes Data Corp., Park Ridge, New Jersey.

Going, L. H. (1967). *J. Am. Oil Chem. Soc.* **44**, 414A–456A.

Going, L. H., and Dobson, R. D. (1966). U.S. Patent 3,253,927.

Greenwell, B. A. (1981). *J. Am. Oil Chem. Soc.* **58**, 206–207.

Gutcho, M. (1979). "Edible Oils and Fats, Recent Developments," Food Technol. Rev. No. 49. Noyes Data Corp., Park Ridge, New Jersey

Handschumaker, E., and Hoyer, H.G. (1964). Canadian Patent 698,516.

Harris, R. D. (1974). Canadian Patent 945,476.

Hastert, R. C. (1981). *J. Am. Oil Chem. Soc.* **58**, 169–174.

Hustedt, H. H. (1976). *J. Am. Oil Chem. Soc.* **53**, 390–392.

Japikse, C. H. (1969a). Canadian Patent 815,481.

Japikse, C. H. (1969b). U.S. Patent 3,425,482.

Katzer, A., Strecker, L., and Fal, U. (1974). *Tluszcze Jadalne* **15**, 165–184.

Kidger, D.P. (1966). U.S. Patent 3,244,536.

Kriz, E. F., and Oszlanyi, A.G. (1976). U.S. Patent 3,985,911.

Latondress, E.G. (1981). *J. Am. Oil Chem. Soc.* **58**, 185–187.

Lesieur, B. (1976). *J. Am. Oil Chem. Soc.* **53**, 413–416.

Liebing, H., and Lau, J. (1974). *Seifen Öle Fette Wachse* **100**, 467–469.

Linteris, L. L., and Thompson, S. W. (1958). *J. Am. Oil Chem. Soc.* **35**, 28–32.

List, G. R., Mounts, T. L., Warner, K., and Heakin, A. J. (1977). *J. Am. Oil Chem. Soc.* **55**, 277–279.

Litchfield, C. (1971). *J. Am. Oil Chem. Soc.* **48**, 467–472.

Madsen, J., and Als, G. (1969). *Seifen Öle Fette Wachse* **95**, 593.

Mag, T. K. (1980). U.S. Patent 4,230,630.

Mattil, K. F. (1964). *In* "Bailey's Industrial Oil and Fat Products" (D. Swern, ed.), 3rd ed., pp. 846–862. Wiley (Interscience), New York.

Naudet, M. (1974). *Rev. Fr. Corps Gras* **21**, 35–43.

Neumunz, G. M. (1978). *J. Am. Oil Chem. Soc.* **55**, 396A–398A.

Ney, K. H. (1964). *Fette, Seifen, Anstrichm.* **66**, 512–517.

Norris, F. A. (1964). *In* "Bailey's Industrial Oil and Fat Products" (D. Swern, ed.), 3rd ed., pp. 741–744. Wiley (Interscience), New York.

Ohlson, R., and Svensson, C. (1976). *J. Am. Oil Chem. Soc.* **53**, 8–11.

Pardun, H., Kroll, E., and Werber, O. (1968). *Fette, Seifen, Anstrichm.* **70**, 531–536.

Patterson, H.B.W. (1976). *J. Am. Oil Chem. Soc.* **53**, 339–341.

Rocquelin, G., Sergiel, J.-P., Martin, B., Leclerc, J., and Cluzan, R. (1971). *J. Am. Oil Chem. Soc.* **48**, 728–732.

Seiden, P. (1967). U.S. Patent 3,298,837.

Sergiel, J.-P. (1973). *Rev. Fr. Corps Gras* **20**, 137–141.

Sreenivasan, B. (1978). *J. Am. Oil Chem. Soc.* **55**, 796–805.

Sullivan, F. E. (1976). *J. Am. Oil Chem. Soc.* **53**, 358–360.

Svensson, C. (1976). *J. Am. Oil Chem. Soc.* **53**, 443–445.

Teasdale, B. F. (1966). Canadian Patent 726,140.

Teasdale, B. F. (1975). *In* "Oilseed and Pulse Crops in Western Canada—A Symposium" (J.T. Harapiak, ed.), pp. 551–585. Western Co-operative Fertilizers Ltd., Calgary, Alberta.

Thomas, A. E. (1978). *J. Am. Oil Chem. Soc.* **55**, 830–833.

Unilever Limited (1968). British Patent 1,121,662.

Wettström, R. (1972). *In* "Rapeseed" (L.-A. Appelqvist and R. Ohlson, ed.), pp. 218–248. Elsevier, Amsterdam.

Wiedermann, L. H. (1968). *J. Am. Oil Chem. Soc.* **45**, 515A–560A.

Wiedermann, L. H. (1978). *J. Am. Oil Chem. Soc.* **55**, 823–829.

Wiedermann, L. H. (1981). *J. Am. Oil Chem. Soc.* **58**, 159–166.

10

Current Consumption of Low Erucic Acid Rapeseed Oil by Canadians

M. VAISEY-GENSER

High and Low Erucic Acid Rapeseed Oils

I. INTRODUCTION

Low erucic acid, low glucosinolate rapeseed has been developed in Canada. The oil derived from this rapeseed has been officially called "canola" oil by its producers since 1980. This oil is Canada's major edible oil. In 1981, canola oil accounted for close to 45% of the Canadian production of deodorized oils which provides the basic stock for margarine, shortening, and salad oil manufacture. In contrast, 26% of the domestically deodorized oil was from soybeans (Statistics Canada, 1982). Rapeseed was introduced to Canadian agriculture during the 1950s. Until 1966, the amounts entering the food system were too small to be listed independently in food production statistics. The past 15 years have seen rapeseed oil achieve its present dominance in the Canadian food supply. The objective of this chapter is to describe, as accurately as possible, how much rapeseed oil is eaten by Canadians as a basis for considering its influence on the nutritive status of the people. Thomas et al. (1981) have presented data from the United Kingdom which suggest that a population's average depot fat composition reflects the fatty acid composition of its average dietary fat, at least in terms of *trans* fatty acids, "higher" fatty acids (C_{20} and C_{22}) and "lower" fatty acids (C_{14}–C_{17}).

A. Methods for Assessing Food Consumption

1. DOMESTIC DISAPPEARANCE DATA

Statistics on the per capita annual disappearance of food provide time series information about the availability of food. A residual approach is used in these calculations. That is, annual balance sheets are prepared which show the supply and disposition of each food item. Net supply is determined by subtracting exports and ending stocks from the gross supply which consists of imports, beginning stocks, and production estimates. Division by the July 1 population results in a per capita value. The net food figure essentially represents the supply of food leaving the wholesale level; as such, it may substantially overestimate actual food consumption. No allowance is made for waste in stores, households, or food service establishments. Nevertheless these statistics are now labeled "Apparent Per Capita Food Consumption in Canada" (Statistics Canada, 1981).

2. FAMILY FOOD EXPENDITURE SURVEYS

Since 1953 there have been continuing surveys of urban family food expenditures using the diary system. The data are classified by income group, family size, region, and other stratifications. While the surveys were originally done to provide information for the Consumer Price Index, in later years, quantities of food purchased, as well as dollars spent, have been recorded. The quantitative records provide an opportunity for some cross-sectional analyses of food purchases, at least for 1969, 1974, and 1978 where population samples were comparable (Statistics Canada, 1980; Robbins and Barewal, 1981).

3. DIETARY SURVEYS

Between 1970 and 1972, the dietary intake of a population sample of more than 15,000 persons was collected in Canada by the 24 hr recall system. The data from this survey provide a reasonable estimate of food eaten by various physiological groups in different regions of the country (Health and Welfare Canada, 1975). Although recall data are not precise, they at least provide a closer estimate of food consumed than domestic disappearance data and a comparison between the two sets of data permits some estimate of waste between the wholesale level and the plate.

B. Statistics for Estimating Fat Intake

1. DOMESTIC DISAPPEARANCE OF FAT

The fat contribution from the approximately 150 foods included in the food disappearance data is calculated by multiplying the per capita disappearance by the appropriate factor. Generally the food compositon values now used are from the "Nutrition Canada Survey Food Nutrient Conversion File" (Health and Welfare Canada, 1977), which are based on USDA data (Watt and Merrill, 1963) updated and modified to Canadian novelties. Apparent per capita fat consumption is estimated by summing the fat in all food disappearance estimates. A summary of the nutritive value of the food available for consumption between 1960 and 1975 has recently been published (Robbins and Barewal, 1981) which has minimized the discrepancies in counting systems and waste factors that have been used through the years.

Recently, however, Statistics Canada has revised its method for reporting the available consumption of the fats and oils commodity group. For 1979 and thereafter, sales data have been used to represent production rather than considering beginning and year-end stocks (Statistics Canada, 1981). Imports and exports are considered in completing the calculation of domestic disappearance of fats. Nevertheless, estimates from 1979 on are not entirely consistent with those of earlier years.

2. PRODUCTION AND SALES DATA ON FATS AND OILS

Time series data are available describing the total domestic production and sales of deodorized fats and oils (Statistics Canada, 1982). Information on the species from which these products are derived is incomplete. When a species is processed by only one or two companies, the production statistics are withheld for that species to comply with the secrecy requirements of the Statistics Act.

Production and sales data indicate the amounts of deodorized fats and oils that are used in the major visible products: margarine, shortenings, and salad oil. As yet, no distinctions are made among market forms within these categories. For example, production and sales of soft and brick margarines are not separated. The shortenings category includes products tailored for specific purposes such as cake baking and pastry baking as well as both solid and liquid forms of heavy-duty shortenings used for institutional frying. The salad oil category includes oil used for both salad dressings and frying. In contrast, U.S. statistics publish the amounts of fats and oils used annually in mayonnaise, potato chips, frozen French fries, and mellorine-type products (United States Department of Agriculture, 1977).

II. LEVEL OF USE OF FATS AND OILS

A. Fats in the Food Supply

1. FOOD FATS AND OILS

In total, the Canadian supply of tablespreads, cooking fats, and salad oils has increased from 22 to 24 kg per capita annually, during the 30 years between 1950 and 1980. The most significant changes within this period have occurred in the pattern of supply of specific products (Fig. 1). For example, the apparent consumption of butter has dropped by more than half. Margarine use, in turn, has increased. However, tablespread consumption in total has dropped by almost 25%, likely reflecting the decrease in bread consumption (Anonymous, 1979); it is also possible that some of the tablespread use in baking has been replaced by shortenings. Shortening use has increased to the point where it is the largest single contributor to the apparent consumption of food fats and oils. The marked upswing in shortening use is partly a result in the increase of fried foods made possible by the development of the fast-food industry, and by the increase in commercial baking. Salad oils showed the largest proportional gain, increasing from less than 1 kg per capita to close to 4 kg in 1980. Salad oils are used by the consumer for frying and some baking but the largest increase likely is a result of salads having been established as part of the food habit.

The apparent use of tablespreads, cooking fats, and salad oils in Canada over the years has been similar in total quantity to that reported for the United States (Rizek et al., 1974; U.S. Department of Agriculture, 1977). However, a comparison of the data on individual product categories shows

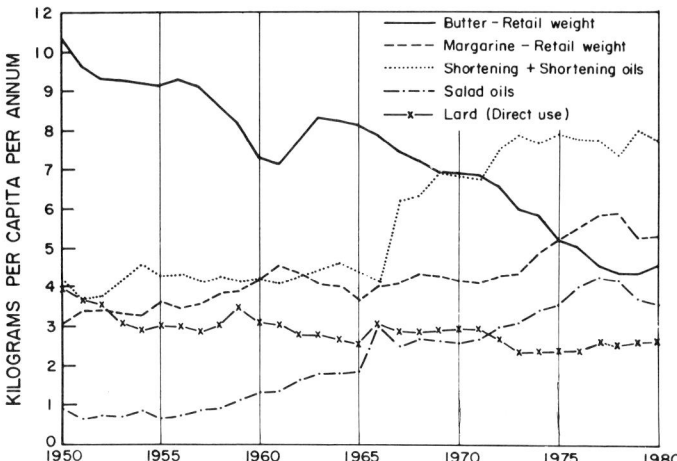

Fig. 1. Canadian apparent consumption of food fats and oils (Hassan and Karamchandani, 1977, 1980; Statistics Canada, 1979; 1980 figures preliminary).

some distinct novelties between the two North American populations. Canadian tablespread consumption, particularly that of butter, has consistently exceeded that of the United States, while the introduction of margarine has been much slower. In 1975, the per capita consumption of butter in Canada was twice that of the United States. The use of margarine surpassed that of butter in 1958 in the United States (Rizek et al., 1974) but this shift did not take place in Canada until 1976 (Fig. 1). Similarly, the United States reported consumption of lard has decreased more rapidly; by 1975 it was half of that reported for Canada, 1.4 kg vs. 2.9 kg per capita, respectively. Between 1950 and 1970 the per capita United States consumption of shortenings and salad oils was substantially greater than that of Canada but this gap has since narrowed (U.S. Department of Agriculture, 1977; Fig. 1).

2. APPARENT CONSUMPTION OF DIETARY FATS

The food fats in the Canadian food supply are shown again in Table I, but here they are expressed strictly as fat, as opposed to the retail weight of products (Fig. 1) and on a daily per capita basis, together with other sources of dietary fat. Table I is limited to the period beginning with 1966 to reflect those years in which rapeseed oil has made up a measurable part of the food supply. During this time, the total dietary fat available for daily consumption increased some 4–8 g per person, coincidental with minor increases in fat from manufactured products, plant sources, and the meat group, and decreases in milk sources. Tablespreads, shortenings, and salad oils accounted for 40–43% of the fat in the Canadian food supply, which is very much the same as in the United States (Rizek et al., 1974).

TABLE I

Canadian Apparent Consumption of Dietary Fats (g/capita/day)

	Food fats and oils[a]					Fat in foods[b]			
Calendar year	Butter (Retail weight x 0.81)	Marg-arine	Lard[c]	Short-ening	Salad oil	Cereals, fruits, veg., pulses, nuts	Meat, fish, poultry, eggs	Milk and milk prod.	Total dietary fat available
1966	17.9	9.0	8.4	15.9	8.2	7.2	54.2	19.6	141
1967	17.0	9.2	7.9	17.1	6.9	7.5	57.2	19.6	142
1968	16.6	9.5	7.9	17.5	7.5	7.1	57.4	19.3	143
1969	15.8	9.7	8.1	19.0	7.3	7.3	57.8	19.4	145
1970	15.9	9.4	8.0	19.0	7.2	7.2	59.4	19.0	145
1971	15.8	9.3	8.0	18.7	7.4	7.3	63.0	19.0	149
1972	15.0	9.7	7.5	20.6	8.4	7.4	61.6	19.5	150
1973	13.7	9.9	6.7	21.7	8.6	8.1	56.0	20.0	145
1974	13.4	10.8	6.7	21.2	9.5	8.2	56.0	20.0	146
1975	11.7	11.6	5.8	21.6	9.7	9.9	55.6	19.3	145
1976	11.3	12.2	6.8	21.3	11.2	8.2	59.7	18.0	149
1977	10.1	13.0	8.2	20.8	12.0	8.0	58.4	18.2	149
1978	9.9	13.2	7.9	20.5	11.6	9.0	57.6	18.3	148
1979	9.9	11.9	8.3	22.2	10.3	9.0	57.5	18.4	147
1980	10.0	12.0	8.0	21.2	9.8	—[d]	—	—	—

[a] Adapted from Statistics Canada 1967–1979, 1979, 1981.

[b] Adapted from Robbins and Barewal, 1981; Statistics Canada, 1981.

[c] Direct use except for 1966, and 1977–1980, when indirect use in margarine and shortenings was not reported.

[d] Data not available.

B.   Waste of Fat in the Food Supply

1.   TOTAL FAT INTAKE

Domestic disappearance data are only approximations of food intake because no allowances are made for waste past the wholesale level; for this reason such data yield information only on "apparent" food consumption. Food consumption data from the 1970–1972 Nutrition Canada Survey (Health and Welfare Canada, 1975) provided an estimate of the actual fat intake by different physiological groups of Canadians (Table II). Fat intake ranged from 31 g per day by infants to 154 g per day by males in the 20–39 year age group. The fat intake for the population can be estimated by calculating a mean that is weighted by the proportions of the population in each

TABLE II

Mean Daily Fat Intake of Canadians

Physiological group[a]	Average daily fat intake 1970–1972[b] (g)	Population weighting factors 1971[c]	Portion of daily fat consumed by physiological group (g)
Infants	31	0.0165	0.5
MF 1–4 yr	70	0.0677	4.7
MF 5–11 yr	96	0.1481	14.2
MF 12–19 yr	147	0.0823	12.1
F	100	0.0793	7.9
M 20–39 yr	154	0.1413	21.8
F	89	0.1388	12.4
M 40–64 yr	118	0.1218	14.4
F	75	0.1233	9.2
M 65+	89	0.0363	3.2
F	63	0.0446	2.8
		1.0000	103.3

[a] M = male; F = female.
[b] Nutrition Canada Survey (Health and Welfare Canada, 1975).
[c] Statistics Canada (1978).

physiological group. In this way the mean daily intake of fat for the Canadian population for 1970–72 was estimated as 103.3 g (Table II). This is substantially less than the 149 g of fat in the 1971 food supply (Table I), suggesting that approximately 30% of available dietary fat is wasted, rather than consumed.

2. WASTE OF FAT BY FOOD CATEGORY

Table III shows the fat supplied by each food group in the 1971 food supply together with the weighted mean fat intake for the same food groups from the 1970–1972 Nutrition Canada Survey. The differences between these two sets of data provide the opportunity to identify the sources of fat waste. For example, it is evident that 27.9 g (44%) of the fat supplied by the meat group was wasted. As the total waste of fat was 45.7 g, the balance (17.8 g) must be accounted for from the products of food fats and oils. Table IV summarizes such an accounting.

Of the 59.2 g of food fats and oils in the 1971 food supply, 22.8 g appear to have been consumed in prepared foods; this amount represented the increase in fat intake from that available in cereals, fruits, and vegetables, and "other" foods such as mixed dishes (Tables III and IV). Food fats and oils

TABLE III

Comparison of Food Sources of Fat at Retail and Intake Levels (g/capita/day)

Food groups	Fat available at retail level (apparent consumption)[a] 1971	Fat consumed (Nutrition Canada Survey weighted means)[b] 1970–1972	Food fats apparently added in preparation (by difference)
Meat, fish, poultry, eggs	63.0	35.1	
Cereals	2.1	15.6[c]	13.5[c]
Fruits, vegetables, potatoes	0.9	6.3[c]	5.4[c]
Milk and milk products	19.0	19.0	
Butter	15.8	11.3	
Margarine	9.3[c]	4.8[c]	
Lard, shortening, salad oil	34.1[c]	2.5[c]	
Other foods (pulses, nuts, cocoa, mixed dishes)	4.8	8.7[c]	3.9[c]
	149.0	103.3	22.8

[a] Robbins and Barewal (1981).
[b] Calculated from Nutrition Canada food consumption data (Health and Welfare Canada, 1975) using the factors shown in Table II.
[c] Possible sources of rapeseed oil.

would be introduced in preparing such items as baked products of flour, French-fried potatoes and assorted casserole items. Specific amounts of tablespreads and salad oils were consumed as such. However, 9 g or 36% of the tablespreads supplied remained available for preparing other foods. Similarly, 66% of the salad oil supply was available for use in cookery. These quantities together with all the lard and shortening made a total of 40.6 g of food fats and oils in the food supply available for food preparation, yet only 22.8 g were consumed in prepared products. Thus, the ratio of fat intake:fat available for food preparation was 22.8/40.6 or 0.562. Applying this ratio permitted estimates of the use as opposed to the supply of each product in prepared foods. The final waste factors by product were calculated as 16% of tablespreads, 44% of lard and shortenings, and 28% of salad oils (Table IV). The high waste of lard and shortening is understandable considering that substantial amounts of used fats can be discarded by institutional and commercial frying operations. Application of these waste factors permits the adjustment of apparent food consumption data for food fat products from years other than 1970–1972 which was the only period represented by the Nutrition Canada Survey.

TABLE IV

Estimated Waste of Food Fats and Oils (1970–1972)[a] (g/person/day)

Fat statistics	Table-spreads[b]	Lard	Shorten-ing	Salad oil	Fats in prepared foods	Total food fats
Apparent consumption	25.1	8.0	18.7	7.4	—	59.2
Reported intake	16.1	—	—	2.5	22.8	41.4
Apparent use in food preparation (by difference)	9.0	8.0	18.7	4.9	40.6	—
Estimated intake in prepared foods[c]	5.0	4.5	10.5	2.8	22.8	—
Total intake	21.1 (84)[d]	4.5 (56)[d]	10.5 (56)[d]	5.3 (72)[d]	—	41.4 (70)[d]
Waste	4.0 (16)[d]	3.5 (44)[d]	8.2 (44)[d]	2.1 (28)[d]	—	17.8 (30)[d]

[a] Derived from data in Tables I and III.
[b] Data for butter and margarine combined assuming interchangeable use.
[c] Derived for each product by: apparent use in food preparation × (22.8/40.6).
[d] Represents proportion of apparent consumption in percent.

III. SOURCES OF DOMESTICALLY DEODORIZED FATS AND OILS

A. The Total Edible Oil Production

The total production of edible oils has increased by 85% over the past 15 years (Tables V through VII). At the same time the proportion of vegetable oil used has increased to 88%, largely at the expense of marine oils (Fig. 2). Grading systems that favor leaner hogs and beef have held the amounts of lard and tallow relatively constant so that their proportion has decreased gradually over time.

Within vegetable sources, rapeseed and soybean have steadily replaced other vegetable oils. Together they accounted for 50% of all edible oils in the 1966–1970 period; this proportion increased to an average of 68% in the last 5 years (Fig. 2). When viewed as 5-year averages, rapeseed oil use has exceeded that of soybean throughout the last decade, and by 1981 accounted for 45% of all domestically deodorized oils (Statistics Canada, 1982). The increase in rapeseed oil use was interrupted by a drop in production in 1974 which was related to the conversion from high to low erucic acid canola varieties of rapeseed. Early canola cultivars had a lower yield than their high erucic acid prototypes.

TABLE V

Source of Fats and Oils Used in the Domestic Production of Canadian Shortening Oils[a] (1000 metric tonnes)

Calendar year	Vegetable								Marine		Animal			Total all species
	Rape-seed[b]	Soy-bean	Palm	Sun-flower	Coco-nut	Cotton-seed	Palm kernel	Total[c]	Herr-ing	Total[d]	Lard	Tallow	Total[e]	
1966	17.0	29.4	7.0	x[f]	x	x	x	76.5	x	x	x	x	x	114.9
1967	17.6	32.1	5.7	x	12.1	x	4.3	85.8	8.5	8.5	12.9	20.1	33.0	127.4
1968	20.9	32.0	5.4	3.8	11.1	x	4.3	89.6	6.7	7.4	11.3	20.7	33.3	130.4
1969	22.2	37.6	9.4	5.3	11.5	6.6	4.6	103.1	9.3	10.4	9.6	20.1	31.0	144.1
1970	19.8	44.9	7.3	2.6	13.2	9.0	4.5	106.1	5.0	5.5	11.5	23.5	35.5	147.1
1971	27.6	34.0	9.1	1.6	13.4	6.6	4.2	100.3	4.6	4.6	15.4	23.3	39.0	144.0
1972	33.6	32.9	16.3	2.1	16.5	5.3	4.6	114.8	5.2	5.7	14.5	24.9	39.5	160.1
1973	40.6	38.4	16.1	1.5	17.6	5.7	5.1	127.9	5.4	5.7	12.9	23.0	36.4	170.0
1974	29.7	54.5	9.1	1.9	10.5	4.4	3.9	117.0	2.0	2.9	19.4	22.4	43.0	162.8
1975	22.1	46.3	25.6	0.2	16.2	4.3	5.0	122.7	3.2	3.4	13.1	23.9	37.6	163.7
1976	21.5	50.0	30.4	x	18.0	2.7	x	133.0	1.7	1.9	8.2	20.5	30.0	164.9
1977	32.8	42.7	24.2	x	x	x	x	129.2	x	x	x	x	x	163.7
1978	35.8	47.2	x	2.8	x	x	x	123.6	x	x	x	x	x	161.8
1979	55.8	x	x	x	x	x	x	150.4	x	x	x	x	x	201.5
1980	65.5	x	x	x	x	x	x	154.4	x	0	x	x	x	206.6
1981	72.8	44.8	10.3	x	x	x	x	156.9	0	0	x	x	x	209.1[g]

[a] Statistics Canada (1982).
[b] Prior to 1966 rapeseed oil was included under "other vegetable oils."
[c] Includes peanut, corn, and others.
[d] Includes seal, whale, and others.
[e] Includes all types of oleo.
[f] Confidential.
[g] Estimated.

TABLE VI

Source of Fats and Oils Used in the Domestic Production of Canadian Margarine Oils[a] (1000 metric tonnes)

Calendar year	Vegetable					Marine		Animal		Total all species
	Rapeseed[b]	Soybean	Palm	Corn	Total[c]	Herring	Total[d]	Lard	Total[e]	
1966	x[f]	26.1	x	x	50.1	x	x	x	x	66.3
1967	16.4	24.4	x	x	48.2	15.0	15.0	1.4	1.4	64.5
1968	14.9	23.3	3.1	x	46.3	16.2	17.8	1.7	1.7	65.8
1969	19.0	23.9	3.9	x	50.3	17.4	18.2	0.7	0.7	69.3
1970	18.8	27.0	2.5	2.6	51.6	10.2	11.7	1.3	1.3	64.5
1971	21.5	18.5	3.1	3.0	47.0	11.0	11.5	1.6	1.6	60.0
1972	31.2	19.0	4.3	3.8	59.1	7.3	7.6	1.1	1.2	67.9
1973	34.4	26.9	4.0	3.7	69.5	6.4	6.4	0.4	0.6	76.5
1974	28.8	41.0	4.1	4.7	79.2	2.9	3.0	2.0	2.2	84.5
1975	33.8	39.8	6.2	5.9	87.0	3.4	3.7	1.8	2.8	93.4
1976	31.9	50.0	6.9	7.2	97.3	1.4	1.6	1.7	1.9	100.7
1977	35.0	53.4	x	x	103.0	x	x	x	x	105.2
1978	39.9	53.9	x	x	111.6	x	x	x	x	114.0
1979	44.0	55.5	x	x	119.0	x	x	x	x	122.7
1980	38.6	61.1	x	x	120.5	x	x	x	x	123.0
1981	46.0	62.4	3.6	x	127.3	x	x	x	x	129.7[g]

[a] Statistics Canada (1982).
[b] Prior to 1966 rapeseed was included under "other vegetable oils."
[c] Includes coconut, cottonseed, palm kernel, peanut, sunflower, and others.
[d] Includes seal, whale, and others.
[e] Includes all types of oleo and tallow.
[f] Confidential
[g] Estimated.

241

TABLE VII

Source of Vegetable Oils Used in the Domestic Production of Canadian Salad and Cooking Oils[a] (1000 metric tonnes)

Calendar year[b]	Rapeseed	Soybean	Palm	Sunflower	Cottonseed	Total all species[c]
1966	x[d]	13.6	x	x	x	57.7
1967	12.2	13.2	x	12.5	x	50.5
1968	17.2	11.2	xx[e]	16.0	x	55.9
1969	20.4	8.4	x	14.9	0.4	55.1
1970	20.7	11.3	xx	10.0	2.1	55.5
1971	23.9	12.6	—[f]	7.9	1.1	57.8
1972	31.6	14.0	xx	7.9	0.9	66.8
1973	33.0	13.4	—	10.4	0.1	69.2
1974	32.9	22.6	0.1	7.4	0.4	76.7
1975	38.9	19.5	1.5	x	0.4	80.5
1976	47.3	21.3	1.1	x	0.7	94.5
1977	53.5	20.4	x	x	x	101.7
1978	56.0	x	x	x	x	99.8
1979	69.2	x	—	x	—	111.1
1980	79.1	x	—	x	—	116.8
1981	96.4	13.2	2.7	x	—	128.2

[a] Statistics Canada (1982).
[b] Prior to 1966 salad oils were not listed by species.
[c] Includes coconut, palm kernel, peanut, corn, and other oils.
[d] Confidential.
[e] Less than .04 metric tonnes.
[f] Nil.

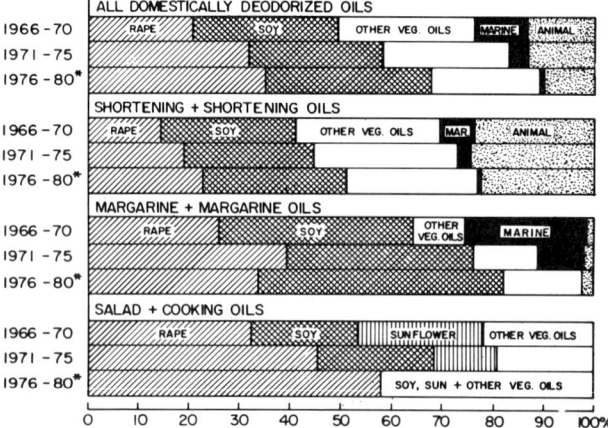

Fig. 2. Sources of Canadian domestically deodorized fats and oils, by product category. (Five-year averages derived from data in Tables V, VI, and VII.)

B.   Shortening and Shortening Oils

The manufacture of shortenings occupies the largest share of the total edible oil production; however, the proportion so used has decreased from 52 to 45% between the late 1960s and the 1980s as the use of both margarine and salad oils has increased. Shortening formulations have traditionally used animal fats and continue to do so. However, the marine oil that was used in the 1960s has now been largely replaced by vegetable oils. From 1976 to 1980, vegetable oils accounted for 77% of all shortening oils that were processed (Fig. 2).

A wide variety of vegetable oil sources contribute to shortenings reflecting the responsiveness of the product category to world supply and local prices (Table V). Coconut and palm kernel oils have been used in apparently constant quantities; cottonseed oil use has been decreasing while palm oil use increased to very substantial amounts in the mid-1970s. Sunflower oil has been used to a minor and varying extent; it is expected to rise with the current increase in local production. However, over the past 15 years soybean and rapeseed oils have been major contributors to shortening blends; in 11 of the 14 years for which statistics are available, soybean has been the largest single source. Rapeseed oil use exceeded that of soybean oil in 1972 and 1973 just prior to the drop in production that was experienced with the conversion to low erucic acid varieties; it took 6 years for rapeseed oil to exceed its 1973 level of use in shortening products.

Concern about the functional properties of low erucic acid rapeseed oil may have limited its growth in some segments of the shortening industry. While reducing erucic acid in rapeseed oil has improved its nutritional potential, the genetically altered oil exhibits novel crystal behavior following hydrogenation; the fine crystal structure established in processing can grow increasingly coarser on storage. The extent of this polymorphic shift appears to be inversely related to the erucic acid content of the oil (Persmark and Bengtsson, 1976). While a coarser crystal structure is desirable for pie crusts and Danish pastries, cakes and breads made with plastic fat require fine fat crystals to provide the creamy texture needed for good volume in the finished products (Knightly, 1981). Recrystallization in hydrogenated low erucic acid rapeseed oil can be avoided by combining it with other oil sources, or limited by the use of crystal modifying additives such as sorbitan tristearate (Loewen, 1980).

Although canola rapeseed shortenings perform well in frying (Stevenson et al., 1981) manufacturers of fried snack foods have been reluctant to use high linolenic acid oils like soybean and rapeseed, due to concerns about reversion flavors in finished products. This reluctance has been offset in recent years by more product demonstration as well as by pricing pressures. Also processors of rapeseed oil are finding that less arduous conditions may be used to hydrogenate canola varieties of rapeseed. Canola seed is lower in glucosinolates than the original rapeseed, thus, the drift of sulfur from the seed to the oil during extraction is virtually eliminated. Residual sulfur in oil "poisons" the nickel catalyst. With catalyst life extended, more moderate hydrogenation conditions can be used with canola oils; the benefits are reduced cost and improved product quality (El-Shattory et al., 1981).

C. Margarine and Margarine Oils

Margarine oils have increased from 26 to 29% of the edible oil production during the last 15 years, reflecting the growing acceptance of this tablespread by consumers. Vegetable oils have contributed to margarine composition throughout this period so successfully that their proportion has increased from 75 to 98% (Fig. 2). The increase has been largely due to the replacement of marine oils, which in the early 1960s accounted for close to 25% of margarine production.

Soybean and rapeseed are the main sources of oil for Canadian margarines; in the 5 years between 1976 and 1980 these two oils accounted for 48 and 33% of margarine oils, respectively (Fig. 2). The use of rapeseed oil in margarine increased steadily between 1966 and 1973. However, in 1974, coincidental with the introduction of low erucic acid varieties, the proportion of rapeseed oil used in margarines dropped and subsequent increases were gradual (Table VI). The susceptibility of hydrogenated low erucic acid rapeseed oil to recrystallization is more troublesome in tablespreads than shortenings (deMan, 1978). To avoid the hazard of a grainy, gritty texture developing in their margarines, many manufacturers prefer to use canola oil as one of two or three oils in the margarine base stocks. This practice limits the use of canola oil in margarine and maintains a dependency on imported oil sources.

Recently, sunflower oil margarines have been introduced to the Canadian market. While hydrogenated sunflower oil also is susceptible to crystal growth on storage (Vane, 1981), the high content of linoleic acid offers the opportunity for manufacturers to claim a nutritional advantage. Previously, only corn oil margarines have so qualified. Neither canola rapeseed oil nor soybean oil, when hydrogenated, contains the 25% level of cis, cis-polyunsaturated fatty acids that is required to permit a nutritionally slanted label claim or advertising statement (Health and Welfare Canada, 1979).

D. Salad and Cooking Oils

By 1981, salad and cooking oils represented a 27% segment of the total domestic edible oil production. This product category is derived entirely from vegetable oils with rapeseed, soybean, and sunflower seed as the main sources (Fig. 2). Oils from other sources such as corn, cottonseed, peanut, and olive are on the Canadian market; however, their individual quantities have not been reported consistently by Statistics Canada (Table VII).

The rapeseed component of salad and cooking oils has increased steadily throughout the past 15 years apart from the short supply in 1974 (Table VII). By 1981, it accounted for 71% of the entire domestic production of salad oils (Statistics Canada, 1982). The popularity of rapeseed oil in this product category may be credited to its good functional properties and ease of processing.

As a frying oil, low erucic acid rapeseed compares well with soybean oil in smoke point, frying stability, and finished product quality (Vaisey-Genser and Eskin, 1979). For salad dressings, rapeseed oil has the necessary bland flavor and cold stability. Both soybean and rapeseed oils are high in linolenic acid, which is expected to threaten their flavor stability during storage. Soybean processors routinely partially hydrogenate and winterize soybean oil to improve its flavor stability. This processing sequence has not proved necessary for rapeseed oil. Despite its higher linolenic acid content, commercially processed rapeseed oil compares well in flavor stability with partially hydrogenated soybean oil (Eskin and Frenkel, 1977).

Research from the baking industry suggests that liquid oils will increasingly replace plastic fats in the coming years. If this trend becomes established, the salad and cooking oil segment of the edible oil production can be expected to show marked increases (Knightly, 1981).

IV. AVAILABILITY OF RAPESEED OIL TO CANADIANS (1966–1980)

A. Apparent Consumption of Rapeseed Oil

The amounts of rapeseed oil that have apparently been consumed by the Canadian population over time can be estimated by considering the proportion of rapeseed oil in annual supplies of product categories (Tables V through VII) together with the apparent consumption of margarine, shortenings, and salad oils (Table I). The results of such calculations are shown in Table VIII. Since 1967, the amount of rapeseed oil in the food supply has increased from 6.4 to 17.1 g per capita daily in 1980, a gain of 167%. The 1980 apparent consumption of low erucic acid rapeseed oil by the Canadian population thus represented 28% of the fat from food fat products and approximated 12% of the total dietary fat available (Table I).

TABLE VIII

Canadian Apparent Consumption of Rapeseed Oil

Calendar year	% Production from rapeseed oil[a]			Apparent consumption of rapeseed oil[b] (g/capita/day)				Apparent consumption of erucic acid from rapeseed oil	
	Marg-arine	Short-ening	Salad oil	Marg-arine	Short-ening	Salad oil	Total	% Erucic acid[c]	g/capita/day
1966	x[d]	14.8	x	—[e]	2.4	—	—	26.0	—
1967	25.5	13.8	24.2	2.3	2.4	1.7	6.4	26.0	1.66
1968	22.7	16.0	30.8	2.2	2.8	2.3	7.3	26.0	1.90
1969	27.3	15.4	37.1	2.7	2.9	2.7	8.3	26.0	2.16
1970	29.1	13.4	37.2	2.7	2.6	2.7	8.0	26.0	2.08
1971	35.8	19.2	41.2	3.4	3.6	3.0	10.0	26.0	2.60
1972	45.9	21.0	47.3	4.5	4.4	4.0	12.8	9.5	1.22
1973	45.0	23.9	47.7	4.5	5.2	4.1	13.8	9.7	1.34
1974	34.1	18.2	42.9	3.7	3.9	4.0	11.6	5.6	0.65
1975	36.2	13.5	48.3	4.2	2.9	4.7	11.8	3.1	0.37
1976	31.7	13.0	50.1	3.9	2.8	5.6	12.3	2.3	0.28
1977	33.3	20.0	52.6	4.3	4.2	6.3	14.8	1.6	0.24
1978	35.0	22.1	56.1	4.6	4.6	6.5	15.7	1.3	0.20
1979	35.9	27.7	62.3	4.3	6.2	6.4	16.9	1.3	0.22
1980	31.4	31.7	67.7	3.8	6.7	6.6	17.1	1.1	0.19

[a] Derived from Tables V, VI, and VII.
[b] Derived as follows: apparent consumption of product (Table I) x % production of product from rapeseed oil.
[c] Average erucic acid content in rapeseed oil (Daun, 1981).
[d] Confidential.
[e] Data not available.

B. Apparent Consumption of Rapeseed Oil Corrected for Waste

The apparent consumption of rapeseed oil makes no allowance for the waste experienced between sale to the retailer and use by the consumer. Applying the waste factors calculated earlier in this chapter (Table IV) makes it possible to estimate more closely the per capita daily intake of rapeseed oil, as opposed to what was available. The results of such corrections are illustrated in Fig. 3. It may be noted that, as rapeseed oil in the food supply increased, the difference between apparent consumption and estimated in-, take became greater. This difference was accentuated particularly by the substantial increase in shortening supplies over the years (Fig. 1); shortenings appear to have the greatest waste of the three visible fat product categories (Table IV).

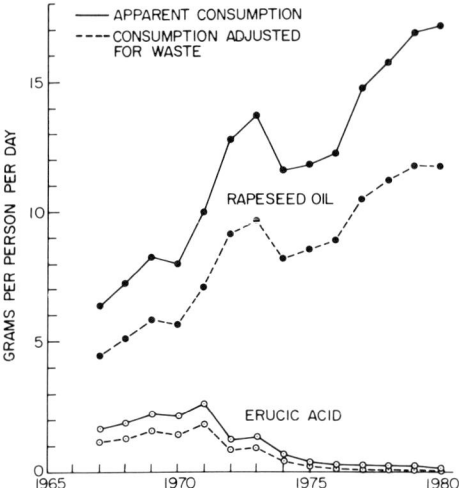

Fig. 3. Apparent consumption of rapeseed oil and erucic acid by the Canadian population. (See Table VIII; correction factors for waste in Table IV.)

Allowing for waste in all products the intake of rapeseed oil by the Canadian population in 1980, was estimated as 11.7 g per capita daily (Fig. 3), representing 12% of the expected total intake of dietary fat (Table I).

C. Apparent Consumption of Docosenoic Acids (22:1)

1. ERUCIC ACID

The level of consumption of rapeseed oil by a population is of interest nutritionally as it affects the fatty acid composition of the total diet. The traditional rapeseed oils were unique among edible vegetable oils because of their high erucic acid (22:1) content. The dietary intake of erucic acid was of particular concern in the early 1970s when evidence suggested that it was linked to cardiac lipidosis and necrosis in some experimental animals (Abdellatif, 1972). These findings prompted Canadian growers to begin converting their rapeseed production to low erucic acid varieties in the 1971 crop year. A steady decrease in the erucic acid content of the annual production took place from 1972 to 1980 (Table VIII). As a result the apparent consumption of erucic acid reached its highest level in 1971 and has declined fairly steadily since, despite substantial increases in the supply of rapeseed oil (Fig. 3). Lately, the supply of rapeseed oil has been entirely from low erucic acid canola cultivars. Correction of the available erucic acid for waste in food preparation, using the factors shown in Table IV, sug-

TABLE IX

Apparent Consumption of Docosenoic Acids

Calendar year	Apparent consumption of docosenoic acids from rape and herring oils (g/capita/day)	Docosenoic acid consumption corrected for waste[b] (g/capita/day)		
		Rape	Herring	Total
1967	2.45	1.17	0.59	1.76
1968	2.68	1.32	0.60	1.92
1969	3.04	1.52	0.66	2.18
1970	2.60	1.47	0.39	1.86
1971	3.15	1.83	0.43	2.26
1972	1.63	0.87	0.30	1.17
1973	1.71	0.94	0.26	1.20
1974	0.80	0.46	0.11	0.57
1975	0.57	0.26	0.14	0.40
1976	0.37	0.20	0.07	0.27
1977	0.24	0.17	x[c]	0.17
1978	0.20	0.15	x	0.15
1979	0.22	0.15	x	0.15
1980	0.19	0.13	x	0.13

[a] Erucic acid (Table VIII) plus cetoleic acid from partially hydrogenated herring oil (22:1 = 24.1%, Beare-Rogers et al., 1971) used in margarine and shortening (Tables I, V, and VI).
[b] Derived using factors in Table IV.
[c] Herring oil use not reported for 1977–1980.

gests that the daily intake of erucic acid by Canadians has ranged from a high of 1.83 g per capita in 1971 to 0.13 g per capita in 1980 (Fig. 3, Table IX).

2.   OTHER DOCOSENOIC ACIDS

Marine oils, like rapeseed oil, contain docosenoic acids (Eaton et al., 1977). Partially hydrogenated herring oil containing 24% cetoleic acid, a positional isomer of erucic acid, was also shown to be associated with cardiac lipidosis in young rats (Beare-Rogers et al., 1971). In 1973, the Health Protection Branch of the Canadian government required food manufacturers to restrict the content of C_{22} monenoic fatty acids in processed foods to 5% of the total fatty acids present (Anonymous, 1973). This goal had been achieved by 1972 by the combined effect of replacing high with low erucic acid rapeseed oil and reducing the use of herring oil (Table IX). By 1976, the herring oil in margarines and shortenings was less than 1.5%; precise amounts have not since been reported. Correction of the known available supply of docosenoic acids for waste suggests that their intake has dropped from a high of 2.26 g per capita daily in 1971 to 0.13 g per capita daily in 1980 (Table IX).

D. Margin of Safety in the Intake of Docosenoic Acids

In 1971, Beare-Rogers and co-workers estimated that there was no myocardial lipidosis in young rats fed 3% of calories as erucic acid. This value provides an approximate "no effect" level for gauging the margin of safety for docosenoic acids that exists in the food supply. The average caloric intake of the Canadian population, weighted by physiological group, has been estimated as 2353 calories in 1970–1972 (Robbins and Barewal, 1981); 3% represents 7.84 g docosenoic acids [(2353 × 0.03)/9 = 7.84]. To reach a 100-fold safety factor would limit the "acceptable daily intake" (ADI) of docosenoic acids to 0.0784 g for the average person. This amount would reduce to 1.12 mg/kg for the average 70 kg male, which is appreciably lower than the ADI of 8.5 mg/kg suggested by calculations of the British Industrial Biologists Research Association in 1972 (Anonymous, 1972). When comparing a no-effect level of 7.84 to the apparent consumption of docosenoic acids (corrected for waste), which is shown in Table IX, it may be seen that in 1980 there was a 60-fold margin of safety (7.84/0.13). Such estimates of safety margins should be considered in light of the questions that have been raised about the reliability of the rat as an experimental animal for docosenoic acid toxicity testing (see also Chapters 11 and 21).

The consumption estimates for docosenoic acids that are listed in Table IX should be viewed with caution. The values for marine oils may well be underestimated, since no allowance has been made for sources other than herring, and even herring oil data are excluded from 1977–1980. Seal oil is known to contain 3–5% docosenoic acid (Eaton et al., 1977). On the other hand, the intake of erucic acid is likely overestimated considering that no allowance has been made for the possibility that some erucic acid may convert to behenic acid (22:0) in the process of hydrogenating rapeseed oil. Behenic acid apparently does not produce the pathology associated with erucic acid (Nolen, 1981). Data from the literature suggest that 0–22% of erucic acid may be hydrogenated to behenic acid in processing rapeseed oil. The amount of conversion depends on the type of process used and the firmness desired in the finished product. Selective hydrogenation, which has been more widely used than the nonselective process, forms appreciably less behenic acid (Swindells, 1970). Margarines are desirably less firm than shortenings and so are hydrogenated more briefly, particularly in the case of soft tub-type products which are increasing in popularity. Formulations for margarine and baking shortenings often combine oil fractions hydrogenated to differing degrees of firmness. Where low erucic acid rapeseed oil is used as one of two or three components, it may be only lightly hydrogenated to avoid later recrystallization problems (Teasdale, 1975). However, in designing cube-type shortenings for heavy duty frying, there will be more extensive hydrogenation to assure stability. Selectively hydrogenated solid frying fats may have 10–12% of the erucic acid converted to behenic acid. Absence of adequate data on the proportions of the various products within the shortening category precludes correction of erucic acid consumption for hydrogenation effects.

V. DEMOGRAPHIC VARIABLES AFFECTING RAPESEED OIL CONSUMPTION

Average values for the apparent consumption of food by a population should not infer perfect distribution across all members. The differences in food fat intake among physiological groups and income levels are illustrated in Fig. 4 which summarizes some of the food consumption data from the Nutrition Canada Survey of 1970–1972 (Myres and Kroetsch, 1978). The design of the survey did not permit precise statistical interpretation. However, it is evident that males consumed appreciably more food fats than females, in all age groups. The differences in fats eaten were most dramatic between sexes, and were more marked among age groups than among income levels.

Men between 20 and 39 years showed the highest intake of fats in the Nutrition Canada Survey (Table II). Vaisey *et al.* (1973) carried out a controlled metabolic study in which young adults consumed a "typical" Canadian diet except that all tablespread, cooking fat, and salad oil were entirely rapeseed oil. The young males (19–28 years) in this study consumed an

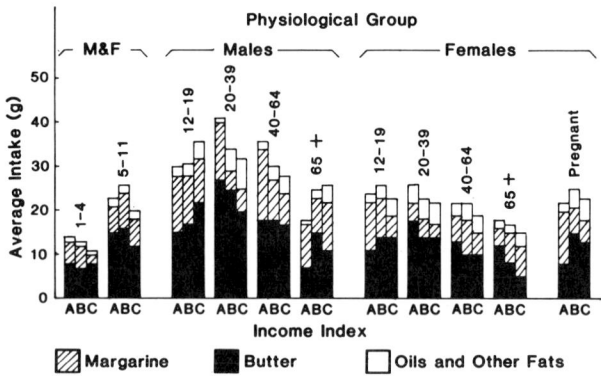

Fig. 4. Average daily intake of food fats and oils by physiological group and income index: A = lowest income (*n* = 1515); B = low income (*n* = 3034); C = all others (*n* = 7729). Adapted from Myres and Kroetsch (1978).

average of 136 g of fat daily of which 79 g or 58% was estimated to be from rapeseed oil. If this percentage is applied to the 154 g fat intake of the 20–39 year male group surveyed in 1970–1972 (Table II) then 90 g rapeseed oil may represent the upper intake possible in the highest fat-consuming segment of the population.

Urban family food expenditure data suggest that the use of margarine and shortening increases as income decreases (Statistics Canada, 1980). This observation is confirmed by the Nutrition Canada Survey data in Fig. 4 where it can be seen that the intake of food fats by young adult males was highest in the lowest income category. The trend toward increasing food fat intake with decreasing income was also apparent among adult women; however, the opposite was observed among adolescent boys and elderly men. Considering that rapeseed may often be the oil of choice for lower-priced house brands of margarine and salad oil, it is likely the rapeseed oil intake is higher among lower income families.

The current economic situation favors increasing use of low erucic acid rapeseed oil by consumers facing higher food prices and by food manufacturers. It is reasonable to forecast that the Canadian consumption of low erucic acid rapeseed oil will continue to rise. World prices favor the development of technology to process local products and pressures on the food budget encourage consumers to seek the best buys in the supermarket.

REFERENCES

Abdellatif, A. M. M. (1972). *Nutr. Rev.* **30**, 2–6.
Anonymous (1972). *Inf. Bull.—Br. Ind. Biol. Res. Assoc.* **11**, 272–273.
Anonymous (1973). *Food Can.* **33**, 38.
Anonymous (1979). "Fleischman Consumer Panel Report on Consumer Purchases of Bakery Products in Canada." Standard Brands, Montreal, Quebec.
Beare-Rogers, J. L., Nera, E. A., and Heggtveit, H. A. (1971). *Can. Inst. Food Technol. J.* **4**, 120–124.
Daun, J. (1981). "Grain Research Laboratory Annual Crop Survey." Board of Grain Commission, Winnipeg, Manitoba.
deMan, J. M. (1978). *Can. Inst. Food Sci. Technol. J.* **11**, 194–203.
Eaton, C. A., Ackman, R. G., and Brodie, P. F. (1977). *Fish. Mar. Serv. Tech. Rep.* **724**. Fisheries and Environment Canada, Ottawa.
El-Shattory, Y., deMan, L., and deMan, J. M. (1981). *Can. Inst. Food Sci. Technol. J.* **14**, 53–58.
Eskin, N. A. M., and Frenkel, C. (1977). *Actes Congr. Mond.—Soc. Int. Etude Corps Gras, 13th, 1976* Sect. A, pp. 1–9.
Hassan, Z. A., and Karamchandani, D. (1977). "Handbook of Food Expenditures, Prices and Consumption," Agric. Can. Publ. No. 77/13. Supply and Services Canada, Ottawa, Ontario.
Hassan, Z. A., and Karamchandani, D. (1980). "Handbook of Food Expenditures, Prices and Consumption," Agric. Can. Publ. No. 80/4. Supply and Services Canada, Ottawa, Ontario.

Health and Welfare Canada (1975). "Nutrition Canada Food Consumption Patterns Report." Supply and Services Canada, Ottawa, Ontario.

Health and Welfare Canada (1977). "Nutrition Canada Survey Food Nutrient Conversion File." Supply and Services Canada, Ottawa, Ontario.

Health and Welfare Canada (1979). "Health Protection and Food Laws." Supply and Services Canada, Ottawa, Ontario.

Knightly, W. H. (1981). *Cereal Chem.* **58**, 171–174.

Loewen, D. K. (1980). *ISF/AOCS World Congr., 1980* Abstract 323.

Myres, A. W., and Kroetsch, D. (1978). *Can. J. Public Health* **69**, 208–221.

Nolen, G. A. (1981). *J. Am. Oil Chem. Soc.* **58**, 31–37.

Persmark, U., and Bengtsson, L. (1976). *Riv. Ital. Sostanze Grasse* **53**, 307–311.

Rizek, R. L., Friend, B., and Page, L. (1974). *J. Am. Oil Chem. Soc.* **51**, 244–250.

Robbins, L., and Barewal, S. (1981). "The Apparent Nutritive Value of Food Available for Consumption in Canada 1960–75," Agric. Can. Publ. No. 180/6. Supply and Services Canada, Ottawa, Ontario.

Statistics Canada (1978). "1976 Census of Canada. Population: Demographic Characteristics Five-year Age Groups," Cat. No. 92.823. Supply and Services Canada, Ottawa, Ontario.

Statistics Canada (1967–1979). "Apparent Per Capita Food Consumption in Canada," Cat. No. 32–226. Supply and Services Canada, Ottawa, Ontario.

Statistics Canada (1979). "Livestock and Animal Product Statistics." Cat. No. 23–203. Supply and Services Canada, Ottawa, Ontario.

Statistics Canada (1980). "Urban Family Food Expenditure 1978," Cat. No. 62–548. Supply and Services Canada, Ottawa, Ontario.

Statistics Canada (1981). "Apparent Per Capita Food Consumption in Canada 1979. Part II," Cat. No. 32-230. Supply and Services Canada, Ottawa, Ontario.

Statistics Canada (1982). "Oils and Fats," Cat. No. 32–006. Annual Supply and Services Canada, Ottawa, Ontario.

Stevenson, S., Vaisey-Genser, M., Eskin, N. A. M., Tassos, L., and Fyfe, B. (1981). *Can. Inst. Food Sci. Technol. 24th Ann. Meet.* Abstract No. 89.

Swindells, C. E. (1970). *Can. Inst. Food Technol. J.* **3**, 171–175.

Teasdale, B. F. (1975). In "Oilseed and Pulse Crops in Western Canada—A Symposium" (J. T. Harapiak, ed.), pp. 551–585. Western Co-operative Fertilizers Ltd., Calgary, Alberta.

Thomas, L. H., Jones, P. R., Winter, J. A., and Smith, H. (1981). *Am. J. Clin. Nutr.* **34**, 877–886.

U. S. Department of Agriculture (1977). "U.S. Fats and Oils Statistics," Econ. Res. Serv. Stat. Bull. No. 574. USDA, Washington, D.C.

Vaisey, M., Latta, M., Bruce, V. M., and McDonald, B. E. (1973). *Can. Inst. Food Sci. Technol. J.* **6**, 142–147.

Vaisey-Genser, M., and Eskin, N. A. M. (1979). "Canola Oil—Properties, Processes and Food Quality," Publ. No. 55. Rapeseed Assoc. Can. Winnipeg, Manitoba.

Vane, B. (1981). M.Sc. Thesis, Department of Foods and Nutrition, University of Manitoba, Winnipeg.

Watt, B. K., and Merrill, A. L. (1963). U. S. Dep. Agric., Agric. Hand. **8**.

11

The Problems Associated with the Feeding of High Erucic Acid Rapeseed Oils and Some Fish Oils to Experimental Animals

F. D. SAUER AND J. K. G. KRAMER

High and Low Erucic Acid Rapeseed Oils
Copyright © 1983 by Academic Press Canada

I. INTRODUCTION

Rape belongs to the *Brassica* genus of crops which also includes mustard turnips, cabbage, rutabagas, broccoli, and kale. The English word "rape" as it applies to the oilseed forms of rape is derived from the Latin "rapum" meaning turnip. Although rapeseed crops are a fairly recent introduction into North America, they were among the first crops domesticated by early man. As reviewed by Singh (1958) reference is made to rapeseed cultivation in ancient Sanskrit writings as early as 2000–1500 BC. Rapeseed was introduced into Japan from China and Korea about the year 0 AD. Rapeseed oil has been the preferred cooking oil in these Asian countries as well as in Pakistan since ancient times. The two rapeseeds, i.e., *Brassica campestris* and *B. napus*, also have a long history of use in Europe. Records indicate that there was extensive cultivation of rapeseed in Middle and Northern Europe about the thirteenth century when the oil was widely used both in foods as cooking oil and also as lamp oil (Appelqvist, 1972).

The early high erucic acid containing rapeseed oils which have been used for cooking and in food products since ancient times do not appear to have had any nutritional or health problems associated with their use. At least no records have been found in early writings to suggest the occurrence of problems related to these oils.

II. GROWTH PERFORMANCE WITH HEAR OILS

The first reports of adverse nutritional effects with high erucic acid rapeseed oils, commonly abbreviated as HEAR oils, came in the 1940s, during a period in time when the relative merits of butter against margarine were being debated. Boer and his colleagues, over 40 years ago, noted that the growth rate in young rats was poorer with HEAR oil than with diets in which butter was substituted for rapeseed oil (Boer *et al.*, 1947). These results were readily confirmed and extended by Deuel who noted that not only did butterfat outperform HEAR oil in rat growth trials, but so did a number of other vegetable oils which included corn, cottonseed, olive, peanut, and soybean oils (Deuel et al., 1944, 1948b). Deuel was aware that HEAR oils were poorly digested by the rat and suspected that erucic acid (cis 22:1 n-9) was the principal culprit in the poor performance obtained with this oil.

Scientists from the Unilever Laboratories confirmed the observation of Deuel that erucic acid retards the growth rate in young rats. In a series of experiments that were conducted in the mid-1950s, Thomasson and Boldingh showed that weight gains in young rats decreased in proportion to the erucic acid content of the diet and that growth could be retarded equally by trierucin added to peanut oil as by HEAR oil (Thomasson and Boldingh, 1955). It is worthwhile to explain briefly the experimental protocol of Thomasson and Boldingh. In the first experiment, HEAR oil (with an erucic acid content of 50%) was incorporated into rat diets in increasing quantities from 10 calorie % to 73 calorie % (Thomasson, 1955a). As expected, the growth rate of 3 week old rats decreased in proportion to the HEAR oil content of the diet. In their next experiment, they fed different fat mixtures at 30 calorie % of the diet and were able to show that trierucin mixed into peanut oil in a half and half mixture had the same growth retarding effect as the HEAR oil with 50% erucic acid (Thomasson and Boldingh, 1955). If this HEAR oil was fed at extreme levels, i.e., at 73 calorie % of the diet, the rats died after about 16 days. Yet, somewhat paradoxically, Thomasson (1955b) found that HEAR oil fed at 50 calorie % of the diet prolonged the life of rats by 20–25% over that of the rats that were fed butterfat (669 vs. 545 days). Booth et al. (1972) observed no significant differences in mortality of Fischer male and female rats fed for 18 months 20% by weight soybean oil, HEAR oil, or crambe seed oil.

III. THE DIGESTIBILITY OF HEAR OILS

A. In Man

For results on the digestibility of HEAR oil in man, one can look to some early studies reported from the U.S. Department of Agriculture by Holmes (1918). Holmes found the average digestibility of HEAR oils in man to be 98.8%. This value was confirmed some 30 years later by Deuel and his collaborators who reported 99.0% digestibility for HEAR oil and 96.5% digestibility for cottonseed oil in human subjects (Deuel et al., 1949).

B. In Experimental Animals

Deuel observed a rather interesting species difference with respect to rapeseed oil digestibility. Whereas HEAR oil was almost completely digestible in man, in adult female rats HEAR oil was only 77% digestible when crude and 82% digestible when refined (Deuel et al., 1948a). In contrast to the rat, HEAR oil is also quite readily digested in swine with apparent digestibility values similar to those observed in man (Paloheimo and Jahkola, 1959). Some other laboratory animals such as rabbits and guinea pigs also have low digestibility coefficients for HEAR oil (Carroll, 1957). Although there is general agreement that laboratory rats digest high levels of HEAR oil rather poorly, there is some evidence to suggest that even among different strains of rats, the ability to digest HEAR oil can vary. Thus the Wistar rat, a smaller albino rat than the Sprague–Dawley rat, had a digestibility coefficient of 83% for HEAR oil (a 37% erucic acid HEAR oil was fed at 20% by weight of the diet) while the Sprague–Dawley rat fed the same oil had a digestibility coefficient of only 65% (Beare et al., 1960).

C. Digestibility of Free Fatty Acids and Their Esters in the Rat

The digestibility experiments reported here were generally carried out with standard rat diets containing approximately 20% protein and 4–5% mineral mixture by weight (see Chapter 13). It is well recognized, however, that changes in the salt mixture, particularly with divalent metal ions, have considerable influence on the digestibility coefficients of fatty acids (Carroll and Richards, 1958). Extreme protein concentrations can also affect fat digestibility (Carroll and Richards, 1958). These points were emphasized by Astorg (1980) who obtained near quantitative absorption of erucic and brassidic acids when calcium salts were omitted from the diet.

In a series of experiments, Carroll (1958) found that the coefficient of digestibility decreased with increasing chain length for both saturated and monounsaturated fatty acids (Table I). Interestingly, the fatty acid binding protein (cytosolic protein) involved in the translocation of fat across the intestinal cells shows a similar specificity for fatty acids, i.e., a decreasing affinity with increasing chain length (Ockner et al., 1972).

When Carroll (1958) fed fatty acids as the methyl or ethyl esters, these esters were more efficiently absorbed than the free fatty acids by the rat. However, for fatty acids in the form of triglycerides (TG) the effect was opposite for saturated and monounsaturated fatty acids. A saturated TG was less digestible than the fatty acid itself, while a TG composed of a monounsaturated fatty acid was absorbed better than its free fatty acid (Carroll and Richards, 1958). Triglyceride fatty acids also influence absorption, particularly with the poorly digested fatty acids. For example, the absorption coefficient of erucic acid was increased from 72% when fed as ethyl erucate to 80% when ethyl erucate was mixed with soybean oil (Ziemlanski et al., 1973a). Even the absorption of 22:1 in HEAR oil was improved by mixing

TABLE I

Coefficient of Digestibility[a] of Fatty Acids in the Rat

Fatty acid	Coefficient of digestibility as		
	Free fatty acid	Ester	Triglyceride
10:0	100		
12:0	86		
14:0	64		
16:0	39		22
18:0	23		14
22:0	7		
18:1n-9	74	96	99
20:1	65		
22:1n-9	51	59	63
24:1n-9	20		

[a] Coefficient of digestibility = (lipid ingested − lipid excreted)/ lipid ingested × 100. Results taken from Carroll, 1958; Carroll and Richards, 1958.

HEAR oil with soybean oil (73 vs. 80%). These results suggest that higher concentrations of polyunsaturates may improve micelle formation or enhance fat transport across the intestinal cell, thereby resulting in increased uptake of poorly absorbed fatty acids.

In addition to the type of fatty acid, the position of the fatty acid in the TG will affect its digestibility. Maximum absorption is obtained when a poorly absorbed fatty acid is esterified at the 2-position of the TG molecule (Tomarelli et al., 1968; Filer et al., 1969) and there are unsaturated fatty acids in the 1,3-position (Mattson and Streck, 1974). A poorly absorbed acid like behenic acid (22:0) can be made partly digestible by feeding it in the form of 2-behenoyl dilinolein (Mattson and Streck, 1974).

The generally poor digestibility of HEAR oil, particularly in rats of the Sprague–Dawley strain, is probably caused by a number of factors. First, there is evidence to indicate that pancreatic acid lipase hydrolyzes eicosenoic and erucic acid esters more slowly than esters of oleic acid, with relative rates of 61, 72, and 100% for the esters of 22:1, 20:1, and 18:1, respectively (Brockerhoff and Jensen, 1974). Second, 22:1 in the HEAR oil TG is almost entirely in the 1,3-position with less than 5% in the 2-position (Brockerhoff and Yurkowski, 1966). The 2-position is occupied primarily by 18:1, 18:2, or 18:3. Since pancreatic lipase cleaves fatty acids at the 1,3-position of the TG molecule, it follows that the bulk of 22:1 in HEAR oil is released as the free fatty acid during digestion. The residual monoglyceride which contains primarily 18:1, 18:2, or 18:3 would be readily absorbed but the free erucic and gondoic (cis 20:1 n-9) acids would not be. It therefore seems probable that the specific triglyceride isomerism of HEAR oil, with 22:1 in the 1,3-position, contributes to the low digestibility of the oil. Direct support for this is found in interesterification experiments with HEAR oils (Rocquelin et al., 1971). The apparent digestibility of a HEAR oil with 45% erucic acid was increased from 81 to 91% by interesterification which increased the apparent digestibility of the erucic acid from 73 to 85%. Interesterification of HEAR oils probably improves the apparent digestibility by increasing the proportion of 2-monoerucin in the TG mixture which would be digested better than the free erucic acid. The same reasoning can be used to explain the observation that trierucin is better digested than free erucic acid (Carroll and Richards, 1958) or HEAR oils (Sergiel and Gabucci, 1980).

Finally, it seems clear that the poor digestibility of HEAR oil in the rat can be attributed to the 22:1 content of the oil (Table I). In these HEAR oils, 22:1 comprised 40–50% of the fatty acids. This is in marked contrast to the newly developed low erucic acid rapeseed (LEAR) oils (<5% 22:1 by definition) which are well digested by rats with a digestibility of 96% which is equal to that of peanut oil (Rocquelin and Leclerc, 1969).

IV. MYOCARDIAL LIPIDOSIS IN RATS

A. Background

In 1970, an undesirable property of HEAR oils was observed which caused considerable concern with health regulatory agencies and in the scientific community. At that time, the Unilever scientists found that in rats fed a HEAR oil containing 49% 22:1 at 50 calorie % of the diet, there was a rapid and severe fat infiltration of the myocardium (Abdellatif and Vles, 1970). The heart became pale after 1 day on this diet and light cream in color after 3–6 days. Appropriately prepared sections of heart tissue showed that the myocardial color changes were the result of a massive fat infiltration (see Fig. 2, Chapter 12). The same cardiac fat infiltration was present when trierucin was fed in amounts equivalent to that present in HEAR oil. This made it quite clear that it was the 22:1 acid in HEAR oil which triggered this response. These findings were readily confirmed by other investigators (Houtsmuller et al., 1970; Beare-Rogers et al., 1971,1972a; Rocquelin 1972; Jaillard et al., 1973; Kramer et al., 1973; Ziemlanski et al., 1973b).

B. The Nature of the Lesion

The histology of the fat infiltrated rat heart has been thoroughly studied (Charlton et al., 1975; Engfeldt and Brunius, 1975a; Ziemlanski et al., 1973b). The fat droplets are round to oval, ranging in size up to 8 μm and are arranged in rows along the longitudinal axis of the fiber. The fat droplets apparently are without surrounding membrane and often are in close contact with muscle mitochondria. At times these mitochondria appear deformed from the encroachment by fat droplets.

Enlarged mitochondria are occasionally noted in heart sections examined by the electron microscope. This has given rise to the term "megamitochondria" (Vodovar et al., 1973; Engfeldt and Brunius, 1975b), which may be somewhat misleading since moderate mitochondrial enlargement may be a fairly common occurrence (Schiefer et al., 1978; Bodak et al., 1979).

The physiological effects of this pronounced myocardial fat infiltration are not entirely clear. Berglund (1975) reported no electrocardiographic anomalies with male rats fed a HEAR oil diet when he used standard leads I, II, and III. Apparently there were no changes in the cardiac conductive system in spite of the gross morphological changes. On the other hand, the contractile force in hearts isolated from rats fed HEAR oil is decreased (Eeg-Larsen, quoted by Berglund, 1975; Ten Hoor et al., 1973).

Ten Hoor and co-workers (1973) measured heart function in rats during maximum lipidosis, i.e., 3 days after feeding a HEAR oil containing diet. Heart function was measured two ways, one, in isolated left ventricular papillary muscle and the other with a heart–lung preparation in which the work load put on the heart could be adjusted as desired. With all measured parameters (i.e., maximal isometric contractile force, maximal developed tension and rate of tension development, and left ventricular stroke work), the hearts from rats that were put on a 50 calorie % HEAR oil diet had poorer contractile properties than did the hearts from the control group that received sunflower oil. Similarly, when rats were reared on an essential fatty acid deficient diet, the contractile force of isolated papillary muscle from heart was weaker than that from the control group that received linoleic acid, i.e., sunflower oil supplement. These authors suggest that in both instances the decreased contractile force of heart muscle may be related to an impaired mitochondrial function and a decreased rate of ATP synthesis.

The myocardial lipidosis has been associated with an increase in TG while the level of phospholipid and cholesterol remained fairly constant (Houtsmuller et al., 1970; Beare-Rogers et al., 1971, 1972b; Rocquelin, 1972). Lipidosis has been detected as early as 3 hr after feeding HEAR oil (Ziemlanski et al., 1973b). The high levels of cardiac free fatty acids in rats fed HEAR oils reported by the early investigators have been ascribed to inappropriate lipid extraction techniques which permitted extensive autolysis of the cardiac lipids (Kramer and Hulan, 1978). The cardiac lipid, more specifically the cardiac TG determination by chemical means, is not as sensitive an indicator as the histological techniques of oil red O staining or electron microscopy. Even the recently developed method that uses the Iatroscan cannot detect increased levels of cardiac TG (Kramer, 1980) which could be detected by oil red O staining (Kramer et al., 1973, 1979). Therefore for sensitivity, specific staining techniques and electron microscopic examination are preferable. However, with the latter methods, the possibility of selecting areas of myocardial lipidosis could become a problem unless a large number of sections are investigated.

C. Factors That Affect Myocardial Lipidosis in the Rat

1. CHAIN LENGTH

Myocardial fat infiltration in the rat heart is directly related to long chain monoenoic fatty acids with 20 or more carbon atoms. The docosenoic (22:1) fatty acids are more damaging than the eicosenoic (20:1) fatty acids (Beare-Rogers et al., 1972a). Presumably, the tetracosenoic (24:1) fatty acids would be even more effective in producing myocardial lipidosis, although this has not been tested experimentally.

2. POSITIONAL AND GEOMETRICAL ISOMERS OF DOCOSENOIC ACID

The position of the double bond appears to have little effect in producing myocardial lipidosis in the rat, if similar levels of fatty acids are compared (Table II). Generally, little differences can be detected histologically or chemically. In two studies, purified positional isomers were fed at the same level in the diet, and it appeared that cetoleic acid resulted in slightly less severe lipidosis when measured histologically (Beare-Rogers et al., 1972a; Svaar, 1982). An attempt to compare erucic and cetoleic acid using HEAR oil and partially hydrogenated fish oil (Astorg and Cluzan, 1977) must be interpreted with caution, since the latter oil no longer contains the pure isomer (Ackman et al., 1971b; Lambertsen et al., 1971) and the digestibility of the trans isomers formed is less than that of the cis isomers. A comparison of the geometric isomers shows that erucic acid (cis 22:1 n-9) is more effective than brassidic acid (trans 22:1 n-9) in producing myocardial lipidosis, measured as either the total lipid or the docosenoic acid content in cardiac TG (Rocquelin et al., 1975). When each isomer was fed to male rats for 3 days, erucic acid comprised about 35% of cardiac TG while brassidic acid was present at only about 10%. These differences were accounted for mainly by the difference of digestibility between the two isomers which were 83 and 46% for erucic and brassidic acids, respectively.

The cis and trans forms of cetoleic acid have not been tested in a controlled experiment. In one study, unhydrogenated fish oil was compared to partially hydrogenated fish oil with a similar level of 22:1 (Opstvedt et al., 1979). Partially hydrogenated fish oil, which contains trans isomers, again gave a lower level of cardiac 22:1 in rats than unhydrogenated fish oil even

TABLE II

Myocardial Lipidosis Caused by Different Docosenoic Fatty Acids

Diet	Duration (days)	% 22:1 in oil	Histological staining[a]	Heart triglycerides mg/g	Heart triglycerides % 22:1	Reference
H$_2$-Fish oil[b]	3	14.8	++	7.1	14.4	Astorg and Cluzan, 1977
HEAR oil/peanut oil	3	10.9	+ to ++	10.5	14.3	
H$_2$-Fish oil	7	14.8	+	5.6	—	Astorg and Cluzan, 1977
HEAR oil/peanut oil	7	14.2	+ to ++	7.0	—	
Fish oil/sunflower oil	4	10.9	+ to ++++	4.7	18.0	Opstvedt et al., 1979
HEAR oil/lard/sunfl. oil	4	11.6	+ to ++++	5.5	16.6	
H$_2$-Fish oil	3	29.7	++ to +++	12.9	18.2	Astorg and Cluzan, 1977
HEAR oil/peanut oil	3	27.2	+ to ++	17.1	25.8	
H$_2$-Fish oil	7	29.7	++ to +++	10.6	—	Astorg and Cluzan, 1977
HEAR oil/peanut oil	7	27.2	+ to ++	15.9	—	
Cetoleic acid	7	30.7	+ to +++	(40.8)[c]	(9.1)	Beare-Rogers et al., 1972a
Erucic acid	7	30.4	++ to ++++	(39.0)	(9.8)	
Cetoleic acid	6	40	Grade 2.3	—	—	Svaar, 1982
Erucic acid	6	40	Grade 3.5	—	—	

[a] A grading system was used in which increasing number of pluses represents increasing severity of myocardial lipidosis.

[b] H$_2$ represents partial hydrogenation of herring or capelin fish oil.

[c] Values given in parentheses represent the total cardiac lipids measurements rather than that of the cardiac TG, which were not reported.

though the dietary concentration of 22:1 was the same. However, the level of cardiac TG was similar when the two oils were fed.

3. OTHER FATTY ACIDS IN THE DIET

Increasing the level of saturated fatty acid in diet containing HEAR (Beare et al., 1963) or LEAR (Farnworth et al., 1982) oils improves growth in the rat. However, higher levels of dietary saturated fatty acids do not lessen the effect on myocardial lipidosis (Beare-Rogers et al., 1972a). On the other hand, a lower severity of myocardial lipidosis was reported in rats fed higher levels of linoleic acid (Astorg and Cluzan, 1977). How linoleic acid was effective is not quite clear since increased levels of dietary polyunsaturated fatty acids would be expected to increase the absorption of 22:1 and thus lead to increased lipidosis.

4. SEX, STRAIN, AND AGE

Myocardial lipidosis develops in both male and female albino rats when fed diets which contain the same quantity of docosenoic fatty acids (Beare-Rogers et al., 1971; Kramer et al., 1973). It has been suggested that lipidosis disappears slightly more rapidly in female rats than in male rats (Engfeldt and Brunius, 1975b).

All the strains of rats so far examined have shown myocardial lipidosis. This includes the albino Wistar (Abdellatif and Vles, 1970, 1973; Rocquelin, 1972; Ziemlanski et al., 1973b), the albino Sprague–Dawley (Beare-Rogers et al., 1971 and 1972a; Kramer et al., 1973), and the hooded Chester Beatty rat (Kramer et al., 1979). It was not possible to distinguish between the severity of myocardial lipidosis in the Sprague-Dawley and Chester Beatty rats by either specific fat staining or gravimetrically (Kramer et al., 1979). No study has been reported in which the two albino rats were compared with respect to lipidosis.

Myocardial fat infiltration in rats fed HEAR oil appears to be age related. Thus weanling rats are most susceptible, while rats older than 12 weeks appear to be more resistant. Four- and 12-week-old rats had a 100% incidence, while the incidence in 32-week-old rats dropped to 20% (Beare-Rogers and Nera, 1972).

D. General Conclusion

Many authors have suggested that cardiac fat infiltration is caused solely by the erucic acid present in HEAR oil, and have thus described erucic acid as toxic to experimental animals (Lancet, 1974; Mulky, 1979; Nolen, 1981). Probably a more realistic approach would be to view erucic acid as one of a number of fatty acids that are poorly metabolized by rats (and other laboratory animals) and that, if fed in large quantities, lead to heart lesions. Almost three decades ago, Wilgram and associates (1954) reported fat droplets, myocardial degeneration, and areas of focal necrosis in the hearts of rats that were fed high fat and choline deficient diets. The myocardial lesions were induced by a combination of choline deficiency and 35% lard or beef fat in the diet. Synthetic TG, particularly trilaurin (or ethyl laurate), produced this myocardial damage when fed at 25–40% of the dietary fat. As early as 1955, Hartroft had suggested that the accumulation of fat in these hearts may be related to impaired myocardial oxidative capacity. This subject is discussed in Chapter 14.

V. MYOCARDIAL NECROSIS IN RATS

A. Background

Myocardial damage affecting the muscle fibers was observed when high levels of HEAR oil were fed to young Sprague–Dawley male rats for an extended period of time. The lesion described by Roine et al. (1960) consisted of a generalized myocardial inflammation with cloudy swelling of muscle fibers and reduced striations. Areas of plasma cell, lymphocytes, and histiocyte infiltration were present. There were small necrotic foci and variable numbers of fibroblasts. According to the authors the condition resembled toxic myocarditis. No organ other than the heart appeared to be affected.

It was well known in 1960, and probably overlooked by later investigators, that any number of high fat diets alone or in combination with low choline (Kesten et al., 1945; Wilgram et al., 1954; Hartroft, 1955; Williams and Oliver, 1961) or low protein (Highman and Daft, 1951; Williams and Aronsohn, 1956) produced identical myocardial lesions in male rats to those recorded by Roine et al. (1960). In addition, diets deficient in thiamine (Ashburn and Lowry, 1944), cysteine (Williams and Aronsohn, 1956), vitamin E (Dessau et al., 1954), potassium (French, 1952; Perdue and Phillips, 1952), and magnesium (Lowenhaupt et al., 1950) also produced essentially the same type of heart lesion. These heart lesions were most evident in the young male rodent (Wilgram et al., 1954; Hartroft, 1955; Williams, 1960). Roine et al. (1960) by not referring to this early literature left the impression that the induction of myocardial necrosis in rodents was a unique property of HEAR oil. This, in turn, led later investigators to ascribe the problem solely to the erucic acid component of HEAR oil. It seems in retrospect that it would have been more realistic to investigate these myocardial lesions in the general context of the effects of high fat diets in the rodent.

TABLE III

Myocardial Necrosis Reported in Male Rats Fed High High Erucic Acid Rapeseed (HEAR) and Control Oils

% 22:1	% fat	% as HEAR	% 22:1 in HEAR	Strain of male rat[a]	Length of feeding (weeks)	HEAR %	HEAR n	Control %	Control n	Control Type	References
15.8	35		~45[b]	SD	5.1	100	7	0	6	Soybean	Roine et al., 1960
12.5	30	25	50.1	W	24	100	12	0	12	Sunflower	Abdellatif and Vles, 1973
8.7	21		41.6	SD	30	96	25	39	26	Peanut	Svaar and Langmark, 1980
7.8	20		38.9	W	18	94	17	50	16	Corn	Umemura et al., 1978
7.6	20		38.1	SD	16	82	11	0	12	Lard/corn (3/1)	Beare-Rogers et al., 1972b
7.6	20		38.1	SD	16	100	15	0	17	Lard/corn (3/1)	Beare-Rogers et al., 1974
7.6	15		50.6	W	8.6	93	15	0	16	Peanut	Rocquelin et al., 1973
7.6	15		50.6	W	8.6	40	5	0	7	Peanut	Rocquelin et al., 1974
7.6	15		50.6	W	25.7	100	5	13	8	Peanut	Rocquelin et al., 1974
7.1	15		47.6	W	30.3	89	9	14	14	Butter	Ziemlanski et al., 1972
7.1	15		47.6	W	52	90	19	8	24	Butter	Ziemlanski et al., 1972
6.8	20		34.2	SD	16	40	15	0	15	Corn	Hung et al., 1977
6.7	15		44.7	W	26	90	10	20	10	Peanut	Rocquelin and Cluzan, 1968
6.6	20	15	44.0	SD	32	88	8	29	7	Sunflower	Abdellatif and Vles, 1973
6.5	19.5	14.6	44.8	W	26	92	12	33	12	Sunflower	Vles et al., 1976
6.4	25		25.6	SD	10–20	25	4	25	4	Lard/corn (3/1)	Ackman add Loew, 1977

6.4	25		25.6	W	10–20	50	4	25	4	Lard/corn (3/1)	Ackman and Loew, 1977
6.2	20		30.9	SD	16	80	10	40	10	Soybean	Clandinin and Yamashiro, 1980
6.2	20		30.9	SD	16	90	10	10	10	Soybean	Clandinin and Yamashiro, 1980
6.2	20		30.9	SD	28	100	10	0	10	Soybean	Clandinin and Yamashiro, 1980
5.9	20		29.4	RN	10	75	8	0	8	Lard/corn (3/1)	Beare-Rogers and Nera, 1972
5.8	20		28.8	W	25	100	14	64	14	Soybean	McCutcheon et al., 1976
5.1	20		25.5	SD	16	100	24	33	24	Corn	Kramer et al., 1979
4.7	10		47.0	na	13	83	24	21	24	Sunflower	Ziemlanski, 1977
4.7	20		23.3	SD	16	72	29	7	56	Lard/corn (3/1)	Beare-Rogers et al., 1974
4.5	20	15	30.2	SD	16	53	15	27	15	Soybean	Nolen, 1981
4.5	20		22.3	SD	16	70	10	40	10	Corn	Kramer et al., 1973
4.1	15	7.95	51.3	W	16	100	6	0	6	Peanut	Astorg and Cluzan, 1976
3.3	20		16.4	SD	16	70	20	20	20	Lard/corn (3/1)	Beare-Rogers et al., 1974
3.1	15		20.6	SD	10	70	10	60	10	Soybean	Vogtmann et al., 1975
1.7	15	n.a.	n.a.	W	30.3	96	24	8	25	Fat–oil mixture	Ziemlanski et al., 1974

[a] SD, Sprague–Dawley; W, Wistar; RN, *Rattus norvegicus*; n.a., not available.
[b] Approximate level of 22:1 in HEAR oils available at that time.

TABLE IV

Myocardial Necrosis in Male Rats Fed Decreasing Levels of Fat in the Form of High Erucic Acid Rapeseed (HEAR) Oils

Diet			Strain of male rat[a]	Length of feeding (weeks)	Heart lesions					Ref.[e]
% 22:1	% fat as HEAR	% 22:1 in HEAR			HEAR		Control			
					%	n	%	n	Type	
15.75	35	~45[b]	SD	5.1	100	7	0	6	Soybean	1
11.25	25				100	7	—[c]	—	—	
6.75	15				0	6	0	9	Soybean	
3.38	7.5				0	5	—	—	—	
4.7	10	47.0	W	13	83	24	21	24	Sunflower	2
2.35	5				61	23	13	24	Sunflower	
8.74	21	41.6	SD	30	96	25	39	26	Peanut	3
4.37	10.5				61	26	12[d]	26	Peanut	

[a] SD, Sprague–Dawley; W, Wistar.
[b] Approximate level of 22:1 in HEAR oils available at that time.
[c] No corresponding control was fed.
[d] Peanut oil fed at 4.3% of the diet.
[e] References: 1, Roine et al. (1960); 2, Ziemlanski (1977); 3, Svaar and Langmark (1980).

B.   Results of Testing HEAR Oils

As seen in Table III, numerous investigators were able to confirm the find-ings of Roine *et al.* (1960). HEAR oils, fed at a high level in the diet (15–30% by weight) to different strains of young male rats for an extended period of time (9–52 weeks), caused a greater incidence of myocardial necrosis than other oils, fats, or fat–oil mixtures. The erucic acid content of the HEAR oils tested ranged from 16 to 51%. The results in Table III are arranged in the order of decreasing content of erucic acid in the diet. The differences in experimental protocol and histological interpretation by the different inves-tigators make detailed comparisons difficult but it is clear from the data in Table III that there is no correlation between the level of 22:1 and the inci-dence of heart lesions. However, when the data are taken from single exper-iments, several effects are apparent. When HEAR oil was the only fat added to the diet (Table IV), the incidence of myocardial necrosis decreased as the level of HEAR oil was reduced. The decrease in heart lesions could be attrib-uted either to a decreased level of fat or erucic acid in the diet.

In order to eliminate one of the variables, i.e., the level of fat in the diet, HEAR oil was diluted with either sunflower oil (Abdellatif and Vles, 1973) or a 3 to 1 lard/corn oil mixture (Beare-Rogers *et al.*, 1972b), and the level of fat in the diet was maintained at a constant level. In both experiments, as the proportion of HEAR oil increased, so did the incidence of heart lesions

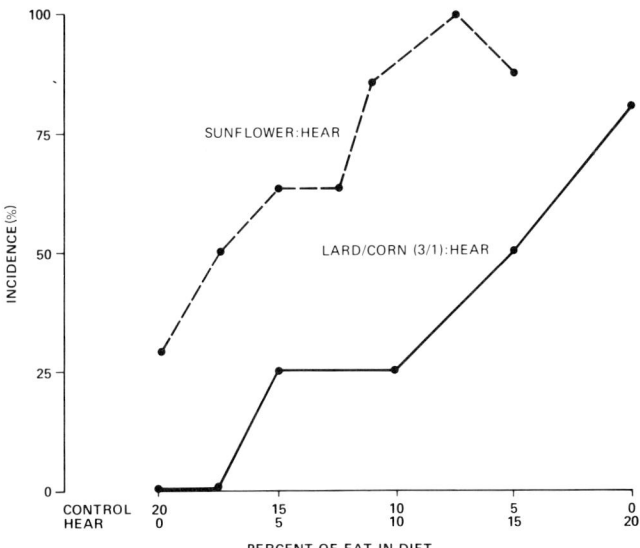

Fig. 1.   Relationship between incidence of myocardial lesions and level of HEAR oil in the diet when mixed in various proportion with sunflower oil (data from Abdellatif and Vles, 1973) or a lard/corn oil (3/1) mixture (data from Beare-Rogers *et al.* 1972b).

(Fig. 1). It is, however, interesting to observe that whereas with lard/corn oil control diets there was a zero incidence of heart lesions, when sunflower oil diets were used as controls, the heart lesion incidence was 29%. This suggests that myocardial necrosis can be reduced quite effectively by saturated fatty acids, while a mixture of polyunsaturated fatty acids, as in sunflower oil, is not as effective. It must be kept in mind, however, that in these two experiments both the rat strain and the experimental feeding period were different. Wistar rats were fed the sunflower oil mixtures for 32 weeks, while the Sprague–Dawley rats were fed the lard/corn oil mixtures for only 16 weeks.

C. The Cardiopathogenicity of Docosenoic Acids

1. ERUCIC ACID

It is quite evident from the results in Table III that male rats fed diets containing high levels of HEAR oil will develop myocardial necrosis. These heart lesions can also be produced by feeding an equal mixture of trierucin and sunflower oil at 60 calorie % (Abdellatif and Vles, 1973) with a level of 22:1 in the oil of about 42% (Table V). Subsequently, this finding was confirmed when rats were fed olive oil interesterified with erucic acid to give a mixture which contained 30% 22:1 (Beare-Rogers, 1975).

Scientists from the University of Guelph (McCutcheon et al., 1976) repeated the earlier study by Abdellatif and Vles (1973) and found that erucic acid will increase heart lesions irrespective of the source of erucic acid, i.e., whether from a HEAR oil or from a non-*Brassica* source, such as nasturtium seed (*Tropaeolum majus*) (Table V). They reported that removal of linolenic acid (18:3) from a simulated HEAR oil which contained 28.7% 22:1 resulted in a significant reduction in the severity of heart lesions. On the other hand, increasing the level of linoleic acid (18:2) in a HEAR oil or a simulated HEAR oil had no apparent effect. These results suggest that there may be an interaction of erucic and linolenic acid, or that linolenic acid predisposes the heart to lesions.

2. COMPARISON OF POSITIONAL ISOMERS OF DOCOSENOIC ACIDS

The two most common naturally occurring *cis* docosenoic acids are erucic acid (*cis* 22:1 *n*-9) found in the seed oil of the Brassica family and cetoleic acid (*cis* 22:1 *n*-11) found mainly in marine oils (Ackman et al., 1971a). The cetoleic acid in fish oils is derived by oxidation of the corresponding alcohol present in small crustacea which form part of the food chain for marine life (Nenenzel, 1970; Pascal and Ackman, 1976). To date, no long-term studies have been reported in which these two docosenoic fatty acids were fed in purified form. When a comparison was made between a HEAR

TABLE V

Myocardial Necrosis in Male Rats Reproduced by Erucic Acid

Description	Fatty acids					Incidence		Severity[a]				Ref.[a]
	22:1	18:3	18:2	18:1 (trans)	Saturates	%	n	1	2	3	4	
HEAR/sunflower (5/1)	41.8	8.4	21.9	12.7	7.2	100	12	5	2	3	2	1
Trierucin/sunflower (1/1)	42.0	0.2	33.8	13.2	6.5	100	12	5	3	3	1	
Sunflower	—	0.3	64.6	23.8	11.0	0	12	0	0	0	0	
HEAR	28.8	7.9	17.9	28.6	5.8	100	14	11	14	10		2
Nasturtium/oils[b] (high 18:3)	29.9	8.3	19.0	28.4	6.5	100	15	15	12	11		
Nasturtium/oils[c] (low 18:3)	28.7	—	42.1	16.0	6.0	93	14	7	10	7		
HEAR/oils[d] (high 18:2)	26.0	5.9	36.1	17.1	5.9	100	15	12	15	6		
Nasturtium/oils[e] (high 18:2)	28.5	6.5	40.8	10.0	6.3	100	15	14	14	11		
Soybean	—	8.4	50.7	25.4	15.3	64	14	5	7	4		
Trierucin (low 18:2)	38.3	—	1.7	50.3	7.7	100	9	3	4	2		3
Trierucin (high 18:2)	39.4	—	32.5	18.0	8.4	100	9	5	2	2		
Tribrassin (low 18:2)	38.6	—	1.3	50.8	7.4	33	9	3	0	0		
Tribrassin (high 18:2)	38.1	—	34.8	18.4	8.6	67	9	5	1	0		
Triolein (low 18:2)	—	—	1.8	89.4	7.5	33	9	3	0	0		
Triolein (high 18:2)	—	—	32.0	60.2	7.9	100	9	8	1	0		
Trielaidin (low 18:2)	—	—	1.7	51.0 (39.4)	7.3	100	9	9	0	0		
Trielaidin (high 18:2)	—	—	29.4	18.6 (41.0)	8.7	100	9	8	1	0		

[a] Reference for severity categories of heart lesions: 1, Abdellatif and Vles (1973): mild (1), moderate (2), definite (3), and severe (4) cellular and/or fibrotic scars; 2, McCutcheon et al. (1976): fresh (1) and old (2) myocardial necrosis and microvascular alterations (3); 3, Astorg and Levillain (1979): histiocytes (1), granules (2), and granules and necrosis (3).

[b] Olive (30%), safflower (18%), and linseed (12%).

[c] Safflower (45%) and olive (10%).

[d] Safflower (34%) and linseed (1%).

[e] Safflower (50%) and linseed (10%).

TABLE VI

Comparison of Myocardial Necrosis Induced by Different Positional Isomers of Docosenoic Fatty Acids in Male Rats

Level of dietary fat		% 22:1 in dietary oil	Heart lesions		Ref.[a]
% test oil	% other fat or oil		%	n	
2.5% HEAR	17.5% lard/corn oil (3/1)	4.8	0	12	1
2.5% H$_2$-HEAR	17.5% lard/corn oil (3/1)	4.4	0	12	
2.5% H$_2$-fish oil	17.5% lard/corn oil (3/1)	3.9	30	10	
5% HEAR	15% lard/corn oil (3/1)	9.5	25	12	1
5% H$_2$-HEAR	15% lard/corn oil (3/1)	8.8	9	11	
5% H$_2$-fish oil	15% lard/corn oil (3/1)	7.8	20	10	
10% HEAR	10% lard/corn oil (3/1)	19.1	25	12	1
10% H$_2$-HEAR	10% lard/corn oil (3/1)	17.6	8	12	
10% H$_2$-fish oil	10% lard/corn oil (3/1)	15.7	40	10	
15% HEAR	5% lard/corn oil (3/1)	28.6	50	12	1
15% H$_2$-HEAR	5% lard/corn oil (3/1)	26.4	58	12	
15% H$_2$-fish oil	5% lard/corn oil (3/1)	23.3	60	10	
H$_2$-fish oil[b]	Lard/corn oil (3/1)	4.6	30	20	2
H$_2$-redfish oil[b]	Lard/corn oil (3/1)	4.5	30	20	
15% H$_2$-fish oil	5% corn oil	16.7	30	20	
15% redfish oil	5% corn oil	16.0	45	20	
4.2% HEAR	10.8% peanut oil	14.2	83	6	3
15% H$_2$-fish oil		14.8	33	6	
13% H$_2$-fish oil	2% corn oil	13.1	50	6	
7.95% HEAR	7.05% peanut oil	27.2	100	6	3
15% H$_2$-fish oil		29.7	67	6	
13% H$_2$-fish oil	2% corn oil	25.6	33	6	
25% HEAR		25.6	25	4	4
25% H$_2$-fish oil		19.7	50	4	
25% HEAR		25.6	50	4	5
25% H$_2$-fish oil		19.7	25	4	
21% HEAR		41.6	96	25	6
21% H$_2$-fish oil		15.1	14	22	

[a] References: 1, Beare-Rogers *et al.* (1972b); 2, Ackman (1974); 3, Astorg and Cluzan (1976); 4, Ackman and Loew (1977) (Sprague–Dawley); 5, Ackman and Loew (1977) (Wistar); 6, Svaar and Langmark (1980).

[b] Ratio of fish oil to lard/corn oil mixture not specified.

oil and a fish oil, both of which were partially hydrogenated to the same iodine value (76–78), the incidence of heart lesions was found to be similar (Beare-Rogers *et al.*, 1972b). From the results in Table VI it appears that there is reasonably good agreement among most research workers that the heart lesion incidence is quite similar for oils that contain equivalent amounts of cetoleic or erucic acids (Beare-Rogers *et al.*, 1972b; Ackman and Loew, 1977). A Norwegian study, in which the heart lesion incidence produced by HEAR oil was greater than that produced by partially hydrogenated fish oil (Svaar and Langmark, 1980), is difficult to interpret because the docosenoic isomer concentration was almost three times higher in the HEAR oil than in the fish oil. In one study, the heart lesion incidence appears to be greater when HEAR oil was fed instead of partially hydrogenated fish oil (Astorg and Cluzan, 1976). Recently Svaar (1982) reported results of feeding trials with erucic and cetoleic acids, however, the trial only lasted 1 week. The severity of lesions was greater with erucic acid than with cetoleic acid, but because of the short duration of the experiment these results will need to be confirmed.

In summary, it appears that the positional isomers of docosenoic acid are equivalent in their cardiopathogenic properties, and to date there is no convincing evidence that real differences exist when fish oils or rapeseed oils, with the same concentration of docosenoic fatty acid isomers, are fed to rats.

3. COMPARISON OF GEOMETRIC ISOMERS OF DOCOSENOIC ACIDS

There has been a great interest in comparing the cardiopathogenic response of geometric isomers of docosenoic acids because much of the HEAR oils and fish oils, which contain docosenoic acids, are used after partial hydrogenation in the form of margarines and shortenings. During partial hydrogenation, numerous geometric and positional isomers of docosenoic acids are formed in partially hydrogenated HEAR oils (Conacher and Page, 1972) and fish oils (Ackman *et al.*, 1971b; Conacher *et al.*, 1972; Lambertsen *et al.*, 1971). Two basic approaches have been used in these investigations; some have tested the pure isomers in synthetic fatty acid mixtures while others have tested partially hydrogenated HEAR oils.

Experiments in which rats were fed diets containing similar levels of erucic (*cis* 22:1 *n*-9) and brassidic (*trans* 22:1 *n*-9) acids, show that the *cis*-docosenoic acid isomer is more cardiopathogenic than the trans isomer (Astorg and Levillain, 1979). However, the cis isomer may be much better absorbed than the trans isomer, as indicated by their respective digestibility coefficients. Rocquelin *et al.* (1975) found that the trans isomer was less digestible (46%) than the cis isomer (83%), and this was shown by the fact that the rat hearts contained 1 mg/g of the trans isomer and 18 mg/g of the cis isomer. Therefore, it is quite possible that if one allows for the difference in digestibility of the two isomers, the cardiopathogenicity may be identical. This is also supported by experiments in which HEAR oil is partially hydrogenated. In partially hydrogenated HEAR oil there are a mixture of positional and geometric isomers of docosenoic acid (Conacher and Page, 1972). Since in partially hydrogenated HEAR oil the position of the docosenoic acids on the triglyceride molecule is not altered from its original 1,3-position (Brockerhoff and Yurkowski, 1966), one may assume that the digestibility should not be altered. A number of investigations have shown that partially hydrogenated HEAR oil is as cardiopathogenic as the corresponding unhydrogenated HEAR oil (Beare-Rogers et al., 1972b and 1974; Rocquelin et al., 1974; Ziemlanski et al., 1972), and that the cardiac lesion incidence is not lowered unless the HEAR oil is diluted (Beare-Rogers et al., 1972b; Ziemlanski et al., 1972). Of course if the hydrogenation of the HEAR oil is carried to completion, the heart lesion incidence is reduced (Nolen, 1981). This probably is of no physiological significance since the behenic acid (22:0) which forms on complete hydrogenation of docosenoic acids is almost undigestible (Mattson and Streck, 1974). The same is expected to hold true for cetoleic acid (cis 22:1 n-11) and its corresponding trans isomer when fully hydrogenated. Although no long-term study has been reported in which the pure isomers were compared, a comparison of unhydrogenated and partially hydrogenated fish oil with a similar level of docosenoic acid indicated that the cardiopathogenic response was similar (Ackman, 1974).

Therefore, one may conclude that all the docosenoic acid isomers, whether geometric or positional, are about equally cardiopathogenic in rats, provided that allowance is made for differences in digestibility.

D. Differences in Rats

1. EFFECT OF AGE

The early experiments in which myocardial necrosis was induced by a variety of dietary deficiencies clearly established that the young, postweaned male rat is the most sensitive (Wilgram et al., 1954; Wilgram and Hartroft, 1955; Williams, 1960; Scott et al., 1962; Rings and Wagner, 1972). It appears that the areas of focal myocardial necrosis and fibrosis proliferate throughout the life of the rat and thus became more abundant in the older animal. Kaunitz and Johnson (1973) showed that rats that died naturally between 400 to 500 days of age had a 26% incidence of heart lesions, while those that died after 900 days of age had an 80% incidence. It therefore becomes clear that it is good experimental practice to use 3- to 4-week-old weanling rats in experiments designed to test the cardiopathogenic response of fats and oils, and not to extend the trial unnecessarily, so that the lesion incidence in the control rats stays as low as possible. As seen in Table III, all studies reported to date, in which the cardiopathogenicity of rapeseed oils were tested, were designed with this in mind.

2. EFFECT OF STRAIN

In studies with HEAR oils mainly two strains of albino rats, Sprague–Dawley and Wistar, have been used. There appears to be little difference in heart lesion response to HEAR oil between these two strains. Hulan et al. (1977) observed that male Chester Beatty (hooded) rats were resistant to the formation of myocardial necrosis. A subsequent study showed that this strain of rat did not develop a higher incidence of heart lesion with HEAR oil (26% 22:1) than with corn oil (Kramer et al., 1979). These results were confirmed (Clandinin and Yamashiro, 1980), and suggest that the cardiac lesions are strain specific.

3. LESION INCIDENCE IN MALE AND FEMALE RATS

It was known from the early literature that the male rat was more susceptible to myocardial necrosis than the female rat (Wilgram et al., 1954; Hartroft, 1955). In 1968, Rocquelin and Cluzan tested both male and female rats, and observed that the male rat heart was more affected by high levels of dietary HEAR oil than the female rat (Table VII). Kramer et al. (1973) found only a few scattered focal accumulations of mononuclear cells and minute foci of myocardial degeneration in female rats, which is in marked contrast to the frequent occurrence of overt myocardial necrosis found in male rats fed HEAR oils. This has been confirmed by other investigators (Engfeldt and Brunius, 1975b; Nolen, 1981; Vogtmann et al., 1975). In an Indian study in

TABLE VII

Comparison of Myocardial Necrosis in Male and Female Rats Fed High Erucic Acid Rapeseed (HEAR) Oils

Diet			Strain of rat[a]	Length of feeding (weeks)	Heart lesions				Ref.[b]
					Males		Females		
% 22:1	% fat	% 22:1 in HEAR			%	n	%	n	
6.71	15	44.7	W	26	90	10	70	10	1
4.46	20	22.3	SD	16	70	10	0	10	2
3.09	15	20.6	SD	10	70	10	10	10	3
4.53	20[c]	30.2	SD	16	53	15	7	15	4

[a] SD, Sprague–Dawley; W, Wistar.
[b] References: 1, Rocquelin and Cluzan (1968); 2, Kramer et al. (1973); 3, Vogtmann et al. (1975); 4, Nolen (1981).
[c] HEAR oil comprised 15%, soybean oil 5%.

which only female rats were used to test the cardiopathogenicity of mustard oil with 47% erucic acid, no fibrotic lesions were observed (Bhatia et al., 1979).

VI. IS MYOCARDIAL LIPIDOSIS LINKED TO MYOCARDIAL NECROSIS?

The evidence to date suggests that these two types of heart lesions are not causally linked. At first it was believed that myocardial necrosis does not occur without the preceding early lipidosis, since this was the sequence of events observed in male rats fed high levels of HEAR oil (Abdellatif and Vles, 1970, 1973; Abdellatif, 1972). However, the evidence indicates that myocardial lipidosis can occur without the subsequent occurrence of myocardial necrosis. For example, female albino rats develop myocardial lipidosis when fed diets containing HEAR oils but do not develop myocardial necrosis (Charlton et al., 1975; Kramer et al., 1973). A further example may be cited. Male rats from the Sprague–Dawley and Chester Beatty strains were observed to have similar myocardial lipidosis but myocardial necrosis was evident only in the Sprague–Dawley strain (Kramer et al., 1979). In another experiment, male rats were fed HEAR oil for 1 week so that maximum cardiac lipidosis was produced, and then they were fed a control diet for the remaining 9 weeks. These rats did not develop myocardial necrosis (Beare-Rogers and Nera, 1972). It appears that myocardial necrosis need not follow myocardial lipidosis.

On the other hand, myocardial necrosis occurs in male rats even in the absence of myocardial lipidosis. Kramer et al. (1973) observed that when male rats were fed a diet that contained either lard, corn oil, or a low erucic acid rapeseed oil, either no, or very little myocardial lipidosis developed, but still the rats showed myocardial necrosis after 16 weeks on test. This indicates that myocardial necrosis can develop independently of the early myocardial fat accumulation.

VII. EFFECTS ON OTHER TISSUES WHEN HEAR OIL IS FED TO RATS

A. Adrenals

Carroll (1951) noted that the adrenals from rats fed a diet with 25% by weight HEAR oil were enlarged and pale and that the cholesterol content of these glands was increased three- to fourfold. Since normal adrenal function is critical for the survival of the animal, this original observation by Carroll has since then been thoroughly investigated by a number of different investigators. It is the erucic acid in HEAR oil that causes the increased adrenal cholesterol level (Carroll, 1953). Of a number of fatty acids tested, only nervonic acid (24:1 n-9) gave a similar increase in the adrenal cholesterol level. When rat diets contained 15% by weight purified erucic acid, the adrenal lipids contained eicosenoic and erucic acid esters of cholesterol and triglycerides. These two fatty acids were also incorporated into adrenal phospholipids, but to a lesser degree (Carroll, 1962). The extent to which adrenal glands from HEAR oil fed rats are functionally impaired is not entirely clear, although there are some well documented observations. Adrenocorticotropin treatment is known to deplete adrenal cholesterol (Sayers *et al.*, 1944). There is a simultaneous increase in the cholesterol ester hydrolase activity (Shima *et al.*, 1972), which depletes the adrenal cholesterol ester concentration. It is postulated that the adrenal enzyme, cholesterol ester hydrolase, is activated by a cyclic AMP dependent protein kinase that phosphorylates the inactive form of the enzyme to the active form (Garren *et al.*, 1971; Trzeciak and Boyd, 1974). Hydrolysis of the cholesterol esters releases cholesterol, which is subsequently utilized for steroid synthesis.

When rats are exposed to stress, there is an adrenal cholesterol ester depletion, similar to that observed with adrenocorticotropin treatment. Rats fed a diet with 10% olive oil when cold stressed for 30 min at 4°C showed a 35% decrease in cholesterol ester concentration. When the rats were fed a diet containing ethyl erucate in corn oil, and then cold stressed, the cholesterol ester depletion was much less, i.e., only 17% (Walker and Carney, 1971). This suggests that there may be a suboptimal adrenal response to stressful situations in rats fed diets with erucic acid.

It appears there are two main reasons for the accumulation of cholesterol esters in rats that are fed HEAR oil or erucic acid containing diets. First, the cholesterol ester hydrolase fails to increase in activity when these rats are stressed, while in control rats, when stressed, the enzyme doubles its activity (Beckett and Boyd, 1975). Second, cholesteryl erucate, which accumulates in the adrenals of rats fed HEAR oil, is only slowly hydrolyzed by the enzyme, i.e., at 25–30% of the rate of cholesteryl oleate. This may be very significant, since there is considerable cholesteryl erucate accumulation in the adrenal glands of rats fed diets high in erucic acid, i.e., this ester may constitute 29–35% of the total (Carroll, 1962; Walker and Carney, 1971). In addition, in these rats there was an accumulation of 8% cholesteryl eicosenoate. In agreement with this evidence of impaired adrenal function, the results indicate that plasma levels of one of the adrenal hormones, corticosterone, are lower in these rats than in control rats when exposed to an environmental stress (Walker and Carney, 1971; Budzynska-Topolowska *et al.*, 1975).

Carney *et al.* (1972) made the interesting observation that prostaglandin production was depressed in the adrenals from rats given a 10% HEAR oil diet. Cholesteryl arachidonate is hydrolyzed during sterol synthesis liberating free arachidonic acid which is converted to prostaglandins. This process is stimulated by adrenocorticotropin. As observed by these authors, the HEAR oil diet depressed adrenal prostaglandin E_2 and $F_{2\alpha}$ synthesis by 40 and 37%, respectively. Thus there is some evidence that erucic acid and probably other long chain monoenoic fatty acids and their isomers can interfere with normal adrenal gland metabolism. This, in turn, may decrease the resistance of the animal to stress.

There is evidence to support this. Carroll and Noble reported as early as 1952 that young rats on a 25% HEAR oil diet survived cold stress less well than their littermates which were fed a margarine or olive oil diet. However, he concluded that there were no indications that poor adrenal function contributed to the deaths from cold stress (Carroll and Noble, 1952). This subject was briefly revived two decades later. A report from Canada indicated that 6 week-old male Sprague–Dawley rats had a high mortality when fed a HEAR oil (29% erucic acid) or partially hydrogenated herring oil (31% cetoleic acid) diet (Beare-Rogers and Nera, 1974). The survival rate of rats fed the partially hydrogenated herring oil was similar to that of the HEAR oil fed group which of course is an indication that the 22:1 isomers are potentially troublesome for the rat regardless of whether the source is marine oil or HEAR oil. The mortality correlated with the time of myocardial fat infiltration, but may also reflect a decreased response to cold stress by the adrenal gland. In another report from Canada no deaths were noted in 6-week-old male Sprague–Dawley rats fed a HEAR oil with 24% erucic acid and cold stressed (Hulan *et al.*, 1976). On the other hand, a Swedish report indicated a high mortality in 4 week-old male Sprague–Dawley rats fed a HEAR oil with 42% erucic acid while a better survival rate was observed with either a peanut oil, a LEAR oil (6% 22:1) or a partially hydrogenated marine oil (3% 22:1) (Darnerud *et al.*, 1978). Quite likely the discrepancy is due to a difference in the erucic acid level of the HEAR oils (24 vs. 29 vs. 42%) and in the age of the rat. Carroll and Noble in 1952 had already noted that older rats survive cold stress much better than younger rats. The later studies are of limited value because no attempt was made to measure either adrenal function or the adrenal cholesteryl ester content and relate these to diet and cold stress susceptibility.

B. Effects on the Reproductive System

1. MALE RATS

The effects of erucic acid or HEAR oil diets on the reproductive system of rats has been studied for some time. There were some early reports that when purified erucic acid was fed to young male rats at 15% by weight of the diet (added to a stock diet) for periods of 3–5 months, the rats showed testicular degeneration, reduced spermatogenesis and failure to produce offspring (Carroll and Noble, 1957; Noble and Carroll, 1961). These findings, however, were not duplicated when HEAR oil instead of recrystallized erucic acid was fed to male rats. Wistar rats fed a HEAR oil diet showed no evidence of testicular degeneration although a reduction of tubular size was reported (Beare et al., 1959a). In a thorough investigation, Beare et al. (1959b) showed that in three litters born to the same parents, the HEAR oil supplemented groups had the same number of offspring as the corn oil supplemented groups but the weanling weight was smaller. In a subsequent four generation study, the authors found both weanling number and weights were decreased in the rats fed the HEAR oil supplemented diet (Beare et al., 1961). Obviously, the HEAR oil fed rats were not sterile (in contrast to the rats fed pure erucic acid) in spite of the fact that their total erucic acid intake was similar. Most likely, and as has been suggested (Beare et al., 1959b), this apparent discrepancy can be explained by the fact that HEAR oils contain essential fatty acids, i.e., 15% linoleic acid and 6% linolenic acid, which prevent the testicular degeneration that was reported when a low fat stock diet was supplemented with high levels of a nonessential fatty acid, i.e., erucic acid (Carroll and Noble, 1957). Coniglio et al. (1974) confirmed that there was no testicular degeneration when male rats were fed a diet that contained HEAR oil and that the lipids from the testes of these rats were higher in 18:1, 18:3, 20:1, and 22:6 but lower in 22:5 than from rats fed corn oil.

2. FEMALE RATS

As with the male rat, the early reports that erucic acid somehow interfered with reproduction in the female rat are difficult to assess because the diets that were used may have been low in essential fatty acids (Carroll and Noble, 1957). Decreased numbers of pregnancies, resorption, and pseudopregnancies were observed in female rats fed diets supplemented with either erucic or oleic acids. Somewhat more plausible are the reports that the ovarian cholesterol content is increased in rats fed a HEAR oil containing diet (Carroll and Noble, 1952). When rats were fed ethyl erucate mixed with corn oil which is high in essential fatty acids, there were no noticeable reproductive abnormalities in the females (Walker et al., 1972). As in the adrenal gland, the erucic acid in the ovaries accumulated as the cholesterol ester. Also, the cholesterol esters of 20:4 n-6, 22:4 n-6, and 24:1 n-9 were increased in the ovaries from these rats. An interesting observation by these authors was that rats fed an olive oil containing diet with no erucic acid accumulated appreciable quantities of esterified erucic acid, i.e., 3.1% of the cholesterol ester fraction and 1.4% of the phospholipid fraction. It was suggested that rat ovarian tissue has a propensity for fatty acid chain elongation, i.e., 18:1 to 22:1 and 24:1. Therefore, erucic acid is a normal constituent of body tissues.

VIII. THE MYOCARDIAL TOLERANCE TO HEAR OIL IN THE DIET BY SPECIES OTHER THAN THE RAT

A. Pigs

The pig is generally accepted and widely used as an animal model for human disease. The domestic pig has found favor with scientists in the study of spontaneous and induced atherosclerosis and particularly the pathogenesis of control of this condition. For a review on the suitability and limitations of the pig as an animal model see the reference by Dodds (1982).

1. MYOCARDIAL LIPIDOSIS

Pigs respond quite differently from rats to a diet that contains HEAR oils. There is a clear species difference. The myocardial lipid infiltration is far less noticeable in the piglet and usually only detectable by chemical stains. While HEAR oil may cause the myocardial triglyceride concentration to increase more than sixfold above control values in the rat, no such increase is present in the piglet (Opstvedt et al., 1979). So far no one has been able to show an increase in the myocardial lipids of pigs fed HEAR oils by gravimetric determination (Aherne et al., 1976; Kramer et al., 1975; Opstvedt et al., 1979; Seher et al., 1979). On the other hand, HEAR oil fed to pigs does appear to give an increased response to oil red O staining in the myocardium. Thus by using this very sensitive staining technique, a mild grade lipidosis was recorded in 3-week-old piglets that were fed diets that contained lard, refined fish oil, partially hydrogenated fish oil, or HEAR oil (Table VIII). In comparison the rat showed more severe lipidosis when fed the same diets (Opstvedt et al., 1979). The Norwegian scientists repeated their study with essentially the same results. Partially hydrogenated fish oil or HEAR oil when included in the diet of piglets gave rise to mild myocardial lipidosis which was visible in histological sections stained with oil red O (Svaar et al., 1980). Similar results were observed with mini pigs (Sus scrofa) (Beare-Rogers and Nera, 1972). Myocardial lipid accumulation has also been observed in older pigs (76–80 days of age) fed a HEAR oil diet (Vodovar et al., 1973). These authors describe ultrastructural changes in the hearts of pigs fed this diet for 45 days which included an abnormally high number of mitochondria as well as enlarged mitochondria, i.e., 2–4 times normal size, which were called "megamitochondria." The use of this term

TABLE VIII

Comparison of Myocardial Lipidosis in Pigs and Rats as Determined Histologically by Oil Red O Staining[a]

Animal	Diet[b]	% incidence of lipidosis[c]						n
		0	1	2	3	4	5	
Pig	16% lard	91	4	—	4	—	—	23
	20.5% fish oil	50	50	—	—	—	—	6
	16% H$_2$—fish oil	78	4	9	9	—	—	23
	16% HEAR oil	45	18	18	18	—	—	11
Rat	16% lard	89	11	—	—	—	—	9
	20.5% fish oil	—	11	22	44	22	—	9
	16% H$_2$—fish oil	—	—	—	11	78	11	9
	16% HEAR oil	—	—	—	—	—	100	9

[a] From Opstvedt et al. (1979).
[b] All diets were made up to contain 21% fat with the balance as sunflower oil and lard.
[c] Severity is indicated as: 0 = no lipidosis; 1 to 5 = increasing lipidosis.

has since been criticized as has the observation made by these authors that the lipid droplets in the myocardium are surrounded by membrane (Schiefer et al., 1978).

The occurrence of myocardial lipidosis in piglets fed diets that contain HEAR oil is by no means a consistent finding. Canadian scientists have shown that in pigs fed a diet containing 15% HEAR oil with an erucic acid content of 21%, there was no histological evidence of lipidosis, and there were not any dietary related lesions in muscle, liver, or spleen (Aherne et al., 1975). A second study reported a year later by the same group with a HEAR oil of 34% erucic acid also failed to show evidence of myocardial lipidosis (Aherne et al., 1976).

2. MYOCARDIAL NECROSIS

The areas of focal myocarditis and myocardial necrosis present in pig hearts has been discussed by a number of investigators. These lesions are generally small and consist of individual muscle cell necrosis with some cells showing vacuoles and disintegrating myofibrils (Svaar et al., 1980). These small foci of necrosis may show mononuclear cell infiltrations (Friend et al., 1975a). Endocardial calcification has been reported in some pig hearts although this is apparently not related to the presence of vegetable oils in the diet. This degenerative calcification appeared in the left atrial endocardium (Friend et al., 1975b). It has been suggested that focal myocarditis and necrosis may be the result of ascarid infection, caused by Ascaris suum larval migration through heart tissue (Aherne et al., 1975). In any case, most of the investigators agree that the mild myocarditis observed in pigs up to 1 year of age is not related to diet and that the lesions are observed with just about equal frequency in pigs fed the low fat control diets as in pigs fed up to 20% vegetable oil in the diet. The type of oil fed, i.e., HEAR oil or other vegetable oils, did not appear to influence the incidence of myocarditis (Roine et al., 1960; Aherne et al., 1975, 1976; Friend et al., 1975b, 1976; Bijster et al., 1979; Svaar et al., 1980). All these studies were done with commercial swine breeds, i.e., Yorkshire, Crossbreds, German Landrace, and Norwegian Landrace.

One group of French investigators appears to be in disagreement with these results. In examining heart sections from pigs fed HEAR oils with the electron microscope, Vodovar et al. (1977) observed abnormal and enlarged mitochondria as well as irregularities in the Z lines and intercalated disks of the myofibrils (Vodovar et al., 1973). In a later study, these workers reported the presence of greatly enlarged mitochondria (5–15 times) with a variety of intramitochondrial inclusions (Vodovar et al., 1977). At later stages, i.e., after the HEAR oil diet had been fed for over 60 days, mitochondrial degeneration was reported. According to these authors, the ultrastructural mitochondrial changes were seen regularly in pigs fed rapeseed oil, rarely in pigs fed other vegetable oils and were absent in pigs fed a low fat diet. Their results, however, are in conflict with the many and extensive studies that indicate that focal myocarditis occurs in older pigs irrespective of diet. These myocardial changes should not be attributed to the feeding of rapeseed oil unless and until these studies are confirmed in other laboratories.

B. Primates

1. MYOCARDIAL LIPIDOSIS

Unfortunately relatively few studies have been reported to date on the effects of HEAR oils in nonhuman primates. In a recent study, 11 laboratory born and reared cynomolgus monkeys (*Macaca fascicularis*) were fed balanced diets that were supplemented with 25% of a 3:1 mixture of lard/corn oil, a HEAR oil (25% erucic acid), or a partially hydrogenated herring oil (23% cetoleic acid) for about 120 days (Table IX). The 22:1 containing diets produced myocardial lipidosis (+ + +) while the lard/corn oil diet produced mild to moderate myocardial lipidosis (+ to + +) (Schiefer et al., 1978). Similar results were recorded for skeletal muscle. On the other hand, there was no apparent interference with either electrical conductivity or papillary muscle contractile force (plotted as length against tension, g/cm^2) (Loew et al., 1978). In a repeat experiment of similar design, the monkeys were killed at 6, 12, 18, 24, and 30 months (Table IX). The results were similar except in the second experiment both the herring oil and the lard/corn oil group had a

TABLE IX

Myocardial Necrosis in Monkeys

Test oil	% Fat in diet	% 22:1 in oil	Strain[a]	Sex	Time on diet (weeks)	Heart lesions Affected	Heart lesions Examined	Ref.[b]
Lard/corn oil (3/1)	20	0	S.s.	n.a.	10	0	1	1
HEAR/lard/corn (10/7.5/2.5)	20	14.7				1	3	
HEAR oil	20	29.4				2	2	
Peanut oil	20	0	M.r.	M	52	0	8	2
H$_2$-Peanut oil	20	0				0	8	
Mustard oil	20	40–44				4	8	
Lard/corn oil (3/1)	25	<0.1	M.f.	2M2F	17.3	2	4	3
H$_2$-Fish oil	25	19.7		2M1F		2	3	
HEAR oil	25	25.6		2M2F		2	4	
Peanut oil	10	0	M.r.	M	18	0	9	4
Mustard oil	5	40–44				0	9	
Mustard oil	10	40–44				0	9	
Lard/corn oil (3/1)	25	<0.1	M.f.	2M1F	26	0.31[c]	3	5
H$_2$-Fish oil	25	20.8		2M1F	26	0.40	3	
Lard/corn oil (3/1)	25	<0.1		1M2F	52	1.31	3	
H$_2$-Fish oil	25	20.8		1M2F	52	1.50	3	
Lard/corn oil (3/1)	25	<0.1		3M	78	0.94	3	
H$_2$-Fish oil	25	20.8		3M	78	0.99	3	
Lard/corn oil (3/1)	25	<0.1		2M1F	104	0.00	3	
H$_2$-Fish oil	25	20.8		1M2F	104	0.16	3	
Lard/corn oil (3/1)	25	<0.1		1M2F	130	0.33	3	
H$_2$-Fish oil	25	20.8		2M1F	130	0.61	3	
Survey of 312 primate hearts			S.s.	28M20F	—	48	180	6
			M.f.	23M23F	—	46	80	
			M.m.	1M4F	—	5	28	
			M.a.	4M3F	—	7	24	

[a] S.s., *Saimiri sciureus* (squirrel monkey); M.r., *Macaca radiata*; M.f., *Macaca fascicularis*; M.m., *Macaca mulatta*; M.a., *Macaca assamensis*.

[b] References: 1, Beare-Rogers and Nera (1972); 2, Gopalan *et al.* (1974); 3, Ackman and Loew (1977); 4, Shenolikar and Tilak (1980); 5, Schiefer (1982); 6, Qureshi (1979).

[c] Myocarditis index, number of foci of inflammation in all sections/number of section examined.

+++ myocardial lipidosis after 6 and 12 months. Thereafter, in the lard/corn oil group the myocardial lipidosis decreased in intensity (0 to ++), while the fish oil group retained a ++ to +++ rating throughout the 30 month trial (Schiefer, 1982). When the heart tissue from these monkeys was examined with the electron microscope, monkeys that were fed a 22:1 enriched diet had enlarged and irregularly shaped mitochondria, sometimes with amorphous material present in the matrix. Some deterioration of mitochondrial cristae was observed. The lipid droplets in the myocardium were not surrounded by membrane (Schiefer et al., 1978). It was not stated whether similar ultrastructural changes were present in the control group of monkeys that were fed the lard/corn oil enriched diets, when these monkeys had a myocardial lipidosis with +++ severity.

An earlier study done with squirrel monkeys (Saimiri sciureus) showed the presence of myocardial lipid droplets after 1 week on a diet which contained either HEAR oil or a lard/corn oil mixture (Beare-Rogers and Nera, 1972). After 10 weeks on the diet, more lipidosis was seen in the squirrel monkeys fed HEAR oil than in the ones that received the lard/corn oil mixture.

The Indian Council of Medical Research investigated the problem of feeding high levels of mustard oil to adult monkeys (Macaca radiata). The mustard oil contained 40–44% erucic acid and in the first trial was fed at the level of 20% by weight of the diet (Gopalan et al., 1974). The trial lasted 1 year during which the monkeys showed no particular abnormalities. There were occasional ventricular extrasystoles detected by electrocardiography, but these were apparently not diet related. Serum lipids and serum enzymes (SGOT and SGPT) remained unchanged. At postmortem, the hearts from the mustard oil fed group had the yellowish discoloration associated with myocardial lipidosis. When examined with the light microscope, the outstanding feature in these hearts was the formation of vacuoles in the sarcoplasm, which, although this was not stated, probably was the result of lipid droplet accumulation. In a repeat study from the same institution (Shenolikar and Tilak, 1980) the mustard oil intake was reduced to 10% and 5% by weight of the diet. The hearts from monkeys fed mustard oil for 18 weeks apparently still showed yellowish discoloration, but there was no evidence of vacuole formation when heart sections were examined histologically. No organs other than the heart showed any diet related abnormalities in either the first or second experiment.

Some conclusions can be drawn from the few studies that have been done with monkeys fed a diet containing docosenoic fatty acids from HEAR, mustard or fish oil. It seems that in response to high fat diets, and in particular to fats with long chain monoenoic fatty acids, there is more lipid accumulation in monkey hearts than in pig hearts. This follows from the results recorded with specific fat stains, or from measuring the fat accumulation by gravimetric means. It is possible to detect fat accumulation gravimetrically in monkey hearts (Ackman, 1980), but not in pig hearts (Kramer et al., 1975; Seher et al., 1979; Opstvedt et al., 1979). The monkey also responds differently from the rat to high fat diets. The rat heart becomes lipidotic only when the fat in the diet is high in docosenoic fatty acids. The monkey heart, on the other hand, shows lipidosis in response to high fat diets even in the absence of docosenoic fatty acids. In any case, myocardial lipidosis in the monkey is never as severe as in the rat. While the monkey showed a maximum 1.9-fold increase in the cardiac triglycerides at 6 months (Ackman, 1980), the rat showed a 12-fold increase in 1 week (Kramer et al., 1979) when HEAR oil containing a comparable level of docosenoic acid was fed. There is another difference between the two species. The rat, when fed a HEAR oil enriched diet, shows a rapid accumulation of cardiac lipids followed by a rapid decline (Beare-Rogers et al., 1971; Kramer et al., 1973; Rocquelin et al., 1973). This does not appear to be the case in the monkey in which definite myocardial lipidosis appears to persist for periods up to 30 months when fed a diet containing fish oil (Ackman, 1980).

2. MYOCARDIAL NECROSIS

It is difficult to assess the relationship of HEAR oils and other oils high in docosenoic acid content to the development of focal myocardial degenerative lesions in the monkey. In a recent study, a series of 312 hearts were selected at random from monkeys used in unrelated toxicological studies (Qureshi, 1979). The monkeys, which included squirrel (*Saimiri sciureus*) cynomolgus (*Macaca fascicularis*), rhesus (*Macaca mulatta*) and assam (*Macaca assamensis*) monkeys were of both sexes. Chronic interstitial myocarditis was found in 34% of the monkeys, approximately evenly distributed in males and females (Table IX). The lesions varied from slight necrosis to myocarditis with focal accumulation of lymphocytes, mononuclear cells, plasma cells, and some eosinophiles. Inflammation of the myocardium was distributed throughout the heart. These lesions, which occur frequently in primates, apparently are not related to bacterial, viral, or parasitic infections, but may be related to, and precipitated by, stress (Qureshi, 1979; Soto et al., 1964).

Similar lesions have also been reported in monkeys fed a high fat diet. In an experiment with cynomolgus monkeys that were fed diets containing 25% HEAR oil (25% 22:1), partially hydrogenated herring oil (23% 22:1) or a lard/corn oil mixture, myocardial lesions, specifically inflammatory foci, were found in monkeys from all three dietary groups (Ackman and Loew, 1977; Schiefer et al., 1978) (Table IX). In this experiment the diets were fed for a total of 120 days. In a follow-up study the herring oil diet was tested against the lard/corn oil control diet for longer periods of time. Here again the lesions, i.e., the focal myocarditis, was observed in both groups of monkeys, which were serially slaughtered at 6, 12, 18, 24, and 30 months (Table IX). What was apparent from this study was that the wild-caught monkeys have a higher incidence of inflammatory foci than their laboratory bred counterparts (Schiefer, 1982).

In marked contrast to these studies is the report from the Indian Council of Medical Research (Gopalan et al., 1974). In 4 of 8 monkeys (Macaca radiata) fed for 1 year a diet that contained 20% by weight mustard oil (40–44% erucic acid) there was sarcoplasmic vacuolation, swelling of myofibrils, focal areas of mononuclear cell infiltration, and areas of fibrosis. The fibrotic areas, for some unexplained reason, were limited to the right ventricular myocardium, with the left ventricular myocardium, in all but one animal of this group, showing no fibrotic changes (Gopalan et al., 1974). In this study the control groups, which were fed either peanut oil or partially hydrogenated peanut oil, "exhibited an essentially normal histology." This is somewhat surprising in view of the fact that about 30% of monkeys examined at random showed focal myocardial lesions (Qureshi, 1979). In a follow-up study from the Indian Council of Medical Research (Shenolikar and Tilak, 1980) monkeys were fed diets contained only 5% and 10% mustard oil. These diets were fed for a period of 18 months. According to the authors, the hearts from all monkeys (27 in total) fed either the mustard oil or the peanut oil diet were now free of vacuoles or fibrotic lesions. The authors also report on respiratory studies with heart muscle homogenates and isolated mitochondria. They observed that mitochondrial respiration was impaired only in the left ventricle while fibrotic lesions were present only in the right ventricle of the monkeys that were fed the mustard oil diet. From this they concluded that a causal relationship between impaired respiration and muscle degeneration does not seem probable (Shenolikar and Tilak, 1980).

What seems clear from the research that has been done to date is that primates tolerate fairly high levels of docosenoic acids (erucic or cetoleic) without showing an increased incidence of focal myocarditis that can be attributed to diet. A summary of published data is presented in Table IX. There appears to be little correlation between the incidence of myocardial lesions and the 22:1 content of the diet. Only one experiment with mustard oil (Gopalan et al., 1974) seems to suggest that 22:1 caused heart lesions in the monkey. These experiments with extreme levels of 22:1 in the diet (8.4%) have never been repeated to date. Since results from other laboratories appear to be different, a confirmatory observation would be desirable. An excellent approach would be to summate the area of damage per heart as has been done with rats (Vles et al., 1976, 1978). This permits a quantitative approach to the degree of heart damage caused by different diets which can then be assessed statistically for significance.

IX. THE INVOLVEMENT OF HEALTH AGENCIES IN LOWERING THE ERUCIC ACID CONTENT OF RAPESEED OILS

The Food and Drug Directorate of the Department of National Health and Welfare in Ottawa first expressed concern about the erucic acid in HEAR oils in 1956. The Department initially decided to remove the oil from the food chain, but then eased this restriction when the Canadian Committee on Fats and Oils reported that, based on available evidence, the use of rapeseed oil with its then rather limited use, was no hazard to human health. The committee however urged an expansion of research in this area (Reynolds, 1975). It was from 1956 on that a serious research effort was initiated in Canada in order to produce low erucic acid cultivars of rapeseed and to amass data on the potential biological hazards of erucic acid in the food chain. Within 4 years from this date the world's first cultivar of rapeseed with a low erucic acid content was grown (Stefansson et al., 1961).

In 1960 the first report appeared that a 70 calorie % HEAR oil diet caused myocarditis and myocardial necrosis in rats (Roine et al., 1960). Curiously, no one observed the myocardial lipidosis, which is readily detectable by the naked eye, until a decade later when the scientists from Unilever reported on this disturbing property of HEAR oils (Abdellatif and Vles, 1970). The observation that, in rats, myocardial fat accumulation is found when HEAR oil diets are used was fully reported on and discussed at the International Conference on Rapeseed and Rapeseed Products which was held in Ste. Adèle, Quebec, in 1970. This Conference concluded with an official conference statement indicating that it was "prudent" to gradually effect a changeover to new, low erucic acid rapeseed varieties, although "no hazard to human health had ever been attributed to HEAR oils throughout its long history as a staple component of the human diet" (Migicovsky, 1970). In spite of this well reasoned and deliberately low keyed approach to the changeover from HEAR oils to low erucic acid rapeseed (LEAR) oils, some countries were quite alarmed about the potential health hazards of erucic acid. In 1972 the Italian government banned the sale of edible oils with over 10% erucic acid, and then changed this limit to 15% erucic acid. During that period, the chairman of an Italian vegetable oil manufacturing firm was arrested and his firm's oil stocks seized because the oils contained 46% erucic acid (Parkes, 1974).

In the early 1970s the Canadian Food and Drug Directorate scientists established that in rats the zero effect level of myocardial fat accumulation is 5% HEAR oil by weight of the diet or 10% of calorie intake. The HEAR oil tested contained 33% 22:1. At this level of HEAR oil intake, there was no myocardial fat infiltration detected by oil red O stain, but as the HEAR oil in the diet was increased from 5% to 10, 15, or 20% of the diet, there was definite lipid accumulation in the myocardium (Beare-Rogers *et al.*, 1971). The zero effect level of this HEAR oil fed at 5% of the diet is equivalent to an oil fed at 20% by weight of the diet that contains about 8% erucic acid. Based on these findings and on a report by an Expert Committee on Long-Chain Fatty Acids, Health and Welfare Canada, in December of 1973 restricted the maximum content of C_{22} monoenoic fatty acids in processed edible fats and oils to 5% of the total fatty acids present (News Release, No. 1973-76, Health and Welfare Canada, June 29, 1973; Consolidated Regulations of Canada, 1978, Vol. VIII, C870, B.09.022). In 1976, The Council of European Communities ruled that the maximum level of erucic acid in oils and fats be not greater than 10% as of July 1, 1977, and that this level be fixed at not greater than 5% by July 1, 1979 (Off. J. Eur. Commun., 1976, L202, 35-37, 28.7.1976). Most other countries that are large users of rapeseed oil and mustard oil, and are not included above, generally do not have specific regulations covering the permissible erucic acid content in fats and oils. It is to be expected that among the large users such as India, Pakistan, China, and Japan the trend toward the increased use of low erucic acid rapeseed oils will take effect.

In Western Europe and North America, the high erucic acid rapeseed oils have been phased out and replaced by low erucic acid rapeseed oils. The difference in composition is shown in Table I, Chapter 17. It is clear from this that the new low erucic acid rapeseed (LEAR) oils are very different from the older HEAR oils. Their docosenoic and eicosenoic acid concentration has been sharply decreased and their 16:0, 18:0, 18:1, and 18:2 content has increased. It is quite obvious that the nutritional properties of these two oils will be different, yet many scientists in describing their experiments still state that "rapeseed oil" was used without bothering to define if the oil was a LEAR oil or a HEAR oil. In view of the significant differences in fatty acid composition of these two types of oils, this is an improper omission and makes it unnecessarily difficult for the reader to interpret the experiments.

In North America, the use of HEAR oil (~40% erucic acid) as food for human consumption is permitted only in the fully hydrogenated, or superglycerinated fully hydrogenated form, to an iodine value of 4 or less (*Federal Register*, 1977). As such it has use as a stabilizer and thickener component for peanut butter at a maximum concentration of 2%, and in cake mix formulations (as an emulsifier) in shortening at a maximum concentration of 4% of the shortening or 0.5% of the cake mix. For this use the fully hydrogenated HEAR oil has been granted GRAS (generally recognized as safe) status by the U. S. Food and Drug Administration.

In summary, one may conclude that physiological, pathological, and nutritional effects of feeding HEAR oils to experimental animals have been studied quite thoroughly. Most of the adverse effects can now be ascribed to erucic acid, which is present in HEAR oils at levels as high as 50%. There is no question that erucic acid, with its multitude of nutritional, biochemical, and physiological effects, will interest scientists from various disciplines for many years. One might, however, hope that future experiments will be designed to test the purified fatty acid as a component of synthetic fat mixtures rather than HEAR oils. This would be scientifically advantageous. It is well recognized that there are complex nutritional interactions of fatty acids. Different HEAR oils differ in their composition of all their fatty acids, not just erucic acid. This creates a problem in the interpretation of results, particularly where different laboratories and different oils are involved. This could be circumvented by testing the effects of erucate with standardized diets that contain properly defined fat mixtures.

X.   CURRENT REGULATIONS ON PERMISSIBLE ERUCIC ACID LEVELS

Because the rat is highly susceptible to myocardial lipidosis when fed diets that contain 22:1, some health regulatory agencies deemed it prudent to limit the content of these acids in the human diet.

Member nations of the European Economic Community have had in effect since July 1, 1977, the directive that the level of erucic acid in oils, fats, or mixtures thereof, intended for human consumption, is not to exceed 10%. As of July 1, 1979, the erucic acid level, calculated on the total level of fatty acids in the fat, is not to exceed 5%. These regulations also apply to compound foodstuffs to which oils or fats are added and the overall fat content exceeds 5%. [*Official Journal of the European Communities* **L 202**, 35–37 (1976); *Journal Officiel de la République Française* **110**, No. 187, 3045–3047 (1978); *Bundesgesetzblatt (Rechtsvorschriften)* **31**, 782 (1977).]

The Canadian regulations state that cooking oil, margarine, salad oil, simulated dairy product, or shortening may not contain more than 5% C_{22} monoenoic fatty acid calculated as a proportion of the total fatty acids contained in the product. Both erucic and cetoleic acids are included in this regulation. [Health and Welfare Canada, News Release 1973–1976, (1973); Consolidated Regulations of Canada Vol. VIII, C870, B.09.022 (1978).]

In Sweden, by voluntary agreement, the limit of C_{22} monoenoic fatty acid is from 3 to 4% (Christophersen *et al.*, 1976).

For the remaining rapeseed oil consuming countries there do not appear to be any regulations that deal specifically with the erucic content of the oil. There is, however, a general trend in all countries to change from HEAR oils to the low erucic acid rapeseed oils.

ACKNOWLEDGMENT

This work is Contribution No. 1092 from the Animal Research Center, Ottawa, Ontario, Canada.

REFERENCES

Abdellatif, A. M. M. (1972). *Nutr. Rev.* **30**, 2–6.
Abdellatif, A. M. M., and Vles, R. O. (1970). *Nutr. Metab.* **12**, 285–295.
Abdellatif, A. M. M., and Vles, R. O. (1973). *Nutr. Metab.* **15**, 219–231.
Ackman, R. G. (1974). *Lipids* **9**, 1032–1035.
Ackman, R. G. (1980). "A Report to Fisheries and Oceans Canada," PDR Contract No. 08SC-01532-9-0244.
Ackman, R. G., and Loew, F. M. (1977). *Fette, Seifen,* Anstrichm. **79**, 15–24, 58–69.
Ackman, R. G., Epstein, S., and Eaton, C. A. (1971a). *Comp. Biochem. Physiol. B* **40B**, 683–697.
Ackman, R. G., Hooper, S. N., and Hingley, J. (1971b). *J. Am. Oil Chem. Soc.* **48**, 804–806.
Aherne, F. X., Bowland, J. P., Christian, R. G., Vogtmann, H., and Hardin, R. T. (1975). *Can. J. Anim. Sci.* **55**, 77–85.
Aherne, F. X., Bowland, J. P., Christian, R. G., and Hardin, R. T. (1976). *Can. J. Anim. Sci.* **56**, 275–284.
Appelqvist, L.-A. (1972). *In* "Rapeseed" (L.-A. Appelqvist and R. Ohlson, eds.), pp. 1–8. Elsevier, Amsterdam.
Ashburn, L. L., and Lowry, J. V. (1944). *Arch. Pathol.* **37**, 27–33.
Astorg, P.-O. (1980). *Ann. Nutr. Aliment.* **34**, 625–640.
Astorg, P.-O., and Cluzan, R. (1976). *Ann. Nutr. Aliment.* **30**, 581–602.
Astorg, P.-O., and Cluzan, R. (1977). *Ann. Nutr. Aliment.* **31**, 43–68.
Astorg, P.-O., and Levillain, R. (1979). *Ann. Nutr. Aliment.* **33**, 643–658.
Beare, J. L., Murray, T. K., Grice, H. C., and Campbell, J. A. (1959a). *Can. J. Biochem. Physiol.* **37**, 613–621.
Beare, J. L., Gregory, E. R. W., and Campbell, J. A. (1959b). *Can. J. Biochem. Physiol.* **37**, 1191–1195.
Beare, J. L., Murray, T. K., and Campbell, J. A. (1960). *Can. J. Biochem. Physiol.* **38**, 187–192.
Beare, J. L., Gregory, E. R. W., Morison Smith, D., and Campbell, J. A. (1961). *Can. J. Biochem. Physiol.* **39**, 195–201.
Beare, J. L., Campbell, J. A., Youngs, C. G., and Craig, B. M. (1963). *Can. J. Biochem. Physiol.* **41**, 605–612.
Beare-Rogers, J. L. (1975). *In* "Modification of Lipid Metabolism" (E.G. Perkins and L. A. Witting, eds.), pp.43–57. Academic Press, New York.
Beare-Rogers, J. L., and Nera, E. A. (1972). *Comp. Biochem. Physiol.* **41B**, 793–800.
Beare-Rogers, J. L., and Nera, E. A. (1974). *Lipids* **9**, 365–367.
Beare-Rogers, J. L., Nera, E. A., and Heggtveit, H. A. (1971). *Can. Inst. Food Technol. J.* **4**, 120–124.
Beare-Rogers, J. L., Nera, E. A., and Craig, B. M. (1972a). *Lipids* **7**, 46–50.
Beare-Rogers, J. L., Nera, E. A., and Craig, B. M. (1972b). *Lipids* **7**, 548–552.
Beare-Rogers, J.L., Nera, E. A., and Heggtveit, H. A. (1974). *Nutr. Metab.* **17**, 213–222.
Beckett, G. J., and Boyd, G. S. (1975). *Eur. J. Biochem.* **53**, 335–342.
Berglund, F. (1975). *Acta Med. Scand. Suppl.* 585, 47–49.

Bhatia, I. S., Sharma, A. K., Gupta, P. P., and Ahuja, S. P. (1979). *Indian J. Med. Res.* **69**, 271–283.

Bijster, G. M., Timmer, W. G., and Vles, R. O. (1979). *Fette, Seifen, Anstrichm.* **81**, 192–194.

Bodak, A., Dutrieux, J. M., Moravec, M., Guillemot, H., and Hatt, P. Y. (1979). *Ann. Biol. Anim. Biochim. Biophys.* **19**, 523–536.

Boer, J., Jansen, B. C. P., and Kentie, A. (1947). *J. Nutr.* **33**, 339–358.

Booth, A. N., Robbins, D. J., Gumbmann, M. R., Gould, D. H., Tallent, W. H., and Wolff, I. A. *J. Am. Oil Chem. Soc.* **49**, 304A.

Brockerhoff, H., and Jensen, R. G. (1974). "Lipolytic Enzymes," Chapter IV. Academic Press, New York.

Brockerhoff, H., and Yurkowski, M. (1966). *J. Lipid Res.* **7**, 62–64.

Budzynska-Topolowska, J., Ziemlanski, S., and Kochman, E. (1975). *Ann. Nutr. Aliment.* **29**, 33–43.

Carney, J. A., Lewis, A., Walker, B. L., and Slinger, S. J. (1972). *Biochim. Biophys. Acta* **280**, 211–214.

Carroll, K. K. (1951). *Endocrinology* **48**, 101–110.

Carroll, K. K. (1953). *J. Biol. Chem.* **200**, 287–292.

Carroll, K. K. (1957). *Proc. Soc. Exp. Biol. Med.* **94**, 202–205.

Carroll, K. K. (1958). *J. Nutr.* **64**, 399–410.

Carroll, K. K. (1962). *Can. J. Biochem. Physiol.* **40**, 1115–1122.

Carroll, K. K., and Noble, R. L. (1952). *Endocrinology* **51**, 476–486.

Carroll, K. K., and Noble, R. L. (1957). *Can. J. Biochem. Physiol.* **35**, 1093–1105.

Carroll, K. K., and Richards, J. F. (1958). *J. Nutr.* **64**, 411–424.

Christophersen, B. O., Svaar, H., Langmark, F. T., Gumpen, S. A., and Norum, K. R. (1976). *Ambio* **5**, 169–173.

Charlton, K. M., Corner, A. H., Davey, K., Kramer, J. K. G., Mahadevan, S., and Sauer, F. D. (1975). *Can. J. Comp. Med.* **39**, 261–269.

Clandinin, M. T., and Yamashiro, S. (1980). *J. Nutr.* **110**, 1197–1203.

Conacher, H. B. S., and Page, B. D. (1972). *J. Am. Oil Chem. Soc.* **49**, 283–286.

Conacher, H. B. S., Page, B. D., and Chadha, R. K. (1972). *J. Am. Oil Chem. Soc.* **49**, 520–523.

Coniglio, J. G., Grogan, W. M., and Harris, D. G. (1974). *Proc. Soc. Expt. Biol. Med.* **146**, 738–741.

Darnerud, P. O., Olsen, M., and Wahlström, B. (1978). *Lipids* **13**, 459–463.

Dessau, F. I., Lipchuck, L., and Klein, S. (1954). *Proc. Soc. Exp. Biol. Med.* **87**, 522–524.

Deuel, H. J., Movitt, E., Hallman, L. F., and Mattson, F. (1944). *J. Nutr.* **27**, 107–121.

Deuel, H. J., Cheng, A. L. S., and Morehouse, M. G. (1948a). *J. Nutr.* **35**, 295–300.

Deuel, H. J., Greenberg, S. M., Straub, E. E., Jue, D., Gooding, C. M., and Brown, C. F. (1948b). *J. Nutr.* **35**, 301–314.

Deuel, H. J., Johnson, R. M., Calbert, C. E., Gardner, J., and Thomas, B. (1949). *J. Nutr.* **38**, 369–379.

Dodds, W. J. (1982). *Fed. Proc., Fed. Am. Soc. Exp. Biol.* **41**, 247–256.

Engfeldt, B., and Brunius, E. (1975a). *Acta Med. Scand. Suppl.* **585**, 15–26.

Engfeldt, B., and Brunius, E. (1975b). *Acta Med. Scand. Suppl.* **585**, 27–40.

Farnworth, E. R., Kramer, J. K. G., Thompson, B. K., and Corner, A. H. (1982). *J. Nutr.* **112**, 231–240.

Federal Register (1977). **42**, 48335–48336.

Filer, L. J., Mattson, F. H., and Fomon, S. J. (1969). *J. Nutr.* **99**, 293–298.

French, J. E. (1952). *Arch. Pathol.* **53**, 485–496.

Friend, D. W., Corner, A. H., Kramer, J. K. G., Charlton, K. M., Gilka, F., and Sauer, F. D. (1975a). *Can. J. Anim. Sci.* **55**, 49–59.

Friend, D. W., Gilka, F., and Corner, A. H. (1975b). *Can. J. Anim. Sci.* **55**, 571–578.

Friend, D. W., Kramer, J. K. G., and Corner, A. H. (1976). *Can. J. Anim. Sci.* **56**, 361–364.

Garren, L. D., Gill, G. N., Masui, H., and Walton, G. M. (1971). *Recent Prog. Horm. Res.* **27**, 433–478.

Gopalan, C., Krishnamurthi, D., Shenolikar, J. S., and Krishnamachari, K. A. V. R. (1974). *Nutr. Metab.* **16**, 352–365.

Hartroft, W. S. (1955). *Fed. Proc., Fed. Am. Soc. Exp. Biol.* **14**, 655–660.

Highman, B., and Daft, F. S. (1951). *Arch. Pathol.* **52**, 221–229.

Holmes, A. D. (1918). *U.S., Dep. Agric. Bull.* **687**.

Houtsmuller, U. M. T., Struijk, C. B., and Van der Beek, A. (1970). *Biochim. Biophys. Acta* **218**, 564–566.

Hulan, H. W., Kramer, J. K. G., Mahadevan, S., Sauer, F. D., and Corner, A. H. (1976). *Lipids* **11**, 6–8.

Hulan, H. W., Kramer, J. K. G., and Corner, A. H. (1977). *Can. J. Physiol. Pharmacol.* **55**, 258–264.

Hung, S., Umemura, T., Yamashiro, S., Slinger, S. J., and Holub, B. J. (1977). *Lipids* **12**, 215–221.

Jaillard, J., Sezille, G., Dewailly, P., Fruchart, J. C., and Bertrand, M. (1973). *Nutr. Metab. 15*, 336–347.

Kaunitz, H., and Johnson, R. E. (1973). *Lipids* **8**, 329–336.

Kesten, H. D., Salcedo, J., and Stetten, DeW. (1945). *J. Nutr.* **29**, 171–177.

Kramer, J. K. G. (1980). *Lipids* **15**, 651–660.

Kramer, J. K. G., and Hulan, H. W. (1978). *J. Lipid Res.* **19**, 103–106.

Kramer, J. K.G., Mahadevan, S., Hunt, J. R., Sauer, F. D., Corner, A. H., and Charlton, K. M. (1973). *J. Nutr.* **103**, 1696–1708.

Kramer, J. K. G., Friend, D. W., and Hulan, H. W. (1975). *Nutr. Metab.* **19**, 279–290.

Kramer, J. K. G., Hulan, H. W., Trenholm, H. L., and Corner, A. H. (1979). *J. Nutr.* **109**, 202–213.

Lambertsen, G., Myklestad, H., and Braekkan, O. R. (1971). *J. Am. Oil Chem. Soc.* **48**, 389–391.

Lancet (1974) **2**, 1359–1360.

Loew, F. M., Schiefer, B., Laxdal, V. A., Prasad, K., Forsyth, G. W., Ackman, R. G., Olfert, E. D., and Bell, J.M. (1978). *Nutr. Metab.* **22**, 201–217.

Lowenhaupt, E., Schulman, M. P., and Greenberg, D. M. (1950). *Arch. Pathol.* **49**, 427–433.

Mattson, F. H., and Streck, J. A. (1974). *J. Nutr.* **104**, 483–488.

McCutcheon, J. S., Umemura, T., Bhatnagar, M. K., and Walker, B. L. (1976). *Lipids* **11**, 545–552.

Migicovsky, B. B. (1970). *Proc. Int. Conf. Sci., Technol. Market. Rapeseed Rapeseed Prod., 1970*, pp. 556–560.

Mulky, M. J. (1979). *Ind. J. Nutr. Diet.* **16**, 241–250.

Nenenzel, J. C. (1970). *Lipids* **5**, 308–319.

Noble, R. L., and Carroll, K. K. (1961). *Recent Prog. Horm. Res.* **17**, 97–118.

Nolen, G. A. (1981). *J. Am. Oil Chem. Soc.* **58**, 31–37.

Ockner, R. K., Manning, J. A., Poppenhausen, R. B., and Ho, W. K. L. (1972). *Science* **177**, 56–58.

Opstvedt, J., Svaar, H., Hansen, P., Pettersen, J., Langmark, F. T., Barlow, S. M., and Duthie, I. F. (1979). *Lipids* **14**, 356–371.

Paloheimo, L., and Jahkola, B. (1959). *J. Sci. Agric. Soc. Finl.* **31**, 212–214.

Parkes, C. (1974). *New Sci.* **62**, 339–340.

Pascal, J.–C., and Ackman, R. G. (1976). *Chem. Phys. Lipids* **16**, 219–223.

Perdue, H. S., and Phillips, P. H. (1952). *Proc. Soc. Exp. Biol. Med.* **81**, 405–407.

Qureshi, S. R. (1979). *Vet. Pathol.* **16**, 486–487.

Reynolds, J. R. (1975). MSc. Thesis, University of Saskatchewan, Saskatoon.

Rings, R. W., and Wagner, J. E. (1972). *Lab. Anim. Sci.* **22**, 344–352.

Rocquelin, G. (1972) *C.R. Hebd. Seances Acad. Sci. Ser. D* **274**, 592–595.

Rocquelin, G., and Cluzan, R. (1968). *Ann. Biol. Anim., Biochim., Biophys.* **8**, 395–406.

Rocquelin, G., and Leclerc, J. (1969). *Ann. Biol. Anim., Biochim. Biophys.* **9**, 413–426.

Rocquelin, G., Sergiel, J.–P., Martin, B., Leclerc, J., and Cluzan, R. (1971). *J. Am. Oil Chem. Soc.* **48**, 728–732.

Rocquelin, G., Sergiel, J.–P., Astorg, P. O., and Cluzan, R. (1973). *Ann. Biol. Anim. Biochim. Biophys.* **13**, 587–609.

Rocquelin, G., Sergiel, J. P., Astorg, P. O., Nitou, G., Vodovar, N., Cluzan, R., and Levillain, R. (1974). *Proc. Int. Rapskongr., 4th, 1974,* pp. 669–683.

Rocquelin, G., Juaneda, P., Peleran, J. C., and Astorg, P. O. (1975). *Nutr. Metab.* **19**, 113–126.

Roine, P., Uksila, E., Teir, H., and Rapola, J. (1960). *Z. Ernährungswiss.* **1**, 118–124.

Sayers, G., Sayers, M. A., Fry, E. G., White, A., and Long, C. N. H. (1944). *Yale J. Biol. Med.* **16**, 361–392.

Schiefer, B. (1982). *In* "Nutritional Evaluation of Long-Chain Fatty Acids in Fish Oil" (S. M. Barlow and M. E. Stansby, eds.), pp. 215–243. Academic Press, New York (in press).

Schiefer, B., Loew, F. M., Laxdal, V. A., Prasad, K., Forsyth, G. W., Ackman, R. G., and Olfert, E. D. (1978). *Am. J. Pathol.* **90**, 551–564.

Scott, R. F., Imai, H., Goodale, F., Lee, K. T., and Morrison, E. S. (1962). *Exp. Mol. Pathol.* **1**, 1–14.

Seher, A., Arens, M., Krohn, M., and Petersen, U. (1979). *Fette, Seifen, Anstrichm.* **81**, 181–187.

Sergiel, J. P., and Gabucci, L. (1980). *Reprod. Nutr. Develop.* **20**, 1415–1427.

Shenolikar, I. S., and Tilak, T. B. G. (1980). *Nutr. Metab.* **24**, 199–208.

Shima, S., Mitsunaga, M., and Nakao, T. (1972). *Endocrinology* **90**, 808–814.

Singh, D. (1958). "Rape and Mustard," Indian Central Oilseeds Committee, Bombay.

Soto, P. J., Beall, F. A., Nakamura, R. M., Kupferberg, L. L. (1964). *Arch. Pathol.* **78**, 681–690.

Stefansson, B. R., Hougen, F. W., and Downey, R. K. (1961). *Can. J. Plant Sci.* **41**, 218–219.

Svaar, H. (1982). *In* "Nutritional Evaluation of Long–Chain Fatty Acids in Fish Oil" (S. M. Barlow and M. E. Stansby, eds.), pp. 163–184. Academic Press, New York (in press).

Svaar, H., and Langmark, F. T. (1980). *Acta Pathol. Microbiol. Scand., Sect. A* **88**, 179–187.

Svaar, H., Langmark, F. T., Lambertsen, G., and Opstvedt, J. (1980). *Acta. Pathol. Microbiol. Scand., Sect. A* **88**, 41–48.

Ten Hoor, F., van de Graaf, H. M., and Vergroesen, A. J. (1973). *Recent Adv. Stud. Card. Struct. Metab.* **3**, 59–72.

Thomasson, H. J. (1955a). *J. Nutr.* **56**, 455–468.

Thomasson, H. J. (1955b). *J. Nutr.* **57**, 17–27.

Thomasson, H. J., and Boldingh, J. (1955). *J. Nutr.* **56**, 469–475.

Tomarelli, R. M., Meyer, B. J., Weaber, J. R., and Bernhart, F. W. (1968). *J. Nutr.* **95**, 583–590.

Trzeciak, W. H., and Boyd, G. S. (1974). *Eur. J. Biochem.* **46**, 201–207.

Umemura, T., Slinger, S. J., Bhatnagar, M. K., and Yamashiro, S. (1978). *Res. Vet. Sci.* **25**, 318–322.

Vles, R. O., Bijster, G. M., Kleinekoort, J. S. W., Timmer, W. G., and Zaalberg, J. (1976). *Fette, Seifen, Anstrichm.* **78**, 128–131.

Vles, R. O., Bijster, G. M., and Timmer, W. G. (1978). *Arch. Toxicol., Suppl.* **1**, 23–32.

Vodovar, N., Desnoyers, F., Levillain, R., and Cluzan, R. (1973). *C.R. Hebd. Seances Acad. Sci. Ser. D* **276**, 1597–1600.

Vodovar, N., Desnoyers, F., Cluzan, R., and Levillain, R. (1977). *Biol. Cell.* **29**, 37–44.

Vogtmann, H., Christian, R., Hardin, R. T. and Clandinin, D. R. (1975). *Int. J. Vitam. Nutr. Res.* **45**, 221–229.

Walker, B. L., and Carney, J. A. (1971). *Lipids* **6**, 797–804.

Walker, B. L., Atkinson, S. M., Zehaluk, C. M., and Mackey, M. G. (1972). *Comp. Biochem. Physiol.* **42B**, 619–625.

Wilgram, G. F., and Hartroft, W. S. (1955). *Br. J. Exp. Pathol.* **36**, 298–305.

Wilgram, G. F., Hartroft, W. S., and Best, C. H. (1954). *Br. Med. J.* **2**, 1–5.

Williams, W. L. (1960). *Yale J. Biol. Med.* **33**, 1–14.

Williams, W. L., and Aronsohn, R. B. (1956). *Yale J. Biol. Med.* **28**, 515–524.

Williams, W. L., and Oliver, R. I. (1961). *Anat. Rec.* **141**, 97–107.

Ziemlanski, S. (1977). *Bibl. "Nutr. Dieta"* **25**, 134–157.

Ziemlanski, S., Opuszynska, T., and Krus, S. (1972). *Pol. Med. J.* **11**, 1625–1633.

Ziemlanski, S., Okolska, G., Cieslakowa, D., and Kucharczyk, B. (1973a). *Pol. Med. Sci. Hist. Bull.* **15**, 453–460.

Ziemlanski, S., Rosnowski, A., and Opuszynska-Freyer, T. (1973b). *Acta Med. Pol.* **14**, 279–290.

Ziemlanski, S., Opuszynska, T., Bulhak-Jachymczyk, B., Olszewska, I., Wozniak, E., Krus, S., and Szymanska, K. (1974). *Pol. Med. Sci. Hist. Bull.* **15**, 3–10.

12

Cardiopathology Associated with the Feeding of Vegetable and Marine Oils

A. H. CORNER

I. INTRODUCTION

Roine *et al.* (1960) were the first to describe myocardial lesions in male rats and in pigs fed high levels of high erucic acid rapeseed (HEAR) oil. Subsequent studies indicate that there are variations in the incidence and type of lesions based on the experimental animal and oil fed. The principal

293

High and Low Erucic Acid Rapeseed Oils
Copyright © 1983 by Academic Press Canada
ISBN 0-12-425080-7

lesions described are myocardial lipidosis and necrosis. Most of these studies have been conducted with the rat, but work has also been done employing a variety of other experimental animals, including gerbils, monkeys, pigs, ducks, chickens, turkeys, guinea pigs, and rabbits. In retrospect the choice of the male rat as an experimental animal for cardiopathological studies was perhaps unfortunate. Lesions consisting of myocardial necrosis (death of myocardial cells) followed by removal of the necrotic cellular debris by macrophages and repair by replacement fibrosis are commonly reported in studies where rapeseed and marine oils are fed to male rats. However, myocardial necrosis in rats and other rodents has been associated with a variety of other experimental conditions.

Willens and Sproul (1938) in describing spontaneous cardiovascular disease in the rat report that fibrosis of the myocardium was one of the most common findings, being present in approximately 60% of all rats examined. Ashburn and Lowry (1944) described myocardial necrosis of both auricular and ventricular walls in thiamine deficient rats. Choline deficient rats fed ethyl laurate died of congestive heart failure due to widespread interstitial myocarditis. There was necrosis of individual myocytes in all four chambers (Kesten et al., 1945). French (1952) described primary degeneration of myocardium which progressed to necrosis in rats fed a potassium deficient diet. He also demonstrated lipidosis in muscle fibers at the periphery of the lesions. Perdue and Phillips (1952) found the incidence of myocardial necrosis and fibrosis in potassium deficient rats greater when the diet contained 20% corn oil than when it contained 5% corn oil. Wilgram et al. (1954) produced myocardial necrosis in rats by feeding choline deficient diets that were either fat free or contained various sources of fat (35% lard, beef fat, corn oil, coconut oil, or synthetic triglycerides). These workers also demonstrated lipid droplets in degenerating myocytes. Female rats were almost completely resistant to the development of cardiac damage by choline deficiency. The injection of pituitary growth hormone and testosterone into choline deficient female rats produced lesions of choline deficiency (Wilgram and Hartroft, 1955).

Williams and Aronsohn (1956) showed that mice fed low protein diets developed two types of lesions: (1) myocardial necrosis and fibrosis, and (2) deposits of mineral ceroid. Williams (1960) found that mice fed high fat, low protein diets developed cardiovascular disease characterized more by small areas of myocardial fibrosis than frank necrosis. Choline supplementation of the diet produced more severe lesions. Carroll et al. (1965) described atrial thrombosis and atrial and ventricular myocardial necrosis in mice fed high fat low protein diets. These workers also described the presence of fat droplets within the lesions in oil red O stained sections. They reported similar lesions (necrosis) in the hearts of older control mice. Thomas et al. (1968)

also produced massive atrial thrombosis in young adult mice fed diets of purified high fat low protein with and without choline supplementation. Lostroh (1958) produced calcification of the interventricular septum and right ventricle of mice by the injection of hydrocortisone. The male was less affected than the female and the concurrent administration of testosterone to females alleviated the condition. Selye (1958) using an active corticoid and a variety of vegetable and animal fats (corn oil, peanut oil, olive oil, pork, and chicken fat) produced myocardial necrosis and nephrocalcinosis in male Sprague–Dawley rats in the short period of 6 days.

Rings and Wagner (1972) reported on the incidence of spontaneous dystrophic cardiac calcification in DBA/2 mice. Ayers and Jones (1978) state that focal areas of myocardial necrosis and/or fibrosis are occasionally observed in a wide variety of species including the dog, cat, rat, nonhuman primates, and reptiles, and appear with increasing frequency with advancing age. They report that such lesions are particularly common in the rat in which there is both an age and sex interaction, the male being more severely affected and at an earlier age. These observations of the influence of sex on the incidence of myocardial necrosis in rats were extended by Hulan et al. (1977c). They found that the incidence of myocardial lesions in entire and castrated female and castrated male rats was similar but significantly lower than in entire male rats when fed diets of corn oil or a low erucic acid rapeseed (LEAR) oil. Hulan et al. (1977a) have also shown that the strain and source of the rats used to study the effect of feeding various dietary oils can influence the incidence of myocardial necrosis and fibrosis.

The Agriculture Canada group have conducted numerous studies on the cardiopathology in rats fed various rapeseed and other vegetable oils. Hulan et al. (1977b) found the incidence of these myocardial lesions to vary from 41 to 42% when the rats were fed soybean oil. Kramer et al. (1979a) reported the incidence of myocardial lesions in soybean oil fed rats to vary from 46 to 60%. The incidence of similar lesions in corn oil fed rats was reported to be 23% (Hulan et al., 1977a), 27% (Hulan et al., 1977c), 33% (Kramer et al., 1979b), 38% (Hulan et al., 1977b), and 40% (Kramer et al., 1973). Thirty-five percent of rats fed safflower oil (Kramer et al., 1975), 29 and 42% of rats fed olive oil (Hulan et al., 1976b) and 16 and 30% of rats fed lard (Hulan et al., 1976b) developed myocardial necrosis and fibrosis. In three different experiments Kramer et al. (1975) reported the incidence of myocardial necrosis and fibrosis in rats fed commercial rat chow was found to be 10, 30, and 55%.

It can be seen from the above that myocardial necrosis may occur in rats subjected to many different nutritional and metabolic conditions and is influenced by many additional factors such as age, sex, and strain. Obviously this information must be taken into consideration in the interpretation of experimental data. The purpose of this chapter is to provide a general over-view of the types of lesions reported following the feeding of rapeseed and marine oils and to give an opinion as to their relative importance.

II. RATS

A. Myocardial Lipidosis

Myocardial cells metabolize large quantities of fatty acids for energy pro-duction and an accumulation of fat globules within myocardiocytes is con-sidered by pathologists to be an indication of injury to the cell. The causes of cellular injury leading to intracellular lipid accumulation or lipidosis, as it is termed, are diverse but include hypoxia, high fever, toxins, or infection with toxin-producing microorganisms. Droplets or globules of triglyceride appear as vacuoles within the cytoplasm because of the solubility of the lipid in the solvents used in preparation of the histological sections. The lipid can be demonstrated in frozen sections stained with such fat-soluble dyes as oil red O. As indicated previously, lipid accumulation in myocardiocytes associ-ated with necrotic lesions has been reported in rats with potassium (French, 1952) or choline deficiency (Wilgram et al., 1954), and in mice fed high fat, low protein diets (Carroll et al., 1965).

Abdellatif and Vles (1970a) using frozen sections stained with Sudan red were the first to describe the presence of intracellular fat globules in the myocardium, skeletal muscles, and adrenals of rats fed HEAR oil. These workers noted pale hearts, skeletal muscles, and adrenals in rats fed HEAR oil but not in control animals fed sunflower oil. The myocardial lipidosis in the rats fed HEAR oil was present after 1 day of feeding, reached a peak at 6 days, was greatly diminished at 4 and 8 weeks, and had almost disappeared after 16 weeks. The feeding of glyceryl trierucate to rats produced a similar lipidosis.

Workers at the Food and Drug Directorate in Ottawa in a series of papers (Beare-Rogers et al., 1971, 1972a, b; Beare-Rogers and Nera, 1972) con-firmed the findings of Abdellatif and Vles (1970a). These workers demon-strated lipidosis in the myocardium of rats fed HEAR oil, partially hydrogen-ated HEAR oil, and partially hydrogenated herring oil and correlated this with the amount of docosenoic acid in cardiac lipids. They concluded that cardiac tissue of young rats cannot readily oxidize docosenoic acids from either a vegetable or marine source.

The development and regression of cardiac lipidosis following the long term administration of HEAR oil has been described by Ziemlanski et al. (1974). Charlton et al. (1975) found cardiac lipidosis in rats fed a variety of rapeseed oils to be more severe in the ventricular walls and interventricular

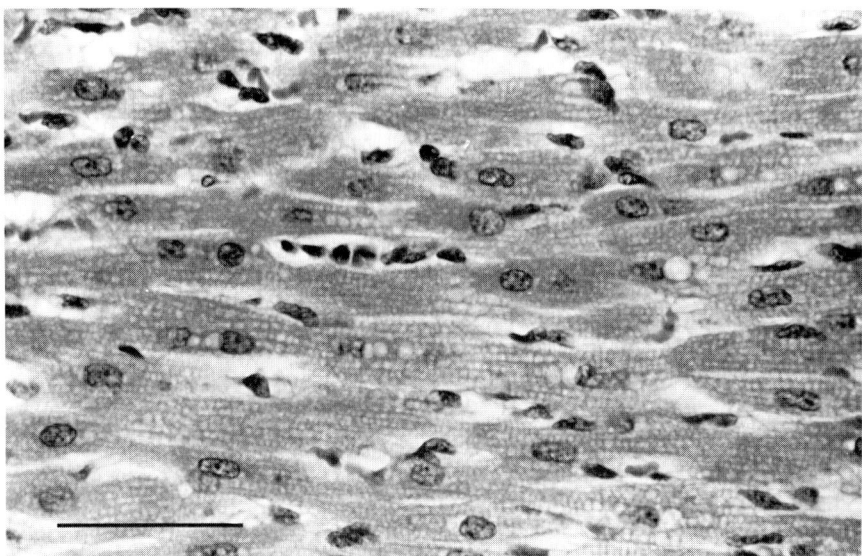

Fig. 1. Myocardium of female rat fed HEAR oil (22.3% erucic acid) and killed on day 7. Vacuoles in myocardial fibers. H&E. ×600. Bar = 50 μm. Courtesy *Can. J. Comp. Med.*

septum than in the atrial walls. The lipid globules were 1–8 μ in diameter in linear arrays within the muscle fibers. Rats fed HEAR oil (22.3% erucic acid) and killed on days 3, 7, and 14 or rats fed LEAR oil (cv. Span) (4.3% erucic acid) and killed on day 7 had vacuoles in the myocardial fibers in sections stained with hematoxylin and eosin (Fig. 1). The degree of lipid accumulation varied according to the type of rapeseed oil (erucic acid content) and the duration of the feeding. Cardiac lipidosis was marked in rats fed HEAR oil (22.3% erucic acid) after 3, 7, and 14 days, moderate in rats fed LEAR oil (cv. Span) (4.3% erucic acid), and very slight in rats fed LEAR oil (cv. Oro) (1.6% erucic acid). The degree of lipidosis decreased with time and after 28 days on diet no lipid was detected in the myocardium of rats fed LEAR oils (cv. Span and cv. Oro). Lipidosis was present in the hearts of rats fed HEAR oil for long periods; some globules were still present at 112 days.

The gross and histological appearance of hearts from rats fed oils of varying erucic acid content is illustrated in Fig. 2. It will be noted that the hearts from rats fed oils high in erucic acid are paler and contain more lipid on histological examination. The hearts from rats fed low or no erucic acid oils are difficult to distinguish grossly. Lipidosis is not demonstrable histologically in the hearts of rats fed LEAR oil (cv. Tower) (0.5% erucic acid) or soybean oil (no erucic acid).

An electron microscopic study of the early stages of myocardial lipidosis in rats fed HEAR oils revealed lipid droplets in the myocardial cells as early as 3 hr (Ziemlanski et al., 1973). These lipid globules were approximately 1 μ in diameter and were always situated in the immediate vicinity of the sarcoplasmic reticulum. After 6 hr, lipid globules had increased in number and size and were noted not only between sarcomeres but also near the nucleus. The globules were more numerous and larger at 24 and 48 hr. Their number and size caused mechanical deformation of the sarcomeres, mito-chondria, and nuclei, but apart from this distortion these elements were not otherwise damaged. No limiting membrane surrounded the lipid globules.

In another study Ziemlanski et al. (1975) reported on the ultrastructural findings in the myocardium of rats fed rapeseed oils for 1, 4, and 8 weeks. The myocardium of control rats fed sunflower oil had minor changes in the form of dilated channels of endoplasmic reticulum and the presence of a few fat globules. Similar changes were seen in the myocardium of rats fed LEAR oil, but were followed after 4 and 8 weeks by mitochondriosis with com-pression and atrophy of sarcomeres. Lipidosis as indicated by the presence of vacuoles was most severe in the myocardium of rats fed HEAR oil for 1 week. Mitochondrial changes were similar to those seen in the rats fed LEAR oil.

Engfeldt and Brunius (1975a, 1975b) reported fat droplets up to 5 μ in diameter arranged in rows along the longitudinal axis of the fibers. On elec-tron microscopic examination the lipid droplets were closely associated with mitochondria, and in sections cut longitudinally to the muscle fibers the droplets were often located in areas corresponding to the Z lines. Like the Polish workers they found no limiting membrane around the droplets. They reported impressions in mitochondria caused by the lipid, but other-wise all cell organelles were normal. Engfeldt and Gustafsson (1975) studied the effects of feeding HEAR oil to germ-free rats and concluded that the development of myocardial lipidosis was not influenced by the presence or absence of a normal intestinal flora. Yamashiro and Clandinin (1980) also conducted ultrastructural studies on the myocardium of rats fed HEAR oil and reported distortion of mitochondria by lipid droplets, a lamellar ar-rangement of mitochondrial cristae, and some separation of intercalated discs.

B. Myocardial Necrosis and Fibrosis

1. GENERAL DESCRIPTION OF FOCAL MYOCARDIAL NECROSIS AND FIBROSIS

It should be stressed that lesions encountered in the hearts of rats are similar whether the rats are fed rat chow, various control oils such as corn, peanut, olive, and soybean, or the various marine and rapeseed oils. As will

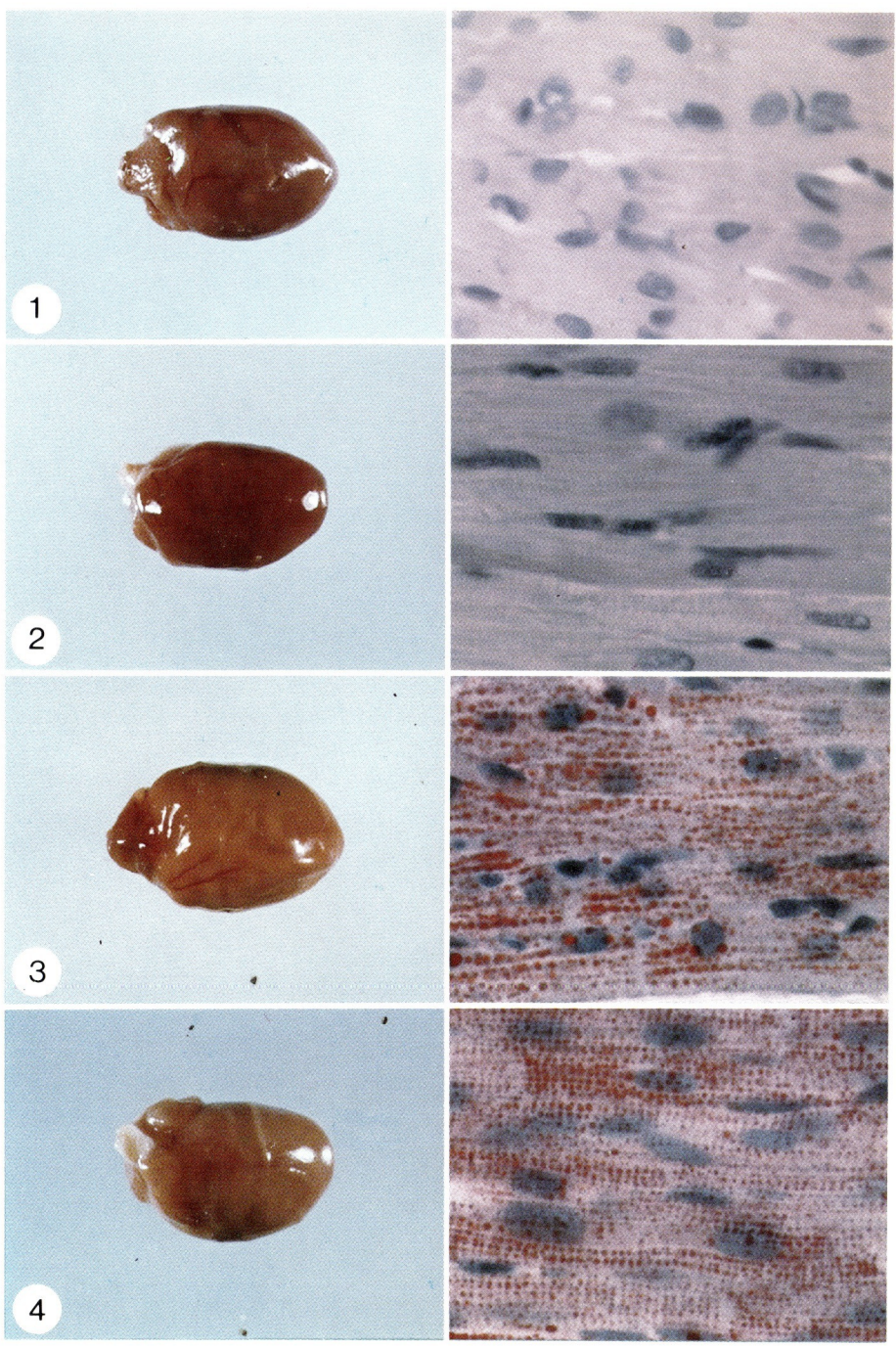

Fig. 2. Gross and histological appearance (oil red O staining) of hearts from rats fed diets containing 20% by weight of the following oils: 1. soybean oil, 2. low erucic acid rapeseed oil (0.5%, 22:1), 3. high erucic acid rapeseed oil (28.8%, 22:1), and 4. mustard seed oil (54.1%, 22:1).

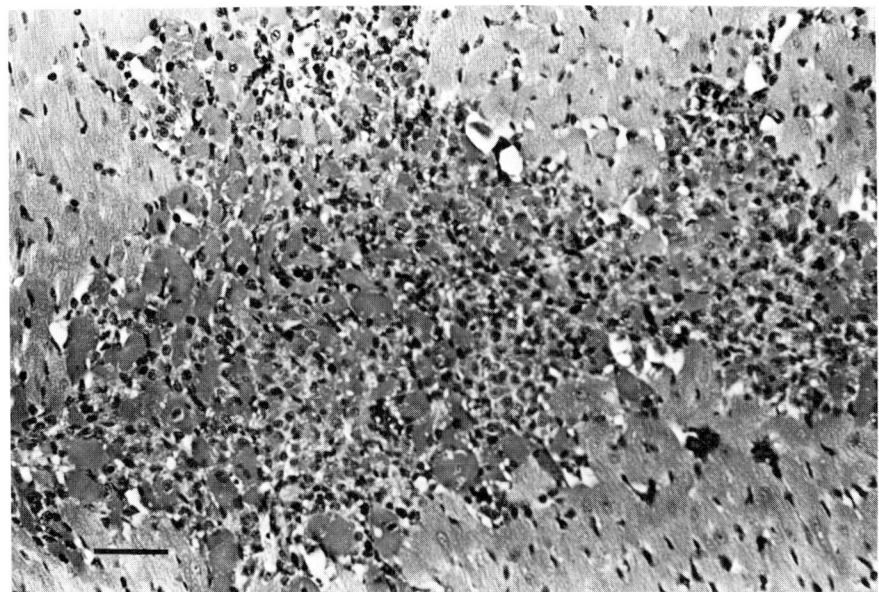

Fig. 3. Myocardium of male rat fed LEAR oil (cv. Tower) (0.3% erucic acid) and killed on day 112. Focus of necrosis. Note hyalinized fibers with pyknotic nuclei and accumulation of a few macrophages. H&E. ×214. Bar = 50 μm.

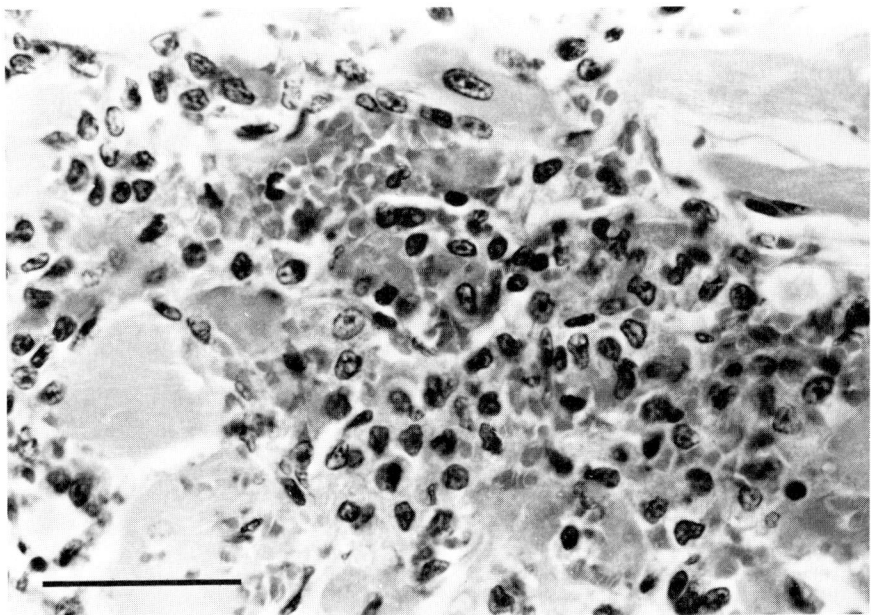

Fig. 4. Myocardium of male rat fed HEAR oil (22.3% erucic acid) and killed on day 112. Focal necrosis and hemorrhage. H&E. ×544. Bar = 50 μm. Courtesy *Can. J. Comp. Med.*

be seen from the following review, the description of the long-term cardiac lesions encountered in rats fed various oil diets varies from author to author. There is general agreement that necrosis of the myocardium occurs followed by removal of the necrotic muscle cells by various phagocytic elements and replacement of the lost muscle by a fibrous connective tissue scar. The areas of necrosis and fibrosis vary from small foci 10 μ in diameter involving one fiber to large oval or irregularly shaped regions approximately 2 mm in diameter. They are randomly distributed in the ventricular walls, interventricular septum, and papillary muscles. Lesions are frequently found close to the epicardial surface of both ventricles but no areas of necrosis or fibrosis are found in the atrial walls. Acute and chronic lesions sometimes are present in the same heart, suggesting that the focal necrosis does not occur at one time but is continuous or recurrent during at least part of the feeding trials. Acute lesions (Fig. 3) consisted of small or large foci in which the myocardial fibers are hyalinized, swollen, fragmented, and the muscle nuclei pyknotic. There is hemorrhage in addition to myocardial necrosis in some foci (Fig. 4). Early acute lesions contain only a few macrophages in addition to small numbers of polymorphonuclear leukocytes and eosinophils (Fig. 5), but in more pronounced lesions there are many closely packed macrophages and a few fibroblasts (Fig. 6). In a few chronic lesions there are

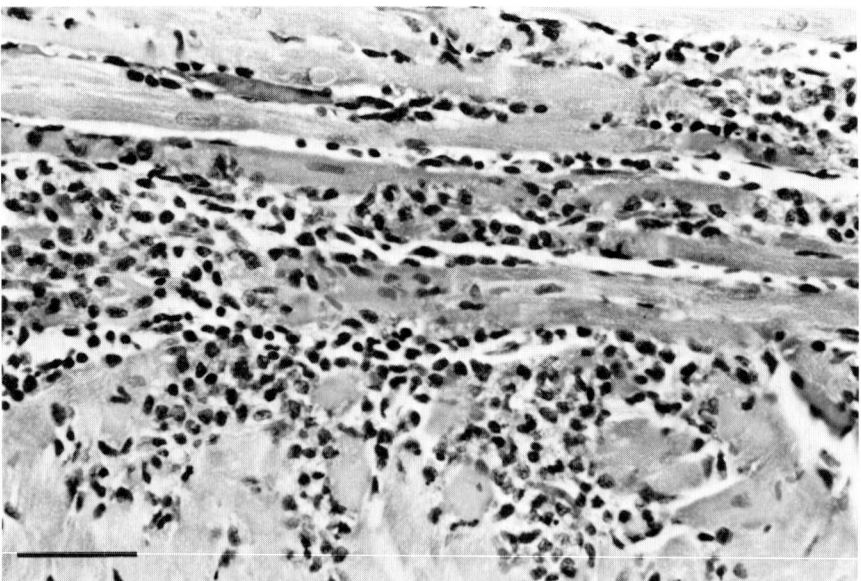

Fig. 5. Myocardium of male rat fed soybean molecular distillate and killed on day 112. Myocardial necrosis with accumulation of macrophages. H&E. ×340. Bar = 50 μm.

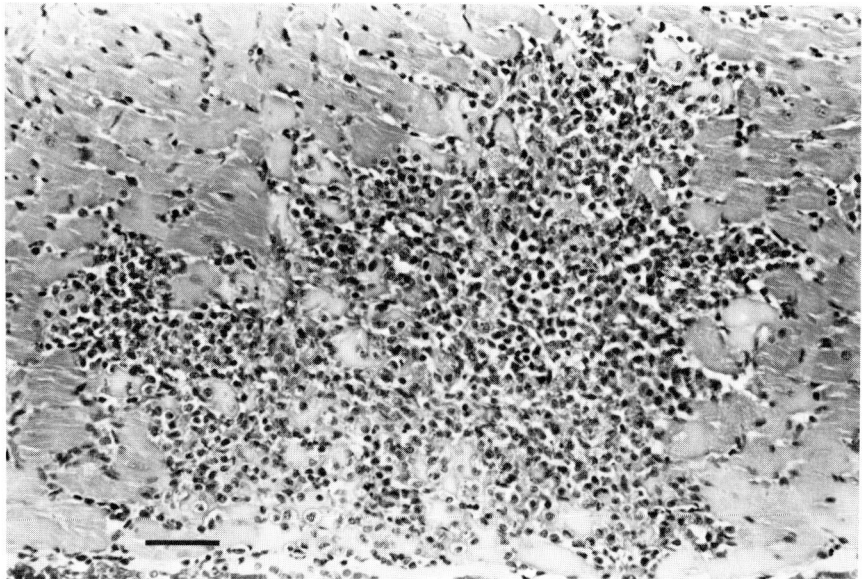

Fig. 6. Myocardium of male rat fed corn oil and killed on day 112. Myocardial necrosis with large numbers of macrophages. H&E. ×214. Bar = 50 μm.

scattered hyalinized muscle fibers. Chronic lesions are round or linear areas of fibrous connective tissue (Fig. 7), some of which contain macrophages with brown intracytoplasmic granules. Those granules were stained blue by the Prussian blue technique indicating the presence of hemosiderin.

There have been several hypotheses proposed to explain the etiology of the focal myocardial necrosis and fibrosis seen in male rats following the feeding of various rapeseed and marine oils. Several workers felt that erucic acid alone was responsible for the production of the focal necrosis (Abdellatif and Vles, 1973; Engfeld and Brunius, 1975b; Astorg and Cluzan, 1976). The development of new cultivars of rapeseed led to the production of new low erucic acid rapeseed. The fact that the oil from the new rapeseed cultivars also produced a high incidence of myocardial lesions coupled with the fact that similar necrotic lesions were found in the hearts of rats fed "control" oils (Kramer et al., 1973) led to the conclusion that factors other than, or in addition, to erucic acid were responsible for the increased incidence of lesions. Studies were directed toward the isolation and identification of a cardiopathogenic toxin in LEAR oil and soybean oil without success (Kramer et al., 1975, 1979a). Several studies suggested that the increased incidence of myocardial necrosis was the result of a fatty acid imbalance for the growing male albino rat (Hulan et al., 1977b; Kramer et al., 1973, 1975). A statistical evaluation of much of the published data (Trenholm et

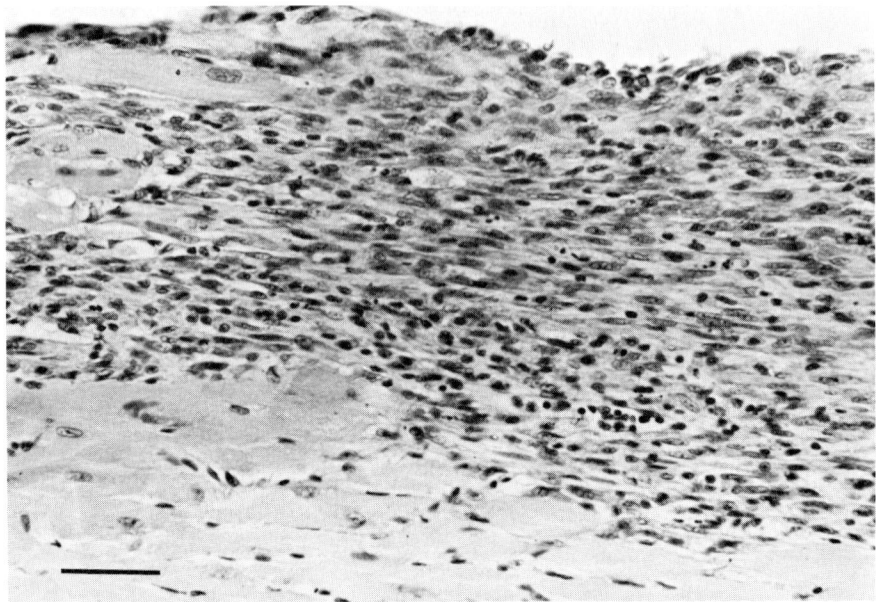

Fig. 7. Myocardium of male rat fed HEAR oil (22.3% erucic acid) and killed on day 112. Part of a fibrous connective tissue scar replacing necrotic myocardium. H&E. ×272. Bar = 50 μm.

al., 1979) and recent experimental evidence (Kramer *et al.*, 1981) strengthen this belief.

2. METHODS EMPLOYED AND LESIONS REPORTED BY VARIOUS
   GROUPS OF WORKERS

Roine *et al.* (1960) were the first to report myocardial lesions in male rats fed HEAR oil. These workers conducted histological examinations on thyroid, heart, liver, spleen, kidneys, adrenal, stomach, small intestine, large intestine, aorta, and striated muscle and found lesions only in the myocardium. They described lesions that they felt resembled toxic myocarditis (interstitial edema, with variable numbers of fibroblasts, histiocytes, lymphocytes, plasma cells and neutrophils, cloudy swelling, and loss of striation of muscle fibers). Small necrotic foci were also reported. The number of sections examined from each heart or the plane of the sections was not described. Lesions were found in rats receiving 70 calorie % HEAR oil only, and not in rats fed 50 calorie % HEAR oil or less, or in the soybean oil fed controls. The number of rats per treatment was relatively low (5–9) and the duration of the feeding trial relatively short (6–7 weeks).

The Unilever Research Laboratory group (Abdellatif and Vles, 1970a,

1973) in their early work gave little detail of their methods of histological examination beyond that of routine fixation and staining. They described the presence of fibrotic scars and histiocytic foci in the myocardium of rats fed rapeseed oil for 16 weeks. No lesions were demonstrated in control rats fed sunflower oil. Later, Vles et al. (1976) described a detailed procedure for the histomorphometric evaluation of rat hearts. They removed the auricles and divided the ventricles transversely into a large central part and smaller basal and apical portions. From the large central portion, five different sites 0.8 mm apart were sectioned. From each site two successive sections were stained with Harris's hematoxylin and two with Masson's trichrome. One site was similarly sampled from the apical and basal portions making a total of 28 sections per heart. In some experiments only the 20 sections from the large central part were examined. Three types of lesions were recorded: (1) foci of mononuclear cells, fibroblasts, and sarcolemmal cells, (2) small necrotic areas with loose connective tissue, and (3) focal scars. The size of these lesions was assessed by measuring the horizontal Feret diameter and the longest diameter. Lesions in control rats fed sunflower oil were indistinguishable from those in the hearts of rats fed LEAR oils and were referred to as "spontaneous" or "background" lesions.

Ziemlanski et al. (1972) fed HEAR oil, hydrogenated HEAR oil, and a number of experimentally produced margarines to rats and produced heart lesions, which they referred to as histiocytic granulomas. These lesions consisted of small necrotic foci, surrounded by histiocytic granulomas and focal fibrosis. They found no myocardial lesions in rats fed a standard diet (7% fat) for 7 and 12 months.

Agriculture Canada workers (Charlton et al., 1975) examined three sections from each heart consisting of a central section from the apex to the base including the interventricular septum and the atrial and ventricular walls. Two additional sections were made parallel to and equidistant from the central section in the middle of each half of the heart. All sections were routinely stained with hematoxylin–eosin. Myocardial lesions in rats fed various rapeseed oils included scattered focal interstitial accumulations of mononuclear cells and foci of myocardial necrosis and fibrosis. Similar lesions were found in the hearts of rats fed the control diets. These workers recorded the number of unequivocal foci of myocardial necrosis and fibrosis found in three sections of each heart as an index of severity in a large number of studies (Charlton et al., 1975; Hulan et al., 1976a, b, 1977a–d; Hunsaker et al., 1977; Kramer et al., 1973, 1975, 1979a, b). These studies have shown that it is possible to rank the various vegetable oils in order of their effect on the incidence and severity of focal myocardial necrosis and fibrosis in the male albino rat.

Beare-Rogers and Nera (1977) removed the hearts from the rats under ether anesthesia and fixed them in buffered formalin. The heart was transversly cut into four portions and sections were stained with hematoxylin–eosin. Lesions noted were described as cardiac necrosis or fibrosis. The lesions were graded 1 to 4 but no description of the grading system was given.

Engfeldt and Brunius (1975b) in their studies on the cardiopathogenic effects of rapeseed oil on rats serially sectioned the whole heart and, in addition, conducted ultrastructural studies. The lesions, aside from the lipidosis mentioned previously, consisted of accumulations of macrophages, myolysis, proliferation of fibroblasts, and finally scarring. They reported the occasional small myocardial lesion in control rats and concluded that these lesions were a "normal" finding. Ultrastructural studies revealed replacement of muscle fibers by scar tissue, but examination of muscle fibers surrounding these lesions revealed no specific alterations apart from lipid droplets. The different cell organelles were reported to have normal structure and the myofibrils of the heart were not affected.

McCutcheon et al. (1976) studying various rapeseed oils and oil blends conducted their examination of rat hearts in a similar manner to that of Charlton et al. (1975), but cut each of the three blocks of tissue serially so that they examined a total of 27 sections per heart. These authors recorded three types of lesions: (1) "Microvascular alteration." This lesion was characterized by the presence of many varying sized vacuoles in walls of small and minute blood vessels. Stenosis of the lumen and necrobiotic changes in the smooth muscle cells of the vascular walls were sometimes noted. (2) "Fresh myocardial necrosis." Foci of myocardial necrosis were frequently found close to the epicardium of both ventricular walls, usually accompanied by congested minute blood vessels and/or extravasation of erythrocytes. In "fresh necrotic lesions" the muscle fibers frequently contained vacuoles and were undergoing hyaline degeneration, basophilic degeneration, and sarcolysis. (3) "Old myocardial necrosis." Foci of relatively old myocardial necrosis were characterized by remarkable mesenchymal cell proliferation with some fibroblastic differentiation. They reported finding all three types of lesions in all dietary groups. In a paper dealing primarily with the histopathology of the hearts of rats fed rapeseed oils (Umemura et al., 1978) this same group explained the focal myocardial necrosis as a secondary development to circulatory disturbances. In addition to the three types of lesions described earlier (McCutcheon et al., 1976) they reported finding intravascular lipid droplets up to 20 μ in diameter, some of which had aggregates of platelets adhering to their surface. They postulate that these lipid droplets as well as the microvascular alterations could sufficiently alter local hemodynamics to lead to the development of the necrotic lesions. It is unfortunate that these workers did not use stains to show that the vacuolation noted in the vessel walls was due to the presence of lipid and not the result of fixation procedures. This group also report intravascular fat globules in the hearts of pigs fed rapeseed meals, but found no necrotic changes associated with them (Umemura *et al.*, 1977). The pigs were killed by electrocution and bled by severing the jugular vein. It is conceivable that their method of killing and bleeding could lead to the development of fat emboli. They did not describe their method of killing rats. This group is the only one to date to describe vascular lesions in hearts from rats fed rapeseed oils.

Rocquelin and Cluzan (1968) described the lesions encountered in hearts of rats fed rapeseed oils as a type of myocarditis characterized by changes ranging from simple edema with separation of myocardial fibers and mild mononuclear cell infiltration to massive histiocytic infiltration replacing necrotic muscle fibers. More recently, this group (Cluzan *et al.*, 1979) devised a grading system where in addition to the number of lesions present they judge the size of the lesion and the cellular response to it. The diameter of the lesion was judged as follows: (a) 1–5 myocardiocytes, (b) 6–20 myocardiocytes, and (c) >20 myocardiocytes. The granulomatous reaction or cellular response was graded as follows: (a) 3–4 macrophages, (b) 5–20 cellular elements, and (c) >20 cellular elements. Lesions difficult to interpret were classified as "doubtful."

III. SWINE

Roine *et al.* (1960) reported on the study of a wide variety of tissues from swine fed soybean oil and HEAR oil. They found hyperplasia of thyroid epithelium in pigs fed soybean oil or HEAR oil. There was a mild interstitial myocarditis in pigs fed oil diets, but not in those fed a conventional diet.

Beare-Rogers and Nera (1972) fed small numbers of miniature 10-day-old piglets HEAR oils for 1 week. Each pig heart was sliced laterally through the apex, and the ventral portion examined histologically. There were stainable fat droplets in the myocardium, but not more chemically discernible fat than in controls fed lard/corn oil. Three-week-old commercial piglets fed HEAR oils reacted in a similar manner.

The Agriculture Canada group in a large study involving 180 pigs of both sexes fed soybean oil or low erucic acid rapeseed oil (4.3% 22:1) at 10 and 20% levels for periods of 1, 4, and 16 weeks (Friend *et al.*, 1975a). They prepared five sections, one from each of the right and left ventricular walls, interventricular septum, and right and left atrial walls. A trace lipidosis was demostrated by oil red O staining in some hearts of all treatment groups at most time intervals. The results indicated a slightly greater number of pigs fed the 20% fat diets than those fed the 10% fat diets had lipidosis. Lipidosis was present in the myocardium of more pigs fed 20% low erucic acid rapeseed oil (18 of 36 pigs) than in the hearts of pigs fed soybean oil (13 of 36 pigs). Minute focal interstitial infiltrations of mononuclear cells were found in all experimental groups including initial controls (pigs killed at day 0 to provide reference tissues). Foci of overt myocardial necrosis as seen in the rat were not encountered and there was no relationship between the lesions found and the diets fed. In a second study (Friend et al., 1975b) 72 boars were divided into groups and fed a control diet unsupplemented and supplemented with 20% soybean oil or 20% HEAR oil for 16 weeks. Small foci of myocardial necrosis accompanied by mononuclear cell infiltrations were encountered in only five pigs (one boar fed the control diet, one fed soybean, and three fed HEAR oil). Cardiac lipidosis was not demonstrated in any of the 72 boars. In a third study by this group (Friend et al., 1976) 60 boars were fed a control diet unsupplemented and supplemented with 20% corn oil, 20% low erucic acid rapeseed oil, or 20% HEAR oil. A number of boars in each dietary group had cardiac lesions characterized by focal areas of myocardial necrosis accompanied by an inflammatory response of mononuclear cells and eosinophils. Focal interstitial infiltrations of mononuclear cells and eosinophils were also encountered. Fibrous connective tissue was evident in older lesions resulting in scar formation. Statistically there was no significant difference in lesion incidence between diet groups.

Aherne et al. (1975) examined tissues from 112 pigs fed four diets including HEAR oil, low erucic acid rapeseed oil, or soybean oil at the 15% level, and a no oil diet. They gave no detailed description of their method of sampling the heart or the number of sections examined. Myocardial lipidosis was not demonstrated by oil red O staining in any of the pigs in this experiment. Focal areas of interstitial myocarditis with an infiltration of eosinophils, lymphocytes, and plasma cells were present in 35.8% of the hearts examined and were found in hearts from all dietary groups. Foci of myocardial degeneration were found in only six pigs (three had been fed soybean oil, two low erucic acid rapeseed oil, and one the no oil control diet). They postulated that the foci of myocardial degeneration may have been caused by the migration of Ascaris suum larvae. However, Friend et al. (1976) questioned whether Ascaris suum migration was the cause of the lesion. In a second experiment utilizing 80 boars and gilts, various groups were fed three low erucic acid rapeseed oil, one HEAR oil, or a no fat control diet. Myocardial necrosis was seen in only five pigs (Aherne et al., 1976). Three of these animals had been fed the control diet, one the HEAR oil, and the other one of the low erucic acid rapeseed oil diets. Again no myocardial lipidosis was detected by oil red O staining.

Opstvedt et al. (1979) conducted short term studies feeding piglets up to 22 days with HEAR oil (48% 22:1), fish oil (14.6% 22:1), partially hydrogenated fish oil (14.3% 22:1), and a lard/sunflower oil control. They demonstrated some stainable fat globules in the myocardium of piglets fed HEAR oil for 8–13 days but not before or after. After 10 days of feeding, a mild to moderate cardiac lipidosis was found in piglets fed diets containing 2% or more 22:1 fatty acids, but there were no significant differences among those fed the HEAR oil, refined fish oil or partially hydrogenated fish oil. The same diets fed to rats produced a lipidosis five times greater than that seen in the piglets. These workers felt that the moderate lipidosis produced in the piglet was due to a higher metabolic capacity for handling docosenoic fatty acids.

In a longer term experiment the same group (Svaar et al., 1980) studied the cardiopathological effects of feeding HEAR oil, fish oil, partially hydrogenated fish oil, partially hydrogenated soybean oil, and lard in 40 female Norwegian Landrace piglets for periods up to 1 year. They cut the hearts in transverse sections from the apex to the base, usually studying five to six sections per heart. A mild cardiac lipidosis was found in a few pigs fed only partially hydrogenated fish oil and HEAR oil for 1, 5, and 27 weeks. Minor heart lesions consisting of small foci of individual muscle cell necrosis, focal accumulation of leukocytes and macrophages, and small areas of fibrosis were found after 1 week and more frequently after 6 months and 1 year. They found no relationship between the incidence and severity of the heart lesions and any particular type of fat in the diet.

Bijster et al. (1979) fed 68 castrated pigs a variety of oil diets for a period of 17 weeks. They sampled swine hearts by using a homemade clamp and cutting the hearts into seven even transverse slices. Specimens for histological examination were taken from similar preselected areas of all hearts. A total of 28 sites were examined from each heart. These sections were examined using the same morphometric techniques described for rats (Vles et al. 1976). Lesions consisted of foci of myolysis or necrosis with macrophage or leukocyte infiltration. Lesions in pigs fed rapeseed oils with various levels of erucic acid did not differ from lesions in pigs fed either soybean oil or a low fat diet.

Vodovar et al. (1973) conducted an ultrastructural study on pigs fed HEAR oil and peanut oil. He described the presence of lipid droplets in a large number of myocardial cells after 7 days feeding of HEAR oil. While the number of cells containing lipid droplets varied from section to section they were more numerous in sections of left ventricle taken midway between apex and base. The extent of lipidosis was considerably reduced after 45 days, and after 90 and 180 days cells containing lipid droplets were rare. These workers described alterations in the mitochondria with particular reference to their size, number, and degeneration. The value of this study is reduced in that results found in the controls were not described. In another ultrastructural study (Vodovar et al., 1977) they utilized diets with and without vegetable oils [HEAR, LEAR (cv. Primor), sunflower, and peanut oils]. At 0, 6, and 15 days of feeding the appearance and size of mitochondria in myocardiocytes were the same in pigs fed all diets. There was an increase in both number and size of mitochondria at 30 days. There was a further increase at 45 days. At 60 days they found mitochondria 5−15 times greater in size than normal. Mitochondrial inclusions including glycogen, lipid residues, and undetermined substances were also present. After 60 days, mitochondria had degenerative changes which eventually led to their lysis. These degenerative changes became more severe with the length of time on diet. The mitochondrial changes were not common in the sunflower or peanut oil fed pigs, and were rare but apparently present in those fed the nonfat diet.

IV. MONKEYS

Focal areas of idiopathic myocardial necrosis and/or fibrosis are occasionally observed in nonhuman primates (Ayers and Jones, 1978). Soto *et al.* (1964) studied the hearts of 20 rhesus monkeys imported from India and found 18 with varying degrees of myocarditis. The myocarditis, regardless of its severity, was usually focal but occasionally interstitial and extensive. The inflammatory response was characterized by the presence of lymphocytes, large mononuclear cells, and occasional plasma cells and eosinophils. Necrosis was seen only when the myocarditis was marked. They were unable to isolate bacterial or viral agents from the hearts or brains, and found no serological evidence of encephalomyocarditis or Coxsackie virus. Chronic interstitial myocarditis was diagnosed in 34% of 312 monkeys of various species (Qureshi, 1979). This idiopathic myocarditis was similar to that reported by Soto *et al.* (1964). The high incidence of such lesions in the hearts of monkeys must be taken into consideration when using this animal to study the cardiopathological effects of various foods and points out the need to have an adequate number of control animals as well as principals.

Beare-Rogers and Nera (1972) fed squirrel monkeys *(Saimiri sciureus)* HEAR oil at 10% and 20% of the diet for 1- and 10-week periods. Each monkey heart was sliced laterally through the apex and the ventral portion was examined histologically. Lipidosis was demonstrated in the hearts of all monkeys at 1 week including those fed the control diet. Lipidosis was still present in one of three monkeys fed HEAR oil at the 10% level and in two of two monkeys fed the same oil at the 20% level at 10 weeks. They reported the presence of cardiac fibrosis in both monkeys fed HEAR oil at 20% for 10 weeks but did not describe the lesions.

Researchers at the National Institute of Nutrition in India fed three groups of eight adult male monkeys *(Macaca radiata)* 20% mustard oil, peanut oil, or hydrogenated peanut fat for over a year (Gopalan *et al.*, 1974). Five representative pieces of heart muscle, except from the apical region, were rountinely examined. Animals fed mustard oil had vacuolation of the sarcoplasm of the right and left ventricular myocardium. No specific fat stains were employed in this study. Four of the eight monkeys in the group fed mustard oil exhibited various degrees of myocardial fibrosis. These changes were not found in the hearts of monkeys fed peanut oil or hydrogenated peanut fat. Shenolikar and Tilak (1980) from the same institute fed three groups of nine male monkeys *(Macaca radiata)* 5 and 10% mustard oil and 10% peanut oil. Histological examination of the hearts failed to reveal vacuolation or fibrosis in any of the dietary groups as previously reported (Gopalan *et al.*, 1974). Macroscopically the hearts of the monkeys fed mustard oil were yellowish indicating lipidosis, but again fat stains were not employed.

Kramer *et al.* (1978a, b) fed male and female *Macaca fascicularis* 20% by weight LEAR oil (cv. Tower) or soybean oil for 24 weeks. Samples were taken from the interventricular septum and from right and left ventricle including the papillary muscle. All groups of monkeys had mild myocardial lesions including focal interstitial collections of inflammatory cells, occasional focal scars, and groups of swollen interstitial fibroblasts. These lesions were judged to represent nonspecific focal myocardial injury that had no relationship to the nature of the diet fed.

Loew *et al.* (1978) and Schiefer *et al.* (1978) reported on an experiment where 11 monkeys *(Macaca fascicularis)* which were fed diets containing 25% by weight HEAR oil, partially hydrogenated herring oil, or a lard/corn oil control diet for 4 months. These authors used three horizontal levels of sections: one close to the base of the heart, one approximately midline between base and apex, and one close to the apex. Myocardial lipidosis was demonstrated in all groups including the lard/corn oil fed controls. The lipidosis was mild to moderate in the lard/corn oil fed controls, severe in those fed partially hydrogenated herring oil and most severe in those fed HEAR oil. Small foci of histiocytic infiltration were found in the myocardium of some animals of all dietary groups. Ultrastructural studies confirmed the lipidosis seen in oil red O stained sections. The size of myocardial mitochondria was increased in animals fed partially hydrogenated herring oil and HEAR oil over that of the control group. Mitochondria contained amorphous material or vacuoles and there was loss and distortion of cristae.

V. POULTRY

Relatively little work has been conducted utilizing poultry as experimental animals. Abdellatif and Vles (1970b) were the first to investigate the feeding of rapeseed oils to ducklings. In this and in a subsequent study (Abdellatif and Vles, 1971), they found that feeding HEAR oil (50% erucic acid) at a rate of 40 calorie % of the diet or above produced mortality in young Pekin ducklings. Grossly they found pale hearts, hydropericardium, and mottling of the liver. Microscopically there was vacuolation of the myocardium and skeletal muscle suggesting lipid accumulation. Edema and disintegration of the muscle fibers was found in association with the lipidosis. Leukocytic infiltration accompanied these changes especially in the birds that died. Fatty metamorphosis of the liver was accompanied by an increase in the fibrous connective tissue elements and by bile duct hyperplasia.

Ratanasethkul et al. (1976) fed chickens, ducks, and turkeys high and low erucic acid rapeseed oil, soybean oil, and a lard/corn oil control diet. All the ducks and some of the chickens fed the HEAR oil diet (36% erucic acid) died with hydropericardium and ascites. Myocardial lipidosis was present in all species fed the HEAR oil and the severity of the lipidosis (as judged by oil red O staining) decreased with time on diet. In ducks, they found thickening of the epicardium and myocardial fibrosis. Granulomas characterized by giant cells and histiocytic infiltration were present in some of the hearts of turkeys fed the HEAR oil.

VI. CONCLUSIONS

1.   Myocardial lipidosis is produced in the rat by feeding various marine and rapeseed oils. This lipidosis reaches a peak at 6–7 days and declines thereafter. The severity and duration of this lesion are directly related to the docosenoic acid content of the diet.

2.   Focal myocardial necrosis followed by reparative fibrosis is a spontaneous idiopathic lesion in the male rat. The incidence and severity of this lesion can be influenced by the feeding of various marine and vegetable oils and per se are not related to the erucic acid content of the oil. It has been possible to repeatedly rank various oils according to their effect on the severity and incidence of these cardiac lesions.

3.   The feeding of HEAR oils to swine produces a lipidosis only slightly more severe than that produced by control oils.

4.   Degenerative lesions seen in the hearts of swine fed various rapeseed oils bear no relationship to the diets fed.

5.   Oils containing docosenoic acid produce a cardiac lipidosis in monkeys which is more severe than that produced when control oils are fed.

6.   Degenerative lesions in the hearts of monkeys fed various marine and vegetable oils bear no relationship to the diets fed.

7.   Poultry when fed high erucic acid rapeseed oils develop myocardial lipidosis, myocardial degeneration, hydropericardium and ascites. The occurrence and severity of these lesions would appear to be related to the amount of erucic acid in the diet.

8.   A number of workers have described an increase in number and size of myocardial mitochondria and various mitochondrial degenerative changes following the feeding of docosenoic acid containing oils to rats,

swine, and monkeys. These changes appear to be related to the concentration of docosenoic acid in the diet but their significance is not yet known.

REFERENCES

Abdellatif, A. M. M., and Vles, R. O. (1970a). *Nutr. Metab.* **12**, 285–295.
Abdellatif, A. M. M., and Vles, R. O. (1970b). *Nutr. Metab.* **12**, 296–305.
Abdellatif, A. M. M., and Vles, R. O. (1971). *Nutr. Metab.* **13**, 65–74.
Abdellatif, A. M. M., and Vles, R. O. (1973). *Nutr. Metab.* **15**, 219–231.
Aherne, F. X., Bowland, J. P., Christian, R. G., Vogtmann, H., and Hardin, R. T. (1975). *Can. J. Anim. Sci.* **55**, 77–85.
Aherne, F. X., Bowland, J. P., Christian, R. G., and Hardin, R. T. (1976). *Can. J. Anim. Sci.* **56**, 275–284.
Ashburn, L. L., and Lowry, J. V. (1944). *Arch. Pathol.* **37**, 27–33.
Astorg, P.-O., and Cluzan, R. (1976). *Ann. Nutr. Aliment.* **30**, 581–602.
Ayers, K. M., and Jones, S. R. (1978). *In* "Pathology of Laboratory Animals" (K. Benirschke, F. M. Garner and T. C. Jones, eds), Vol. I., pp. 1–69. Springer-Verlag, Berlin and New York.
Beare-Rogers, J. L., and Nera, E. A. (1972). *Comp. Biochem. Physiol. B* **41B**, 793–800.
Beare-Rogers, J. L., and Nera, E. A. (1977). *Lipids* **12**, 769–774.
Beare-Rogers, J. L., Nera, E. A., and Heggtveit, H. A. (1971). *Can. Inst. Food Technol. J.* **4**, 120–124.
Beare-Rogers, J. L., Nera, E. A., and Craig, B. M. (1972a). *Lipids* **7**, 46–50.
Beare-Rogers, J. L., Nera, E. A., and Craig, B. M. (1972b). *Lipids* **7**, 548–552.
Bijster, G. M., Trimmer, W. G., and Vles, R. O. (1979). *Fette, Seifen, Anstrichm.* **81**, 192–194.
Carroll, R. B., Clower, B. R., and Williams, W. L. (1965). *Arch. Pathol.* **80**, 391–396.
Charlton, K. M., Corner, A. H., Davey, K., Kramer, J. K. G., Mahadevan, S., and Sauer, F. D. (1975). *Can. J. Comp. Med.* **39**, 261–269.
Cluzan, R., Suschetet, M., Rocquelin, G., and Levillain, R. (1979). *Ann. Biol. Anim., Biochim., Biophys.* **19**, 497–500.
Engfeldt, B., and Brunius, E. (1975a). *Acta Med. Scand., Suppl.* **585**, 15–26.
Engfeldt, B., and Brunius, E. (1975b). *Acta Med. Scand., Suppl.* **585**, 27–40.
Engfeldt, B., and Gustafsson, B. (1975). *Acta Med. Scand., Suppl.* **585**, 41–46.
French, J. E. (1952). *Arch. Pathol.* **53**, 485–496.
Friend, D. W., Corner, A. H., Kramer, J. K. G., Charlton, K. M., Gilka, F., and Sauer, F. D. (1975a). *Can. J. Anim. Sci.* **55**, 49–59.
Friend, D. W., Gilka, F., and Corner, A. H. (1975b). *Can. J. Anim. Sci.* **55**, 571–578.
Friend, D. W., Kramer, J. K. G., and Corner, A. H. (1976). *Can. J. Anim. Sci.* **56**, 361–364.
Gopalan, C., Krishnamurthi, D., Shenolikar, I. S., and Krishnamachari, K. A. V. R. (1974). *Nutr. Metab.* **16**, 352–365.
Hulan, H. W., Kramer, J. K. G., Mahadevan, S., Sauer, F. D., and Corner, A. H. (1976a). *Lipids* **11**, 6–8.
Hulan, H. W., Kramer, J. K. G., Mahadevan, S., Sauer, F. D., and Corner, A. H. (1976b). *Lipids* **11**, 9–15.
Hulan, H. W., Kramer, J. K. G., and Corner, A. H. (1977a). *Can. J. Physiol. Pharmacol.* **55**, 258–264.
Hulan, H. W., Kramer, J. K. G., and Corner, A. H. (1977b). *Lipids* **12**, 951–956.
Hulan, H. W., Kramer, J. K. G., Corner, A. H., and Thompson, B. (1977c). *Can. J. Physiol. Pharmacol.* **55**, 265–271.

Hulan, H. W., Thompson, B., Kramer, J. K. G., Sauer, F. D., and Corner, A.H. (1977d). *Can. Inst. Food Sci. Technol. J.* **10**, 23–26.

Hunsaker, W. G., Hulan, H. W., Kramer, J. K. G., and Corner, A. H. (1977). *Can. J. Physiol. Pharmacol.* **55**, 1116–1121.

Kesten, H. D., Salcedo, J., and Stetten, De W. (1945). *J. Nutr.* **29**, 171–177.

Kramer, J. K. G., Hulan, H. W., Corner, A. H., Thompson, B. K., Holfeld, N., and Mills, J. H. L. (1979a). *Lipids* **14**, 773–780.

Kramer, J. K. G., Hulan, H. W., Mahadevan, S., Sauer, F. D., and Corner, A. H. (1975). *Lipids* **10**, 511–516.

Kramer, J. K. G., Hulan, H. W., Procter, B. G., Dussault, P., and Chappel, C. I. (1978a). *Can. J. Anim. Sci.* **58**, 245–256.

Kramer, J. K. G., Hulan, H. W., Procter, B. G., Rona, G., and Mandavia, M. G. (1978b). *Can. J. Anim. Sci.* **58**, 257–270.

Kramer, J. K. G., Hulan, H. W., Trenholm, H. L., and Corner, A. H. (1979b). *J. Nutr.* **109**, 202–213.

Kramer, J. K. G., Mahadevan, S., Hunt, J. R., Sauer, F. D., Corner, A. H., and Charlton, K. M. (1973). *J. Nutr.* **103**, 1696–1708.

Kramer, J. K. G., Farnworth, E. R., Thompson, B. K., and Corner, A. H. (1981). *Prog. Lipid Res.* **20**, 491–499.

Loew, F. M., Schiefer, B., Laxdal, V. A., Prasad, K., Forsyth, G. W., Ackman, R. G., Olfert, E. D., and Bell, J. M. (1978). *Nutr. Metab.* **22**, 201–217.

Lostroh, A. J. (1958). *Proc. Soc. Exp. Biol. Med.* **98**, 84–88.

McCutcheon, J. S., Umemura, T., Bhatnagar, M. K., and Walker, B. L. (1976). *Lipids* **11**, 545–552.

Opstvedt, J., Svaar, H., Hansen, P., Pettersen, J., Langmark, F. T., Barlow, S. M., and Duthie, I. F. (1979). *Lipids* **14**, 356–371.

Perdue, H. S., and Phillips, P. H. (1952). *Proc. Soc. Exp. Biol. Med.* **81**, 405–407.

Qureshi, S. R. (1979). *Vet. Pathol.* **16**, 486–487.

Ratanasethkul, C., Riddell, C., Salmon, R. E., and O'Neil, J. B. (1976). *Can. J. Comp. Med.* **40**, 360–369.

Rings, R. W., and Wagner, J. E. (1972). *Lab. Anim. Sci.* **22**, 344–352.

Rocquelin, G., and Cluzan, R. (1968). *Ann. Biol. Anim. Biochim., Biophys.* **8**, 395–406.

Roine, P., Uksila, E., Teir, H., and Rapola, J. (1960). *Z. Ernährungswiss.* **1**, 118–124.

Schiefer, B., Loew, F. M., Laxdal, V., Prasad, K., Forsyth, G., Ackman, R. G., and Olfert, E. D. (1978). *Am. J. Pathol.* **90**, 551–564.

Selye, H. (1958). *Proc. Soc. Exp. Biol. Med.* **98**, 61–62.

Shenolikar, I. S., and Tilak, T. B. G. (1980). *Nutr. Metab.* **24**, 199–208.

Soto, P. J., Beall, F. A., Nakamura, R. M., and Kupferberg, L. L. (1964). *Arch. Pathol.* **78**, 681–690.

Svaar, H., Langmark, F. T., Lambertsen, G., and Opstvedt, J. (1980). *Acta Pathol. Microbiol. Scand. Sect. A* **88**, 41–48.

Thomas, H. M., Williams, W. L., and Clower, B. R. (1968). *Arch. Pathol.* **85**, 532–538.

Trenholm, H. L., Thompson, B. K., and Kramer, J. K. G. (1979). *Can. Inst. Food Sci. Technol. J.* **12**, 189–193.

Umemura, T., Yamashiro, S., Bhatnagar, M. K., Moody, D. L., and Slinger, S. J. (1977). *Res. Vet. Sci.* **23**, 59–61.

Umemura, T., Slinger, S. J., Bhatnagar, M. K., and Yamashiro, S. (1978). *Res. Vet. Sci.* **25**, 318–322.

Vles, R. O., Bijster, G. M., Kleinekoort, J. S. W., Timmer, W. G., and Zaalberg, J. (1976). *Fette, Seifen, Anstrichm.* **78**, 128–131.

Vodovar, N., Desnoyers, F., Levillain, R., and Cluzan, R. (1973). *C. R. Hebd. Sciences Acad. Sci., Ser. D* **276**, 1597–1600.

Vodovar, N., Desnoyers, F., Cluzan, R., and Levillain, R. (1977). *Biol. Cell.* **29**, 37–44.

Wilgram, G. F., and Hartroft, W. S. (1955). *Br. J. Exp. Pathol.* **36**, 298–305.

Wilgram, G. F., Hartroft, W. S., and Best, C. H. (1954). *Br. Med. J.* **2**, 1–5.

Willens, S. L., and Sproul, E. E. (1938). *Am. J. Pathol.* **14**, 177–199.

Williams, W. L. (1960). *Yale J. Biol. Med.* **33**, 1–14.

Williams, W. L., and Aronsohn, R. B. (1956). *Yale J. Biol. Med.* **28,** 515–524.

Yamashiro, S., and Clandinin, M. T. (1980). *Exp. Mol. Pathol.* **33**, 55–64.

Ziemlanski, S., Opuszynska, T., and Krus, S. (1972). *Pol. Med. J.* **11**, 1625–1633.

Ziemlanski, S., Rosnowski, A., and Opuszynska, T. (1973). *Acta Med. Pol.* **14**, 279–290.

Ziemlanski, S., Rosnowski, A., and Ostrowski, K. (1975). *Pol. Med. Sci. Hist. Bull.* **15**, 123–132.

Ziemlanski, S., Opuszynska, T., Bulhak-Jachymczyk, B., Olszewska, I., Wozniak, E., Krus, S., and Szymanska, K. (1974). *Pol. Med. Sci. Hist. Bull.* **15**, 3–10.

13

The Composition of Diets Used in Rapeseed Oil Feeding Trials

E. R. FARNWORTH

I. INTRODUCTION

Fat provides about 40 calorie % of the typical North American diet (Nutrition Canada, 1977) of which approximately 15 calorie % can be derived from vegetable oils (Enig et al., 1978). Not surprisingly, considerable effort has been spent in testing vegetable oils for their nutritional properties and safety. As in most safety testing procedures, laboratory animals are used, the most common of which is the albino rat. What perhaps has not been properly appreciated is the fact that there are significant interspecies differences in fat metabolism (see Chapter 14), and that results obtained with rats may not be applicable to other species, including man. Nevertheless, by using a

315

High and Low Erucic Acid Rapeseed Oils
Copyright © 1983 by Academic Press Canada
All rights of reproduction in any form reserved.
ISBN 0-12-425080-7

TABLE I

Percent Composition of Experimental Diets

Ingredients (% by weight)	Diets									
	1	2	3	4	5	6	7	8[a]	9	10
Oil or fat	20	5	20	20	20	20	20	15	37.8	16.6
Protein										
Casein	20	20	20	20	25	20	20	18	32.6	11.3
Flour										
Soybean meal										20.9
Methionine										
Yeast										
Carbohydrate										
Sucrose	20	20	20	20	15			24		
Dextrose						37	50.5			
Starch	30	45	30	30	30			37	24	41
Cellulose	5	5	5	6	6	18.5	5			7.7
Agar-agar								2		
Mineral mix	4	4	4	3	3	3.5	3.5	4	4.8	2.1
Vitamin mix	1	1	1	1	1	1	1	—[e]	0.6	0.4
References[b]	1–11	6,8, 9	12,14, 15	15	13,15, 16	17,19	18	20–23	24,25	26

[a] 1 kg diet combined with 500 g water.
[b] Other diets contained 30.1, 15.7, 7.2n, and 0% oil.
[c] Control diet contained 4.3% oil and 48.2% starch.
[d] Composed of wheat 9%, wheat bran 5%, barley 12%, oats 10%, soybean 33.4%, herring meal 4%, linseed meal 3%, meatmeal 3%, alfalfa meal 3%.
[e] Vitamins added to dry basal.
[f] Vitamin A and D given as supplement.
[g] Included as part of sucrose.

large number of different species, and through an understanding of the metabolic differences that exist between these species, it is possible to obtain a thorough knowledge of both the nutritional and possible toxicological properties of vegetable oils. In all of these testing procedures, it is implicit that the fats and oils to be tested must be included in an otherwise nutritionally adequate diet. It must also be kept in mind that nutritional as well as physical limits determine the maximum amount (and therefore the safety margin) of oil that can be included in the test diet.

The following sections outline the diets used by investigators in animal feeding trials involving low erucic acid rapeseed (LEAR) oil and other vegetable oils. Problems involved with using high fat diets, especially the consequences in terms of changes in nutrient requirements, are also discussed.

Used in Rat Feeding Trials

					Diets							
11	12	13	14	15	16	17	18	19[b]	20	21	22	23[c]
18.1	27	15	15	27	10	21	21	49.6	20	20	15	21
											82.4[d]	
26.2	27	30	25	27	30	20	22	18	19.75	27		19.6
								18				
	0.12	0.12	0.12	0.12	0.12				0.25		0.2	0.4
								12				
	33	15	10	10	20	52.76			18.9	46		20
44.6	8	35	45	31	35		48.5		30			31.5
8.4						1	1		5	3		1
2.3	4	4	4	4	4	5	5	2.4	5	4	1.9	5
0.4	0.88	0.88	0.88	0.88	0.88	0.24	2.5	—[f]	1.1	—[g]	0.5	1.5
26,27	28	28	29	29	30	31,32	33	34	35,36	37	38	39

[h] References: 1, Kramer et al. (1973); 2, Kramer et al. (1975); 3, Kramer et al. (1979a); 4, Kramer et al. (1979b); 5, Hulan et al. (1976); 6, Hulan et al. (1977a); 7, Hulan et al. (1977b); 8, Hulan et al. (1977c); 9, Charlton et al. (1975); 10, Farnworth et al. (1982b); 11, Ackman (1974); 12, Beare-Rogers and Nera (1972); 13, Beare-Rogers and Nera (1977); 14, Beare-Rogers et al. (1972); 15, Beare-Rogers et al. (1974); 16, Beare-Rogers et al. (1979); 17, McCutcheon et al. (1976); 18, Hung et al. (1977); 19, Umemura et al. (1978); 20, Rocquelin and Cluzan (1968); 21, Rocquelin et al. (1973); 22, Astorg and Cluzan (1976); 23, Astorg and Levillain (1979); 24, Abdellatif and Vles (1970); 25, Abdellatif and Vles (1973); 26, Vles et al. (1976); 27, Vles et al. (1978); 28, Ziemlanski et al. (1974); 29, Ziemlanski et al. (1975a); 30, Ziemlanski et al. (1975b); 31, Engfeldt and Brunius (1975a); 32, Engfeldt and Brunius (1975b); 33, Engfeldt and Gustafsson (1975); 34, Roine et al. (1960); 35, Clandinin and Yamashiro (1980); 36, Yamashiro and Clandinin (1980); 37, Nolen (1981); 38, Vogtmann et al. (1975); 39, Svaar and Langmark (1980).

II. DIETARY PROTOCOLS

Published results comparing the cardiopathogenic effects of feeding LEAR and other vegetable oils to laboratory animals have originated primarily from seven research teams, viz., Agriculture Canada, Ottawa, Canada; Health and Welfare Canada, Ottawa, Canada; University of Guelph, Guelph, Canada; Institut National de la Recherche Agronomique, Dijon, France; Unilever Research, Vlaardingen, The Netherlands; Institute of Food and Nutrition, Warsaw, Poland; and the Karolinska Institute, Huddinge, Sweden. These investigators have fed LEAR oil alone or in combination with other oils and fats to a variety of animals including the rat, pig, monkey, duck, mouse, and rabbit.

A. Rat Diets

Diets for rat feeding trials have varied in composition depending on the investigator and the nature of the study. The lipid content of the diets has ranged from 5 to 49.6% (Table I). Casein with few exceptions has been the only protein source, but its proportion of the diet has not been constant. Sugars, starches, and cellulose have provided the carbohydrate in the diets.

Table II lists the ingredients of vitamin mixes used in rat feeding trials. Table II also includes the National Academy of Sciences (NAS) requirements (National Academy of Sciences, 1972) based on a 3600 kcal/kg diet (5% fat) and extrapolated requirements for a 4500 kcal/kg diet (20% fat). Most investigators fed vitamin mixtures in excess of the 1972 NAS requirements because of the relatively high caloric density of high fat diets.

Table III shows the composition of mineral mixes, NAS requirements, and extrapolated requirements. All groups except the Unilever group have used commercial mineral mix formulations. Agriculture Canada and Health and Welfare Canada changed the mineral mix for rats in 1974 in order to more closely approximate the copper, manganese, and zinc requirement specified by NAS. The level of supplementation of some of the minerals does not appear to conform to the 1972 NAS requirements. In the future, high fat diets should possibly contain the extrapolated NAS requirements.

1. TECHNICAL PROBLEMS ASSOCIATED WITH HIGH FAT DIETS

Working with diets high in fat presents technical as well as nutritional problems. Rancidity is one major concern for diets that can contain up to 27% fat. Most investigators have stored excess feed at low temperatures, while others have also added antioxidants. Low erucic acid rapeseed oil presents a unique situation of having a relatively high 18:3 and total unsaturated fatty acid content and thus being more susceptible to oxidative deterioration, but also having a relatively high tocopherol content (Mordret and Helme, 1974), which is known to be a natural antioxidant. Many investigators have used commercially refined oil which has antioxidants included while other groups have added additional antioxidants (Kramer et al., 1973; Svaar et al., 1980).

The separation of dietary ingredients is a problem inherent with mixtures having a large liquid content. As was pointed out, the inclusion of cellulose (Alexander and Mattson, 1966) or sawdust (Thomasson et al., 1970) is necessary to prevent the liquid fat from separating from the dry components of

TABLE II

Composition of Vitamin Mixtures Used in Rat Feeding Trials

Ingredients (mg/kg diet)	NAS[a]	NAS[b]	Mixture											
			1	2	3	4	5	6	7	8	9	10	11	12
p-Aminobenzoic acid	NR[c]		100	20	—	20	500	62	—	—	100	100	300	100
Biotin	NR		0.2	0.5	0.2	0.5	0.2	2	0.2	0.2	0.3	0.4	1.0	0.3
Choline chloride[d]	750	938	1275	1000	956	1000	666	1240	1407	1291	1	2000	2000	3000
Folic acid	NR		2	1	1	1	1	10.5	1	0.9	10	2	20	0.3
Inositol	NR		500	1000	100	1000	200	1240	105	96	100	100	1000	2000
Niacin	15	19	50	50	50	22.5	50	62	21	19	50	40	200	20
Calcium pantothenate	8	10	30	20	22	12	10	70	21	19	50	40	100	20
Pyridoxine-HCl	7	8.75	10	5	10	10	4	28	2.1	1.9	15	5	20	4
Riboflavin	2.5	3.1	10	10	10	3.75	4	28	6.3	5.8	20	8	20	6.7
Thiamine-HCl	1.25	1.56	10	10	10	1.9	4	74	6.3	5.8	20	5[e]	50[e]	4
DL-α-Tocopherol[f]	35	44	91	240	200	58	15	15[g]	84	77	150	50	500	100
Vitamin A[h]	0.6	0.75	0.8	8	11	0.9	0.6	0.6[g]	33	30	4.5	5.8	7.4	4.12
Vitamin B12	0.005	0.006	0.02	0.01	0.02	0.0075	0.03	—	0.02	0.02	0.03	0.03	0.02	0.15
Vitamin C	NR		—	—	—	—	—	—	—	—	20	—	1000	100

(Continued)

TABLE II (CONTINUED)

Ingredients (mg/kg diet)	NAS[a]	NAS[b]	Mixture 1	2	3	4	5	6	7	8	9	10	11	12
Vitamin D (IU/kg)[j]	1000	1250	1000	4000	1000	1500	500	500[g]	210920	193800	5000	2000	4000	12000
Vitamin K	0.05	0.06	5	1	1	0.075	1	5	1.0	0.9	10	5	10	3
References[j]			1–13	14	15	16	17–20	21,22	23,24	23	25–27	28,29	30	31

[a] NAS, National Academy of Sciences (1972); requirements based on 3600 kcal/kg.
[b] Extrapolated values based on 4500 kcal/kg.
[c] NR, no requirement level established.
[d] Choline chloride or choline chloride equivalent.
[e] As thiamine mononitrate.
[f] As acetate.
[g] Given biweekly.
[h] retinal or retinol equivalent.
[i] Values given in weight units converted to IUs (1 μg = 40 IU).
[j] References: 1. Kramer et al. (1973); 2. Kramer et al. (1975); 3. Kramer et al. (1979a); 4. Kramer et al. (1979b); 5. Hulan et al. (1976); 6. Hulan et al. (1977a); 7. Hulan et al. (1977b); 8. Hulan et al. (1977c); 9. Charlton et al. (1975); 10. Farnworth et al. (1982b); 11. Beare-Rogers et al. (1972); 12. Beare-Rogers et al. (1974) 13. Beare-Rogers et al. (1979); 14. McCutcheon et al. (1976); 15. Hung et al. (1977); 16. Umemura et al. (1978); 17. Rocquelin and Cluzan (1968); 18. Rocquelin et al. (1973); 19. Astorg and Cluzan (1976); 20. Astorg and Levillain (1979); 21. Abdellatif and Vles (1970); 22. Abdellatif and Vles (1973); 23. Vles et al. (1976); 24. Vles et al. (1978); 25. Ziemlanski et al. (1974); 26. Ziemlanski et al. (1975a); 27. Ziemlanski et al. (1975b); 28. Engfeldt and Brunius (1975a); 29. Engfeldt and Brunius (1975b); 30. Engfeldt and Gustafsson (1975); 31. Nolen (1981).

the diet (oil rises to the top). Most diets used in rat feeding trials contain cellulose (Table I).

2. NUTRITIONAL PROBLEMS ASSOCIATED WITH HIGH FAT DIETS

The nutritional adequacy of the diets listed in Table I may be summarized by the fact that no nutrient deficiency symptoms have been reported by any investigators using the diets in spite of the fact that some feeding trials lasted for up to 1 year. Generally, the diets were readily accepted by the test animals, and normal growth patterns have been reported except when large quantities of high erucic acid rapeseed oils were fed to rats (Roine et al., 1960). When diets with high caloric densities are fed, nutritional problems such as lowered food intake and changes in nutrient requirements must be considered.

It is well known that the rat consumes food to satisfy its need for calories (Sibbald et al., 1956 and 1957). Therefore, to ensure the maximum intake of diet, ad libitum feeding regimes have been used almost exclusively; Roine et al. (1960) and Svaar and Langmark (1980) being exceptions. The group at Dijon, France, have reported adding water to their diets to cut down on feed wastage. But again, this may be a method of increasing feed consumption, since Rogers and Harper (1965) reported increased feed consumption by rats receiving diets containing added water.

The NAS nutrient requirements are generally accepted as standards for rat feeding trials (National Academy of Sciences, 1972). However, diets used in early feeding trials may not meet 1972 dietary standards, since the nutrient standards have changed over the years. A comparison between the first NAS requirements published in 1962 and the second published in 1972 shows the addition of chromium, L-arginine, L-asparagine, and L-glutamic acid to the list of essential nutrients, and a definition of the vitamin D requirement (National Academy of Sciences, 1962, 1972). In addition, the levels of 12 other nutrients were changed. However, of these 12 only the amino acids, vitamin E ,and pyridoxine requirements might be expected to affect lipid metabolism. The nutrient data in Table II show that with few exceptions the NAS requirements for vitamins have been met. Although the mineral mixes described in Table III generally met 1972 NAS requirements, the level of zinc and copper used in most of the earlier diets tended to be low; selenium was not added by most investigators.

The use of NAS requirements as standards for rat diets containing high fat levels may not be entirely valid, since the NAS requirements are based on a diet containing 5% fat (3600 kcal/kg ME). The caloric density of a typical rat diet increases by 20% as the fat level is raised from 5 to 20%. Indeed, the requirements are given with the caution that nutrient levels in diets of different caloric densities can be obtained by extrapolation so that a constant nutrient-to-calorie ratio is maintained. It can be seen from the data in Table II

TABLE III

Composition of Mineral Mixtures Used in Rat Feeding Trials

Salt mix	NAS[a]	NAS[b]	USP XIV	USP XVII	USP XVIII	Williams–Briggs modified	Bernhart–Tomarelli[c]	Hubbell–Mendel–Wakeman	Custom mixtures				
									1	2[d]	3	4	5
Aluminum (mg/kg)	NR[e]		0.2	—	—	—	—	0.19	5.7	—	—	—	—
Calcium (%)	0.5	0.6	0.442	0.611	0.764	0.623	0.900	0.434	0.529	0.312	0.339	0.545	0.15
Chlorine (%)	0.05	0.06	0.424	0.338	0.426	0.347	0.074	0.190	0.756	0.214	0.07	0.948	0.71
Cobalt (mg/kg)	NR	—	—	0.23	0.30	—	—	—	0.86	—	—	0.576	—
Copper (mg/kg)	5	6.3	0.8	4.9	6.1	5.2	6.5	7.2	4.5	5.2	5.7	3	9.3
Fluorine (mg/kg)	NR	—	9.2	—	—	—	—	9.0	2.14	—	—	—	—
Iodine (mg/kg)	0.15	0.19	1.2	24.2	30	0.59	0.22	1.2	1.71	0.17	0.18	24.2	1.8
Iron (mg/kg)	35	44	107	217	280	25.2	37.3	47.2	152	28.4	30.9	161	58.6
Magnesium (%)	0.04	0.05	0.066	0.046	0.27	0.046	0.060	0.019	0.05	0.05	0.05	0.039	—
Manganese (mg/kg)	50	62	2.6	52.1	65	50.1	46.4	2.3	52.9	65.6	71.4	65.6	38.1
Phosphorus (%)	0.4	0.5	0.267	0.354	0.444	0.399	0.746	0.100	0.409	0.117	0.127	0.332	—
Potassium (%)	0.18	0.23	0.655	0.448	0.0009	0.383	0.268	0.239	0.473	0.23	0.25	0.703	—
Selenium (mg/kg)	0.04	0.05	—	—	—	—	—	—	—	—	—	—	—

Sodium (%)	0.05	0.06	0.122	0.219	0.274	0.211	0.076	0.063	0.218	0.064	0.070	0.259	0.46
Sulfur (mg/kg)	NR		410.5	770.5	581	644.4	500.5	90.6	116	—	—	1.53	26
Zinc (mg/kg)	12	15	—	5	6.2	11.8	17.1	—	3.23	15.5	16.9	4.74	—
References[f]	1–3,11		4–10,12, 16,34	17,18	19–21	13–15,22, 23,36	24–27	28,29	30	30,31	32,33	35	

[a] NAS, National Academy of Sciences (1972); requirements based on 3600 kcal/kg.

[b] Extrapolated values based on 4500 kcal/kg.

[c] Modified by Clandinin and Yamashiro (1980) to contain, in addition (mg/kg) Mn 77.5; Se 0.06.

[d] Also includes (mg/kg) Ni 1.13; Mo 0.89; As 0.08; B 0.3; Br 7.86.

[e] NR, no requirement level established.

[f] References: 1, Kramer et al. (1973); 2, Beare-Rogers and Nera (1972); 3, Beare-Rogers et al. (1972); 4, Kramer et al. (1975); 5, Kramer et al. (1979a); 6, Kramer et al. (1979b); 7, Hulan et al. (1976); 8, Hulan et al. (1977a); 9, Hulan et al. (1977b); 10, Hulan et al. (1977c); 11, Charlton et al. (1975); 12, Farnworth et al. (1982b); 13, Beare-Rogers and Nera (1977); 14, Beare-Rogers et al. (1974); 15, Beare-Rogers et al. (1979); 16, Engfeldt and Gustafsson (1975); 17, Engfeldt and Brunius (1975a); 18, Engfeldt and Brunius (1975b); 19, McCutcheon et al. (1976); 20, Hung et al. (1977); 21, Umemura et al. (1978); 22, Clandinin and Yamashiro (1980); 23, Yamashiro and Clandinin (1980); 24, Rocquelin and Cluzan (1968); 25, Rocquelin et al. (1973); 26, Astorg and Cluzan (1976); 27, Astorg and Levillain (1979); 28, Abdellatif and Vles (1970); 29, Abdellatif and Vles (1973); 30, Vles et al. (1976); 31, Vles et al. (1978); 32, Ziemlanski et al. (1974); 33, Ziemlanski et al. (1975a); 34, Nolen (1981); 35, Roine et al. (1960); 36, Ackman (1974).

and III that most investigators exceeded the 1972 NAS requirements for this reason.

Clandinin and Yamashiro (1980) have recently pointed out that if the methionine per calorie ratio is calculated for rat diets containing 20% casein and 20% fat (4500 kcal/kg ME), the diet is apparently deficient in methionine. However, these workers were not able to demonstrate a consistent improvement in weight gains when rats were fed the diet supplemented with methionine. An experiment carried out in this laboratory indicated that methionine supplementation did improve weight gains but the choline level of the diet had no effect. Neither the choline nor the methionine status of the diet affected the incidence or severity of heart lesions (Farnworth et al., 1982a).

Calculation of the nutrient to calorie values of other known lipotropic factors for diets containing 20% fat indicate that the diets listed in Table I exceed 1972 NAS requirements. It has been suggested that all nutrient requirements be expressed as a function of energy intake (Crampton, 1964). However, until more data is available on the requirements of rats fed high fat diets, the NAS requirements extrapolated to 4500 kcal/kg should be used.

B. Pig Diets

Experiments to test the effects of feeding rapeseed oil to pigs follow rat experiments in frequency, perhaps due to the suitability of the pig as a model for human health and nutrition studies (Bustad et al., 1966; Dodds, 1982). Compared to rat diets, a great diversity of feedstuffs have been used to formulate pig diets (Table IV).

The inclusion of fat or oil in the diets of pigs is not common practice. This is perhaps due to the reported reduction in growth and increase in energy required per unit gain for baby pigs that received diets containing added fats (Peo et al., 1957; Asplund et al., 1960; Eusebio et al., 1965). However, in spite of reduced feed consumption by animals fed high fat diets, no evidence of adverse affects on growth rate was reported by investigators comparing rapeseed and other vegetable oils. The control diets included in the studies reported by Aherne et al. (1975, 1976) and Friend et al. (1975a,b, 1976) contained no added fat. These investigators took care to maintain constant crude protein, calcium, and phosphorus levels in their control and experimental diets. No increases in vitamins or minerals were made when diets contained added oil (Table V). However, since these nutrients were in excess of the NAS requirements for growing swine, the adequacy of the high caloric density diets is probably not in doubt (National Academy of Sciences, 1973).

TABLE IV

Composition of Experimental Diets Used in Pig Feeding Trials

Ingredients (% by weight)	NAS[a]	1	2	3	4	5	6	7[b]	8	9	10	11	12
Oil or fat	—	10	20	5	10	20	3.5	13	15	15	17.5	4	8
Protein	22												
Alfalfa meal													
Barley		37.5	37.5	36.75	36.75	37.5			49	50	39	72	64
Corn								30.5					
Herring meal									11.2				
Meat meal													
Oats											15		
Rapeseed meal													
Skim milk							96.3	26					
Soybean meal		20	20	23.95	25.65	21.5			11.3	20	26	21.25	25
Wheat		8	8	8	8	8		30.5	12	11.2			
Methionine											0.04		
Carbohydrate	—												
Cellulose		2	4			4							
Corn starch		16	4	21.2	14.5	4							

(Continued)

TABLE IV (CONTINUED)

Ingredients (% by weight)	NAS[a]	Diets											
		1	2	3	4	5	6	7[b]	8	9	10	11	12
Mineral mix	—[e]	0.5	0.5	0.1	0.1	0.5		—	—[d]	—[d]	3.5	—[d]	—[d]
CaCO$_3$		1	1	0.7	0.7	1.0				1.2			
Ca Phos.		2.5	2.5	3.3	3.3	2.5				1.2			
NaCl (iodized)		0.5	0.5	0.5	0.5	0.5			0.5	0.4	0.47		
Vitamin mix	—[e]	2	2	0.5	0.5	0.5		[c]	1	1	0.45	2.75	3
Emulsifier							0.2						
References[f]		1	2	2	2	2, 3	4	5	6	7	8	9	9

[a] NAS, National Academy of Sciences (1973); requirements for 5–10 kg pigs.
[b] 1150 g diet combined with 1500 g water.
[c] Vitamins A and D given weekly.
[d] Vitamin and mineral mix combined.
[e] For mineral and vitamin requirement see Table V.
[f] References: 1, Friend et al. (1975a); 2, Friend et al. (1975b); 3, Friend et al. (1976); 4, Beare-Rogers and Nera (1972); 5, Roine et al. (1960); 6, Aherne et al. (1975); 7, Aherne et al. (1976); 8, Svaar et al. (1980); 9, Petersen et al. (1979).

TABLE V

Composition of Vitamins and Minerals in the Diets Used in Pig Feeding Trials

Ingredients	NAS[a]	Diets						
		1	2	3	4	5	6	7
Vitamins (mg/kg diet)								
Choline chloride	1467	220	220	2207		55.7	1.5	15
Folic acid	NR[b]					1.7		
Niacin	22	13.2	13.2	13.2		50.5	44	9
Calcium pantothenate	14	8.8	8.8	8.8		22.2	28	16.3
Pyridoxine-HCl	1.5						3	
Riboflavin	3	4.4	4.4	4.4			6	
Thiamin-HCl	1.3						2.6	
DL-α-Tocopherol (acetate)	11	5.1	11	11		11	100	6
Vitamin A (IU/kg)	2200	5000	5000	5000	24000[c]	4400	2800	12000
Vitamin B_{12}	0.022	0.022	0.022	0.022		0.020	0.044	0.023
Vitamin D (IU/kg)	220	500	500	500	4800[c]	550	400	1500
Minerals (% of diet)								
Calcium	0.8	0.9[d]	1.05[d]	0.9[d]		0.72[d]	0.75	7.2
Cobalt (mg/kg)	NR					2.8		0.36
Copper (mg/kg)	(6)[e]	11	22	22		24.6		34.5
Iodine (mg/kg)	(0.2)	1	2	2		1.7	2.2	7.2
Iron (mg/kg)	(80)	111	222	222		294.1	48	72
Magnesium (mg/kg)	(400)							300
Manganese (mg/kg)	(20)	45	90	90		76.2	60	85.8

(Continued)

TABLE V (CONTINUED)

Ingredients	NAS[a]				Diets			
		1	2	3	4	5	6	7
Phosphorus	0.60	0.8[d]	0.91[d]	0.8[d]		0.59[d]	0.18	1.95
Potassium	(0.26)							
Selenium (mg/kg)	(0.1)						0.1	
Sodium	NR	0.2	0.2	0.2		0.16		1.8
Zinc (mg/kg)	(50)	111	222	222		88.5	51	143
References[f]		1	2	2, 3	4	5, 6	7	8

[a] NAS, National Academy of Sciences (1973); requirements for 5–10 kg pigs.
[b] NR - no requirement stated.
[c] Vitamins A and D given weekly.
[d] Percent of total diet by analysis.
[e] Numbers in parentheses are suggested levels.
[f] References: 1, Friend et al. (1975a); 2, Friend et al. (1975b); 3, Friend et al. (1976); 4, Roine et al. (1960); 5, Aherne et al. (1975); 6, Aherne et al. (1976); 7, Svaar et al. (1980); 8, Petersen et al. (1979).

C. Nonhuman Primate Diets

Nonhuman primates have been used to test the cardiopathogenicity of vegetable oils. The diets that have been used in these experiments are listed in Table VI and VII. It should be recognized that the nutritional requirements of the various nonhuman primates have not been studied in detail. More research, particularly on mineral requirements and differences in nutrient requirements by the different species of monkeys, is necessary to overcome the general lack of information on the effect nutrition has on more subtle parameters of health (National Academy of Science, 1972). Nevertheless, from the examples of adequate diet given in the NAS requirements, it is apparent that most of the diets represented in Table VI and VII were nutri-

TABLE VI

Composition of Experimental Diets Used in Monkey Feeding Trials

Ingredient (% by weight)	NAS[a]	Diets					
		1	2	3	4	5	6
Oil or fat	$(0.25)^b$	20	20	20	25	5	10
Protein	(3)						
Alfalfa					1		
Bengalgram				40		60	60
Casein			20	3			
Oats					19.57		
Skim milk powder		14		10	9		
Soybean meal		30			31		
Wheat				18	4	30	25
Wheat bran					4.68		
Yeast		2.8			2		
Carbohydrate	—						
Alfa cel			4				
Corn starch		17		4			
Sucrose		10	50				
Mineral mix	—[c]	4	4	4	0.75	4	4
CaCO₃					1.25		
CaPO₄					1.25		
Vitamin mix	—[c]	2.2	2	1	0.5	1	1
Banana extract (ml/kg)			15				
References[d]		1	2	3	4–6	7	7

[a] NAS, National Academy of Sciences (1972); requirements for growing 3 kg rhesus monkey.

[b] Number in parentheses given in g/kg body weight; fat as linoleic acid.

[c] For mineral and vitamin requirement see Table VII.

[d] References: 1, Kramer et al. (1978); 2, Beare-Rogers and Nera (1972); 3, Gopalan et al. (1974); 4, Ackman and Loew (1977); 5, Ackman et al. (1977); 6, Loew et al. (1978); 7, Shenolikar and Tilak (1980).

330

TABLE VII

Composition of Vitamin and Mineral Mixtures Used in Monkey Feeding Trials

		Diets			
Ingredients	NAS[a]	1	2	3	4
Vitamins (mg/kg diet)					
p-Aminobenzoic acid	NR	110	100	1000	—
Biotin	0.01	0.44	0.40	1	0.57
Choline	Req.	1238	1125	150	930
Folic acid	(0.04)	1.98	1.80	10	2.3
Inositol	NR	110	100	1000	—
Niacin	(2)	93.5	85	80	114
Ca pantothenate	Req.	66	60	60	12
Pyridoxine-HCl	0.5–5	22	20	10	15
Riboflavin	0.03	22	20	16	1.7
Thiamin-HCl	0.33	22	20	10	1.7
DL-α-Tocopherol	0.83	484	440	2^b	66
Vitamin A (IU/kg)	400	19800	18000	1500^b	23000
Vitamin B_{12}	0.07	0.028	0.025	0.02	0.004
Vitamin C	25	990	900	200	25^c
Vitamin D (IU/kg)	25	2337	2125	400^b	1430
Vitamin K	0.0001	49.5	45.0	200	6
Minerals (mg/kg diet)					
Aluminum		—	—	0.2	—
Calcium (%)	150	0.558	0.558	0.440	0.984
Chlorine (%)		0.407	0.407	0.423	0.182
Copper		0.305	0.305	9.9	5.3
Iodine		2.45	2.45	1.2	—[d]
Iron		11.3	11.3	127	34.6
Magnesium (%)	40	0.040	0.040	0.066	0.040
Manganese		4.9	4.9	2.05	507
Phosphorus (%)		0.354	0.354	0.211	0.249
Potassium (%)		0.741	0.741	0.649	0.128
Sodium (%)		0.262	0.262	0.121	0.117
Sulfur		56	56	417	0.002
Zinc	Req.	0.479	0.479	—	11.2
References[e]		1	2	3, 4	5–7

[a] NAS, National Academy of Sciences (1972) requirements for growing 3 kg rhesus monkey in units/kg body weight; NR, no requirement stated; numbers in brackets, tentative requirement; Req., required but level not established; minerals in g/kgBW.

[b] Given weekly.

[c] Vitamin C in drinking water to supply 25 mg/kg body weight/day.

[d] Iodized salt included in diet.

[e] References: 1, Kramer et al. (1978); 2, Beare-Rogers and Nera (1972); 3, Gopalan et al. (1974); 4, Shenolikar and Tilak (1980); 5, Ackman and Loew (1977); 6, Ackman et al. (1977); 7, Loew et al. (1978).

tionally sound. Evidence of this is found in the observation by most research workers that the monkeys grew well and remained healthy throughout the feeding trial (Beare-Rogers and Nera, 1972; Ackman and Loew, 1977; Loew *et al.*, 1978; Kramer *et al.*, 1978; Schiefer, 1982). Exceptions to this are found in the experiments by Gopalan *et al.* (1974) and Shenolikar and Tilak (1980) who used adult monkeys (*Macaca radiata*) and found that the monkeys had difficulty in adjusting to the experimental diets during the first 5 weeks. During this time they exhibited loss of appetite and diarrhea. Once adapted, however, food intake improved and body weights were well maintained throughout the experiment. Some losses were experienced with the monkeys receiving the mustard oil diets. One developed signs of acute scurvy, another anorexia, and loss of body weight and two died after 52 weeks on experiment.

III. CONCLUSIONS

Feeding trials using a variety of animals have been an integral part of the research efforts to establish the nature of the cardiopathic properties of rapeseed and other vegetable oils. The experimental diets out of necessity have contained high levels of oil or fat. Since nutrient requirements are usually established for low fat diets, the adequacy of some nutrients could be questioned for high fat diets. However, the variety of dietary constituents and vitamin and mineral supplements precludes the possibility that the lack of one particular nutrient is affecting the experimental results.

ACKNOWLEDGMENT

Animal Research Centre Contribution No. **1093**.

REFERENCES

Abdellatif, A. M. M., and Vles, R. O. (1970). *Nutr. Metab.* **12**, 285–295.
Abdellatif, A. M. M., and Vles, R. O. (1970). *Nutr. Metab.* **15**, 219–231.
Ackman, R. G. (1974). *Lipids* **9**, 1032–1035.
Ackman, R. G., and Loew, F. M. (1977). *Fette, Seifen, Anstrichm.* **79**, 15–24, 58–69.
Ackman, R. G., Eaton, C. A., Loew, F. M., Sipos, J. C., and Hancock, D. (1977). *Bibl. Nutr. Dieta* **25**, 170–185.
Aherne, F. X., Bowland, J. P., Christian, R. G., Vogtmann, H., and Hardin, R. T. (1975). *Can. J. Anim. Sci.* **55**, 77–85.
Aherne, F. X., Bowland, J. P., Christian, R. G., and Hardin, R. T. (1976). *Can. J. Anim. Sci.* **56**, 275–284.

Alexander, J. C., and Mattson, F. H. (1966). *Can. J. Biochem.* **44**, 35–43.

Asplund, J. M., Grummer, R. H., and Phillips, P. H. (1960). *J. Anim. Sci.* **19,** 709–714.

Astorg, P. O., and Cluzan, R. (1976). *Ann. Nutr. Aliment.* **30**, 581–602.

Astorg, P. O., and Levillain, R. (1979). *Ann. Nutr. Aliment.* **33**, 643–658.

Beare-Rogers, J. L., and Nera, E. A. (1972). *Comp. Biochem. Physiol.* **41B**, 793–800.

Beare-Rogers, J. L., and Nera, E. A. (1977). *Lipids* **12**, 769–774.

Beare-Rogers, J. L., Nera, E. A., and Craig, B. M. (1972). *Lipids* **7**, 46–50.

Beare-Rogers, J. L., Nera, E. A., and Heggtveit, H. A. (1974). *Nutr. Metab.* **17**, 213–222.

Beare-Rogers, J. L., Gray, L., Nera, E. A., and Levin, O. L. (1979). *Nutr. Metab.* **23**, 335–346.

Bustad, L. K., McClellan, R. O., and Burns, M. P. (1966). "Swine in Biomedical Research." Frayn Printing Co., Seattle, Washington.

Charlton, K. M., Corner, A. H., Davey, K., Kramer, J. K. G., Mahadevan, S., and Sauer, F. D. (1975). *Can. J. Comp. Med.* **39**, 261–269.

Clandinin, M. T., and Yamashiro, S. (1980). *J. Nutr.* **110**, 1197–1203.

Crampton, E. W. (1964). *J. Nutr.* **82**, 353–365.

Dodds, W. J. (1982). *Fed. Proc., Fed. Am. Soc. Exp. Biol.* **41**, 247–256.

Engfeldt, B., and Brunius, E. (1975a). *Acta Med. Scand., Suppl.* **585**, 15–26.

Engfeldt, B., and Brunius, E. (1975b). *Acta Med. Scand., Suppl.* **585**, 27–40.

Engfeldt, B., and Gustafsson, B. (1975). *Acta Med. Scand., Suppl.* **585**, 41–46.

Enig, M. G., Munn, R. J., and Keeney, M. (1978). *Fed. Proc., Fed. Am. Soc. Exp. Biol.* **37**, 2215–2220.

Eusebio, J. A., Hays, V. W., Speer, V. C., and McCall, J. T. (1965). *J. Anim. Sci.* **24**, 1001–1007.

Farnworth, E. R., Kramer, J. K. G., Corner, A. H., and Thompson, B. K. (1982a). *J. Am. Oil Chem. Soc.* **59**, 290A, Abstr. 243.

Farnworth, E. R., Kramer, J. K. G., Thompson, B. K., and Corner, A. H. (1982b). *J. Nutr.* **112**, 231–240.

Friend, D. W., Corner, A. H., Kramer, J. K. G., Charlton, K. M., Gilka, F., and Sauer, F. D. (1975a). *Can. J. Anim. Sci.* **55**, 49–59.

Friend, D. W., Gilka, F., and Corner, A. H. (1975b). *Can. J. Anim. Sci.* **55**, 571–578.

Friend, D. W., Kramer, J. K. G., and Corner, A. H. (1976). *Can. J. Anim. Sci.* **56**, 361–364.

Gopalan, C., Krishnamurthi, D., Shenolikar, I. S., and Krishnamachari, K. A. V. R. (1974). *Nutr. Metab.* **16**, 352–365.

Hulan, H. W., Kramer, J. K. G., Mahadevan, S., Sauer, F. D., and Corner, A. H. (1976). *Lipids* **11**, 9–15.

Hulan, H. W., Kramer, J. K. G., and Corner, A. H. (1977a). *Can. J. Physiol. Pharmacol.* **55**, 258–264.

Hulan, H. W., Kramer, J. K. G., and Corner, A. H. (1977b). *Lipids* **12**, 951–956.

Hulan, H. W., Kramer, J. K. G., Corner, A. H., and Thompson, B. (1977c). *Can. J. Physiol. Pharmacol.* **55**, 265–271.

Hung, S., Umemura, T., Yamashiro, S., Slinger, S. J., and Holub, B. J. (1977). *Lipids* **12**, 215–221.

Kramer, J. K. G., Mahadevan, S., Hunt, J. R., Sauer, F. D., Corner, A. H., and Charlton, K. M. (1973). *J. Nutr.* **103**, 1696–1708.

Kramer, J. K. G., Hulan, H. W., Mahadevan, S., Sauer, F. D., and Corner, A. H. (1975). *Lipids* **10**, 511–516.

Kramer, J. K. G., Hulan, H. W., Procter, B. G., Dussault, P., and Chappel, C. I. (1978). *Can. J. Anim. Sci.* **58**, 245–256.

Kramer, J. K. G., Hulan, H. W., Corner, A. H., Thompson, B. K., Holfeld, N., and Mills, J. H. L. (1979a). *Lipids* **14**, 773–780.

Kramer, J. K. G., Hulan, H. W., Trenholm, H. L., and Corner, A. H. (1979b). *J. Nutr.* **109**, 202–213.

Loew, F. M., Schiefer, B., Laxdal, V. A., Prasad, K., Forsyth, G. W., Ackman, R. G., Olfert, E. D., and Bell, J. M. (1978). *Nutr. Metab.* **22**, 201–217.

McCutcheon, J. S., Umemura, T., Bhatnagar, M. K., and Walker, B. L. (1976). *Lipids* **11**, 545–552.

Mordret, F., and Helme, J. P. (1974). *Proc. Int. Rapskongr, 4th, 1974,* pp. 283–289. Giessen, West Germany.

National Academy of Sciences (1962). "Nutrient Requirements of Laboratory Animals," pp. 51–95. Nat. Acad. Sci., Washington, D.C.

National Academy of Sciences (1972). "Nutrient Requirements of Laboratory Animals," 2nd rev. ed., pp. 29–45, 56–93. Nat. Acad. Sci., Washington, D.C.

National Academy of Sciences (1973). "Nutrient Requirements of Swine," pp. 2–30. Nat. Acad. Sci., Washington, D.C.

Nolen, G. A. (1981). *J. Am. Oil Chem. Soc.* **58**, 31–37.

Nutrition Canada (1977). "Food Consumption Patterns Report." Health and Welfare Canada, Ottawa, Ontario.

Peo, E. R., Ashton, G. C., Speer, V. C., and Catron, D. V. (1957). *J. Anim. Sci.* **16**, 885–891.

Petersen, U., Oslage, H.-J., and Seher, A. (1979). *Fette, Seifen, Anstrichm.* **81**, 177–181.

Rocquelin, G., and Cluzan, R. (1968). *Ann. Biol. Anim. Biochim. Biophys.* **8**, 395–406.

Rocquelin, G., Sergiel, J.-P., Astorg, P. O., and Cluzan, R. (1973). *Ann. Biol. Anim. Biochim. Biophys.* **13**, 587–609.

Rogers, Q. R., and Harper, A. E. (1965). *J. Nutr.* **87**, 267–273.

Roine, P., Uksila, E., Teir, H., and Rapola, J. (1960). *Z. Ernährungswiss.* **1**, 118–124.

Schiefer, B. (1982). *In* "Nutritional Evaluation of Long-Chain Fatty Acids in Fish Oils" (M.E. Stansby and S.M. Barlow, eds.). Academic Press, New York (in press).

Shenolikar, I. S. and Tilak, T. B. G. (1980). *Nutr. Metab.* **24**, 199–208.

Sibbald, I. R., Berg, R. T., and Bowland, J. P. (1956). *J. Nutr.* **59**, 385–392.

Sibbald, I. R., Bowland, J. P., Robblee, A. R., and Berg, R. T. (1957). *J. Nutr.* **61**, 71–85.

Svaar, H., and Langmark, F. T. (1980). *Acta Pathol. Microbiol. Scand., Sect. A* **88**, 179–187.

Svaar, H., Langmark, F. T., Lambertsen, G., and Opstvedt, J. (1980). *Acta Pathol. Microbiol. Scand., Sect. A* **88**, 41–48.

Thomasson, H. J., Gottenbos, J. J., van Pijpen, P. L., and Vles, R. O. (1970). *Proc. Int. Symp. Chem. Technol. Rapeseed Oil Other Cruciferae Oils, 1967* pp. 381–402. Gdansk, Poland.

Umemura, T., Slinger, S. J., Bhatnagar, M. K., and Yamashiro, S. (1978). *Res. Vet. Sci.* **25**, 318–322.

Vles, R. O., Bijster, G. M., Kleinekoort, J. S. W., Timmer, W. G., and Zaalberg, J. (1976). *Fette, Seifen, Anstrichm.* **78**, 128–131.

Vles, R. O., Bijster, G. M., and Timmer, W. G. (1978). *Arch. Toxicol., Suppl.* **1**, 23–32.

Vogtmann, H., Christian, R., Hardin, R. T., and Clandinin, D. R. (1975). *Int. J. Vitam. Nutr. Res.* **45**, 221–229.

Yamashiro, S., and Clandinin, M. T. (1980). *Exp. Mol. Pathol.* **33**, 55–64.

Ziemlanski, S., Opuszynska, T., Bulhak-Jachymczyk, B., Olszewska, I., Wozniak, E., Krus, S., and Szymanska, K. (1974). *Pol. Med. Sci. Hist. Bull.* **15**, 3–10.

Ziemlanski, S., Kucharczyk, B., Ziombski, H., and Szymanska, K. (1975a). *Pol. Med. Sci. Hist. Bull.* **15**, 95–112.

Ziemlanski, S., Rosnowski, A., and Ostrowski, K. (1975b). *Pol. Med. Sci. Hist. Bull.* **15**, 123–132.

14

The Metabolism of Docosenoic Acids in the Heart

F. D. SAUER AND J. K. G. KRAMER

I. SUBSTRATES FOR MYOCARDIAL OXIDATION

In the presence of adequate oxygen supply the healthy heart oxidizes fatty acids, ketone bodies, and pyruvate. Pyruvate is derived from glycogenolysis and glycolysis. Outer-chain glycogen degradation is catalyzed by phosphorylase. Outer-chain glycogen synthesis is catalyzed by a different enzyme, UDP glucose–glycogen synthase. The degradative phosphorylase re-

High and Low Erucic Acid Rapeseed Oils
Copyright © 1983 by Academic Press Canada
All rights of reproduction in any form reserved.
ISBN 0-12-425080-7

action results in the phosphorylitic cleavage of the α-1,4 bond of glycogen with inorganic phosphate to yield a glucose 1-phosphate. The enzyme phosphorylase, which catalyzes this reaction, is regulated similarly to some of the other enzymes that produce substrates for myocardial oxidation, by a phosphorylation–dephosphorylation mechanism (Table I). Specifically, two protein kinases are involved. A cyclic AMP-dependent protein kinase, as the name suggests, is activated by increased cyclic AMP levels. When activated, this enzyme in turn activates a second enzyme, phosphorylase b kinase (inactive), to phosphorylase b kinase (active) with the consumption of 1 mole of ATP. The latter enzyme, when active, converts phosphorylase b to the active phosphorylase a (phosphorylated) with the consumption of another mole of ATP. Both these enzymes are inactivated and dephosphorylated by phosphatases which remove the phosphate group from the enzyme(s).

Glucose breakdown by heart muscle is regulated at several steps. Phosphofructokinase (PFK), which converts fructose 6-phosphate to fructose 1,6-diphosphate, is a key step. PFK is activated by 5'-AMP and phosphate which lower the K_m of the enzyme for fructose 6-phosphate and is inhibited by ATP, ATP•Mg^{2+}, and citrate which, conversely, raise the K_m of the enzyme for fructose 6-phosphate. Whether PFK is also regulated by a dephosphorylation–phosphorylation process similar to that described above is not decided. Brand and Söling (1975) suggest the presence of such a regulatory mechanism for PFK isolated from liver.

The oxidation of pyruvate represents another important regulatory step in

TABLE I

Enzyme Activation by Phosphorylation–Dephosphorylation

Enzyme	Reaction	Enzyme active form
Phosphorylase	$Glycogen_{(n+1)} + Pi^{2-} \rightarrow$ $glycogen_{(n)} + glucose\ 1\text{-}P^{2-}$	phosphorylated
UDP glucose glycogen synthase	$UDP\ glucose + glycogen_{(n)} \rightarrow$ $UDP + glycogen_{(n+1)}$	dephosphorylated
Phosphofructo-kinase	$Fructose\ 6\text{-}phosphate \rightarrow$ $fructose\ 1,6\text{-}diphosphate$	phosphorylated?
Pyruvate dehydrogenase complex	$Pyruvate + CoASH + NAD^+ \rightarrow$ $acetyl\text{-}SCoA + CO_2 + NADH + H^+$	dephosphorylated
Adipose tissue triglyceride lipase	$Triglyceride \rightarrow$ $free\ fatty\ acids + glycerol$	phosphorylated

glucose utilization by heart muscle. The enzyme that catalyzes this reaction, pyruvate dehydrogenase (PDH), is a large multienzyme complex which consists of three enzymes; pyruvate decarboxylase (E_1), dihydrolipoyl trans-acetylase (E_2), and dihydrolipoyl dehydrogenase (E_3). The reaction sequence which decarboxylates pyruvate to acetyl-CoA proceeds via 5 steps as follows:

$$CH_3COCOOH + [TPP]-E_1 \rightleftharpoons [CH_3CHOH-TPP]-E_1 + CO_2 \tag{1}$$
$$[CH_3CHOH-TPP]-E_1 + [LipS_2]-E_2 \rightleftharpoons [CH_3CO-S-LipSH]-E_2 + [TPP]-E_1 \tag{2}$$
$$[CH_3CO-S-LipSH]-E_2 + CoASH \rightleftharpoons CH_3CO-S-CoA + [Lip(SH)_2]-E_2 \tag{3}$$
$$[Lip(SH)_2]-E_2 + FAD-E_3 \rightleftharpoons [LipS_2]-E_2 + FADH_2-E_3 \tag{4}$$
$$FADH_2-E_3 + NAD^+ \rightleftharpoons FAD-E_3 + NADH + H^+ \tag{5}$$

Sum: $CH_3COCOOH + CoASH + NAD^+ \rightleftharpoons CH_3COSCoA + CO_2 + NADH + H^+$

where TPP is thiamine pyrophosphate and $Lip(SH)_2$ and $LipS_2$ are the reduced and oxidized forms of lipoic acid (amide).

The regulation of this reaction sequence is complex. Acetyl-CoA and NADH are inhibitors of pyruvate dehydrogenase competing with CoASH and NAD^+, respectively, for binding sites on the enzyme. In addition, as the concentration of acetyl-CoA increases, the concentration of CoASH decreases, thereby removing one of the reactants in step 3. In addition to this, the pyruvate dehydrogenase complex is regulated by phosphorylation-–dephosphorylation reactions. To carry this out, there is a PDH kinase which catalyzes the phosphorylation of PDH with $ATP\cdot Mg^{2+}$ and inactivates it. The reversal is carried out by a PDH phosphate phosphatase which dephosphorylates the enzyme and activates it. This regulatory step is made even more complex since the kinase and phosphatase reactions are modulated by cellular chemicals. Thus, the kinase reaction is inhibited by pyruvate and ADP. Mg^{2+} and Ca^{2+} activate phosphatase and inhibit kinase which, of course, activates the PDH enzyme.

The regulatory mechanisms that control carbohydrate utilization in heart muscle help to explain many of the observations with perfused heart preparations. In the normal heart with a plentiful oxygen supply, fatty acids and ketone bodies are utilized in preference to carbohydrates. This goes with the finding that in starvation or diabetes where fat mobilization, fatty acid oxidation, and ketone body synthesis are accelerated, glycolysis and pyruvate oxidation in heart muscle are inhibited. Conversely, there is preferential carbohydrate utilization in heart muscle under conditions of increased work load, anoxia, or with the addition of glucose and insulin (Neely et al., 1967, 1970; Mansour, 1963; Morgan and Parmeggiani, 1964; Morgan et al., 1961; Randle et al., 1966; Regen et al., 1964). The finding that carbohydrate utilization is inhibited when fat is oxidized has been explained on the basis of increased acetyl-CoA/CoASH and NADH/NAD^+ ratios (Garland and Randle, 1964; Randle et al., 1966, 1970). In addition, when fatty acids and ketone bodies are oxidized in heart muscle, citrate has been found to accumulate. This in turn could lead to inhibited glycolysis through inhibition of the phosphofructokinase reaction, as described above. At any rate, there seems little doubt that fatty acids, as well as ketone bodies formed during fatty acid oxidation, are of vital importance in the normal metabolism of heart muscle.

II. THE UTILIZATION OF FAT BY HEART MUSCLE

Heart muscle, when functioning normally in the presence of an adequate oxygen supply, utilizes lipids as the preferred energy source (Neely et al., 1972; Opie, 1968, 1969). The free fatty acids (FFA) used as substrate are derived from three sources: (1) from circulating lipoproteins (VLDL) which contain triglycerides, (2) from FFA derived from adipose tissue and are albumin bound in the bloodstream, and (3) by hydrolysis of endogenous triglycerides (TG). Clearly, the lipase reaction is the first reaction of importance in regulating the supply of FFA to the heart. In heart muscle there probably are present a number of different lipases. Lipoprotein lipase, which hydrolyzes the triglycerides present in chylomicrons and VLDL, is located in the endothelial lining of coronary blood vessels and is almost completely released by an infusion of heparin. In addition to utilizing exogenous lipids, heart muscle utilizes endogenous triglycerides. There is an interaction with the rate of FFA influx to the heart from adipose tissue and the rate of endogenous lipolysis, with the former having a sparing action on the latter (Crass, 1972; Crass et al., 1975). Endogenous myocardial lipids are hydrolyzed by a hormone sensitive triglyceride lipase which may well be of lysosomal origin (Wang et al., 1977). It was observed that chloroquine, a lysosomal inhibitor, inhibited endogenous triglyceride lipolysis (Hülsmann and Stam, 1978; Stam et al., 1980).

Heart muscle therefore has at least two triglyceride lipases: one with a pH optimum of pH~5 and probably of lysosomal origin, the other with a pH optimum of pH~7.5 and localized in the soluble as well as particulate fraction (Severson, 1979). Triglyceride lipase appears to be under hormonal control in perfused heart preparations. Catecholamines produce a rise in cyclic AMP which is followed by an increased release of glycerol and a decrease in endogenous triglycerides. The cyclic AMP analogue, dibutyryl cyclic AMP also decreases the concentration of myocardial triglycerides (Christian et al., 1969; Crass et al., 1975; Gartner and Vahouny, 1973; Mayer, 1974). In adipose tissue, triglyceride lipase is phosphorylated to the active form by a protein kinase and dephosphorylated to the inactive state by a phosphoprotein phosphatase. The protein kinase itself is under regulation by modulators and requires cyclic AMP for activation (for a review see Severson, 1979). It has been suggested that the myocardial triglyceride lipase may be regulated in a similar manner (Jesmok et al., 1977). Because the myocardium contains several lipases it is difficult to study the precise regulatory mechanism of each lipase, nevertheless, the general interactions and responses to different modulators are understood.

III. ALTERATIONS IN CARDIAC METABOLISMS WHEN RATS ARE FED DIETS THAT CONTAIN ERUCIC ACID

Since the normal heart depends primarily on fatty acids as a source of metabolic fuel, it is not surprising that this organ should be exquisitely sensitive to changes in the fat consumption of the diet.

When HEAR (high erucic acid rapeseed) oil or trierucin is fed to rats there is a large rise in endogenous cardiac TG levels. The figures given by Stam et al. (1980) are representative. Within 3 days of feeding a HEAR oil or trierucin containing diet, the TG content rose from the normal of 33 μmoles/ g myocardial protein to 300 and 716 μmoles/g, respectively. The fatty acid composition of the accumulated cardiac TG resembles the composition of the oil fed (Kramer et al., 1979). The intramyocardial lipidosis is probably due to a combination of factors, including increased erucic acid transport to the heart via VLDL triglycerides (Thomassen et al., 1979), the slow oxidation rate of erucic acid in the myocardium, and the likelihood that erucic acid inhibits the rate of β-oxidation of other fatty acids (Cheng and Pande, 1975; Christiansen et al., 1977; Christophersen and Bremer, 1972; Heijkenskjöld and Ernster, 1975; Swarttouw, 1974; Vasdev and Kako, 1976). Free fatty acids are the primary source of substrate for heart muscle and therefore have a fast turnover rate. Clearly, any interference with the normal oxidation rates of FFA, no matter how slight, will upset the normal steady state flux and increase the concentration of lipids in the heart. Most enzymes of β-oxidation are less active with the very long chain monoenic fatty acids, such as erucic and gondoic acids, than with the more common shorter chain fatty acids. It may be worthwhile to review how some of the key reactions in myocardial fat oxidation are affected by erucic acid.

Myocardial lipase once isolated is almost totally inactive toward trierucin (Kramer et al., 1973; Mersel et al., 1979) and this has been suggested as a cause for the accumulation of erucic acid containing lipids in heart muscle. This, however, ignores the well-documented observation that in perfused hearts, isolated from rats fed HEAR oil or trierucin, myocardial lipoprotein lipase and the hormone sensitive triglyceride lipase activities are greatly increased (Jansen *et al.*, 1975; Hülsmann *et al.*, 1979). Myocardial lipase was more active in the rats fed HEAR oil diets for 3 or 10 days than in rats fed the stock diet for 3 and 10 days. The difference in lipase activity became more pronounced when glucagon or norepinephrine was added to the heart perfusate (Stam *et al.*, 1980). Since FFA inhibit lipase activity (Crass *et al.*, 1975), it may be that this was overlooked in experiments in which heart homogenates were used which would, of course, contain fairly high levels of FFA. Furthermore, although myocardial lipases are inactive toward trierucin, they most definitely are active toward the endogenous erucic acid containing triglycerides since erucic acid is readily released into the effluent from perfused rat hearts (Stam *et al.*, 1980). It is, therefore, unlikely that the low activity of myocardial lipases for trierucin has much relevance in early myocardial lipidosis.

A more probable contributory factor to myocardial lipidosis when erucic acid is included in the diet lies in the fact that this fatty acid is poorly oxidized by heart mitochondria (for a review, see Sauer and Kramer, 1980). A number of enzymes involved in β-oxidation of fatty acids are inhibited by, or have low activity for, erucic acid. These include long chain fatty acid thiokinase (EC 6.2.1.3), which activates fatty acids to their CoA derivates and is relatively inactive with erucic acid (Kramer *et al.*, 1973; Swarttouw, 1974; Cheng and Pande, 1975), and carnitine acyltransferase (EC 2.3.1.21), which transfers the long chain acyl groups between CoA and carnitine also is less active with erucic acid (Christophersen and Bremer, 1972; Swarttouw, 1974; Cheng and Pande, 1975). Similarly, the long chain acyl-CoA dehydrogenase which dehydrogenates the acyl-CoA chain in the C-2 and C-3 position during β-oxidation also shows reduced activity with erucyl-CoA (Korsrud *et al.*, 1977). Not only are individual enzymes of fatty acid oxidation relatively inactive with erucic acid as substrate, but the overall β-oxidation rates are reduced. There is complete agreement among investigators that oxidation rates with heart mitochondria are much lower with erucylcarnitine as substrate than when the carnitine esters of shorter chain length fatty acids are used as substrates (Christophersen and Bremer, 1972; Kramer *et al.*, 1973; Swarttouw, 1974; Cheng and Pande, 1975).

The mechanism whereby erucic acid inhibits oxidation rates of fatty acids in heart mitochondria has received considerable attention. Erucic acid, as its carnitine ester, may be slow in transferring across the inner mitochondrial membrane via carnitine acyl transferase to reach the mitochondrial matrix space where β-oxidation takes place. This would impair erucic acid oxidation. Evidence for delayed erucylcarnitine entry into heart mitochondria has been presented (Christophersen and Bremer, 1972). What is not explained is how erucylcarnitine inhibits the oxidation of other acylcarnitine esters such as palmitylcarnitine. Erucylcarnitine appears to inhibit palmitylcarnitine oxidation in heart mitochondria while at the same time depressing the overall rate of O_2 uptake. The mechanism for this does not appear to be through an inhibition of citric acid cycle activity (Christophersen and Christiansen, 1975) but rather through a removal of free CoA (as erucyl-CoA) in the mitochondrial matrix which thereby causes a CoA deficit. Alternately, Christophersen and Christiansen (1975) propose that erucyl-CoA may be a competitive inhibitor for the long chain CoA dehydrogenase. With this enzyme inhibited, the oxidation rates of other acyl-CoA esters in heart mitochondria would then be decreased.

It is relevant to point out that this inhibition of fatty acid oxidation is not an exclusive property of erucic acid, but rather a property common to very long chain monoenoic fatty acids. Thus, Christiansen et al. (1977) showed that the carnitine esters of erucic (cis 22:1 n-9), cetoleic (cis 22:1 n-11), brassidic (trans 22:1 n-9), gondoic (cis 20:1 n-9), and oleic (cis 18:1 n-9) acids progressively inhibit rates of β-oxidation and rates of intramitochondrial CoA acylation as the chain length increases from C_{18} to C_{22}. The important difference in these acids is that while oleylcarnitine competes with and inhibits palmitylcarnitine oxidation, the overall respiratory rate is normal, since oleylcarnitine is itself readily oxidized. This is not the case with the 22:1 isomers. Thus, erucyl- and cetoleylcarnitine also inhibit palmitylcarnitine oxidation but in addition are by themselves poorly oxidized substrates; therefore, the overall rate of mitochondrial respiration is inhibited.

Some additional contributory factors to myocardial lipidosis are worth considering. Free fatty acids are transported in blood as albumin bound complexes. Shafrir et al. (1965) have shown that fatty acid affinity for albumin decreases with increasing chain length and that erucic acid has about one-third the affinity for albumin as does palmitate. This may result in an increased fatty acid uptake by tissues when diets rich in erucic acid are fed to experimental animals (Gumpen and Norum, 1973). In addition it is possible that the rate of triglyceride biosynthesis is more rapid in rats fed an erucic acid containing diet. Results have been presented which indicate that during the first week on the test diets the rate of triglyceride synthesis from glycerol 3-phosphate in rats fed an erucic acid containing diet was 64% greater than that in rats fed a corn oil diet (Hung and Holub, 1977).

IV. THE ROLE OF THE PEROXISOMAL SYSTEM IN MYOCARDIAL LIPIDOSIS

Perhaps equally as challenging as the question of why erucic acid causes myocardial lipidosis is the question of why the myocardial lipidosis abates and almost disappears even when experimental animals are maintained on exactly the same erucic acid containing diet that caused the lipidosis in the first place.

Originally, it had been postulated that the enzymes of β-oxidation in the heart in some way might "adapt" to erucic acid oxidation, thereby reducing myocardial lipidosis. From experiments with isolated heart mitochondria there was no evidence of improved erucic acid oxidation rates with prolonged feeding of HEAR oil (Kramer et al., 1973). Similarly, perfused hearts from rats fed HEAR oil or sunflower oil had essentially the same oxidation rates for erucic (and oleic) acids (Stam et al., 1980). There was no indication that HEAR oil feeding improved erucic acid oxidation rates. Recently, it has been proposed that enhanced lipolysis, which occurs during prolonged erucic acid feeding, may contribute to the reduction of myocardial triglycerides (Stam et al., 1980). More important, however, may be the induction of hepatic and myocardial peroxisomal oxidation. A brief summary of this system may be in order.

Lazarow and de Duve (1976) purified peroxisomes from rat liver and found that these were able to oxidize long chain acyl-CoA esters apparently via a β-oxidation mechanism (Lazarow, 1978). This peroxisomal system differs distinctly from the well-characterized β-oxidation system of mitochondria in a number of ways. The first dehydrogenation is carried out by an FAD dependent fatty acyl-CoA oxidase (Lazarow, 1978; Osumi and Hashimoto, 1978) and involves the reduction of O_2 to H_2O_2. The enoyl-CoA hydratase and 3-hydroxyacyl-CoA dehydrogenase activities are carried out by a multifunctional protein (Osumi and Hashimoto, 1979, 1980). Also, the peroxisomal β-ketothiolase has different chromatographic properties and chain length specificity from the mitochondrial enzyme (Krahling and Tolbert, 1981).

The functional properties of this system differ from that in mitochondria in the following respects. Peroxisomal oxidation is not coupled to phosphorylation, i.e., no ATP is generated and peroxisomal oxidation is not inhibited by cyanide (Lazarow and de Duve, 1976). Whereas mitochondrial β-oxidation goes to completion, this is not so with the peroxisomal system, which appears to be inactive with fatty acyl-CoA esters shorter than C_8 (Inestrosa et al., 1979). Peroxisomal β-oxidation is most active with long chain (C_{10}–C_{22}) fatty acyl-CoA substrates and is also very active with erucic acid. Peroxisomes from rat liver contain two carnitine acyltransferases and it has been suggested that their function is to convert acetyl-CoA and octanoyl-CoA to their corresponding carnitine esters for passive diffusion out of these organelles for further metabolism elsewhere in the liver cell (Tolbert, 1981).

That peroxisomal β-oxidation probably played an important role in the metabolism of erucic and cetoleic acids was recognized quite early by the investigators who were concerned about the adverse nutritional effects of these long chain monoenoic fatty acids.

Evidence for the incomplete β-oxidation of erucic and cetoleic acids, a process commonly referred to as chain shortening, has been presented by different workers. In isolated rat hepatocytes, erucic acid (22:1 *n*-9) is chain shortened to gondoic (20:1 *n*-9), oleic (18:1 *n*-9), and hexadecenoic (16:1 *n*-9) acids, presumably by three cycles of peroxisomal β-oxidation (Norseth and Christophersen, 1978). Chain shortening of these long chain monoenes is stimulated by incorporating 0.3% clofibrate [2-(4-chlorophenoxy)-2-methylpropionic acid ethyl ester] into the rat diet for several days (Christiansen, 1978). Clofibrate is known to stimulate peroxisomal proliferation in the liver. Hepatic peroxisomes, as indicated by the presence of marker enzymes (catalase and urate oxidase), also proliferate when rats are fed diets enriched with either HEAR oil or marine oil (Christiansen *et al.*, 1979a). In an elegant series of experiments, Neat *et al.* (1981) isolated peroxisomes from rat liver by Percoll density gradient centrifugation and showed that peroxisomal β-oxidation was induced by feeding the rats diets that contained marine or HEAR oils. Soybean oil also induced peroxisomal activity but to a lesser extent than the 22:1 containing oils. These authors calculated that, in liver, induced peroxisomal β-oxidation rates can approach 30% of that in mitochondria. These results, which showed that rat liver contained an inducible peroxisomal β-oxidation system, were helpful in explaining the transient myocardial lipidosis observed in rats fed HEAR oil diets. The proportion of dietary fat which passes through the liver (estimated to be at least 50%) would therefore be exposed to the peroxisomal chain shortening process and to that extent, at least, reduce the influx of long chain monoenoic fatty acids to the heart. Indeed, it has been shown that in the VLDL-triglycerides, erucic acid is chain shortened and decreased in concentration after a 3 week dietary induction period (Christiansen *et al.*, 1979a).

Although peroxisomal β-oxidation is well characterized in liver, the assumption cannot be made that this process also occurs in other organs such as the heart (Tolbert, 1981). Nevertheless, there is evidence to suggest that chain shortening may also occur in rat heart. For example, cultured rat heart cells when incubated with [1-^{14}C]erucic acid showed an almost linear release of $^{14}CO_2$, but if [14-^{14}C]erucic acid was used there was a lag period of over 12 hours (Pinson and Padieu, 1974). Subcellular particles isolated from rat heart chain shortened [14-^{14}C]-erucic acid to 20:1, 18:1, and 16:1 (Clouet and Bezard, 1978). Chain shortening has also been observed in perfused heart preparations. With hearts from rats fed either HEAR oil or a marine oil containing diets, the chain shortening process was severalfold more active than from control rats fed a peanut oil diet (Norseth, 1979). The products were primarily 18:1 and 20:1. Since the peroxisomal enzyme catalase increased by 85%, the author concluded that 22:1 in the diet induced peroxisomal proliferation in the rat heart. Clofibrate also appears to stimulate peroxisomal activity in the rat heart. Thus, catalase showed a 60% increase in activity with clofibrate added to the diet and the rate of erucic acid oxidation in perfused rat hearts was doubled (Norseth, 1980). Not surprisingly, when clofibrate is fed together with a HEAR oil containing diet to young rats, the fatty infiltration into heart muscle is significantly decreased (Christiansen et al., 1978, 1979b). Recently, evidence has been presented that peroxisomal β-oxidation is present also in human liver and biceps muscle (Bronfman et al., 1979; Shumate and Choksi, 1981).

The viewpoint that the clofibrate effect is entirely through an enhanced peroxisomal β-oxidation has been challenged by Pande and Parvin (1980). These authors point out that the hypotriglyceridemic effect of clofibrate is seen equally in male and female rats although the peroxisomal proliferation is more pronounced in the livers of male rats (Svoboda et al., 1967). Instead, the authors suggest that, since clofibrate triples the carnitine content of the liver, its effect may be due to enhanced, carnitine dependent, mitochondrial oxidation of fatty acids.

V. CARDIAC RESPIRATORY RATES AND OXIDATIVE PHOSPHORYLATION IN RATS FED HEAR OIL CONTAINING DIETS

It has been established quite clearly that erucic acid, as either the carnitine or CoA ester is poorly oxidized by isolated rat heart mitochondria and it is quite likely that the erucyl-CoA (or carnitine) ester may also interfere with the oxidation of shorter chain fatty acyl-CoA esters. There is, however, a current controversy as to whether heart mitochondria isolated from young rats fed a HEAR oil diet are "functionally" impaired, i.e., incapable of normal function even in the absence of erucic acid in the incubation mixture.

Heijkenskjöld and Ernster (1975) observed moderate inhibition with glutamate and succinate oxidation rates in heart mitochondria from rats that received erucic acid. Palmitylcarnitine oxidation, however, was significantly depressed. These authors (in collaboration with B. Chance) were able to show, by use of surface fluorimetry, that erucic acid interfered with oleic acid oxidation in intact, perfused rat hearts. Similar results were obtained by Hsu and Kummerow (1977) and Clandinin (1978). These authors reported that heart mitochondria isolated from rats fed HEAR oil diets or diets with hydrogenated fat had decreased O_2 uptake and ATP synthesis with a variety of substrates, which included oleyl-, erucyl-, and elaidylcarnitine, pyruvate, pyruvate plus malate, and 2-oxoglutarate plus malate. The control rats received diets that contained either corn oil (Hsu and Kummerow, 1977) or soybean oil (Clandinin, 1978). Houtsmuller et al. (1970) reported a two- to threefold decrease in oxygen uptake and ATP synthesis in heart mitochondria from rats fed a HEAR oil diet when compared with the activity of these organelles from rats fed a sunflower seed oil diet. It is of course tempting to speculate that depressed respiratory rates, depressed rates of ATP synthesis and lowered ADP/O ratios (molecules of ADP converted to ATP per atom O consumed) are contributory factors to the transient myocardial lipidosis that follows when erucic acid is fed (Clandinin, 1978).

There is, however, no general agreement that heart mitochondria isolated from rats during the period of maximal lipidosis are, in fact, functionally impaired. Cheng and Pande (1975) compared two diets: one with 50 calorie % HEAR oil (34%, 22:1), and the other with an equal amount of corn oil. Heart mitochondria isolated from rats fed either of these two diets were equally active in substrate oxidation rates (pyruvate, glutamate, palmityl-, or erucyl-CoA) as well as ADP/O ratios and respiratory control. Dow-Walsh *et al.* (1975) obtained results similar to that of Cheng and Pande (1975) and, in addition, in a critical appraisal of earlier work by Houtsmuller *et al.* (1970), expressed concern about the rather low activity of rat heart mitochondria used in these earlier studies. Dow-Walsh *et al.* (1975) observed what appeared to be an increased fragility of heart mitochondria isolated from the HEAR oil fed rats and suggested that the mitochondrial membranes might be weakened because of an altered fatty acid profile of membrane phospholipids (Blomstrand and Svensson, 1974). From studies with intact rat hearts, it appears that myocardial adenine nucleotide concentrations (ATP, ADP, and AMP) are not altered in rats fed diets that contain either olive oil or HEAR oil (Beare-Rogers and Gordon, 1976).

It is difficult at this time to resolve the question of whether *in vivo* myocardial oxidation rates and ATP synthesis are impaired in rats fed high fat, erucic acid containing diets for short periods of time. It is generally accepted that erucyl-CoA (or -carnitine) is poorly oxidized by isolated heart mitochondria and interferes with β-oxidation rates of other acyl-CoA (or -carnitine) esters. It is less clear to what extent this effect is present *in vivo* when rats are fed a diet with HEAR oil or marine oil. Evidence has been presented (Norseth, 1979) that in perfused hearts fatty acid oxidation rates are not significantly different when rats are fed diets that contain marine oil, HEAR oil, or peanut oil. The only observed difference was a tendency for increased di- and triglyceride formation from [U-^{14}C]palmitate in rats fed the diets with 22:1. Similarly, Stam *et al.* (1980) found no differences in erucic acid oxidation rates by perfused heart preparations from rats fed diets with either HEAR oil or sunflower oil added. This argues against a mitochondrial "adaptation" process for erucic acid oxidation and instead supports the idea that heart mitochondria remain functionally unchanged by different dietary fats. Most probably myocardial lipidosis results from an increased rate of influx of long chain monoenoic fatty acids that are of dietary origin. The prolonged feeding of diets that contain these fatty acids in turn induces the hepatic peroxisomal β-oxidation system which, perhaps assisted by the myocardial peroxisomal system, decreases the rate of long chain monoenoic fatty acid influx to the heart mitochondria. This then reduces the extent and severity of myocardial lipidosis.

It is not clear whether ATP production and respiratory control indices and ADP/O ratios are altered in heart mitochondria from rats that have received high levels of 22:1 monoenes. In assessing published data it becomes obvious that a variety of techniques are used by different workers in the isolation of heart mitochondria and that subsequent activities may vary considerably. In the isolation of mitochondria from heart muscle, particularly when lipidosis is present, the possibility of free fatty acid release is always a cause of concern, since free fatty acids are powerful uncouplers of oxidative phosphorylation. Generally, albumin is added to the incubation to bind any fatty acids that may be present but it is not known if this is completely effective. The suggestion has been made (Chan and Higgins, 1978) that the free fatty acid induced uncoupling in mitochondria on aging (i.e., liver mitochondria) may result from increased phospholipase A activity which increases the free fatty acid concentration. Additional studies should be carried out to compare ATP levels in intact hearts from rats fed the appropriate dietary fat, either *in vivo* or in perfused heart preparations using ^{31}P NMR as has already been described (Matthews *et al.*, 1981). This would seem to be the only practical means of circumventing the free fatty acid release during heart tissue manipulations and the ensuing mitochondrial damage.

VI. INTERSPECIES DIFFERENCES IN CARDIAC LIPIDS

A. Cardiac Lipids

The lipid content in pig hearts normally is approximately 2%, which is similar to that observed in rat hearts. Little is known about the cardiac lipid content of monkeys fed a low fat control diet. These species respond differently to experimental diets which contain high levels of fat and are rich in docosenoic acid. In the rat and pig studies (Fig. 1) HEAR oils were fed which contained erucic acid (22:1 *n*-9). In some of the monkey studies fish oils were fed which contained cetoleic acid (22:1 *n*-11) as the main docosenoic acid isomer. The results from both 22:1 isomers are combined because of their similarity in response.

The total lipid content in the rat heart increases from about 2 to 3% when rats are switched to high fat diets, but the cardiac triglyceride content remains unchanged, except when diets that contain 22:1 are fed (Fig. 1). These long chain monoenoic fatty acids cause a rapid accumulation of cardiac triglycerides within the first week which then decline on continued feeding. However, after 16 weeks the level of cardiac triglycerides still remains above that found in rats fed control oils (Fig. 1).

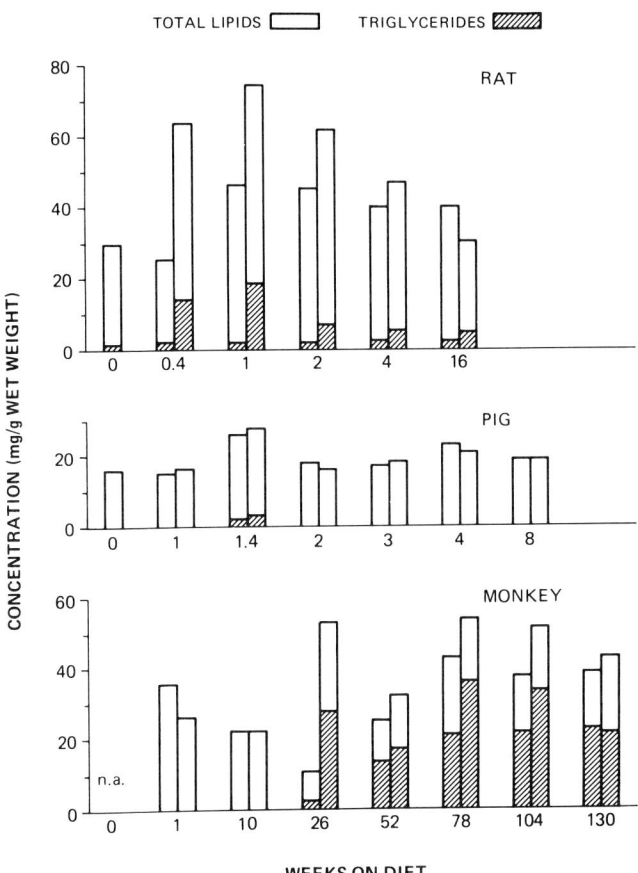

Fig. 1. The concentration (mg/g wet weight) of the total cardiac lipids and the cardiac triglycerides of rats, pigs, and monkeys fed a low fat control diet (time 0) and diets to which a control oil (first bar) or a docosenoic acid containing oil (second bar) was added. The portion of triglycerides in the total lipids are indicated by a hatched bar wherever this information is available. Source of data: rat (Kramer and Hulan, 1978; Kramer et al., 1979); pig, 1.4 weeks (Opstvedt et al., 1979), all other values (Kramer et al., 1975); and monkey, 1 and 10 weeks (Beare-Rogers and Nera, 1972), all other values (Ackman, 1980). Erucic acid was the docosenoic acid in all studies except the monkey data from Ackman (1980) who fed partially hydrogenated fish oil containing mainly cetoleic acid.

On the other hand, the total cardiac lipid content of the pig is not affected by high fat diets, even when the diets contain 22:1. The cardiac triglycerides also remain unaffected by diets with or without 22:1 (Fig. 1, time period 1.4 weeks for pigs).

The cardiac lipids of monkeys appear to increase when they are fed diets rich in fat. This is evident from the results of the total heart lipids of monkeys fed the lard/corn oil (3/1) control diet. The increase in triglycerides appears to be more rapid and more pronounced in monkeys fed diets which contain 22:1. However, after 2.5 years the cardiac lipid content was the same in the two experimental groups (Fig. 1). Unlike the rat or pig, the monkey accumulates high levels of cardiac triglycerides on any diet that is high in fat. In fact, about 50% of the cardiac lipids in monkeys are triglycerides.

These results appear to indicate that cardiac lipids of the pig do not respond to the feeding of high fat diets with or without docosenoic acids. The rat, on the other hand, after the initial acute myocardial lipidosis, appears to adapt after about 1 week to dietary docosenoic acids. The monkey does not appear to adapt well to high fat diets; however, docosenoic acids do not appear to have any additional effects on total heart lipids or heart triglycerides.

B. Fatty Acid Changes

Dietary docosenoic acid (22:1) is incorporated into the cardiac lipids of rats, pigs, and monkeys. When diets containing similar concentrations of 22:1 are fed to these three species, the rat accumulates the greatest amount

TABLE II

Incorporation of Docosenoic Acid into Total Heart Lipids and the Major Lipid Classes in Rats, Pigs, and Monkey

Description	Rat[a]		Pig[b]	Monkey[c]
	During lipidosis	After lipidosis		
Total heart lipids (mg/g wet wt)	74	30	19	43
% 22:1 in:				
Total heart lipids	19.8	6.4	2.2	16.4
Triglycerides	27.4	13.7	2.7	16.4
Phosphatidylcholine		1.8	2.5	
Phosphatidylethanolamine	5.1[d,e]	2.3	11.1	8.3[e]
Diphosphatidylglycerol		3.0	1.4	
Sphingomyelin		5.5	2.0	

[a] Rats were fed HEAR oil (25.5%, 22:1) at 20% by weight of the diet for 1 or 16 weeks (Kramer et al., 1979).

[b] Pigs were fed HEAR oil (22.3%, 22:1) at 20% by weight of the diet for 8 weeks (Kramer et al., 1975). Results of lipid classes from Kramer and Hulan (1977).

[c] Monkeys were fed partially hydrogenated fish oil (20.8%, 22:1) at 25% by weight of the diet for 130 weeks (Ackman, 1980).

[d] Kramer and Hulan (1978).

[e] This number represents the total phospholipids.

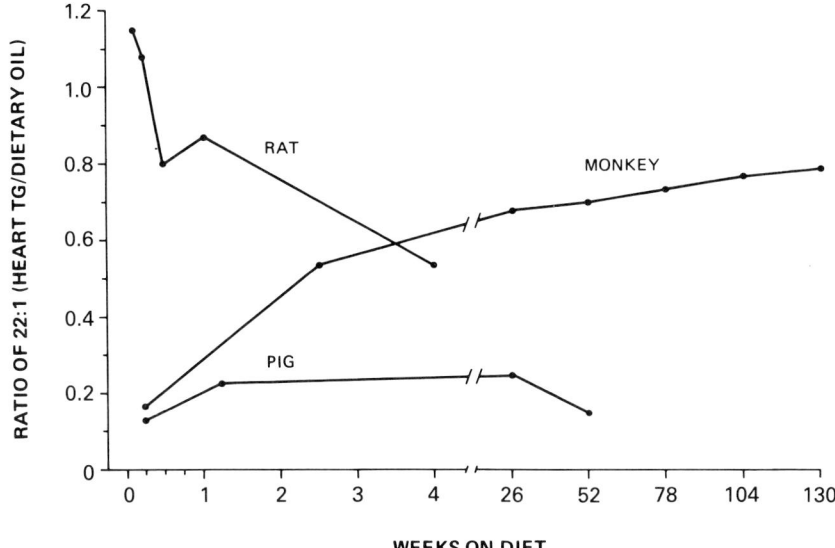

Fig. 2. The concentration of 22:1 in the cardiac triglycerides of rats, pigs, and monkeys is expressed as a ratio to the concentration of 22:1 in the dietary oil fed. Source of data: rat (Kramer and Hulan, 1978; Kramer et al., 1979); pig (Svaar et al., 1980); and monkey, 1 and 10 weeks cardiac total lipids instead of cardiac triglycerides (Beare-Rogers and Nera, 1972), all other values (Ackman, 1980). Erucic acid was the docosenoic acid in all studies except the data from Ackman (1980) who fed partially hydrogenated fish oil containing mainly cetoleic acid to monkeys for 26 to 130 weeks.

of this acid during the period of myocardial lipidosis (Table II). Next to the rat, the monkey appears to incorporate the highest level of 22:1 in the heart. The pig contains the least amount of this acid in the heart. In the rat heart, the highest level of 22:1 occurs in the triglyceride fraction; much less is found in phospholipids. In monkey hearts, the triglycerides contain higher levels of 22:1 than the phospholipids, but the difference is less pronounced than in the rat. On the other hand, in pig hearts, phosphatidylethanolamine shows the highest incorporation of 22:1, while all the other major lipid classes have a similar 22:1 content (Table II).

The change in the 22:1 content with time on diet for the three species is shown in Fig. 2. The concentration of 22:1 in the cardiac triglycerides is compared to the concentration of this acid in the dietary oil and expressed as a ratio. As evident from these results, the 22:1 concentration of the rat heart triglycerides is the same as that of the dietary oil during the first week on the experimental diet. The ratio of 22:1 then decreases on continued feeding to a ratio of approximately 0.5 after 16 weeks. In the pig, the ratio was 0.2 throughout the 1 year study. The monkey shows a tendency to accumulate 22:1 in the cardiac triglyceride fraction. Erucic acid (22:1 *n*-9)

TABLE III

Concentration of C_{22} n-3 Polyunsaturated Fatty Acids in the Cardiac Lipids of Rat, Pig, and Monkey

Description	Rat[a]	Pig[b]	Monkey[c]
Time on diet (weeks)	1.6	8	17.1
% 18:3 in dietary oil	6.2	5.9	7.6
ΣC_{22} n-3 in total lipids[d]	2.8	Trace	0.5
ΣC_{22} n-3 in PE[e]	5.5	Trace	n.a.

[a] Kramer et al. (1979).
[b] Kramer et al. (1975) and Kramer and Hulan (1977).
[c] Ackman and Loew (1977).
[d] ΣC_{22} n-3 refers to the sum of 22:5 n-3 and 22:6 n-3.
[e] PE, phosphatidylethanolamine.

and cetoleic acid (22:1 n-11) essentially have the same effect. The ratio for an erucic acid containing diet has been reported as 0.62 (Ackman and Loew, 1977) while that for a cetoleic acid containing diet has been reported as 0.51 (Ackman and Loew, 1977) and 0.68 (Ackman, 1980). The results with the monkey (Fig. 2) have been obtained with either erucic or cetoleic acids. What is not apparent from Fig. 2 is that the ratio of 22:1 in the heart TG to that in the dietary oil decreases disproportionately as the 22:1 level of the oil is lowered. Thus, for monkeys kept 18 months on experiment, as the dietary docosenoic acid concentration was decreased from 5.2 (Ackman, 1980) to 4.2 and 2.1% (Shenolikar and Tilak, 1980), the ratio dropped from 0.69 to 0.35 and 0.23, respectively.

There also appears to be an interspecies difference in the metabolism of linolenic acid. When rats are fed diets which contain 18:3, triglycerides isolated from heart do not contain appreciable quantities of 18:3, but instead accumulate C_{22} polyunsaturated fatty acids (Table III). In these rat hearts, 22:5 n-3 and 22:6 n-3 accumulate, and in particular in the phosphatidylethanolamine fraction (Kramer, 1980; Kramer et al., 1979). Interestingly, when diets containing the same amount of 18:3 are fed to pigs or monkeys (Table III) almost no 22:5 n-3 and 22:6 n-3 accumulate in the heart. This suggests that there is a significant interspecies difference in 18:3 metabolism.

VII. SOME INTERSPECIES DIFFERENCES IN MYOCARDIAL METABOLISM

The early and acute triglyceride and erucic acid accumulation in heart muscle is a metabolic aberration that is found in rats and other rodents but is not present in pigs or some species of monkeys (Fig. 1). This raises two points of interest. First, why are some species resistant to this acute myocardial lipidosis after eating diets that contain high levels of erucic acid, and second, to what extent is the investigator justified in extrapolating results of erucic acid experiments from one species to another.

It has been recognized for some years that there are species differences in some of the enzymes of fatty acid β-oxidation. Thus, physiochemical and chain length specificity differences have been found in the long chain acyl-coenzyme A dehydrogenase isolated from pig or beef tissue (Hall *et al.*, 1976). Some differences in chain length specificity also appear to be present with the enzyme isolated from sheep or pig liver (Beinert, 1963). Whereas pig liver contained this enzyme with an acyl chain length preference of C_4 to C_{16} and C_6 to C_{16}, the enzyme from sheep liver was more reactive with the shorter acyl chain lengths.

Osmundsen and Bremer (1978) also found interspecies differences as to optimum acyl-carnitine chain lengths when β-oxidation rates were measured with heart mitochondria. In their study, comparisons were made with heart mitochondria from rats, mice, rabbits, cats, frogs, and monkeys (*Cercopithecus aetiops*). These authors carefully distinguished between the polarographic assay method which measures the sum of β-oxidation plus tricarboxylic acid cycle (TCA) oxidation rates, and a spectrophotometric assay with $Fe(CN)_6^{3-}$ as electron acceptor, which selectively measures β-oxidation rates alone. This allowed them to make some interesting observations. In the mouse, rat, and cat, as the fatty acid chain length increased from C_{18} to C_{22} β-oxidation was, as expected, progressively depressed. Surprisingly, however, there was a proportionately larger depression of TCA cycle respiration which exaggerated the *total* respiratory depression. This was not so in the other species, i.e., the pig, rabbit, and sheep. In heart mitochondria from these species, as the β-oxidation rate decreased, the decrease in TCA cycle respiration remained proportional, so that overall respiratory depression was less severe than in the rat, mouse, and cat. This is an important finding and may well help to explain why pigs are relatively resistant to severe myocardial lipidosis when fed diets that are high in erucic acid, while rodents, such as the rat, are extremely susceptible. Unfortunately, monkeys were not included in this experiment. It would have been of considerable interest to determine whether heart mitochondria from monkey show TCA cycle inhibition while metabolizing long chain monoenoic fatty acids. These authors, however, did observe that monkey heart mitochondria exhibit a fatty acid chain length profile for β-oxidation rates that is similar to that of pig heart mitochondria, i.e., there appeared to be no well-defined chain length preference. By contrast, mitochondria from rat, ox, and frog hearts showed a sharply defined chain length preference. In a comparison study with heart mitochondria isolated from rats and pigs, Buddecke *et al.*

(1976) observed that pig heart mitochondria were better able to metabolize erucic acid than were the mitochondria from rat heart. When corrected for relative differences in oxidation rates with oleic acid set at 100%, mitochondria from pig heart had threefold greater erucic acid oxidation rates than did the mitochondria from rat heart. Forsyth *et al.* (1977) compared carnitine acyl transferase activities from rat and monkey (*M. fascicularis*) heart mitochondria. They found that cetoleyl-CoA inhibited the rat heart enzyme but was a good substrate for the enzyme isolated from monkey hearts. They concluded from this that docosenoic acids may be metabolized much more rapidly in the monkey heart than in the rat heart.

It was shown that the human heart mitochondria, incubated within 15 hr after death, metabolized [14-^{14}C]erucic acid more slowly than [10-^{14}C]oleic acid as measured both as $^{14}CO_2$ release and as recovery of perchloric acid soluble intermediates (Clouet et al., 1974). Similar results were obtained in a second experiment in which mitochondria were isolated from human heart auricles by biopsy procedures during surgery. These authors concluded that the rates of erucic acid activation and β-oxidation are decreased similar to that observed in experimental animals (Clouet and Bézard, 1979).

In summary, it seems clear that there are significant interspecies differences in the relative and absolute rates of oxidation of different chain length fatty acids by heart muscle. The large differences in interspecies susceptibility to myocardial lipidosis would appear to be explained reasonably well by the metabolic data now available for some of these species. Obviously, more data on interspecies comparisons are badly needed. Because there are these well-documented interspecies differences, there are some obvious difficulties in extrapolating data from experimental animals to man on the interaction of 22:1 monoenes and myocardial lipidosis.

ACKNOWLEDGMENT

This work is Contribution No. 1094 from the Animal Research Centre, Ottawa, Ontario, Canada.

REFERENCES

Ackman, R. G. (1980). "A Report to Fisheries and Oceans Canada," PDR Contract No. 08SC-01532-9-0244. Fisheries and Oceans Canada, Ottawa.
Ackman, R. G., and Loew, F. M. (1977). *Fette, Seifen, Anstrichm.* **79**, 15–24, 58–69.
Beare-Rogers, J. L., and Gordon, E. (1976). *Lipids* **11**, 287–290.
Beare-Rogers, J. L., and Nera, E. A. (1972). *Comp. Biochem. Physiol.* **41B**, 793–800.
Beinert, H. (1963). *In* "The Enzymes" (P. D. Boyer, H. Lardy, and K. Myrbäck, eds.), 2nd ed., Vol. 7, pp. 447–476. Academic Press, New York.

Blomstrand, R., and Svensson, L. (1974). *Lipids* **9**, 771–780.
Brand, I. A., and Söling, H. D. (1975). *FEBS Lett.* **57**, 163–168.
Bronfman, M., Inestrosa, N. C., and Leighton, F. (1979). *Biochem. Biophys. Res. Commun.* **88**, 1030–1036.
Buddecke, E., Filipovic, I., Wortberg, B., and Seher, A. (1976). *Fette, Seifen, Anstrichm.* **78**, 196–200.
Chan, S. H. P., and Higgins, E. (1978). *Can. J. Biochem.* **56**, 111–116.
Cheng, C.-K., and Pande, S. V. (1975). *Lipids* **10**, 335–339.
Christian, D. R., Kilsheimer, G. S., Pettett, G., Paradise, R., and Ashmore, J. (1969). *Adv. Enzyme Regul.* **7**, 71–82.
Christiansen, R. Z. (1978). *Biochim. Biophys. Acta* **530**, 314–324.
Christiansen, R. Z., Christophersen, B. O., and Bremer, J. (1977). *Biochim. Biophys. Acta* **487**, 28–36.
Christiansen, R. Z., Osmundsen H., Borrebaek, B., and Bremer, J. (1978). *Lipids* **13**, 487–491.
Christiansen, R. Z., Christiansen, E. N., and Bremer, J. (1979a). *Biochim. Biophys. Acta* **573**, 417–429.
Christiansen, R. Z., Norseth, J., and Christiansen, E. N. (1979b). *Lipids* **14**, 614–618.
Christophersen, B. O., and Bremer, J. (1972). *FEBS Lett.* **23**, 230–232.
Christophersen, B. O., and Christiansen, R. Z. (1975). *Biochim. Biophys. Acta* **388**, 402–412.
Clandinin, M. T. (1978). *J. Nutr.* **108**, 273–281.
Clouet, P., and Bézard, J. (1978). *FEBS Lett.* **93**, 165–168.
Clouet, P., and Bézard, J. (1979). *C.R. Acad. Sci. Paris* **288D**, 1683–1686.
Clouet, P., Blond, J.-P., and Bézard, J. (1974). *C.R. Acad. Sci. Paris* **279D**, 1003–1006.
Crass, M. F. (1972). *Biochim. Biophys. Acta* **280**, 71–81.
Crass, M. F., Shipp, J. C., and Pieper, G. M. (1975). *Am. J. Physiol.* **228**, 618–627.
Dow-Walsh, D. S., Mahadevan, S., Kramer, J. K. G., and Sauer, F. D. (1975). *Biochim. Biophys. Acta* **396**, 125–132.
Forsyth, G. W., Carter, K. E., Loew, F. M., and Ackman, R. G. (1977). *Lipids* **12**, 791–796.
Garland, P. B., and Randle P. J. (1964). *Biochem. J.* **91**, 6C–7C.
Gartner, S. L., and Vahouny, G. V. (1973). *Proc. Soc. Exp. Biol. Med.* **143**, 556–560.
Gumpen, S. A., and Norum, K. R. (1973). *Biochim. Biophys. Acta* **316**, 48–55.
Hall, C. L., Heijkenskjöld, L., Bartfai, T., Ernster, L., and Kamin, H. (1976). *Arch. Biochem. Biophys.* **177**, 402–414.
Heijkenskjöld, L., and Ernster, L. (1975). *Acta Med. Scand., Suppl.* **585**, 75–83.
Houtsmuller, U. M. T., Struijk, C. B., and Van der Beek, A. (1970). *Biochim. Biophys. Acta* **218**, 564–566.
Hsu, C. M. L., and Kummerow, F. A. (1977). *Lipids* **12**, 486–494.
Hülsmann, W. C., and Stam, H. (1978). *Biochem. Biophys. Res. Commun.* **82**, 53–59.
Hülsmann, W. C., Geelhoed-Mieras, M. M., Jansen, H., and Houtsmuller, U. M. T. (1979). *Biochim. Biophys. Acta* **572**, 183–187.
Hung, S., and Holub, B. J. (1977). *Nutr. Rep. Int.* **15**, 71–79.
Inestrosa, N. C., Bronfman, M., and Leighton, F. (1979). *Biochem. J.* **182**, 779–788.
Jansen, H., Hülsmann, W. C., van Zuylen-van Wiggen, A., Struijk, C. B., and Houtsmuller, U. M. T. (1975). *Biochem. Biophys. Res. Commun.* **64**, 747–75l.
Jesmok, G. J., Calvert, D. N., and Lech, J. J. (1977). *J. Pharmacol. Exp. Ther.* **200**, 187–194.
Korsrud, G. O., Conacher, H. B. S., Jarvis, G. A., and Beare-Rogers, J. L. (1977). *Lipids* **12**, 177–181.
Krahling, J. B., and Tolbert, N. E. (1981). *Arch. Biochem. Biophys.* **209**, 100–110.
Kramer, J. K. G. (1980). *Lipids* **15**, 651–660.
Kramer, J. K. G., and Hulan, H. W. (1977). *Lipids* **12**, 159–164.
Kramer, J. K. G., and Hulan, H. W. (1978). *Lipids* **13**, 438–445.

Kramer, J. K. G., Mahadevan, S., Hunt, J. R., Sauer, F. D., Corner, A. H., and Charlton, K. M. (1973). *J. Nutr.* **103**, 1696–1708.

Kramer, J. K. G., Friend, D. W., and Hulan, H. W. (1975). *Nutr. Metab.* **19**, 279–290.

Kramer, J. K. G., Hulan, H. W., Trenholm, H. L., and Corner, A. H. (1979). *J. Nutr.* **109**, 202–213.

Lazarow, P. B. (1978). *J. Biol. Chem.* **253**, 1522–1528.

Lazarow, P. B., and de Duve, C. (1976). *Proc. Natl. Acad. Sci. U.S.A.* **73**, 2043–2046.

Mansour, T. E. (1963). *J. Biol. Chem.* **238**, 2285–2292.

Matthews, P. M., Bland, J. L., Gadian, D. G., and Radda, G. K. (1981). *Fed. Proc., Fed. Am. Soc. Exp. Biol.* **40**, 1625 (Abstr. No. 497).

Mayer, S. E. (1974). *Circ. Res.* **35**, Suppl. 3, 129–135.

Mersel, M., Heller, M., and Pinson, A. (1979). *Biochim. Biophys. Acta,* **572**, 218–224.

Morgan, H. E., and Parmeggiani, A. (1964). *Control Glycogen Metab., Ciba Found. Symp., 1963*, pp. 254–270.

Morgan, H. E., Cadenas, E., Regen, D. M., and Park, C. R. (1961). *J. Biol. Chem.* **236**, 262–268.

Neat, C. E., Thomassen, M. S., and Osmundsen, H. (1981). *Biochem. J.* **196**, 149–159.

Neely, J. R., Liebermeister, H., and Morgan, H. E. (1967). *Am. J. Physiol.,* **212**, 815–822.

Neely, J. R., Whitfield, C. F., and Morgan, H. E. (1970). *Am. J. Physiol.,* **219**, 1083–1088.

Neely, J. R., Rovetto, M. J., and Oram, J. F. (1972). *Prog. Cardiovasc. Dis.,* **15**, 289–329.

Norseth, J. (1979). *Biochim. Biophys. Acta* **575**, 1–9.

Norseth, J. (1980). *Biochim. Biophys. Acta* **617**, 183–191.

Norseth, J., and Christophersen, B. O. (1978). *FEBS Lett.* **88**, 353–357.

Opie, L. H. (1968). *Am. Heart J.* **76**, 685–698.

Opie, L. H. (1969). *Am. Heart J.* **77**, 100–122, 383–410.

Opstvedt, J., Svaar, H., Hansen, P., Pettersen, J., Langmark, F. T., Barlow, S. M., and Duthie, I. F. (1979). *Lipids* **14,** 356–371.

Osmundsen, H., and Bremer, J. (1978). *Biochem. J.* **174**, 379–386.

Osumi, T., and Hashimoto, T. (1978). *Biochem. Biophys. Res. Commun.* **83**, 479–485.

Osumi, T., and Hashimoto, T. (1979). *Biochem. Biophys. Res. Commun.* **89**, 580–584.

Osumi, T., and Hashimoto, T. (1980). *Arch. Biochem. Biophys.* **203**, 372–383.

Pande, S. V., and Parvin, R. (1980). *Biochim. Biophys. Acta* **617**, 363–370.

Pinson, A., and Padieu, P. (1974). *FEBS Lett.* **39**, 88–90.

Randle, P. J., Garland, P. B., Hales, C. N., Newsholme, E. A., Denton, R. M., and Pogson, C. I. (1966). *Recent Prog. Horm. Res.* **22**, 1–44.

Randle, P. J., England, P. J., and Denton, R. M. (1970). *Biochem. J.* **117**, 677–695.

Regen, D. M., Davis, W. W., Morgan, H. E., and Park, C. R. (1964). *J. Biol. Chem.* **239**, 43–49.

Sauer, F. D., and Kramer, J. K. G. (1980). *Adv. Nutr. Res.* **3**, 207–230.

Severson, D. L. (1979). *J. Mol. Cell. Cardiol.* **11**, 569–583.

Shafrir, E., Gatt, S., and Khasis, S. (1965). *Biochim. Biophys. Acta* **98**, 365–371.

Shenolikar, I. S., and Tilak, T. B. G. (1980). *Nutr. Metab.* **24**, 199–208.

Shumate, J. B., and Choksi, R. M. (1981). *Biochim. Biophys. Res. Commun.* **100**, 978–981.

Stam, H., Geelhoed-Mieras, T., and Hülsmann, W. C. (1980). *Lipids* **15**, 242–250.

Svaar, H., Langmark, F. T., Lambertsen, G., and Opstvedt, J. (1980). *Acta Pathol. Microbiol. Scand., Sect. A* **88**, 41–48.

Svoboda, D., Grady, H., and Azarnoff, D. (1967). *J. Cell. Biol.* **35**, 127–152.

Swarttouw, M. A. (1974). *Biochim. Biophys. Acta* **337**, 13–21.

Thomassen, M. S., Strom. E., Christiansen, E. N., and Norum, K. R. (1979). *Lipids* **14**, 58–65.

Tolbert, N. E. (1981). *Annu. Rev. Biochem.* **50**, 133–157.

Vasdev, S. C., and Kako, K. J. (1976). *Biochim. Biophys. Acta* **431**, 22–32.

Wang, T.-W., Menahan, L. A., and Lech, J. J. (1977). *J. Mol. Cell. Cardiol.,* **9**, 25–38.

15

The Regulation of Long Chain Fatty Acid Oxidation

S. V. PANDE

I. INTRODUCTION

In the first half of this chapter, I summarize the process of fatty acid β-oxidation as understood today indicating, where possible, the suggested

355

High and Low Erucic Acid Rapeseed Oils
Copyright © 1983 by Academic Press Canada
ISBN 0-12-425080-7

role of the individual steps in the regulatory process; the more integrated control of fatty acid oxidation is described in the second half. The peroxisomal β-oxidation, which has been vigorously studied lately, is included and so is ω-oxidation, but the other minor oxidative routes, the α- and the γ-oxidations, are not being described. For supplementary information I recommend the excellent reviews of Lynen (1954), Fritz (1961), Greville and Tubbs (1968), Green and Allman (1968), Bressler (1970), Wakil and Barnes (1971), Neely and Morgan (1974), Idell-Wenger and Neely (1978), and McGarry and Foster (1980).

II. β-OXIDATION

A. Cellular Transport and the Role of Binding Proteins

Most tissues with the exception of adipose and mammary gland have rather limited stores of triacylglycerol and most of the time such tissues meet their net fatty acid needs by extracting them from the circulation largely from the albumin-free fatty acid complex. The serum concentration of albumin-bound fatty acids fluctuates between intermittent periods of alimentation and generally correlates inversely to the concurrent rates of overall glucose utilization in the body. The rate of uptake of free fatty acids by tissues, particularly liver, correlates to the concentration of albumin-bound free fatty acids in circulation. This led to the assumption initially that passive diffusion accounted for the cellular uptake of fatty acids but experimental evidence now favors the involvement of a rapid carrier mediated transport. It is believed that net free fatty acid export out of cells, usually of adipose, proceeds only when the balance of esterification and lipolysis raises the intracellular concentration of uncomplexed free fatty acid to levels exceeding those in serum and that most tissues are able to import free fatty acid because its rapid utilization maintains intracellular concentration at levels below those in circulation. According to one view the efficient hepatic extraction of albumin-bound fatty acids is facilitated by the presence of albumin receptors on the liver cell surface; presumably, the surface binding of free fatty acid—albumin complex follows dissociation of free fatty acids near the cell surface which then leads to the rapid uptake of free fatty acids (Weisiger et al., 1981). Experiments with cultured chick embryo heart cells, hepatocytes, and adipocytes have shown that at physiological, micro-, and submicromolar concentrations of unbound fatty acids, uptake proceeds by a saturable process, whereas at higher concentrations, fatty acids enter by passive diffusion (Samuel et al., 1976; Paris et al., 1978, 1979; Abumrad et al., 1981).

Studies with Escherichia coli mutants have indicated that the inward transport of fatty acids through the cytoplasmic membrane involves separate mechanisms for the uptake of long and medium chain fatty acids. The up-take of the former proceeds by active transport and requires a specific carrier protein. This protein shows some affinity for medium chain fatty acids also but these acids enter largely by diffusional mechanism. The entire transport process requires further participation of an acyl-CoA synthetase (EC 6.2.1.3) and of a third undefined protein (Nunn and Simons, 1978; Nunn et al., 1979; Maloy et al., 1981).

The possibility of fatty acid uptake being facilitated by intracellular binding proteins has received much attention. Presence of fatty acid binding proteins (MW ~12,000) in various tissues is well documented (Levi et al., 1969; Mishkin et al., 1972; Ockner et al., 1972; Fournier et al. 1978). Moreover, in heart and skeletal muscles myoglobin itself shows appreciable fatty acid binding ability (Gloster and Harris, 1977). In rat liver, fatty acid binding protein constitutes a major protein of the cytosol (Ockner and Manning, 1974); in smaller quantities it is reportedly present also in mitochondria, microsomes (Rustow et al., 1979), and peroxisomes (Appelkvist and Dallner, 1980). These proteins bind a number of ligands having in common a hydrophobic region and their possible role in the metabolism of compounds other than fatty acids has been suggested (Foliot, 1979; Dampsey et al., 1981; Billheimer and Gaylor, 1980). Fatty acid binding protein binds fatty acyl-CoA esters with higher affinity than most other ligands (Mishkin and Turcotte, 1974; Ketterer et al., 1976). The level of fatty acid binding protein and long chain acyl-CoA rises in liver when rats are given clofibrate (Renaud et al., 1978) whereas lower levels of both of these are found in Morris hepatoma cells (Mishkin et al., 1977). Despite speculations that fatty acid binding proteins play some role in fatty acid uptake and its subsequent metabolism (Wu-Rideout et al., 1976; Renaud et al., 1978; Goresky et al., 1978) the exact function and the possible participation of these proteins in the regulation of fatty acid metabolism is unclear. A stimulatory effect of fatty acid binding protein on mitochondrial adenine nucleotide translocase (Chan and Barbour, 1979), acetyl-CoA carboxylase (Lunzer et al., 1977), and fatty acid esterification (Wu-Rideout et al., 1976) has been described but it is not clear to what extent these represent simple *in vitro* protections against the known inhibitory effects of long chain acyl-CoA esters on the reactions examined.

B. Activation and Mitochondrial Transport

Most intracellular routes of fatty acid metabolism require prior activation of these substrates. Activation commonly proceeds according to the reaction: $ATP + fatty\ acid + CoASH = acyl\text{-}CoA + AMP + PP_i$. Several acyl-CoA synthetases, differing on the basis of substrate specificity, distribution in tissues, and intracellular localization are known (for reviews, see Londesborough and Webster, 1974; Groot et al., 1976). Palmitoyl-CoA synthetase, which accepts both long chain unsaturated and saturated fatty acids as substrates, is localized in microsomes (Groot et al., 1976), on the outer side of outer mitochondrial membrane (Pande and Blanchaer, 1970), and in peroxisomes (Krisans et al., 1980). A medium chain acyl-CoA synthetase that shows slight activity with long chain fatty acids as well is present in the matrix of liver (but not in fetal liver mitochondria, Parameswaran and Arinze, 1981) and to some extent in heart mitochondria but is absent in mitochondria of skeletal muscles and adipose tissue (Harper and Saggerson, 1975; Van Tol, 1975). In skeletal muscle, octanoate is activated by the long chain acyl-CoA synthetase of outer mitochondrial membrane and consequently octanoate oxidation shows carnitine dependency unlike that in liver (Van Tol, 1975).

A GTP-dependent acyl-CoA synthetase of broad substrate specificity was once described and characterized from rat liver (Galzigna et al., 1967) but the rates of GTP-dependent activation of fatty acid in most tissues including liver are only a very small fraction of those observed with ATP (Pande and Mead, 1968). The occurrence and the role of GTP-dependent activation in fatty acid metabolism continue to remain uncertain (Groot et al., 1976).

In Escherichia coli an acyl-acyl carrier protein synthetase that uses acyl carrier protein instead of CoA for fatty acid activation has been described (Ray and Cronan, 1976). The hydrocarbon utilizing yeast, Candida lipolytica fabricates two distinct long chain acyl-CoA synthetases: one of them activates fatty acids exclusively for lipid synthesis, while the other one does so for β-oxidation (Numa, 1981). Comparisons of the mitochondrial and microsomal long chain acyl-CoA synthetases of rat liver have shown, however, that the two enzymes are very similar (Philipp and Parsons, 1979; Tanaka et al., 1979).

Whereas the mitochondrial enzymes of β-oxidation reside within the area bound by inner membrane, activation of fatty acids proceeds largely at sites exterior to this membrane. The transport of activated acyl groups across the inner mitochondrial membrane is brought about by a carnitine dependent route (Fritz, 1963; Bremer, 1968; Bressler, 1970). A carnitine acyltransferase localized on the outer aspect of inner membrane utilizes cytosolic free carnitine to convert the cytosolic acyl-CoA to cytosolic acylcarnitine (Fig. 1). A translocase of the inner membrane then moves the acylcarnitine inside in exchange for the simultaneous movement of carnitine in the opposite direction. Another carnitine acyltransferase, situated on the inner side of the inner membrane, utilizes matrix CoA to convert acylcarnitine to acyl-CoA, thus producing the latter in the same compartment where enzymes of the β-oxidation spiral exist (Pande, 1975; Ramsay and Tubbs, 1975; Tubbs and

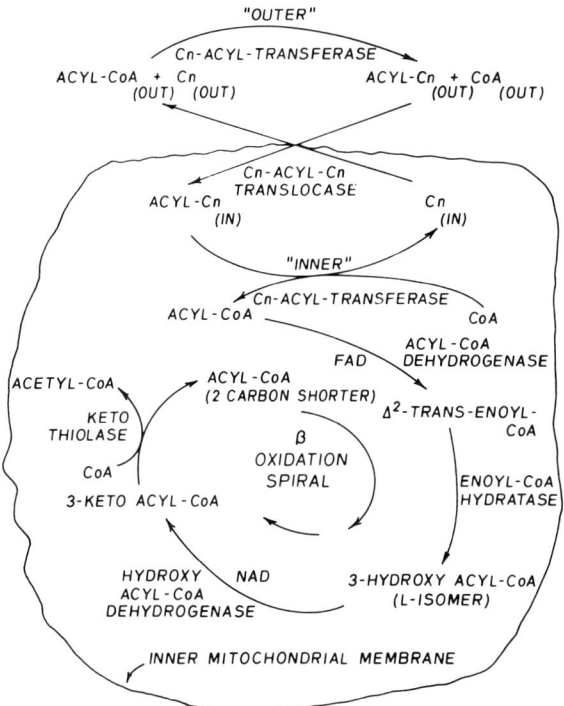

Fig. 1.   Schematic representation of the sequence of reactions involved in the carnitine-dependent transport of acyl group into mitochondria and its subsequent β-oxidation.

Ramsay 1979; Pande and Parvin, 1980c). The presence of carnitine acetyltransferase and carnitine palmitoyltransferase in mitochondria is well established. Most mitochondria show appreciable carnitine octanoyltransferase activity also but whether this activity is due to a third carnitine acyltransferase (Solberg, 1972) or results from the overlapping substrate specificities of carnitine acetyl- and carnitine palmitoyltransferases for medium chain acyl esters (Clark and Bieber, 1981a) is uncertain. There is some evidence that the inner and the outer carnitine palmitoyltransferase of mitochondria are identical proteins; the known differences in their properties presumably arise from their dissimilar membrane association (Bergström and Reitz, 1980; Clark and Bieber, 1981b). Carnitine acetyltransferase and carnitine octanoyltransferase activities are localized also in peroxisomes. Carnitine acylcarnitine translocase of liver and heart mitochondria shows broad substrate specificity but the possible occurrence of more than one translocase of narrower substrate specificity has not been excluded.

C. Mitochondrial System for the Oxidation of Saturated Fatty Acids

Three FAD dependent enzymes of differing chain length specificity together account for the initial dehydrogenation of commonly occurring fatty acids. The electron acceptor for these dehydrogenases is another FAD-containing electron-transfer protein which in turn is linked to the mitochondrial respiratory chain (Beinert, 1963; Wakil and Barnes, 1971). Enoyl-CoA hydratases convert the product of the above reaction, Δ^2-*trans*-enoyl-CoA, to L-3-hydroxyacyl-CoA. Two such enzymes showing preference for long or short chain acyl esters with overlapping activity with medium chain substrates are known (Fong and Schulz, 1977). The corresponding cis isomers are also hydrated by the enzyme but the product then has the D configuration. A NAD dependent hydroxy acyl-CoA dehydrogenase of wide substrate specificity that is specific for the L-antipode produces 3-ketoacyl-CoA from the corresponding hydroxy ester. A CoA dependent 3-oxoacyl-CoA thiolase [acetyl-CoA acyltransferase (EC 2.3.1.16)] cleaves the 3-ketoacyl-CoA producing acetyl-CoA and an acyl-CoA two carbons shorter. Occurrence of multiple forms of thiolases that differ in their substrate specificity, intracellular location, and functions are known. In mitochondria, an enzyme of broad substrate specificity is considered involved in β-oxidation whereas another thiolase, acetyl-CoA acetyltransferase (EC 2.3.1.9) specific for acetoacetyl-CoA, is believed to function only in ketone body metabolism (Middleton, 1973, Staack et al., 1978).

By a repetition of the above four steps of the β-oxidation spiral (Fig. 1), the normally occurring even numbered fatty acids are degraded completely in most tissues to acetyl-CoA; a notable exception is liver where part of the acetoacetyl-CoA derived from the ω-end of fatty acids during β-oxidation escapes further cleavage to acetyl-CoA (Brown et al., 1954; Greville and Tubbs, 1968; Lopes-Cardozo et al., 1975). During the β-oxidation of less frequently occurring odd numbered fatty acids, the ω-terminal portion ends up as propionyl-CoA, instead of acetyl-CoA. The presence of mitochondrial enzymes, a biotin containing propionyl-CoA carboxylase (EC 6.4.1.3), a methylmalonyl-CoA racemase, and a coenzyme B_{12} containing methyl malonyl-CoA mutase (EC 5.4.99.2), allows propionate carbons to enter the citric acid cycle in the form of succinyl-CoA (Wakil and Barnes, 1971).

D. Mitochondrial System for the Oxidation of Unsaturated Fatty Acids

In the degradation of normally occurring unsaturated fatty acids with cis double bonds, intermediates that are not on the direct path of β-oxidation are produced, and their channeling into the β-oxidation spiral entails additional steps. When the double bond is located at odd numbered carbon of the fatty acid, e.g., in oleic and linoleic acids at position 9, chain shortening by β-oxidation produces a Δ^3-cis-enoyl-CoA ester (Fig. 2). A mitochondrial Δ^3-cis-Δ^2-trans-enoyl-CoA isomerase (EC 5.3.3.7) then brings these acyl-CoA esters to the normal path of β-oxidation by converting them to Δ^2-trans-enoyl-CoA (Stoffel et al., 1964; Davidoff and Korn 1965; Struijk and Beerthuis, 1966), a usual β-oxidation intermediate. For the degradation of unsaturated fatty acids having a cis double bond at even numbered carbon atoms, at least three routes need to be considered based on the existence of the relevant enzymatic activities. According to one possibility the usual β-oxidation of such acids can proceed until the cis double bond reaches carbon atom two (see Δ^2-cis-octenoyl-CoA in Fig. 2). (The usual Δ^2-enoyl-CoA esters of β-oxidation spiral have a trans configuration.) As the enoyl-CoA hydratase of the β-oxidation spiral is not specific for the geometric configuration of the double bond in position 2, it hydrates the Δ^2-cis-enoyl-CoA ester but the resulting 3-hydroxyacyl-CoA ester has D-configuration. Conversion of the D- to the L-isomer, a usual β-oxidation intermediate, is made possible by the participation of a mitochondrial epimerase (Stoffel et al., 1964; Stoffel and Caeser, 1965). According to the second possibility, the degradion of Δ^2-cis-enoyl-CoA could involve reduction of the double bond in position two and the resulting saturated acyl-CoA ester would then be a normal substrate of the β-oxidation sequence; the presence of a NADPH-dependent Δ^2-enoyl-CoA reductase (EC 1.3.1.8) that catalyzes such a reaction has been demonstrated in mitochondria (Seubert et al., 1968; Podack and Seubert, 1972; Mizugaki and Uchiyama, 1973). Whereas Δ^4-cis-decenoyl-CoA has been detected as an intermediate of linoleate oxidation even in vivo (Kunau and Lauterbach, 1978) the formation of Δ^2-cis-enoyl-CoA has not been detected. Consequently, the third proposed route excludes Δ^2-cis-enoyl-CoA ester as an intermediate (Kunau and Dommes, 1978) according to which the usual β-oxidation of unsaturated fatty acids with a cis double bond at even numbered carbon proceeds until the formation of a Δ^2-trans-Δ^4-cis-dienoyl-CoA ester (Fig. 2). A mitochondrial NADPH-dependent reductase, initially called 4-enoyl-CoA reductase but more appropriately named 2,4-dienoyl-CoA reductase (Kunau and Bartnik, 1974; Kunau and Dommes, 1978; Borrebaek et al., 1980a), then reduces the double bond in position 4 and Δ^2-trans-enoyl-CoA ester is eventually obtained; whether the latter is formed directly or requires the intermediate participation of Δ^3-cis-Δ^2-trans-enoyl-CoA isomerase has not been ascertained (Kunau and Dommes, 1978). The formation of Δ^2-trans-enoyl-CoA ester allows the acyl group oxidation to proceed via the normal β-oxidation mechanism. Evidence for the possible involvement of the different pathways described above, in the oxidation of certain unsaturated fatty acids, rests largely on in vitro demonstration of the enzymatic reactions involved. Experiments with

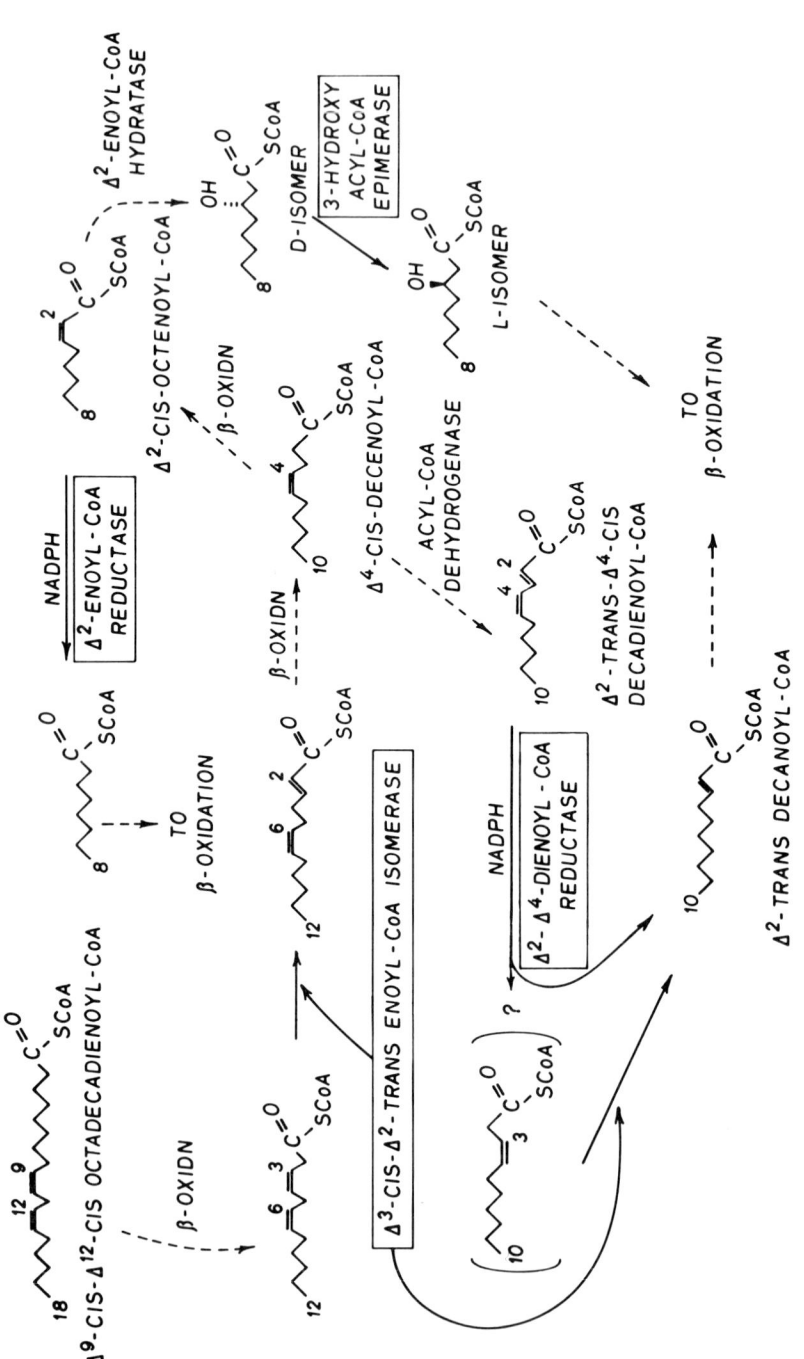

Fig. 2. Reaction sequences implicated in the oxidative degradation of unsaturated fatty acids. Names of enzymes implicated in the oxidation of only unsaturated fatty acids are shown in blocks.

pent-4-enoate have shown that the 2,4-dienoyl-CoA reductase mediated route is the major one involved in the oxidative degradation (Borrebaek *et al.*, 1980a; Hiltunen and Davis, 1981) of this short chain acid. The observation that the activity of 2,4-dienoyl-CoA reductase of liver mitochondria is selectively increased by feeding clofibrate or a high fat diet containing partially hydrogenated marine oils (Borrebaek *et al.*, 1980b) suggest that the 2,4-dienoyl-CoA reductase mediated route participates in the degradation of unsaturated fatty acids *in vivo*.

E. Organization of the Mitochondrial Oxidation System

Mitochondrial oxidation of fatty acids ordinarily proceeds without marked accumulation of intermediates (for references, see Stanley and Tubbs, 1975) and this has led to the speculation that in intact mitochondria the enzymes of β-oxidation might occur organized as a multienzyme complex (Garland and Yates, 1967; Greville and Tubbs, 1968). Some support for this view comes from the observations that the exact number and the substrate specificity of the individual enzymes of the β-oxidation sequence of higher organisms are strikingly similar to those of *E. coli* in which certain enzymes of β-oxidation do occur organized as multienzyme complex (Binstock *et al.*, 1977; O'Brien and Frerman, 1977; Pawar and Schulz, 1981). Clear evidences have not been obtained in support of mechanisms in which, during β-oxidation, either all intermediates accumulate or none do. Thus, under state 3 conditions, in the absence of carnitine, a limited accumulation, mainly of saturated thiol esters, is observed but these do not show a clear precursor–product relationship and behave as though they arise by a "leakage" from the main pathway of true intermediates (Stanley and Tubbs, 1975). The above results would be expected if (1) "leakage" represents slow movement of intermediates from matrix to the extramitochondrial compartment, (2) FAD-linked acyl-CoA dehydrogenase(s) were the rate limiting enzyme(s) of the β-oxidation spiral as appears likely, and (3) the combined activity, and affinity, or both, of the FAD-linked dehydrogenases increased with a decrease in chain length, at least relative to that of the competing carnitine dependent acyl group exporting system, which could otherwise cause substantial intermediate accumulation by diverting them to cytosol. Indeed, substantial accumulation of intermediates as acylcarnitines does occur when oxidation is followed in the presence of carnitine and under state 4 conditions (Lopes-Cardozo *et al.*, 1978). Under conditions of elevated NADH/NAD ratios, accumulation of 3-hydroxyacyl intermediates as carnitine esters has been demonstrated (Bremer and Wojtczak, 1972; Stanley and Tubbs, 1975; Lopes-Cardozo *et al.*, 1978).

F. Peroxisomal Oxidation System

Until recently it was believed that β-oxidation proceeded only in mitochondria. This view changed with the discovery that in germinating castor bean seedlings β-oxidation enzymes are found localized not in mitochondria but in glyoxysomes (Cooper and Beevers, 1969). The presence of β-oxidation machinery in peroxisomes has now been established in liver (Lazarow and de Duve, 1976; Bronfman et al., 1979), in rat brown adipose tissue (Kramar et al., 1978), and there is some indication of its presence in muscles as well (Shumate and Choksi, 1981). The peroxisomal β-oxidation follows the same scheme as that of mitochondria, except that in peroxisomes the initial α,β-desaturation is catalyzed by an oxidase that directly transfers electrons to O_2 forming H_2O_2 unlike that in mitochondria where a dehydrogenase transfers electrons to an electron-transfer flavoprotein. Moreover, whereas mitochondria oxidize long, medium, and short chain acyl groups, the peroxisomes show activity only with acyl groups having more than six carbons. This lack of reactivity with short chain acyl-CoA esters is shared by the different enzymes of the peroxisomal β-oxidation sequence (Osumi and Hashimoto, 1978; Osumi et al., 1980; Miyazawa et al., 1980; Lazarow, 1981) which seem to be distinct from those of the mitochondrial enzymes. Peroxisomes are thus capable of only partial degradation of fatty acids; under optimized conditions in vitro, palmitoyl-CoA undergoes up to five cycles of β-oxidation (Lazarow, 1978). In most studies in vitro, however, rapid oxidation of acyl groups seems limited to the first three cycles only (Thomas et al., 1980). The long chain mono unsaturated fatty acids are oxidized at faster rates than the corresponding saturated acids and it is believed that the peroxisomal oxidation plays an important role in the chain shortening of those fatty acyl groups that otherwise are only poorly oxidized by mitochondria (Hryb and Hogg, 1979). Presence of organelle specific 2,4-dienoyl-CoA reductase and Δ^3-cis-Δ^2-trans-enoyl-CoA isomerase indicates that, like mitochondria, peroxisomes are equipped to degrade unsaturated fatty acids, the usual β-oxidation of which gives rise to intermediates having cis double bonds at even as well as odd numbered carbon atoms (Dommes et al., 1981). Peroxisomes have been also implicated in the oxidation of lignoceric acid (24:0) by rat liver preparations (Kawamura et al., 1981).

From the measurement of enzymatic capacities in vitro, the FAD-dependent acyl-CoA oxidase has been identified as the rate limiting enzyme of the peroxisomal pathway and 3-oxoacyl-CoA ester, one of the products of peroxisomal oxidation, strongly inhibits the acyl-CoA oxidase activity (Osumi et al., 1980). When oxidation is followed under conditions chosen to minimize damage to peroxisomes, the access of added acyl-CoA to the β-oxidation system seems to become a restricted process. Under these conditions, in the absence of Triton X-100, added CoA inhibits the oxidation of palmitoyl-CoA but, surprisingly, not that of erucoyl-CoA. The possibility of CoA inhibition regulating peroxisomal β-oxidation has been suggested (Osmundsen and Neat, 1979). Peroxisomes have enough acyl-CoA synthetase activity to account for their acyl-CoA oxidizing ability (Krisans et al., 1980). They also have a seperate pool of CoA which is available to the thiolase reaction of peroxisomal β-oxidation but not to the acyl-CoA synthetase (Broekhoven et al., 1981). This implies that in peroxisomes, as in mitochondria, a permeability restricting membrane separates the site of β-oxidation from that of fatty acid activation. How activated acyl groups are transported across the peroxisomal membrane is not known. According to one view, such a transport is facilitated by the presence of a fatty acid binding protein (Appelkvist and Dallner, 1980). A possible involvement of ATP in peroxisomal transport has also been suggested (Thomas et al., 1980). Experiments in vitro have not shown any dependency on carnitine although peroxisomes have appreciable carnitine acetyltransferase and carnitine medium chain acyltransferase activities which are greatly enhanced by hypolipidemic drugs that enhance the activities of the enzymes of peroxisomal β-oxidation sequence. However, it is generally assumed that the acetyl and the acyl intermediate products of peroxisomal oxidation leave these organelles as acylcarnitines for further metabolism elsewhere, most likely in mitochondria.

The extent to which the peroxisomes contribute to the oxidation of fatty acids in liver in vivo is controversial. Based on the measured capacity of the peroxisomal pathway under optimized conditions in vitro, the assessment of the contribution of peroxisomal oxidation has varied from a major (exceeding 50% of total) (Lazarow, 1978; Inestrosa et al., 1979) to a relatively minor one (Hryb and Hogg, 1979; Shindo and Hashimoto, 1978; Krahling et al., 1979; Thomas et al., 1980; Neat et al., 1981). Experiments with isolated hepatocytes indicate that the major route of palmitate oxidation is the mitochondrial one with the peroxisomes oxidizing only about 10% of the fatty acid, even after clofibrate administration, which preferentially enhances the peroxisomal pathway (Mannaerts et al., 1979). Estimates using intact perfused rat livers have indicated that palmitate is oxidized exclusively in mitochondria. Peroxisomes were found to oxidize several other fatty acids, particularly the long chain monounsaturated ones, but less than one acetyl unit was formed per mole of added fatty acid. Thus, in intact liver, where competing pathways of fatty acyl group utilization are available, only a small fraction of the peroxisomal oxidative capacity is normally utilized for the β-oxidation (Foerster et al., 1981) of commonly occurring fatty acids.

The ability of peroxisomes to oxidize fatty acids and the activities of the enzymes of this pathway are greatly enhanced following administration of several hypolipidemic compounds (Lazarow and de Duve, 1976; Lazarow, 1977; Shindo and Hashimoto, 1978; Osumi and Hashimoto, 1979; Inestrosa *et al.*, 1979). Starvation and diabetes do not affect the peroxisomal β-oxidation ability (Mannaerts *et al.*, 1979) but a high fat (15% versus 5%) diet, particularly one rich in long chain monounsaturated fatty acids, enhances the peroxisomal oxidation capacity (Neat *et al.*, 1981).

The presence of a cytosolic β-oxidation system has also been described. This system was found capable of only limited chain shortening, preferred long chain acyl-CoA esters as substrate, and was less efficient than that of the mitochondrial β-oxidation system (Fiecchi *et al.*, 1973; Galli-Kienle *et al.*, 1976). The possibility that the cytosolic β-oxidation system was derived from the peroxisomes, during *in vitro* manipulations, appears likely (Osumi and Hashimoto, 1979).

III. ω-OXIDATION

This pathway utilizes free fatty acids as substrates and microsomal enzymes and converts the ω-methyl group to a carboxylic one via the intermediate formation of ω-hydroxy acid (see Spector, 1971 for an earlier account). The resulting dicarboxylic acid then seems to be degraded largely by β-oxidation. Long chain dicarboxylic acids are activated and transported into mitochondria though not as readily as the corresponding monocarboxylic fatty acids (Pettersen, 1973; Pettersen and Aas, 1973). Recent estimates have reconfirmed that the contribution of ω-oxidation to overall fatty acid degradation in liver of starved rats is very minor with only about 5% of the usual fatty acids being subjected to ω-oxidation (Björkhem, 1978; Kam *et al.*, 1978). Starvation and diabetes markedly increase the ω-oxidation; this increase seems to be caused both from an increase in the capacity of this pathway and from increased substrate, free fatty acid, availability (Wada *et al.*, 1971; Björkhem, 1973, 1976). Medium and short chain dicarboxylic acids (Pettersen *et al.*, 1972) and their 3-hydroxy homologues (Greter *et al.*, 1980), the products of ω- and then β-oxidation, are present in normal human urine in traces, but appear in considerable quantities in the urine of ketotic patients. Dicarboxylic aciduria is also seen under conditions of curtailed β-oxidation such as caused by carnitine deficiency (Karpati *et al.*, 1975; Di Donato *et al.*, 1980), congenital acyl-CoA dehydrogenase deficiency (Mantagos *et al.*, 1978; Gregerson *et al.*, 1980), and hypoglycin intoxication (Tanaka *et al.*, 1976). Much evidence indicates a role of the ω-oxidation system in the degradation of fatty acids with structures that block the usual β-oxidation system (see Greter *et al.*, 1980). ω-Oxidation followed by β-oxidation, via the intermediate formation of succinate, provides a route for the conversion of fatty acid carbon to glucose. This was once considered important (Wada and Usami, 1977) but subsequent estimates have shown that the contribution of such a route to glucose synthesis is a very minor one

(Kam et al., 1978; Björkhem, 1978). Nevertheless, dicarboxylic fatty acids, administered to rats, are able somehow to exert an antiketogenic effect (Wada and Usami, 1977; Mortensen, 1981).

IV. INTEGRATED REGULATION OF MITOCHONDRIAL OXIDATION

Evidence so far indicates that substrate and cofactor availability, and product inhibition, constitute the main modes for the regulation of fatty acid oxidation within mitochondria; there is nothing to indicate that substrate flow over the β-oxidation spiral is further regulated by allosteric or covalent modification of the activity of any enzyme of this spiral. As is to be expected for a multistep metabolic pathway, diverse steps contribute to the overall regulation of fatty acid oxidation in different tissues under different conditions. More is known about this process in liver and heart than in other tissues and its brief description follows.

A. In Heart Muscle

In muscles, at resting state, oxidation of fatty acids like that of other substrates is limited primarily by the low rate of energy expenditure that, by limiting the supply of ADP and P_i, restricts mitochondrial oxidative phosphorylation. An elevation in ATP/ADP ratio of mitochondrial matrix, such as at the cessation of exercise in vivo, or at state 3 to state 4 transition in vitro, slows fatty acid oxidation by interaction at several sites, and under these conditions the probable involvement of the following simplistic scenario is indicated by the information available. A rise in mitochondrial ATP/ADP accompanies elevation of mitochondrial NADH/NAD, a curtailed operation of the citric acid cycle, and a decrease in oxaloacetate concentration. The latter limits the activity of citrate synthase and thereby elevates acetyl-CoA/CoA ratios (Hansford, 1980). Acetyl-CoA potently inhibits 3-ketoacyl-CoA thiolase activity, competitively with respect to CoA; it inhibits also the activity of acetoacetyl-CoA thiolase but less strongly (Olowe and Schulz, 1980). Consequently, with increase in acetyl-CoA/CoA the concentration of 3-ketoacyl-CoA and acetoacetyl-CoA, the end products of β-oxidation, would rise momentarily and this in turn would decrease substrate flux over the β-oxidation spiral by severely inhibiting the activity of long chain acyl-CoA dehydrogenase, the first irreversible and rate limiting enzyme of β-oxidation (McKean et al., 1979; Davidson and Schulz, 1981). Acetoacetyl-CoA inhibits the activity of enoyl-CoA hydratase of bovine liver and its possible role in the control of long chain fatty acid oxidation was suggested (Waterson and Hill, 1972). However, this is not so at least in heart, where activity of

long chain enoyl-CoA hydratase is high and acetoacetyl-CoA does not inhibit it (Fong and Schulz, 1977).

The increase in mitochondrial acetyl-CoA/CoA is believed to eventually
contribute to the lowering of free fatty acid uptake that is seen when energy
need of heart decreases relative to free fatty acid availability (Neely and
Morgan, 1974; Idell-Wenger and Neely, 1978). In such situations, fatty acyl-
CoA and fatty acylcarnitine are elevated but corresponding acetyl esters are
elevated even more. The increased acetyl pressure of mitochondria, through
the participation of carnitine acetyltransferases and carnitine acylcarnitine
translocase, is able to tie up cytosolic CoA and carnitine as acetyl esters. If
the resultant decline of free carnitine limits the conversion of fatty acyl-CoA
to fatty acylcarnitine more than the decline of free CoA limits fatty acid
activation then the fatty acyl CoA/CoA ratio would rise in cytosol and this
would further restrict fatty acid activation because fatty acyl-CoA esters re-
versibly inhibit the activity of fatty acyl-CoA synthetase, competitively with
respect to CoA (Pande, 1973; Oram et al., 1975). A buildup of free fatty
acids intracellularly would slow net free fatty acid uptake and this appropriately occurs at a time when free fatty acid availability to heart exceeds its
energy needs (Neely and Morgan, 1974; Idell-Wenger and Neely, 1978).
When heart work is increased, enhanced energy expenditure increases oxygen consumption; the NADH/NAD ratio decreases and so presumably does
that of FADH/FAD. More substrate is pulled over the oxidative route in mitochondria and the levels of tissue acetylcarnitine, acetyl-CoA, and of at least
matrix fatty acyl-CoA, decline, aided in part by relieving the restraints of
high acetyl-CoA/CoA ratio on β-oxidation described above. The resultant
rise in cytosolic free CoA and carnitine then accelerates fatty acid activation
and the formation of fatty acylcarnitine; the level of fatty acylcarnitine goes
up and so does the rate of uptake of free fatty acids. Thus, when the energy
expenditure of heart is increased, acceleration of citric acid cycle oxidation
appears to precede and exceed that of β-oxidation. It should be added that
hormonal and metabolic regulatory mechanisms, involving presumably cyclic AMP-dependent activation of triglyceride lipase and its inhibition by
fatty acyl-CoA esters, exist in heart that allow endogenous triglycerides to be
mobilized when substrate supply from circulation proves insufficient for the
prevailing energy needs (Vary et al., 1981). A relative deficiency of oxygen
as in hypoxic or ischemic myocardium inhibits substrate oxidation, as expected, but somehow β-oxidation is inhibited in preference to the inhibition
of citric acid cycle, prior to it, or both, because under these conditions
acetyl-CoA and acetylcarnitine decline markedly while the levels of fatty
acyl-CoA and fatty acylcarnitine show a large increase; uptake of fatty acids
is reduced because of its intracellular build up owing, most likely, to rise in
fatty acyl-CoA/CoA ratio and in AMP which would decrease the further con-

version of free fatty acids to fatty acyl-CoA esters (Idell-Wenger and Neely, 1978; Whitmer et al., 1978).

B.   In Liver

1.   REGULATION AT ACYLGLYCEROL, ACYLCARNITINE BRANCH POINTS

In liver, cellular energy demands show only minor fluctuations under different conditions. The oxygen consumption of liver rises only modestly in starvation (Exton et al., 1972) and the increased demand of energy and of reducing equivalents of gluconeogenesis under this condition is believed to impose a permissive effect on the pulling of fatty acids toward mitochondria for ketogenesis (Flatt, 1972; Blackshear et al., 1975); it should be noted, however, that in perfusion experiments in vitro, with high enough medium fatty acids, maximal stimulation of ketogenesis is observed without dependence on concurrent gluconeogenesis. Fatty acids are considered to be the major hepatic energy source in the postabsorptive state (Fritz, 1961; Havel et al., 1962) and in this situation the proportion of fatty acids oxidized to CO_2 and to ketone bodies are about similar (Spitzer et al., 1969; Weinstein et al., 1981). In vitro experiments (Mayes and Felts, 1967; Ontko, 1967) have shown that this proportioning varies markedly with the amount of incoming fatty acids; the more fatty acid enters the liver, the more is used for ketogenesis and correspondingly less is oxidized in the citric acid cycle so that the total production of energy remains nearly unchanged.

A major determinant of the mitochondrial fatty acid oxidation normally in liver is the delivery rate of activated fatty acyl groups to the enzymes of the β-oxidation spiral in the matrix. Although the importance of the delivery of free fatty acids to liver is well established in this regard, the fact that at times intracellular lipolysis alone can provide enough free fatty acids for oxidation needs to be appreciated. The observations that livers from diabetic rats continue to produce ketone bodies nearly maximally, even when perfused with medium lacking fatty acids, attests to the ability of intracellular lipolysis to furnish substrates for β-oxidation for appreciable periods, at least in liver, and to its activation in the diabetic state (Krebs et al., 1969; Van Harken et al., 1969); indirect evidences indicate that a cyclic AMP dependent hormone sensitive lipase exists in liver (for references, see Lund et al., 1980).

Much evidence, nevertheless, shows that the acceleration of hepatic fatty acid oxidation under ketogenic conditions entails important intracellular adaptive changes at steps subsequent to fatty acid activation. A set of these adaptations enhance the proportion of extramitochondrial fatty acyl-CoA being directed to the mitochondrial oxidative route over that being channeled for triacylglycerol and lipoprotein synthesis. Opposite changes in the capacities of triacylglycerol and fatty acylcarnitine synthesizing enzymes are seen under a variety of conditions in liver (Bremer et al., 1976; Zammit, 1981; Christiansen and Bremer, 1978), but these changes take some time to manifest and thus seem relevant mainly for the long range adjustment of fatty acid metabolism. The utilization of extramitochondrial fatty acyl-CoA in liver, for both triacylglycerol and fatty acylcarnitine synthesis, seems to proceed at below saturating rates most of the time, inasmuch as under a variety of experimental conditions, any manipulation in rate of one of these two arms of the fatty acyl-CoA utilization branch leads to near reciprocal changes in the rate of the other arm (Heimberg et al., 1978; Williamson, 1979). Considerable discussion has centered around the question of whether a primary control of acylglycerol synthesizing branch in liver then sets the rate of the β-oxidation branch or whether the reverse generally prevails (Mayes and Felts, 1967; Van Tol, 1975; Christiansen, 1979; McGarry and Foster, 1980; Ide and Ontko, 1981). Much indirect circumstantial evidence, however, suggests that synergistic effects brought about by the near simultaneous and coordinate control of the two pathways of cytosolic fatty acyl-CoA disposal occur and that these adjustments then determine the proportioning of fatty acids between β-oxidation and acylglycerol synthesizing routes.

Instances are known in which the partitioning of fatty acyl-CoA utilization for acylglycerol synthesis and for mitochondrial oxidation is influenced by changes in the concentration of the cosubstrates of the two pathways: α-glycerolphosphate and carnitine. Thus, glucagon's enhancement of ketogenesis with simultaneous diminution of acylglycerol synthesis (Heimberg et al., 1978) accompanies a decline of α-glycerolphosphate content (Pilkis et al., 1976; Christiansen, 1979). Addition of glycerol to hepatocytes previously exposed to glucagon enhances α-glycerolphosphate content and acylglycerol synthesis while fatty acid oxidation is diminished (Lund et al., 1980). Although under ketotic conditions the levels of α-glycerolphosphate were initally considered inadequate for acylglycerol synthesis (Tzur et al., 1964; Mayes and Felts, 1967), its concentration normally is not very limiting for acylglycerol synthesis (Christiansen, 1979; Lund et al., 1980) and a lack of simple correlation between α-glycerolphosphate and ketogenic capacity has also been observed (Williamson et al., 1969; McGarry and Foster, 1971b, 1980).

In liver, which has much less carnitine than muscles and heart, the low carnitine concentration becomes a limiting factor for directing fatty acids toward mitochondria for oxidation, especially when the supply of fatty acids is enhanced. Increases in liver carnitine are usually seen under conditions of enhanced ketogenesis (Snoswell and Koundakjian, 1972; Kondrup and Grunnet, 1973; McGarry et al., 1975, 1978; Williamson, 1979) and an increase in the ketogenic capacity of intact liver in vitro is demonstrable on perfusion of liver with carnitine containing medium (McGarry et al., 1975). However, conditions have also been noted where liver carnitine increases but ketogenesis does not (Robles-Valdes et al., 1976; Brass and Hoppel, 1978). A very marked increase in fatty acylcarnitine concentration is invariably seen under conditions of augmented hepatic ketogenesis, and this appears an essential requirement for the enhancement of fatty acid oxidation, insofar as situations where hepatic fatty acid oxidation may increase without elevation of fatty acylcarnitine concentration have not been identified. Inasmuch as under conditions of accelerated fatty acid oxidation in liver, fatty acylcarnitine concentration rises despite enhanced utilization, the possibility that the availability of free fatty acylcarnitine is one of the key factors that limits the rate of mitochondrial fatty acid oxidation is strongly indicated; it should be stressed that only a small fraction of total long chain acylcarnitine would exist in free form in cells owing to its marked adsorption at water-apolar interphases (Pande, 1981). These considerations imply that the mitochondrial carnitine acylcarnitine translocase usually remains subsaturated with respect to the concentration of cytosolic free fatty acylcarnitine. Carnitine acylcarnitine translocase remains subsaturated normally with respect to matrix carnitine concentration as well, and these also rise to further accelerate the rates of the translocase catalyzed exchange diffusion reactions, when increases in total liver carnitine concentration occur (Parvin and Pande, 1979; Pande and Parvin, 1979, 1980a,c). Mitochondrial matrix carnitine concentration changes in parallel with that of total liver carnitine because carnitine acylcarnitine translocase not only catalyzes an exchange diffusion of carnitines but also an equilibrating unidirectional transport that allows adjustment of matrix carnitine concentration in response to changes in cytosolic carnitine concentration (Pande and Parvin, 1979, 1980b). Thus, an increase in liver carnitine favors the oxidation of fatty acids by enhancing their delivery into the mitochondrial matrix.

Evidence that the short-term control of fatty acylglycerol and fatty acylcarnitine synthesizing pathways in liver involves, in addition, modulation of the activities of these two pathways is also available. Thus, preliminary evidence indicates that the activity of acylglycerol synthesizing enzymes is modulated by reversible cyclic AMP dependent phosphorylation (Nimmo and Houston, 1978; Haagsman et al., 1981). Activity of the outer membrane-bound carnitine palmitoyltransferase of liver mitochondria normally remains largely suppressed because the high malonyl-CoA concentration prevailing in livers in the fed state inhibits the carnitine palmitoyltransferase activity, apparently competitively with respect to fatty acyl-CoA. Under conditions of active fatty acid oxidation in liver, the concentrations of fatty acyl-CoA rise whereas those of malonyl-CoA fall, so that the resultant favorable fatty acyl-CoA/malonyl-CoA ratio speeds up fatty acylcarnitine formation. Much evidence consistent with the possible regulatory role of malonyl-CoA

in hepatic mitochondrial fatty acid oxidation is available (McGarry and Foster, 1980). The occasional lack of correlation between hepatic malonyl-CoA
content and the rates of concurrent ketogenesis in certain situations and the
recognition that the inhibitory effect of malonyl-CoA on the outer carnitine
palmitoyltransferase is less marked with mitochondria from livers of starved
as opposed to those from fed rats (Benito and Williamson, 1978; Cook et al.,
1980; Ontko and Johns, 1980; Saggerson and Carpenter, 1981a), owing to
the expression of a malonyl-CoA insensitive form of carnitine palmitoyltransferase on starvation (Bremer, 1981; Saggerson and Carpenter, 1981b),
indicate that additional malonyl-CoA independent mechanisms effectively
contribute to the control of fatty acid oxidation in liver. Although the activity
of mitochondrial outer carnitine palmitoyltransferase of muscles and heart is
also strongly inhibited by malonyl-CoA in vitro (Saggerson and Carpenter,
1981b), whether in these tissues, which lack active acetyl-CoA carboxylase
for malonyl-CoA synthesis, malonyl-CoA inhibition is ever functional for the
regulation of fatty acid oxidation in vivo is unknown, but is considered unlikely because the suppressive effect of carbohydrate on fatty acid oxidation
is less marked in these tissues unlike that in liver (McGarry and Foster,
1980). It should be noted that inasmuch as fatty acyl-CoA esters inhibit acetyl-CoA carboxylase activity, increased fatty acid availability to liver, by rais-
ing cytosolic fatty acyl-CoA, can suppress malonyl-CoA content and thus
bring about full activation of β-oxidation, particularly when liver glycogen
stores are low to presumably further restrict the supply of lipogenic intermediate malonyl-CoA and liver carnitine content is high (McGarry and Foster,
1980).

Much indirect evidence suggests that the distribution of fatty acids at the
acylglycerol, acylcarnitine branch point is subjected to hormonal control.
For example, livers of female rats esterify more fatty acids and oxidize less
than those of male rats. Starvation, anti-insulin serum, alloxan, and glucagon lower acylglycerol synthesis while enhancing ketone body production
and apart from the fact that dibutyryl cyclic AMP reproduces these effects,
details of how hormones bring about these changes are unknown (Heimberg
et al., 1978).

2. REGULATION OF ACETYL-CoA UTILIZATION FOR KETOGENESIS

Although ketogenesis is generally regarded to result from an overproduc-
tion of acetyl-CoA in liver mitochondria and for this rates of β-oxidation are
evidently important, a number of observations suggest that the control of
acetyl-CoA partitioning between ketogenic and nonketogenic routes also
plays an important role. Acetyl-CoA in liver mitochondria is a direct precursor of acetate (Söling et al., 1974), acetylcarnitine, citrate, and acetoacetate
(Lopes-Cardozo et al., 1975). The latter two are the major routes of acetyl-
CoA disposal in liver mitochondria. For the control of acetyl-CoA partition-

ing between citrate and 3-hydroxy-3-methylglutaryl-CoA, substrate availability and product inhibition seem to be the major modes involved and this seems to be accomplished in a number of ways.

The hydroxymethylglutaryl-CoA pathway, the major route involved in acetoacetate production, requires the participation of acetyl-CoA acetyltransferase (EC 2.3.1.9) for the conversion of acetyl-CoA to acetoacetyl-CoA, and of the hydroxymethylglutaryl-CoA synthase (EC 4.1.3.5) which catalyzes the reaction: acetoacetyl-CoA + acetyl-CoA = 3-hydroxy-3-methylglutaryl-CoA + CoA. The activity of acetyl-CoA acetyltransferase is strongly inhibited by acetoacetyl-CoA and CoA, both of which decrease its affinity for acetyl-CoA (Huth et al., 1978). Acetoacetyl-CoA inhibits the activity of 3-hydroxy-3-methylglutaryl-CoA synthase (Reed et al., 1975) also. For both the above steps, acetyl-CoA is a substrate and CoA is a product. One would expect, therefore, an elevation in acetyl-CoA/CoA ratio to favor acetoacetate production and many data support it (Sauer and Erfle, 1966; Lopes-Cardozo et al., 1975; Siess et al., 1976).

Rapid β-oxidation of fatty acids in perfused liver (DeBeer et al., 1974) and in isolated mitochondria (Lopes-Cardozo and Van den Bergh, 1972) has been shown to suppress the operation of citric acid cycle apparently from the elevation of mitochondrial NADH/NAD ratio which restricts oxaloacetate availability for citrate synthase and simultaneously inhibits isocitrate oxidation (Lenartowicz et al., 1976). Considerable support for an earlier postulate that oxaloacetate availability normally determines the rate of citrate synthesis has become available. Thus, because of marked protein binding, the concentration of free, as opposed to total, oxaloacetate in matrix of liver mitochondria is now estimated to be near the K_m of citrate synthase (Siess et al., 1976; Brocks et al., 1980). The antiketogenic effect of alanine (Nosadini et al., 1980) and of 3-mercaptopicolinate, an inhibitor of phosphoenolpyruvate carboxykinase (Blackshear et al., 1975), is believed to be exerted, at least in part, from their ability to raise hepatic oxaloacetate concentration. And, in pyruvate carboxylase deficiency, expected to impair oxaloacetate supply, concentration of ketone bodies is elevated (Saudubray et al., 1976).

Long chain acyl-CoA esters inhibit the activity of isolated citrate synthase specifically (Wieland, 1968; Hsu and Powell, 1975; Caggiano and Powell, 1979) but this effect has not been demonstrated with intact mitochondria and its possible involvement in the control of acetyl-CoA utilization for citrate formation in vivo remains uncertain. Similarly an elevation of palmitoyl-CoA generation at the outside of mitochondrial membrane in vitro increases the relative rates of ketogenesis and β-hydroxybutyrate to acetoacetate ratio, and these events can be rationalized in terms of the known inhibition of mitochondrial adenine nucleotide translocase by long chain acyl-CoA esters (Pande and Blanchaer, 1971; Shug et al., 1971) but whether this inhibition is exerted in intact cells is equivocal (Hansford,

1980). Nevertheless, increases in the hepatic content of long chain acyl-CoA and total CoA are known under ketogenic conditions (Kondrup and Grunnet, 1973; Smith, 1978). The possibility that the ATP inhibition of citrate synthase may be involved in promoting ketone body formation *in vivo* has likewise been considered unlikely (Lopes-Cardozo et al., 1975; Hansford, 1980).

C. Regulatory Steps Implicated under Specialized Conditions

Acids like octanoic and butyric are considerably more ketogenic than a longer chain acid, oleate, not only *in vitro* (Whitelaw and Williamson, 1977; Zaleski and Bryla, 1977) but also *in vivo* (Bach et al., 1977; Mortenson, 1981; Frost and Wells, 1981). Medium and shorter chain fatty acids gain direct access to mitochondria, and in mitochondria of liver presence of an appropriate activating enzyme enables their rapid activation. These acids, therefore, do not require carnitine for their transport into liver mitochondria and they do not serve as substrates for the acylglycerol synthesizing enzymes. The ability of medium chain fatty acids to thus bypass the constraints of cytosolic acyl-CoA partitioning and acylcarnitine transport has generally been assumed to account for their greater ketogenic ability. However, isolated liver mitochondria show much higher oxygen consumption and acetoacetate production rates with saturating concentrations of octanoylcarnitine than with those of palmitoylcarnitine (Lee and Fritz, 1970). Because under these conditions, carnitine acylcarnitine translocase and inner carnitine palmitoyltransferase activities are not rate limiting (S. V. Pande, unpublished), these observations indicate that some step of β-oxidation spiral itself, involved prior to the formation of medium chain acyl-CoA intermediates, most likely acyl-CoA dehydrogenase, limits the ketogenic capacity of long chain fatty acids. Thus, the ability of liver mitochondria to oxidize palmitoyl groups is not limited by the capacity of electron transport oxidative phosphorylation segment as inferred earlier (Pande, 1971); it is limited by the acetyl-CoA producing capacity of β-oxidation just as in mitochondria of heart and skeletal muscles (Pande, 1971). It should be added that although the ketogenic adaptation of starvation and diabetes brings about larger increases in the oxidation of long chain acids by intact liver, increases are seen also with acids like octanoic and butyric (McGarry and Foster, 1971a; Whitelaw and Williamson, 1977). To what extent the latter result from the increases in the capacities of their mitochondrial activation and β-oxidation and from adaptations at the acetyl-CoA partitioning steps is not clear.

Oxidation of relatively longer chain fatty acids, such as C_{20} and C_{22} monoenoic acids of certain oils, follows a pattern distinct from that involved in the oxidation of palmitate. The longer chain acids are poorer substrates for activation, carnitine dependent transport, and for the mitochondrial β-oxidation system. Their presence impedes the usual rapid mitochondrial oxidation of more commonly occurring fatty acids, owing to competition, and this is believed to contribute, in part, to the myocardial lipidosis seen following the ingestion of erucic acid containing rapeseed oil (for review, see Sauer and Kramer, 1980). Evidence indicates that these longer chain acids initially undergo chain shortening in peroxisomes and the resulting intermediates are then oxidized in mitochondria. An adaptive increase in the capacity of peroxisomal oxidative pathway is seen following the feeding of high fat diets particularly of those containing long chain monounsaturated fatty acids (Neat et al., 1981). The peroxisomal system in vitro oxidizes CoA esters of trans monounsaturated fatty acids as fast or faster than the corresponding cis isomers (Neat et al., 1981), whereas the reverse applies for the mitochondrial oxidation (Lawson and Kummerow, 1979); the limiting steps involved, however, have not been identified.

Steps that are not ordinarily rate limiting for fatty acid oxidation become so in the presence of inhibitors of those steps. The same applies for the various nutritional or congenital diseases that cause lack of a cofactor or of an enzyme concerned with the oxidation of fatty acids. The following, admittedly incomplete account, is intended to supplement and to update the information summarized in the review of Osmundsen and Sherratt (1978) on "Inhibitors of β-oxidation."

2-Bromopalmitate, which serves as a substrate for cellular transport and activation but poorly impedes the relatively rapid processing of the usual fatty acids at these steps (Mahadevan and Sauer, 1971; Pande et al., 1971). As ester of CoA, 2-bromopalmitate inhibits the outer carnitine palmitoyltransferase and some segment of β-oxidation (Tubbs et al., 1980). Equally potent and irreversible inhibitor of carnitine palmitoyltransferase is methyl-2-tetradecylglycidate, presumably as CoA-ester (Tutwiler and Ryzlak, 1980). ω-Trimethylaminoacyl esters of carnitine (Tubbs et al., 1980) and sulfobetaines (N-alkyl-N,N-dimethyl-3-ammonio-1-propanesulfonates) (Parvin et al., 1980) selectively inhibit mitochondrial carnitine acylcarnitine translocase and thereby the mitochondrial oxidation of acylcarnitines.

2-Mercaptoacetate, but not 2-mercaptopropionate, in intact mitochondria, inhibits long chain acyl-CoA dehydrogenase but whether 2-mercaptoacetate itself or its metabolite is the inhibitory species is unknown (Bauché et al., 1981).

2-Bromooctanoate, by becoming converted to 2-bromo-3-ketooctanoyl-CoA, irreversibly inhibits the 3-ketothiolase (EC 2.3.1.6) activity (Raaka and Lowenstein, 1979). 4-Bromooctanoate does likewise by becoming converted to 4-bromo-3-ketobutyryl-CoA (Olowe, 1981). Arsenite restrains β-oxidation by inhibiting acetoacetyl-CoA thiolase (EC 2.3.1.9) (Rein et al., 1979). The inhibitory effect of pent-4-enoate on the activities of 3-ketoacyl-

CoA and acetoacetyl-CoA thiolases (Fong and Schulz, 1978) seems to be exerted, at least in part, from the accumulation of penta-2,4-dienoyl-CoA formed during pent-4-enoate metabolism. This inference rests on the following: (1) Borrebaek et al. (1980a) and Hiltunen and Davis (1981) have provided evidence that a dehydrogenase of β-oxidation initially converts pent-4-enoyl-CoA to penta-2,4-dienoyl-CoA. The latter, with the participation of NADPH-dependent 2,4-dienoyl-CoA reductase (see Section II,D), then produces pent-2-enoyl-CoA, the normal β-oxidation of which gives rise to acetyl-CoA and propionyl-CoA. (2) The inhibitory effect of pent-4-enoate on fatty acid oxidation, and also that of methylenecyclopropylacetyl-CoA which has a double bond in position 4 (Van Hoof et al., 1979), is decreased under conditions (clofibrate, high fat diet) that elevate mitochondrial 2,4-dienoyl-CoA reductase activity (Borrebaek et al., 1980a,b). These findings indicate also that 2,4-dienoyl-CoA reductase readily becomes a rate limiting enzyme in the metabolism of certain 2,4-dienoyl-CoA esters.

A dietary deficiency of riboflavin induces inhibition of fatty acid oxidation from depression of various acyl-CoA dehydrogenase activities and presumably also of electron transfer flavoprotein (Hoppel et al., 1979). Hypoglycin A intoxication involves inhibition of short chain acyl-CoA dehydrogenase by intermediary formation of methylencyclopropylacetyl-CoA (Tanaka et al., 1976; Kean, 1976) and a genetic defect in one or more of these dehydrogenases has been implicated as the cause of impaired β-oxidation in certain cases of nonketotic carboxylic aciduria (Mantagos et al., 1978; Gregersen et al., 1980).

A carnitine deficiency, whether congenital, dietary, or related to other causes, impairs fatty acid oxidation and generally elevates muscle triglyceride content (Di Mauro et al., 1980; Engel, 1980). The ethanol induced increase of liver triglyceride is prevented by carnitine administration (Hosein and Bexton, 1975) and carnitine is able to normalize serum triglyceride levels in patients exhibiting hypertriglyceridemia (Maebashi et al., 1978; Bougneres et al., 1979), both effects presumably brought about by the fatty acid oxidation promoting effect of carnitine. These latter findings suggest that at the levels normally prevailing in human liver in vivo, carnitine is one of the factors whose concentration limits the utilization of fatty acids for oxidation, at least in certain situations. Impairment of long chain fatty acid oxidation also results from a deficiency of carnitine palmitoyltransferase, which, although discovered in 1973, is now recognized as the most common cause of hereditary muscle weakness with recurrent myoglobinuria (Di Mauro et al., 1980). Several cases of myopathies, often accompanying lipid storage, have been described, and although some defect in fatty acid oxidation, at steps other than just described above, has been suspected, the exact errors involved remain to be identified (Di Mauro et al., 1980). Hopefully, advancing understanding of metabolism would not only enable this but eventually permit also a better management of such unfortunate abnormalities.

ACKNOWLEDGMENTS

This work was supported by grants from the Medical Research Council of Canada (MT-4264) and the Quebec Heart Foundation.

REFERENCES

Abumrad, N. A., Perkins, R. C., Park, J. H., and Park, C. R. (1981). *J. Biol. Chem.* **256**, 9183–9191.
Appelkvist, E. L., and Dallner, G. (1980). *Biochim. Biophys. Acta 617*, 156–160.
Bach, A., Schirardin, H., Weryha, A., and Bauer, M. (1977). *J. Nutr.* **107**, 1863–1870.
Bauché, F., Sabourault, D., Giudicelli, Y., Nordmann, J., and Nordmann, R. (1981). *Biochem. J.* **196**, 803–809.
Benito, M., and Williamson, D. H. (1978). *Biochem. J.* **176**, 331–334.
Bergström, J. D., and Reitz, R. C. (1980). *Arch. Biochem. Biophys.* **204**, 71–79.
Beinert, H. (1963). *In* "The Enzymes" (P. D. Boyer, H. Lardy, and K. Myrbäck, eds.), 2nd ed., Vol. 7, pp. 447–476. Academic Press, New York.
Billheimer, J. T., and Gaylor, J. L. (1980). *J. Biol. Chem.* **255**, 8128–8135.
Binstock, J. F., Pramanik, A., and Schulz, H. (1977). *Proc. Natl. Acad. Sci. U.S.A.* **74**, 492–495.
Björkhem, I. (1973). *Eur. J. Biochem.* **40**, 415–422.
Björkhem, I. (1976). *J. Biol. Chem.* **251**, 5259–5266.
Björkhem, I. (1978). *J. Lipid Res.* **19**, 585–590.
Blackshear, P. J., Holloway, P. A. H., and Alberti, K. G. M. M. (1975). *Biochem. J.* **148**, 353–362.
Borrebaek, B., Osmundsen, H., and Bremer, J. (1980a). *Biochem. Biophys. Res. Commun.* **93**, 1173–1180.
Borrebaek, B., Osmundsen, H., Christiansen, E. N., and Bremer, J. (1980b). *FEBS Lett.* **121**, 23–24.
Bougneres, P. F., Lacour, B., DiGiulio, S., and Assan, R. (1979). *Lancet* **1**, 1401–1402.
Brass, E. P., and Hoppel, C. L. (1978). *J. Biol. Chem.* **253**, 5274–5276.
Bremer, J. (1968). *In* "Cellular Compartmentalization and Control of Fatty Acid Metabolism" (Fed. Eur. Biochem. Symp. IV), pp. 65–88. Universitetsforslaget, Oslo.
Bremer, J. (1981). *Biochim. Biophys. Acta* **665**, 628–631.
Bremer, J., and Wojtczak, A. B. (1972). *Biochim. Biophys. Acta* **280**, 515–530.
Bremer, J., Bjerve, K. S., Borrebaek, B., and Christiansen, R. (1976). *Mol. Cell. Biochem.* **12**, 113–124.
Bressler, R. (1970). *Comp. Biochem.* **18**, 331–359.
Brocks, D. G., Siess, E. A., and Wieland, O. H. (1980). *Biochem. J.* **188**, 207–212.
Broekhoven, A. V., Peeters, M. C., Debeer, L. J., and Mannaerts, G. P. (1981). *Biochem. Biophys. Res. Commun.* **100**, 305–312.

Bronfman, M., Inestrosa, N. C., and Leighton, F. (1979). *Biochem. Biophys. Res. Commun.* **88**, 1030–1036.

Brown, G. W., Chapman, D. D., Matheson, H. R., Chaikoff, I. L., and Dauben, W. G. (1954). *J. Biol. Chem.* **209**, 537–548.

Caggiano, A. V., and Powell, G. L. (1979). *J. Biol. Chem.* **254**, 2800–2806.

Chan, S. H. P., and Barbour, R. L. (1979). *In* "Membrane Bioenergetics International Workshop" (C. P. Lee, G. Schatz, and L. Ernster, eds.), pp. 521–532. Addison-Wesley, Reading, Massachusetts.

Christiansen, R. Z. (1979). *FEBS Lett.* **103**, 89–92.

Christiansen, R. Z., and Bremer, J. (1978). *In* "Biochemical and Clinical Aspects of Ketone Body Metabolism" (H. D. Söling and D. Claus, eds.), pp. 59–69. Thieme, Stuttgart.

Clark, P. R. H., and Bieber, L. L. (1981a). *J. Biol. Chem.* **256**, 9861–9868.

Clark, P. R. H., and Bieber, L. L. (1981b). *J. Biol. Chem.* **256**, 9869–9873.

Cook, G. A., Otto, D. A., and Cornell, N. W. (1980). *Biochem. J.* **192**, 955–958.

Cooper, T. G., and Beevers, H. (1969). *J. Biol. Chem.* **244**, 3514–3520.

Dampsey, M. E., McCoy, K. E., Baker, H. N., Dimitriadou-Vafiadou, A., Lorsbach, T., and Howard, J. B. (1981). *J. Biol. Chem.* **256**, 1867–1873.

Davidoff, F., and Korn, E. D. (1965). *J. Biol. Chem.* **240**(6), 1549–1558.

Davidson, B., and Schulz, H. (1981). *Fed. Proc., Fed. Am. Soc. Exp. Biol.* **40**(6), Abstr. 292.

DeBeer, L. J., Mannaerts, G., and deSchepper, P. J. (1974). *Eur. J. Biochem.* **47**, 591–600.

Di Donato, S., Peluchetti, D., Rimoldi, M., Bertagnolio, B., Uziel, G., and Cornelio, F. (1980). *Clin. Chim. Acta* **100**, 209–214.

Di Mauro, S., Trevisan, C., and Hays, A. (1980). *Muscle Nerve* **3**, 369–388.

Dommes, V., Baumgart, C., and Kunau, W. H. (1981). *J. Biol. Chem.* **256**, 8259–8262.

Engel, A. G. (1980). *In* "Carnitine Biosynthesis, Metabolism, and Functions" (R. A. Frenkel and J. D. McGarry, eds.), pp. 271–285. Academic Press, New York.

Exton, J. H., Corbin, J. G., and Harper, S. C. (1972). *J. Biol. Chem.* **247**, 4996–5003.

Fiecchi, A., Galli-Kienle, M., Scala, A., Galli, G., and Paoletti, R. (1973). *Eur. J. Biochem.* **38**, 516–528.

Flatt, J. P. (1972). *Diabetes* **21**, 150–153.

Foerster, E. C., Fährenkemper, T., Rabe, U., Graf, P., and Sies, H. (1981). *Biochem. J.* **196**, 705–712.

Foliot, A. (1979). *Actual. Chim. Ther.* **6**, 121–139.

Fong, J. C., and Schulz, H. (1977). *J. Biol. Chem.* **252**, 542–547.

Fong, J. C., and Schulz, H. (1978). *J. Biol. Chem.* **253**, 6917–6922.

Fournier, N., Geoffroy, M., and Deshusses, J. (1978). *Biochim. Biophys. Acta* **533**, 457–464.

Fritz, I. B. (1961). *Physiol. Rev.* **41**, 52–129.

Fritz, I. B. (1963). *Adv. Lipid Res.* **1**, 285–334.

Frost, S. C., and Wells, M. A. (1981). *Arch. Biochem. Biophys.* **211**, 537–546.

Galli-Kienle, M., Cighetti, G., Santaniello, E., Fiecchi, A., and Galli, G. (1976). *Lipids* **11**, 235–240.

Galzigna, L., Rossi, C. R., Sartorelli, L., and Gibson, D. M. (1967). *J. Biol. Chem.* **242**, 2111–2115.

Garland, P. B., and Yates, D. W. (1967). *In* "Mitochondrial Structure and Compartmentation" (E. Quagliariello, S. Papa, E. C. Slater, and J. M. Tager, eds.), pp. 385–399. Adriatica Editrice, Bari.

Gloster, J., and Harris, P. (1977). *Biochem. Biophys. Res. Commun.* **74**, 506–513.

Goresky, C. A., Daly, D. S., Mishkin, S., and Arias, I. M. (1978). *Am. J. Physiol.* **234**, E542–E553.

Green, D. W., and Allman, D. W. (1968). *In* "Metabolic Pathways" (D. M. Greenberg, ed.), Vol. 2, pp. 1–32. Academic Press, New York.

Gregersen, N., Rosleff, F., Kolvraa, S., Hobolth, N., Rasmussen, K., and Lauritzen, R. (1980). *Clin. Chim. Acta* **102**, 179–189.

Greter, J., Lindstedt, S., Seeman, H., and Steen, G. (1980). *Clin. Chem. (Winston-Salem, N.C.)* **26**, 261–265.

Greville, G. D., and Tubbs, P. K. (1968). *Essays Biochem.* **4**, 155–212.

Groot, P. H. E., Scholte, H. R., and Hülsmann, W. C. (1976). *Adv. Lipid Res.* **14**, 75–126.

Haagsman, H. P., De Haas, C. G. M., Geelen, M. J. H., and Van Golde, L. M. G. (1981). *Biochim. Biophys. Acta* **664**, 74–81.

Hansford, R. G. (1980). *Curr. Top. Bioenerg.* **10**, 217–278.

Harper, R. D., and Saggerson, E. D. (1975). *Biochem. J.* **152**, 485–494.

Havel, R. J., Felts, J. M., and van Duyne, C. M. (1962). *J. Lipid Res.* **3**, 297–308.

Heimberg, M., Goh, E. H., Klausner, H. A., Soler-Argilga, C., Weinstein, I., and Wilcox, H. G. (1978). *In* "Disturbances in Lipid and Lipoprotein Metabolism" (J. M. Dietschy, A. M. Gotto, Jr., and J. A. Ontko, eds.), pp. 251–267. Am. Physiol. Soc., Bethesda, Maryland.

Hiltunen, J. K., and Davis, E. J. (1981). *Biochem. J.* **194**, 427–432.

Hoppel, C., Di Marco, J. P., and Tandler, B. (1979). *J. Biol. Chem.* **254**, 4164–4170.

Hosein, E. A., and Bexton, B. (1975). *Biochem. Pharmacol.* **24**, 1859–1863.

Hryb, D. J., and Hogg, J. F. (1979). *Biochem. Biophys. Res. Commun.* **87**, 1200–1206.

Hsu, K. H. L., and Powell, G. L. (1975). *Proc. Natl. Acad. Sci. U.S.A.* **72**, 4729–4733.

Huth, W., Steinmann, R., Holze, G., and Seubert, W. (1978). *In* "Biochemical and Clinical Aspects of Ketone Body Metabolism" (H. D. Söling and C. D. Seufert, eds.), pp. 11–12. Thieme, Stuttgart.

Ide, T., and Ontko, J. A. (1981). *J. Biol. Chem.* **256**, 10247–10255.

Idell-Wenger, J., and Neely, J. R. (1978). *In* "Disturbances in Lipid and Lipoprotein Metabolism" (J. M. Dietschy, A. Gotto, and J. A. Ontko, eds.), pp. 269–284. Am. Physiol. Soc., Bethesda, Maryland.

Inestrosa, N. C., Bronfman, M., and Leighton, F. (1979). *Biochem. J.* **182**, 779–788.

Kam, W., Kumaran, K., and Landau, B. R. (1978). *J. Lipid Res.* **19**, 591–600.

Karpati, G., Carpenter, S., Engel, A. G., Walters, G., Allen, J., Rathman, S., Klassen, G., and Manner, O. A. (1975). *Neurology* **25**, 16–24.

Kawamura, N., Moser, H. W., and Kishimoto, Y. (1981). *Biochem. Biophys. Res. Commun.* **99**, 1216–1225.

Kean, E. A. (1976). *Biochim. Biophys. Acta* **422**, 8–14.

Ketterer, B., Tipping, E., and Hackeney, J. F. (1976). *Biochem. J.* **155**, 511–521.

Kondrup, J., and Grunnet, N. (1973). *Biochem. J.* **132**, 373–379.

Krahling, J. B., Gee, R., Gauger, J. A., and Tolbert, N. E. (1979). *J. Cell. Physiol.* **101**, 375–390.

Kramar, R., Hüttinger, M., Gmeiner, B., and Goldenberg, H. (1978). *Biochim. Biophys. Acta* **531**, 353–356.

Krebs, H. A., Wallace, P. G., Hems, R., and Freedland, R. A. (1969). *Biochem. J.* **112**, 595–600.

Krisans, S. K., Mortensen, R. M., and Lazarow, P. B. (1980). *J. Biol. Chem.* **255**, 9599–9607.

Kunau, W. H., and Bartnik, F. (1974). *Eur. J. Biochem.* **48**, 311–318.

Kunau, W. H., and Dommes, P. (1978). *Eur. J. Biochem.* **91**, 533–544.

Kunau, W. H., and Lauterbach, F. (1978). *FEBS Lett.* **94**, 120–124.

Lawson, L. D., and Kummerow, F. A. (1979). *Lipids* **14**, 501–503.

Lazarow, P. B. (1977). *Science* **197**, 580–581.

Lazarow, P. B. (1978). *J. Biol. Chem.* **253**, 1522–1528.

Lazarow, P. B. (1981). *Arch. Biochem. Biophys.* **206**, 342–345.

Lazarow, P. B., and de Duve, C. (1976). *Proc. Natl. Acad. Sci. U.S.A.* **73**, 2043–2046.

Lee, L. P. K., and Fritz, I. B. (1971). *Can. J. Biochem.* **49**, 599–605.

Lenartowicz, E., Winter, C., Kunz, W., and Wojtczak, A. B. (1976). *Eur. J. Biochem.* **67**, 137–144.

Levi, A. J., Gatmaitan, Z., and Arias, I. M. (1969). *J. Clin. Invest.* **48**, 2156–2167.

Londesborough, J. C., and Webster, L. T., Jr. (1974). *In* "The Enzymes" (P. D. Boyer, ed.), 3rd ed., Vol. 10, pp. 469–488. Academic Press, New York.

Lopes-Cardozo, M., and Van den Bergh, S. G. (1972). *Biochim. Biophys. Acta* **283**, 1–15.

Lopes-Cardozo, M., Mulder, I., VanVugt, F., Hermans, P. G. C., Van den Bergh, S. G., Klazinga, W., and DeVries-Akkerman, E. (1975). *Mol. Cell. Biochem.* **9**, 155–173.

Lopes-Cardozo, M., Klazinga, W., and Van den Bergh, S. G. (1978). *Eur. J. Biochem.* **83**, 629–634.

Lund, H., Borrebaek, B., and Bremer, J. (1980). *Biochim. Biophys. Acta* **620**, 364–371.

Lunzer, M. A., Manning, J. A., and Ockner, R. K. (1977). *J. Biol. Chem.* **252**, 5483–5487.

Lynen, F. (1954). *Harvey Lect. Ser.* **48**, 210–244.

Maebashi, M., Sato, M., Kawamura, N., Imamura, A., and Yoshinaga, K. (1978). *Lancet* **2**, 805–807.

Mahadevan, S., and Sauer, F. (1971). *J. Biol. Chem.* **246**, 5862–5867.

Maloy, S. R., Ginsburgh, C. L., Simons, R. W., and Nunn, W. D. (1981). *J. Biol. Chem.* **256**, 3735–3742.

Mannaerts, G. P., DeBeer, L. J., Thomas, J., and de Schepper, P. J. (1979). *J. Biol. Chem.* **254**, 4585–4595.

Mantagos, S., Genel, M., and Tanaka, K. (1978). *Pediatr. Res.* **12**, 453, Abstr. 538.

Mayes, P. A., and Felts, J. M. (1967). *Nature (London)* **215**, 716–718.

McGarry, J. D., and Foster, D. W. (1971a). *J. Biol. Chem.* **246**, 1149–1159.

McGarry, J. D., and Foster, D. W. (1971b). *J. Biol. Chem.* **246**, 6247–6253.

McGarry, J. D., and Foster D. W. (1980). *Annu. Rev. Biochem.* **49**, 395–420.

McGarry, J. D., Robles-Valdes, C., and Foster, D. W. (1975). *Proc. Natl. Acad. Sci. U.S.A.* **72**, 4385–4388.

McKean, M. C., Frerman, F. E., and Mielke, D. M. (1979). *J. Biol. Chem.* **254**, 2730–2735.

Middleton, B. (1973). *Biochem. J.* **132**, 717–730.

Mishkin, S., and Turcotte, R. (1974). *Biochem. Biophys. Res. Commun.* **60**, 376–381.

Mishkin, S., Stein, L., Gatmaitan, Z., and Arias, I. M. (1972). *Biochem. Biophys. Res. Commun.* **47**, 997–1003.

Mishkin, S., Morris, H. P., Murthy, P. V. N., and Halperin, M. L. (1977). *J. Biol. Chem.* **252**, 3626–3628.

Miyazawa, S., Osumi, T., and Hashimoto, T. (1980). *Eur. J. Biochem.* **103**, 589–596.

Mizugaki, M., and Uchiyama, M. (1973). *Biochem. Biophys. Res. Commun.* **50**, 48–53.

Mortensen, P. B. (1981). *Biochim. Biophys. Acta* **664**, 335–348.

Neat, C. E., Thomassen, M. S., and Osmundsen, H. (1981) *Biochem. J.* **196**, 149–159.

Neely, J. R., and Morgan, H. W. (1974). *Annu. Rev. Physiol.* **36**, 413–459.

Nimmo, H. G., and Houston, B. (1978). *Biochem. J.* **176**, 607–610.

Nosadini, R., Datta, H., Hodson, A., and Alberti, K. G. M. M. (1980). *Biochem. J.* **190**, 323–332.

Numa, S. (1981). *Trends Biochem. Sci.* **6**, 113–115.

Nunn, W. D., and Simons, R. W. (1978). *Proc. Natl. Acad. Sci. U.S.A.* **75**, 3377–3381.

Nunn, W. D., Simons, R. W., Egan, P. A., and Maloy, S. R. (1979). *J. Biol. Chem.* **254**, 9130–9134.

O'Brien, W. J., and Frerman, F. E. (1977). *J. Bacteriol.* **132**, 532–540.

Ockner, R. K., and Manning, J. A. (1974). *J. Clin. Invest.* **54**, 326–328.

Ockner, R. K., Manning, J. A., Pappenhausen, R. B., and Ho, W. K. L. (1972). *Science* **177**, 56–58.

Olowe, Y. (1981). *Fed. Proc., Fed. Am. Soc. Exp. Biol.* **40**(6), Abstr. 285.

Olowe, Y., and Schulz, H. (1980). *Eur. J. Biochem.* **109**, 425–429.

Ontko, J. A. (1967). *Biochim. Biophys. Acta* **137**, 1–12.

Ontko, J. A., and Johns, M. L. (1980). *Biochem. J.* **192**, 959–962.

Oram, J. F., Wenger, J. I., and Neely, J. R. (1975). *J. Biol. Chem.* **250**, 73–78.

Osmundsen, H., and Neat, C. E. (1979). *FEBS Lett.* **107**, 81–85.

Osmundsen, H., and Sherratt, H. S. A. (1978). *Biochem. Soc. Trans.* **6**, 84–88.

Osumi, T., and Hashimoto, T. (1978). *Biochim. Biophys. Res. Commun.* **83**, 479–485.

Osumi, T., and Hashimoto, T. (1979). *J. Biochem. (Tokyo)* **85**, 131–139.

Osumi, T., Hashimoto, T., and Ui, N. (1980). *J. Biochem. (Tokyo)* **87**, 1735–1746.

Pande, S. V. (1971)..*J. Biol. Chem.* **246**, 5384–5390.

Pande, S. V. (1973). *Biochim. Biophys. Acta* **306**, 15–20.

Pande, S. V. (1975). *Proc. Natl. Acad. Sci. U.S.A.* **72**, 883–887.

Pande, S. V. (1981). *Biochim. Biophys. Acta* **663**, 669–673.

Pande, S. V., and Blanchaer, M. C. (1970). *Biochim. Biophys. Acta* **202**, 43–48.

Pande, S. V., and Blanchaer, M. C. (1971). *J. Biol. Chem.* **246**, 402–411.

Pande, S. V., and Mead, J. F. (1968). *Biochim. Biophys. Acta* **152**, 636–638.

Pande, S. V., and Parvin, R. (1979). *In* "Function and Molecular Aspects of Biomembrane Transport" (E. Quagliariello, M. Klingenberg and F. Palmieri, eds.), pp. 287–290. Elsevier/North Holland Biomedical Press, New York.

Pande, S. V., and Parvin, R. (1980a). *Biochim. Biophys. Acta* **617**, 363–370.

Pande, S. V., and Parvin, R. (1980b). *J. Biol. Chem.* **255**, 2994–3001.

Pande, S. V., and Parvin, R. (1980c). *In* "Carnitine Biosynthesis, Metabolism and Functions" (R. A. Frenkel and J. D. McGarry, eds.), pp. 143–155. Academic Press, New York.

Pande, S. V., Siddiqui, A. W., and Gattereau, A. (1971). *Biochim. Biophys. Acta* **248**, 156–166.

Parameswaran, M., and Arinze, I. J. (1981). *Biochim. Biophys. Acta* **672**, 219–223.

Paris, S., Samuel, D., Jacques, Y., Gache, C., Franchi, A., and Ailhaud, G., (1978). *Eur. J. Biochem.* **83**, 235–243.

Paris, S., Samuel, D., Romey, G., and Ailhaud, G. (1979). *Biochimie* **61**, 361–367.

Parvin, R., and Pande, S. V. (1979). *J. Biol. Chem.* **254**, 5423–5429.

Parvin, R., Goswami, T., and Pande, S. V. (1980). *Can. J. Biochem.* **58**, 822–830.

Pawar, S., and Schulz, H. (1981). *J. Biol. Chem.* **256**, 3894–3899.

Pettersen, J. E. (1973). *Biochim. Biophys. Acta* **306**, 1–14.

Pettersen, J. E., and Aas, M. (1973). *Biochim. Biophys. Acta* **326**, 305–313.

Pettersen, J. E., Jellum, E., and Eldjarn, L. (1972). *Clin. Chim. Acta* **38**, 17–24.

Philipp, D. P., and Parsons, P. (1979). *J. Biol. Chem.* **254**, 10785–10790.

Pilkis, S. J., Riou, J. P., and Claus, T. H. (1976). *J. Biol. Chem.* **251**, 7841–7852.

Podack, E. R., and Seubert, W. (1972). *Biochim. Biophys. Acta* **280**, 235–247.

Raaka, B. M., and Lowenstein, J. M. (1979). *J. Biol. Chem.* **254**, 6755–6762.

Ramsay, R. R., and Tubbs, P. K. (1975). *FEBS Lett.* **54**, 21–25.

Ray, T. K., and Cronan, J. E. (1976). *Proc. Natl. Acad. Sci. U.S.A.* **73**, 4374–4378.

Reed, W. D., Clinkenbeard, K. D., and Lane, M. D. (1975). *J. Biol. Chem.* **250**, 3117–3123.

Rein, K. A., Borrebaek, B., and Bremer, J. (1979). *Biochim. Biophys. Acta* **574**, 487–494.

Renaud, G., Foliot, A., and Infante, R. (1978). *Biochem. Biophys. Res. Commun.* **80**, 327–334.

Robles-Valdes, C., McGarry, J. D., and Foster, D. W. (1976). *J. Biol. Chem.* **251**, 6007–6012.

Rüstow, B., Kunze, D., Hodi, J., and Egger, E. (1979). *FEBS Lett.* **108**, 469–472.

Saggerson, E. D., and Carpenter, C. A. (1981a). *FEBS Lett.* **129**, 225–228.

Saggerson, E. D., and Carpenter, C. A. (1981b). *FEBS Lett.* **129**, 229–232.

Samuel, D., Paris, S., and Ailhaud, G. (1976). *Eur. J. Biochem.* **64**, 583–595.

Saudubray, J. M., Marsac, C., Charpentier, C., Cathelineau, L., Besso Leaud, M., and Leroux, J. P. (1976). *Acta Paediatr. Scand.* **65**, 717–724.

Sauer, F. D., and Erfle, J. D. (1966). *J. Biol. Chem.* **241**, 30–37.

Sauer, F. D., and Kramer, J. K. G. (1980). *Adv. Nutr. Res.* **3**, 207–230.

Seubert, W., Lamberts, I., Kramer, R., and Ohly, B. (1968). *Biochim. Biophys. Acta* **164**, 498–517.

Shindo, Y., and Hashimoto, T. (1978). *J. Biochem. (Tokyo)* **84**, 1177–1181.

Shug, A., Lerner, E., Elson, C., and Shrago, E. (1971). *Biochem. Biophys. Res. Commun.* **43**, 557–563.

Shumate, J. B., and Choksi, R. M. (1981). *Biochem. Biophys. Res. Commun.* **100**, 978–981.

Siess, E. A., Brocks, D. G., and Wieland, O. H. (1976). *FEBS Lett.* **69**, 265–271.

Smith, C. M. (1978). *J. Nutr.* **108**, 863–873.

Snoswell, A. M., and Koundakjian, P. P. (1972). *Biochem. J.* **127**, 133–141.

Solberg, H. E. (1972). *Biochim. Biophys. Acta* **280**, 422–433.

Söling, H. D., Graf, M., and Seufert, D. (1974). *Alfred Benzon Symp.* **6**, 695–704.

Spector, A (1971). *Prog. Biochem. Pharmacol.* **6**, 130–176.

Spitzer, J. J., Nakamura, H., Hori, S., and Gold, M. (1969). *Proc. Soc. Exp. Biol. Med.* **132**, 281–286.

Staack, H., Binstock, J. F., and Schulz, H. (1978). *J. Biol. Chem.* **253**, 1827–1831.

Stanley, K. K., and Tubbs, P. K. (1975). *Biochem. J.* **150**, 77–88.

Stoffel, W., and Caeser, H. (1965). *Hoppe-Seyler's Z. Physiol. Chem.* **341**, 76–83.

Stoffel, W., Ditzer, R., and Caesar, H. (1964). *Hoppe-Seyler's Z. Physiol. Chem.* **339**, 167–181.

Struijk, C. B., and Beerthuis, R. K. (1966). *Biochim. Biophys. Acta* **116**, 12–22.

Tanaka, K., Kean, E. A., and Johnson, B. (1976). *N. Engl. J. Med.* **295**, 461–467.

Tanaka, T., Hosaka, K., Hoshimaru, M., and Numa, S. (1979). *Eur. J. Biochem.* **98**, 165–172.

Thomas, J., DeBeer, L. J., de Schepper, P. J., and Mannaerts, G. P. (1980). *Biochem. J.* **190**, 485–494.

Tubbs, P. K., and Ramsay, R. R. (1979). In "Function and Molecular Aspects of Biomembrane Transport" (E. Quagliariello, M. Klingenberg, and F. Palmieri, eds.), pp. 279–286. Elsevier/North Holland Biomedical Press, New York.

Tubbs, P. K., Ramsay, R. R., and Edwards, M. R. (1980). In "Carnitine Biosynthesis, Metabolism and Functions" (R. A. Frenkel and J. D. McGarry, eds.), pp. 207–217. Academic Press, New York.

Tutwiler, G. F., and Ryzlak, M. T. (1980). *Life Sci.* **26**, 393–397.

Tzur, R., Tal, E., Shapiro, B. (1964). *Biochim. Biophys. Acta* **84**, 18–23.

Van Harken, D. R., Dixon, C. W., and Heimberg, M. (1969). *J. Biol. Chem.* **244**, 2278–2285.

Van Hoof, F., Hue, L., and Sherratt, H. S. A. (1979). *Biochem. Soc. Trans.* **7**, 163–165.

Van Tol, A. (1975). *Mol. Cell. Biochem.* **7**, 19–31.

Vary, T. C., Reibel, D. K., and Neely, J. R. (1981). *Annu. Rev. Physiol.* **43**, 419–430.

Wada, F., and Usami, M. (1977). *Biochim. Biophys. Acta* **487**, 261–268.

Wada, F., Usami, M., and Hayashi, T. (1971). *J. Biochem. (Tokyo)* **70**, 1065–1067.

Wakil, S. J., and Barnes, E. M. Jr. (1971). *Compr. Biochem.* **185**, 57–104.

Waterson, R. M., and Hill, R. L. (1972). *J. Biol. Chem.* **247**, 5258–5265.

Weinstein, I., Wasfi, I., and Heimberg, M. (1981). *Biochim. Biophys. Acta* **664**, 124–132.

Weisiger, R., Gollan, J., and Ockner, R. (1981). *Science* **211**, 1048–1051.

Whitelaw, E., and Williamson, D. H. (1977). *Biochem. J.* **164**, 521–528.

Whitmer, J. T., Idell-Wenger, J. A., Rovetto, M. J., and Neely, J. R. (1978). *J. Biol. Chem.* **253**, 4305–4309.

Wieland, O. (1968). *Adv. Metab. Dis.* **3**, 1–47.

Williamson, D. H. (1979). *Biochem. Soc. Trans.* **7**, 1313–1321.

Williamson, D. H., Veloso, D., Ellington, E. V., and Krebs, H. A. (1969). *Biochem. J.* **114**, 575–584.
Wu-Rideout, M.Y. C., Elson, C., and Shrago, E. (1976). *Biochem. Biophys. Res. Commun.* **71**, 809–816.
Zaleski, J., and Bryla, J. (1977). *Arch. Biochem. Biophys.* **183**, 553–562.
Zammit, V. A. (1981). *Trends Biochem. Sci.* **6**, 46–49.

16

The Mechanisms of Fatty Acid Chain Elongation and Desaturation in Animals

H. SPRECHER

I. INTRODUCTION

The pioneering studies by Mead and his colleagues and in Klenk's laboratory, as reviewed by these investigators, established that polyunsaturated fatty acids are made as shown in Fig. 1 (Klenk, 1965; Mead, 1971). Animals convert dietary carbohydrate and protein in part to acetyl-CoA, which can

385

High and Low Erucic Acid Rapeseed Oils

$$
\begin{array}{ccccc}
9 & 6 & 5 & 4 \\
\downarrow & \downarrow & \downarrow & \downarrow
\end{array}
$$

1.　$16{:}0 \rightarrow 16{:}1 \rightarrow 16{:}2 \rightarrow 18{:}2 \rightarrow 18{:}3 \rightarrow 20{:}3 \rightarrow 20{:}4$
　　　$\downarrow$

2.　$18{:}0 \rightarrow 18{:}1 \rightarrow 18{:}2 \rightarrow 20{:}2 \rightarrow 20{:}3$

3.　　　　$18{:}2 \rightarrow 18{:}3 \rightarrow 20{:}3 \rightarrow 20{:}4 \rightarrow 22{:}4 \rightarrow 22{:}5$

4.　　　　$18{:}3 \rightarrow 18{:}4 \rightarrow 20{:}4 \rightarrow 20{:}5 \rightarrow 22{:}5 \rightarrow 22{:}6$

Fig. 1. Pathways for the production of unsaturated fatty acids derived from linolenate (18:3), linoleate (18:2), oleate (18:1), and palmitoleate (16:1).

then be used by acetyl-CoA carboxylase and fatty acid synthetase for the synthesis of palmitic acid. Although substantial amounts of stearic acid are provided in the diet this acid may also be made by chain elongating dietary or newly synthesized palmitic acid. Both palmitic and stearic acid are desaturated at the 9-position to give respectively palmitoleic acid and oleic acid. Each of these acids then serve as the initial unsaturated precursor for the biosynthesis of an independent family of unsaturated fatty acids. Dietary linoleic acid and linolenic acid also each serve as the initial unsaturated precursor for two additional families of unsaturated fatty acids. There is no direct crossover in metabolism between metabolites of these four families of acids. Each series is similar in that the more highly unsaturated metabolites are produced by an alternating series of desaturation and chain elongation reactions. Indeed, as discussed in this chapter, there is an increasing amount of evidence supporting the concept that only four position specific desaturases are required for the synthesis of all unsaturated fatty acids. As shown in Fig. 1 these desaturases introduce double bonds at positions 9, 6, 5, and 4. Although the enzymes required for chain elongating fatty acids have not been studied as extensively, there is a significant amount of data suggesting that microsomes also contain more than one chain elongating system. This evidence, as well as the overall regulation of unsaturated fatty acid biosynthesis in animals, is the topic of this chapter.

II. THE 9-DESATURASE(S)

The enzyme or enzymes that introduce a double bond at position-9 in a fatty acid have been the most extensively studied even though this is the only desaturase that is not required for converting dietary linoleate or linolenate to longer chain (n-6) or (n-3) metabolites. This desaturation, as well as other position specific desaturases, requires molecular oxygen and reducing equivalents as obligatory cofactors (Bloomfield and Bloch, 1958; Stoffel, 1961; Marsch and James, 1962). NADH is the preferred electron donor (Oshino et al., 1974) and the true substrate and product are the acyl-CoA derivatives (Holloway and Holloway, 1974). Desaturation involves the flow of electrons from NADH to cytochrome b_5 (Oshino et al., 1966, 1971; Oshino and Omura, 1973) via NADH-cytochrome b_5 reductase (Jones et al., 1969; Holloway and Wakil, 1970; Holloway, 1971) to the terminal cyanide sensitive 9-desaturase (Oshino, 1972; Shimakata et al., 1972). The desaturase complex has an absolute requirement of lipid for activity (Jones et al., 1969; Holloway, 1971; Enoch et al., 1976).

The 9-desaturase from rat liver is a single polypeptide of 53,000 daltons containing one atom of nonheme iron (Strittmatter et al., 1974). This enzyme from chicken liver microsomes has a molecular weight of 33,600 and the antibody to this enzyme does not cross-react with the 9-desaturase from rat liver (Prasad and Joshi, 1979b). These findings suggest that there are species differences in the enzyme that introduces a double bond at position 9. This hypothesis is supported by several additional findings. Although cyanide inhibits the 9-desaturase from livers of rat, hen, guinea pig, and rat lung, this enzyme from microsomal preparation of rabbit liver, pig thyroid, and bovine adrenocortex is insensitive to cyanide (Hiwatashi et al., 1975). With rat liver microsomes (Paulsrud et al., 1970) or with the purified desaturases (Enoch et al., 1976) the best substrate for desaturation is stearoyl-CoA. Conversely, when saturated fatty acids of varying chain lengths are incubated with hen liver microsomes two peaks of activity are found (Johnson et al., 1969; Brett et al., 1971). One maximum was found for myristic acid while the other was for stearic acid, thus suggesting the presence of two different chain length specific 9-desaturases.

Although animals cannot synthesize linoleate, microsomal preparations from hen and pig liver as well as from goat mammary glands have an enzyme that converts cis-12-octadecenoic acid to linoleate. This desaturation proceeds very slowly when microsomes are used from mouse and rabbit liver and no detectable activity is found with microsomes from rat or hamster liver (Gurr et al., 1972). Indeed, recent studies have shown that a variety of different monoenoic acids of different chain lengths, containing a single cis or trans double bond, will serve as substrates for a 9-desaturase when incubated with rat liver microsomes (Pollard et al., 1980a; Mahfouz et al., 1980). In some cases the two double bonds in the product are conjugated while with other substrates the skipped pattern of unsaturation is initiated. Other substrates gave rise to products containing two or more methylene carbons between the two double bonds. Surprisingly, trans 4-18:1, trans 6-18:1, trans 11-18:1, trans 12-18:1, trans 13-18:1, and trans 14-18:1 are all desaturated at position 9 to give the respective dienoic acids with an inversion of the original double bond to the cis configuration (Mahfouz *et al.*, 1980).

Although the mechanism for introducing a cis double bond at position 9 has not been established the process is stereospecific resulting in removal of only the 9-D and 10-D hydrogens (Schroepfer and Bloch, 1965; Morris *et al.*, 1968). In order to have cis removal of these two hydrogens it is likely that stearoyl-CoA exists in the eclipsed or gauche conformation (Brett *et al.*, 1971). The substrate would thus have the same general configuration as the product, oleyl-CoA. This hypothesis is supported by the finding that oleyl-CoA, DL-*cis*-9,10-epoxy-octadecanoyl-CoA, and DL-*cis*-9,10-methylene-oc-tadecanoyl-CoA are competitive inhibitors of the purified stearoyl-CoA de-saturase while the corresponding trans analogues are noncompetitive inhibitors (Enoch *et al.*, 1976).

The rate of desaturation at position 9 is highly dependent on the nutrition-al and hormonal status of the animal. Fasting depresses the rate of desatura-tion (Oshino and Sato, 1972) and refeeding a high protein diet elevates the level of desaturase activity about fivefold above controls while a sevenfold elevation was observed when animals were refed a high carbohydrate diet. Upon prolonged feeding the rate of desaturation returned to control values. Cycloheximide injection into rats 5 or 16 hr after the initiation of refeeding resulted in a decay of desaturase activity with a half-life of 3–4 hr. Refeeding did not alter the level of cytochrome b_5 or the activity of cytochrome b_5 reductase (Oshino and Sato, 1972). These results are consistent with the dietary regulation of the level of the 9-desaturase with lipogenic diets being the most effective in elevating the amount of this enzyme.

The activity of the 9-desaturase is also regulated by the type of fat includ-ed in the diet. The activity of the 9-desaturase is depressed, by some un-known mechanism, when rats are fed oils containing linoleic acid (Jeffcoat and James, 1977, 1978). This same effect is observed with cultured hepato-cytes isolated from rats fed linoleic acid (Jeffcoat *et al.*, 1979). As noted previously, the 9-desaturase may be viewed as a lipogenic enzyme, the function of which is to desaturate dietary or newly synthesized palmitic or stearic acids for deposition primarily in triglycerides. Indeed it is now well established that the activity of fatty acid synthetase (Volpe and Vagelos, 1976; Flick *et al.*, 1977) and the 9-desaturase are regulated in an almost identical manner by fasting and dietary modification.

This type of coupled control also exists relative to hormonal regulation. The activity of the 9-desaturase is depressed in the alloxan or streptozotocin induced diabetic rat and activity is restored after insulin injection (Gellhorn and Benjamin, 1964, 1965). Administration of agents that block protein syn-thesis prevent the insulin induced restoration of 9-desaturase activity (Gellhorn and Benjamin, 1969). These findings suggest that insulin exerts its effect by promoting the synthesis of one or more of the proteins required for the desaturase reaction.

More recently, it has been shown that insulin exerts its effect solely by increasing the level of the 9-desaturase with no modification in either cytochrome b_5 levels or NADH-cytochrome b_5 reductase activity (Prasad and Joshi, 1979a). When diabetic rats are fed a diet high in either fructose or glycerol the activity of the 9-desaturase is not depressed (DeTomas et al., 1973; Prasad and Joshi, 1979a). Since these two carbohydrates are metabolized by an insulin insensitive pathway (Adelman et al., 1966) it suggests that glycolytic intermediates may also contribute directly in regulating the level of the 9-desaturase. Stearoyl-CoA desaturase activity is negligible in chick embryo liver (Prasad and Joshi, 1979b). When liver explants from chick embryo are cultured in the presence of insulin there is a rapid induction of stearoyl-CoA desaturase activity (Joshi and Aranda, 1979). These findings suggest that insulin plays a primary role in regulating the levels of stearoyl-CoA desaturase with glycolytic intermediates playing a secondary role.

In addition to the above factors that regulate 9-desaturase activity there is now significant evidence suggesting that cytosolic proteins are either directly involved in the desaturase reaction or alternatively mediate the activity of this enzyme. A fatty acid binding protein has been isolated from the 100,000 g supernatant of a liver homogenate and shown to stimulate the activity of the 9-desaturase (Jeffcoat et al., 1976). In addition it has also been shown that catalase stimulates the activity of the 9-desaturase by some unknown mechanism that does not involve breakdown of hydrogen peroxide (Jeffcoat et al., 1978). It has yet to be firmly established whether the cytosolic fatty acid binding proteins (Ockner et al., 1972; Mishkin and Turcotte, 1974; Rustow et al., 1979) are similar to those that mediate the activity of the 9-desaturase.

III. THE 6-DESATURASE(S)

By using an anti-cytochrome b_5 antibody it was established that cytochrome b_5 also was required for the desaturase which introduces a double bond at position 6 in a fatty acid (Okayasu et al., 1977; Lee et al., 1977). Recently a 6-desaturase has been purified from rat liver and shown to be a single polypeptide of 66,000 daltons containing one atom of nonheme iron. NADH, molecular oxygen, cytochrome b_5, cytochrome b_5 reductase and lipid or detergent were all required in order for linoleoyl-CoA to be desaturated (Okayasu et al., 1981).

It is generally thought that a single 6-desaturase acts on each of the unsaturated acids serving as the initial precursor for the biosynthesis of the four families of unsaturated fatty acids. However, if mammalian cells contain a single 6-desaturase it must be capable of desaturating a wide variety of different substrates. The rate of desaturation increases as the chain length of the substrate is extended from 14 to 22 carbons (Pollard et al., 1980b). The double bonds in this homologous series of five even-numbered substrates were always at positions 9 and 12. As with the 9-desaturase, the 6-desaturase is capable of desaturating a number of trans monoenoic acids. (Pollard et al., 1980a). Surprisingly, with rat liver microsomes even small amounts of saturated fatty acids are desaturated at position 6 (Pollard et al., 1980a). Immature rat brain contains an enzyme that desaturates both palmitate and stearate at position 6, but this activity is lost as the animal ages (Cook and Spence, 1973, 1974). If a single 6-desaturase is present in mammalian cells the presence of a cis double bond at position 9 thus is not an absolute prerequisite for enzymatic activity.

The activity of the 6-desaturase is depressed in the diabetic rat and activity is restored upon injection of insulin (Mercuri et al., 1966; Brenner et al., 1968). Diets high in protein stimulate the activity of the 6-desaturase by some unknown mechanism (Inkpen et al., 1969; Castuma et al., 1972).

Cytosolic proteins also stimulate the activity of the 6-desaturase (Catala et al., 1975; Leikin et al., 1979). It appears likely, however, that different cytosolic proteins are involved in mediating the activities of the 9-, 6-, and 5-desaturases (Jeffcoat et al., 1978).

IV. THE 5-DESATURASE(S)

The best evidence that there is a 5-desaturase distinct from a 6-desaturase comes from tissue culture experiments showing that certain cells are able to desaturate fatty acids at the 5 position but not at the 6-position (Dunbar and Bailey, 1975; Maeda et al., 1978). In addition to an acyl-CoA 5-desaturase liver microsomes contain a second 5-desaturase, which converts 1-acyl-2-eicosatrienoyl-sn-glycero-3-phosphorylcholine to the corresponding arachidonyl analogue (Pugh and Kates, 1977). The acyl-CoA 5-desaturase appears to act on a rather large range of different types of substrates. Initially, it was suggested that linoleate could also be converted to arachidonate according to the following pathway: 9,12-18:2 → 11,14-20:2 → 8,11,14-20:3 → 5,8,11,14-20:4 (Stoffel, 1963). Subsequent studies demonstrated that 11,14-20:2 was converted to 5,11,14-20:3 rather than 8,11,14-20:3 (Ullman and Sprecher, 1971). It is now well established that a number of different acids with their first double bond at position 11 are desaturated at position 5 in a variety of tissues (Sprecher and Lee, 1975; Dhopeshwarkar and Subramanian, 1976a; de Alaniz et al., 1976; Albert and Coniglio, 1977; Maeda et al., 1978). The configuration of the double bond at position 5 in these acids has never been established. This may be of interest since it has recently been shown that 5-trans-9,12-18:3 will cure many of the pathological changes observed in essential fatty acid deficiency even though it is not a precursor for prostaglandins and is not metabolized to an acid that is converted to prostaglandins (Houtsmuller, 1981). There is an obvious similarity in structure between 5-trans-9,12-18:2 and 5,11,14-20:3.

In fact a fatty acid does not have to have two or more double bonds in order to serve as a substrate for a 5-desaturase. A number of different cis and trans monoenoic acids are desaturated at position 5 (Lemarchal and Bornens, 1968; Mahfouz and Holman, 1980; Pollard et al., 1980a). When seven different methyl branched isomers of 8,11,14-20:3 were used as substrates for desaturation with rat liver microsomes only the 13, 17, 18, and 19 methyl branched substrates were desaturated at significant rates. The 2, 5, and 10 methyl branched isomers were virtually inactive (Do and Sprecher, 1975). It remains to be determined whether a single acyl-CoA 5-desaturase can act on such a variety of different substrates.

Relatively little is known about the factors controlling the activity of the 5-desaturase. The activity of the 5-desaturase, unlike that of the 9- and 6-desaturases, appears not to be significantly depressed by diabetes (Castuma et al., 1972; Poisson et al., 1979). Diets devoid of fat (Castuma et al., 1972) or high in protein (Inkpen et al., 1969) do not markedly modify the activity of the acyl-CoA 5-desaturase.

V. THE 4-DESATURASE(S)

It is generally assumed that a 4-desaturase is required for converting 7,10,13,16-22:4 to 4,7,10,13,16-22:5 and 7,10,13,16,19-22:5 to 4,7,10,13,16,19-22:6. At present there is no direct evidence supporting this hypothesis. When $[1-^{14}C]7,10,13,16-22:4$ was incubated with testes microsomes no radioactive 4,7,10,13,16-22:5 could be detected even though this preparation was able to desaturate appropriate substrates at positions 5 and 6 (Ayala et al., 1973). Several authors have questioned whether there is a 4-desaturase similar to the acyl-CoA desaturases which introduce double bonds at positions 5, 6, and 9 in fatty acids (Van Golde and Van Den Bergh, 1977; Sprecher and James, 1979).

In the streptozotocin diabetic rat there is a marked elevation in the level of 4,7,10,13,16,19-22:6 in liver lipids which is accompanied by depressed levels of arachidonate (Fass and Carter, 1980). These compositional studies suggest that the pathway for producing this end metabolite derived from linolenate is modified by experimental diabetes.

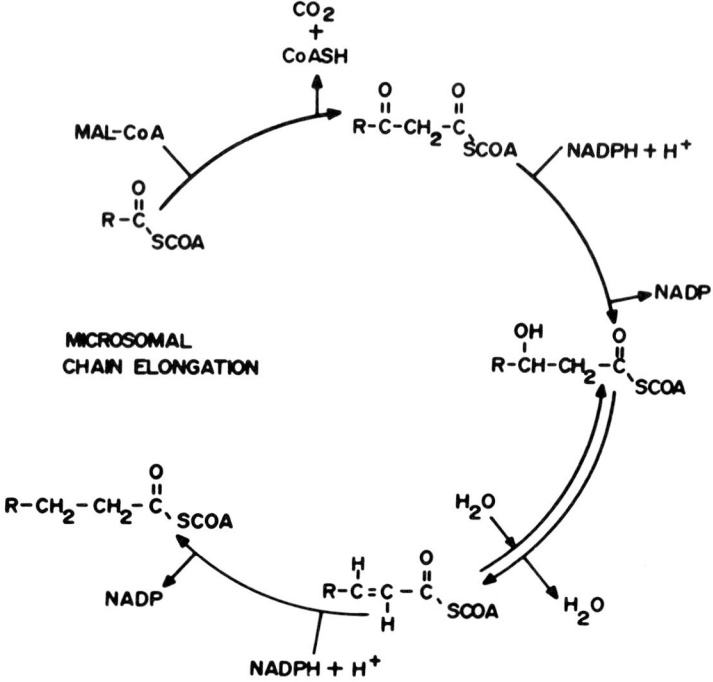

Fig. 2. Pathway for the microsomal chain elongation of fatty acids.

VI. MICROSOMAL FATTY ACID CHAIN ELONGATION

Fatty acids are chain elongated in the microsome according to the reaction sequence shown in Figure 2 (Nugteren, 1965). It is possible to assay three of the four reactions, with microsomes merely by modifying incubation conditions. The condensation reaction can be measured by incubating an acyl-CoA with malonyl-CoA and determining the rate of formation of β-ketoacyl-CoA. β-Ketoacyl-CoA reductase cannot be assayed as a single reaction since when a β-ketoacyl-CoA is incubated with NADPH the fully chain elongated product is produced. β-Hydroxyacyl-CoA dehydrase is assayed by incubating a β-hydroxyacyl-CoA with microsomes and measuring the synthesis of the 2-*trans*-acyl-CoA. When a 2-*trans*-acyl-CoA is incubated in the presence of NADPH it is possible to measure the production of the α,β-saturated product.

In enzymatic studies the rate of condensation depends on the assay conditions. As shown in Fig. 3, with either palmitoyl-CoA or 6,9-octadecadienoyl-CoA as primer, the rate of condensation depends on whether albumin is included in incubations (Bernert and Sprecher, 1977). In the

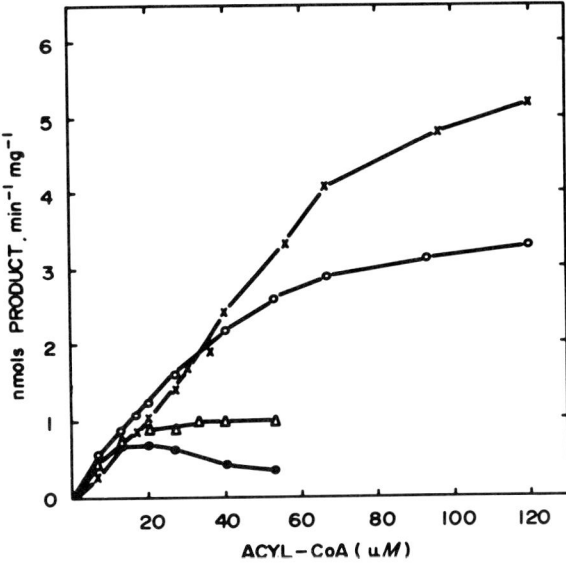

Fig. 3. Rates of β-ketoacyl-CoA synthesis in the presence and absence of bovine serum albumin. 16:0-CoA with albumin (●) and without albumin (○); 6,9-18:2-CoA with albumin (X) and without albumin (Δ). The substrate to albumin molar ratio was 1:2. From Bernert and Sprecher (1977).

absence of albumin maximum rates were found when the primer concentration was 15–20 μM. Acyl-CoA derivatives form micelles when their concentration exceeds 2–5 μM (Barden and Cleland, 1969). In the presence of microsomal protein considerable nonspecific fatty acyl-CoA binding occurs, thus increasing the apparent critical micelle concentration of the primer (Lamb and Fallon, 1972). Failure to find a higher specific activity for condensation, when the primer concentration exceeds 15–20 μM, suggests that micelles are made at this concentration and are inactive as a substrate for condensing enzymes. In fact, as shown in Fig. 4 the rate of condensation is markedly dependent on the amount of albumin included in incubations (Bernert and Sprecher, 1978). At low albumin to primer molar ratios the apparent rate of chain elongation exceeds that of overall chain elongation. Microsomal preparations contain β-ketothiolase activity (Seubert and Podack, 1973). In the presence of low levels of albumin some of the β-ketoacyl-CoA is converted back to the initial primer, thus giving rise to an apparent low rate of condensation. At higher albumin concentrations the rate of condensation slightly exceeds that of overall chain elongation. These findings show that albumin not only mediates the rate of condensation but also protects β-ketoacyl-CoA from cleavage to the original primer. It has not yet been established whether a cytosolic protein or proteins regulate this

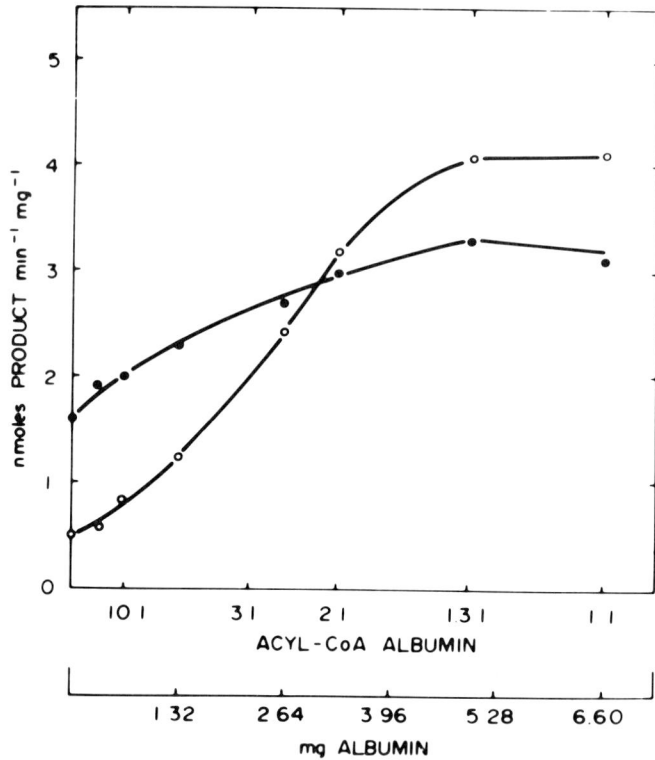

Fig. 4. The effect of increasing concentrations of bovine serum albumin on the rate of condensation (●) and overall chain elongation (○) using palmitoyl-CoA as substrate. From Bernert and Sprecher (1979b).

reaction *in vivo* in a similar way as does albumin in enzymatic experiments.

As shown in Fig. 5 when β-hydroxyacyl-CoA dehydrase was assayed with liver microsomes a biphasic *v/s* curve was obtained (Bernert and Sprecher, 1977). When microsomes were treated with deoxycholate and assayed for dehydrase activity a typical *v/s* curve was obtained. Although the secondary rise in activity could not be recovered following treatment with detergent this procedure served as the initial step for the partial purification of the enzyme. Most of the activity was recovered in the 100,000 g supernatant. This enzyme was purified about 100-fold by ion exchange chromatography using an eluting medium containing Triton X-100 (Bernert and Sprecher, 1979a). The activity of this partially purified enzyme was at a maximum when the amount of Triton X-100 exceeded 0.24 m*M* which is the critical micelle concentration of this nonionic detergent (Robinson and Tanford, 1975). The *v/s* curves of a series of saturated β-hydroxyacyl-CoA deriva-

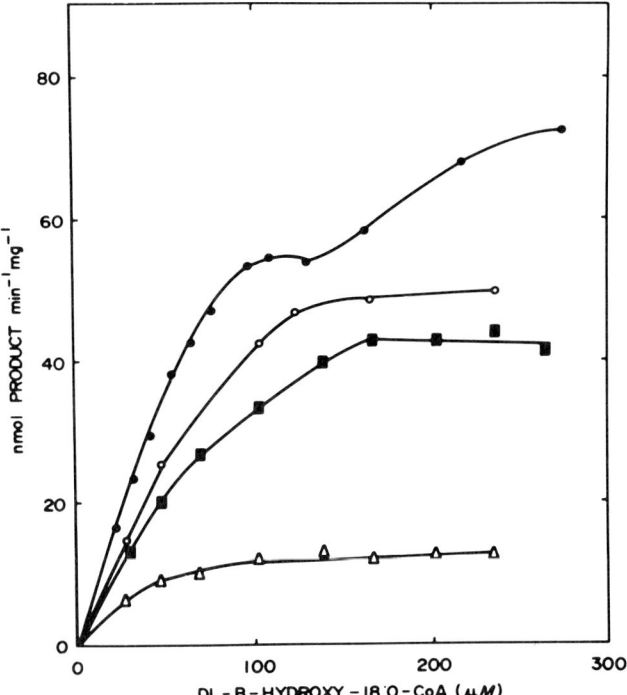

Fig. 5. Kinetics of rat liver microsomal β-hydroxyacyl-CoA dehydrase using β-hydroxy-16:0-CoA with intact microsomes (●) or after treating microsomes with deoxycholate (○). This preparation was then centrifuged at 100,000 g for 1 hour and supernatant (β) and pellet (Δ) were assayed. From Bernert and Sprecher (1979a).

tives, in the presence of optimum amounts of Triton X-100, are shown in Fig. 6. It is apparent that a higher concentration of substrate was required, as the chain length of the primer decreased, before enzymatic activity could be detected. In fact, there was an almost exact correlation between the critical micelle concentration of these substrates with the minimum amount of substrate required in order to detect dehydrase activity. These findings suggest that the active substrate is the micellar form and that substrate micelles are formed even in the presence of Triton X-100 which also was present above its own critical micellar concentration.

2-*trans*-Enoyl-CoA reductase activity, like that for condensation, but unlike that for β-hydroxyacyl-CoA dehydrase, is markedly dependent on the addition of bovine serum albumin (Bernert and Sprecher, 1978). As shown in Fig. 7, even in the presence of optimum amounts of NADPH, 2-*trans*-octadecenoyl-CoA is preferentially hydrated to β-hydroxystearate by rever-

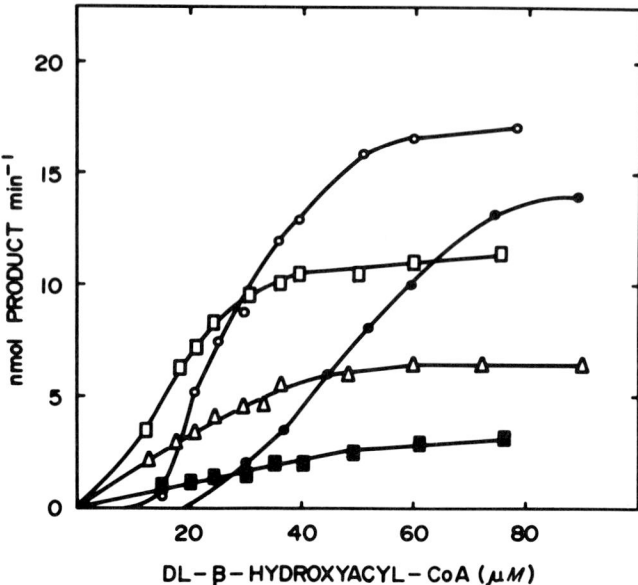

Fig. 6. Rates of dehydration using β-hydroxy-12:0-CoA (●), β-hydroxy-14:0-CoA (○), β-hydroxy-16:0-CoA (□), β-hydroxy-18:0-CoA (△), and β-hydroxy-20:0-CoA (■). From Bernert and Sprecher (1979a).

sal of the β-hydroxyacyl-CoA dehydrase reaction. As the concentration of albumin increases the synthesis of the α,β-saturated product is favored with a decline in reverse dehydrase activity.

It was initially suggested that microsomal chain elongation was carried out by a single multifunctional enzyme in which the primer and malonyl-CoA were both transferred directly to the enzyme (Podack et al., 1974). Several lines of evidence argue against this hypothesis. First, as already noted, β-hydroxyacyl-CoA dehydrase has been partially purified from rat liver microsomes. Second, the CoA derivatives of the products of each of the following reactions have been isolated: overall chain elongation, condensation, β-hydroxyacyl-CoA dehydrase, and 2-trans-enoyl-CoA reductase. When the 2-trans-enoyl-CoA reductase was assayed in the presence of NADPH it was even possible to isolate the CoA derivative of both the α,β-saturated product as well as the β-hydroxy derivative formed by reversal of β-hydroxyacyl-CoA dehydrase (Bernert and Sprecher, 1979b). If the true substrates were covalently linked to the enzyme then it should not be possible to isolate the CoA derivative of component reactions unless the CoASH released from the substrate was tightly bound to the enzyme and transferred to the product via a transacylase.

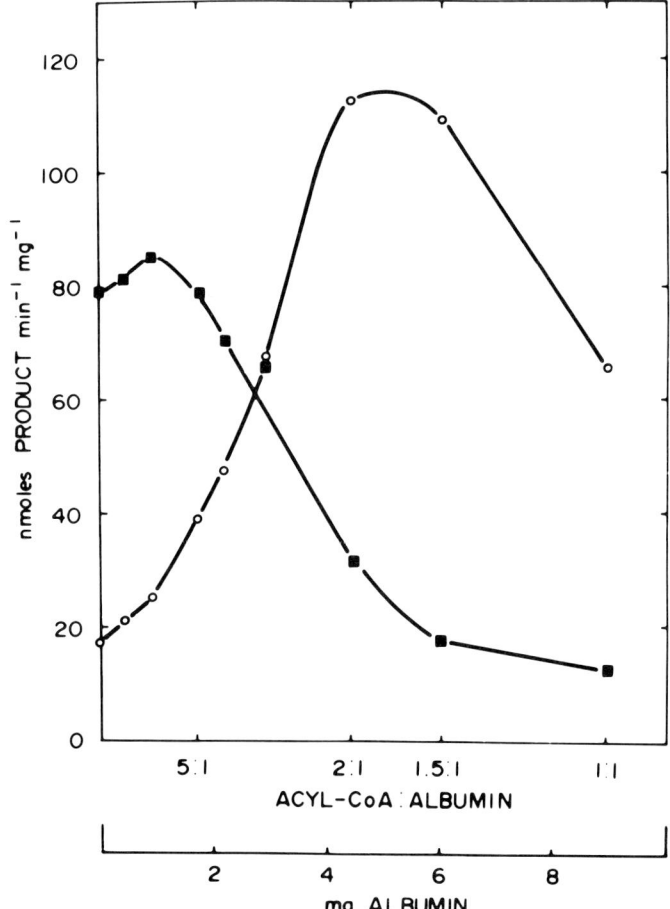

Fig. 7. The effect of increasing concentrations of bovine serum albumin on both the reduction and hydration of 2-*trans*-octadecenoyl-CoA; stearic acid (○), β-hydroxystearic acid (■). From Bernert and Sprecher (1979b).

It is likely that microsomes contain two or more chain elongating systems. Mouse brain microsomes have been shown to contain two (Bourre *et al.*, 1973; Murad and Kiskimoto, 1978) or perhaps even three chain elongating systems (Goldberg *et al.*, 1973). The results in Table I show the rate of overall chain elongation and condensation, using palmitoyl-CoA, 6,9-octadecadienoyl-CoA, and 6,9,12-octadecatrienoyl-CoA as primers as well as rates of the β-hydroxyacyl-CoA and 2-*trans*-enoyl-CoA reductase reaction using the appropriate substrates required for producing stearoyl-CoA and 8,11-eicosadienoyl-CoA (Bernert and Sprecher, 1977). With all three substrates

TABLE I

Rates of Component Reactions in the Microsomal Chain Elongation of Fatty Acids Using Livers from Rats Raised on a Normal Chow or Fat Free Diet[a]

Initial substrate	16:0-CoA			6,9-18:2-CoA			6,9,12-18:3-CoA		
	Normal (N)	Fat-Free (FF)	N/FF	Normal (N)	Fat-Free (FF)	N/FF	Normal (N)	Fat-Free (FF)	N/FF
Chain elongation	0.73±0.01 (2)	2.89~0.23 (2)	0.25	2.79~0.05 (2)	4.00~0.10 (2)	0.70	3.53~0.08 (2)	6.00~0.15 (2)	0.59
Condensation	0.90~0.03 (2)	2.98~0.06 (4)	0.30	3.04~0.02 (2)	4.14~0.04 (4)	0.73	4.29~0.09 (2)	6.79~0.08 (2)	0.63
β-Hydroxyacyl-CoA dehydrase	45.4~1.26 (2)	49.5~2.01 (3)	0.92	45.7~1.08 (2)	49.3~1.60 (3)	0.93			
2-trans Enoyl-CoA reductase	99.6~2.70 (2)	101.8~2.46 (2)	0.98	99.0~1.10 (2)	104.2~3.04 (2)	0.95			

[a] Rates of reactions are expressed as nanomoles of product produced $min^{-1} \cdot$ mg of microsomal protein $^{-1}$ $\pm$ S. E. Values in parentheses are the number of determinations (Bernert and Sprecher, 1977).

the rate of condensation was equal to that of overall chain elongation, thus showing that condensation was rate limiting. The rate of condensation for palmitoyl-CoA was stimulated to a larger extent than was found with the unsaturated primers when microsomes were used from rats raised on a fat free versus a chow diet. These findings coupled with differential inhibition by N-ethylmaleimide for condensation with saturated versus unsaturated primers suggest that rat liver microsomes contain at least two condensing enzymes. One preferentially uses saturated primers while the other is specific for unsaturated acids. The last two reaction rates were both much more rapid than condensation and were not influenced by the dietary history of the animal or substrate modification.

Although desaturation and chain elongation are generally viewed as separate albeit coupled processes recent studies suggest that cytochrome b_5 may be required for both types of reactions (Keyes et al., 1979; Keyes and Cinti, 1980). The addition of malonyl-CoA to microsomes stimulated reoxidation of cytochrome b_5 and this effect was more pronounced when microsomes were used from animals that had been fasted and then refed versus those maintained on a chow diet. In addition, the rate of cytochrome b_5 oxidation was enhanced by including ATP and CoASH in the incubation. Cytochrome b_5 may thus be involved not only in the desaturation reactions but also in microsomal chain elongation as well as in the conversion of 5α-cholest-7-en-3β-ol to cholesta-5,7-dien-3β-ol (Reddy et al., 1977) and in plasmalogen synthesis (Paultauf et al., 1974). It remains to be established whether the flow of electrons through this cytochrome to various enzymes is a mechanism of metabolic regulation.

The rate of microsomal chain elongation, as determined in enzymatic studies, is influenced by structural changes in the substrate (Nugteren, 1965; Ludwig and Sprecher, 1979). The results in Table II compare the rates of overall chain elongation for a series of 18 carbon acids containing either two or three cis double bonds in the skipped pattern of unsaturation. In both series the acid with its first double bond at position 7 was the best substrate for overall chain elongation. The rate of chain elongation also depends on the chain length of the substrate. There was an increase in specific activity of chain elongation as the chain length of the primer increased from 7,10,-14:2 to 7,10-16:2. A maximum rate was found for 7,10-18:2 with virtually no activity for 7,10-20:2.

The presence of cis double bonds in the primer is not mandatory since the following trans monoenoic acids were all chain elongated to 20-carbon trans monoenoic acids: 7-trans-18:1, 8-trans-18:1, 9-trans-18:1, 10-trans-18:1, 11-trans-18:1 and 12-trans-18:1 (Kameda et al., 1980).

The factors altering chain elongation activity have not been as carefully defined as they have for the various desaturases. Fasting does depress the rate of chain elongation of palmitoyl-CoA to stearoyl-CoA and activity is

TABLE II

Rates of Chain Elongation of an Isomeric Series of
Eighteen Carbon Dienoic and Trienoic Acids[a]

Reaction		Rate
4,7-18:2 →	6,9-20:2	0.3
5,8-18:2 →	7,10-20:2	0.5
6,9-18:2 →	8,11-20:2	2.9
7,10-18:2 →	9,12-20:2	5.2
8,11-18:2 →	10,13-20:2	0.6
9,12-18:2 →	11,14-20:2	0.5
10,13-18:2 →	12,15-20:2	0.8
11,14-18:2 →	13,16-20:2	0.2
4,7,10-18:3 →	6,9,12-20:3	0.8
5,8,11-18:3 →	7,10,13-20:3	2.8
6,9,12-18:3 →	8,11,14-20:3	3.8
7,10,13-18:3 →	9,12,15-20:3	4.7
8,11,14-18:3 →	10,13,16-20:3	0.3
9,12,15-18:3 →	11,14,17-20:3	0.2

[a] Rates expressed as nanomoles of product produced $min^{-1} \cdot$ mg of rat liver microsomal protein^{-1} (Ludwig and Sprecher, 1979).

restored upon refeeding (Donaldson et al., 1970; Sprecher, 1974a; Kawashima et al., 1977). The stimulation of chain elongation observed upon refeeding was abolished if rats were injected with either cycloheximide or actinomycin D thus suggesting that chain elongation activity, like that for the 9-desaturase, is regulated by protein synthesis (Kawashima et al., 1977). It remains to be determined whether this effect is confined to only the condensing enzyme(s) or whether the regulation of enzyme synthesis is involved in mediating reaction rates subsequent to the initial and rate limiting reaction.

VII. RETROCONVERSION

In addition to serving as substrates for either desaturation or chain elongation some unsaturated fatty acids undergo partial degradation and the resulting acids are again reesterified into tissue lipids. For example, 4,7,10,13,16-22:5 (Verdino et al., 1964) and 7,10,13,16-22:4 (Sprecher, 1967) were not incorporated into liver lipids to any significant extent when they were fed to rats that had been raised on a fat free diet. Instead they were converted to arachidonate which was incorporated into tissue lipids. This partial degradative process or retroconversion takes place in the mitochondria (Stoffel et *al.*, 1970). The loss of two carbon atoms presumably involves one revolution of the β-oxidation pathway. The resulting acid is then removed by some unknown mechanism and incorporated into tissue lipids. Not all fatty acids are substrates for this process. For example, linoleic acid apparently is not a substrate for retroconversion (Stoffel'*et al.*, 1970) even though 7,10-16:2, the potential product of this reaction, is readily chain elongated to linoleate (Klenk, 1965; Sprecher, 1968). A recent review summarizes those acids which are substrates for retroconversion (Sprecher and James, 1979).

The conversion of 4,7,10,13,16-22:5 to arachidonate involves the loss of not only two carbon atoms but also a double bond. In both mitochondria (Kunau and Bartnik, 1974; Kunau and Dommes, 1978) and peroxisomes (Dommes *et al.*, 1981) this conversion most likely proceeds as follows: 4,7,10,13,16-22:5 → 2-*trans*-4,7,10,13,16-22:6 → 3-*trans*-7,10,13,16-22:5 → 2-*trans*-7,10,13,16-22:5 → 7,10,13,16-22:4 → 5,8,11,14:20:4. This reaction pathway requires the enzyme 2,4-dienoyl-CoA reductase, an NADPH dependent enzyme that converts the 2,4-conjugated system into the 3-enoyl-CoA.

VIII. REGULATION OF UNSATURATED FATTY ACID BIOSYNTHESIS

The types of unsaturated fatty acids found in specific lipids must be determined both by the factors regulating fatty acid and phospholipid biosynthesis. At present little is known about the integrated regulation of these two processes. For example, is arachidonic acid produced at a constant rate from dietary linoleate or is its rate of synthesis altered as the need for this acid varies for phospholipid biosynthesis? Most studies on phospholipid biosynthesis have been confined to determining rates of incorporation of various fatty acids into phospholipid precursors. In a similar way the regulation of unsaturated fatty acid biosynthesis has either been studied by defining how dietary fat modification alters tissue lipid composition or enzymatic studies determining how substrate modification mediates rates of desaturation and chain elongation. In fact the regulation of unsaturated fatty acid biosynthesis may be subdivided into those factors that regulate the production of fatty acids within a metabolic sequence versus interactions that exist between metabolites from two or more of the different families of acids for common enzymes. For example, liver lipids generally contain large amounts of both linoleate and arachidonate but only small amounts of the other (n-6) metabolites. The rate limiting reaction in the (n-6) pathway is desaturation of linoleate to 6,9,12-18:3 (Marcel *et al.*, 1968). Tissue lipids generally contain only low levels of 6,9,12-18:3. The results in Table III show that 6,9,12-18:3 is chain elongated at a rapid rate (Bernert and Sprecher, 1975). Failure to find significant amounts of 6,9,12-18:3 in liver lipids may thus suggest that it is preferentially chain elongated to 8,11,14-20:3 rather than used as a substrate for acylation into phospholipids. The low level of 8,11,14-20:3 found in liver lipids is not readily explained by its rate of desaturation to arachidonate. As shown in Table III the desaturation of 8,11,14-20:3 via an acyl-CoA 5-desaturase proceeds at a slow rate. It is possible that the low level of 8,11,14-20:3 in liver phospholipids could in part be due to conversion of this acid by direct desaturation of 1-acyl-2-eicosa-trienoyl-sn-glycero-3-phosphorylcholine to the corresponding arachidonyl analogue (Pugh and Kates, 1977). The relative roles of these two 5-desaturases in making arachidonate has yet to be quantitated.

The factors regulating the conversion of 8,11,14-20:3 to arachidonate are of considerable interest since prostaglandin E_1 is a potent inhibitor of platelet aggregation (Kloeze, 1969). In contrast thromboxane A_2 derived from arachidonic acid stimulates platelet aggregation (Hamberg and Samuelson, 1974; Hamberg et al., 1975). If the ratio of 8,11,14-20:3 to arachidonate could be increased by dietary supplementation with 8,11,14-20:3 it might be a way to inhibit platelet aggregation by altering the types of acids available for prostaglandin biosynthesis. The levels of 8,11,14-20:3 in tissue lipids of rat (Danon et al., 1975), rabbit (Oelz et al., 1976), and man (Stone et al., 1979) are elevated when 8,11,14-20:3 is included in a balanced diet. The benefits of increasing the levels of 8,11,14-20:3 in platelet lipids is, however, uncertain since it is converted by the platelet primarily to 12-hydroxy-heptadecadienoic acid, which is an inactive metabolite (Needleman et al., 1980). In addition, due to the lack of a fourth double bond, 8,11,14-20:3 cannot be converted to prostacyclin by the artery. The potential role played by 8,11,14-20:3 in mediating platelet aggregation has recently been reviewed in depth (Willis, 1981).

Liver lipids contain only low levels of 7,10,13,16-22:4 and 4,7,10,13,16-22:5. As shown in Table III arachidonate is converted to 7,10,13,16-22:4 at

TABLE III

Rates of Desaturation and Chain Elongation for Acids in the Linoleate Pathway[a]

Reaction		Rate
9,12-18:2 →	6,9,12-18:3	1.0
6,9,12-18:3 →	8,11,14-20:3	4.4
8,11,14-20:3 →	5,8,11,14-20:4	0.8
5,8,11,14-20:4 →	7,10,13,16-22:4	1.2

[a] Rates expressed as nanomoles of product produced $min^{-1} \cdot$ mg of rat liver microsomal protein^{-1} (Bernert and Sprecher, 1975).

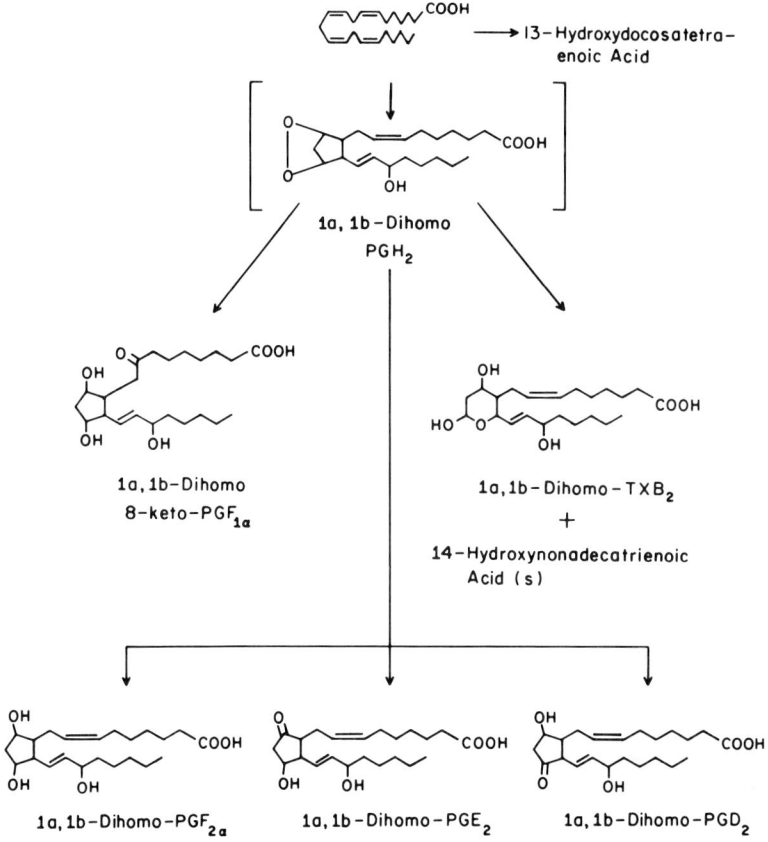

Fig. 8. Metabolites produced when $[1\text{-}^{14}C]7,10,13,16$-docosatetraenoic acid was incubated with rabbit kidney medullary microsomes. From H. Sprecher, M. Van Rollins, F. Sun, A. Wyche, and P. Needleman (unpublished results).

a reasonably rapid rate when compared with other rates of reactions we have measured. Feeding studies have shown that these acids are preferentially converted back to arachidonate rather than being incorporated directly into liver lipids (Verdino et al., 1964; Sprecher, 1967). Although this may be an effective way to control liver lipid fatty acid composition different mechanisms must operate in extrahepatic tissues. The triglycerides of kidney medulla (Comai et al., 1975) and the cholesteryl esters of adrenal gland (Walker, 1970; Vahouny et al., 1979) contain relatively large amounts of 7,10,13,16-22:4. As shown in Fig. 8 this acid is converted by rabbit kidney medullary microsomes into a complete series of prostaglandin analogous to those derived from arachidonic acid (H. Sprecher, M. VanRollins, F. Sun, A. Wyche, and P. Needleman, unpublished results). The physiological properties of these prostaglandins are, however, quite different from those produced from arachidonate. The endoperoxide produced from 7,10,13,16-22:4 did not contract smooth muscle strips and it did not aggregate platelets. The prostacyclin derived from this acid did not inhibit platelet aggregation. The 22-carbon prostacyclin and prostaglandin E_2 both stimulated renomedullary intestitial cell adenylate cyclase although neither compound was as active as were the corresponding prostaglandins produced from arachidonate.

Rat testes lipids contain a high level of 4,7,10,13,16-22:5 (Bieri and Prival, 1965; Carpenter, 1971) and small amounts of both 9,12,15,18-24:4 and 6,9,12,15,18-24:4 (Bridges and Coniglio, 1970a). These acids accumulate in testes lipids even though rat testes are capable of carrying out retroconversion (Bridges and Coniglio, 1970b). Clearly, the mechanism regulating unsaturated fatty acid biosynthesis in extrahepatic tissues differs from the mechanism operative in liver. In addition, little is known about how the liver acts in concert with each specific tissue to establish what acids are available for incorporation into lipids of extrahepatic tissues. For example, both brain (Dhopeshwarkar and Subramanian, 1975; Dhopeshwarkar and Subramanian, 1976b) and testes (Bridges and Coniglio, 1970a; Ayala et al., 1973) are able to desaturate and chain elongate fatty acids. However, the types of acids supplied to these tissues will depend on the dietary history of the animal and undoubtedly liver metabolism. Conversely, the kidney is able to chain elongate fatty acids but apparently does not contain a 9-desaturase (Cinti and Montgomery, 1976; Montgomery and Cinti, 1977).

In addition to those factors regulating the biosynthesis of acids within a family there also are interactions between metabolites from two or more families of unsaturated acids. For example, numerous studies have shown that 5,8,11-20:3 accumulates in tissue lipids when animals are raised on a diet devoid of essential fatty acids. As soon as the diet is fortified with acids of either the (n-3) or (n-6) families the level of 5,8,11-20:3 is depressed with a concomitant increase in the level of either (n-3) or (n-6) metabolites. Competitive feeding experiments led to a hypothesis suggesting that (n-9) metabolites competed with (n-6) or (n-3) acids and that the metabolites from the latter two families of acids were preferred substrates thus effectively preventing oleate from being converted to 5,8,11-20:3 (Mohrhauer and Holman, 1963; Holman, 1964). Competitive enzyme experiments were consistent with this hypothesis (Brenner and Peluffo, 1966). The results in Table IV show rates of reactions in the oleate biosynthetic sequence (Bernert and Sprecher, 1975). As in the linoleate pathway the rate limiting reaction is catalyzed by a 6-desaturase. If linoleate, linolenate, and oleate are all substrates for a common 6-desaturase, then upon including these acids in the diet they might well inhibit production of 5,8,11-20:3 by serving as preferred substrates for a common 6-desaturase. Rates of reaction in the oleate family

TABLE IV

Rates of Desaturation and Chain Elongation for Acids in the Oleate Pathway[a]

Reaction		Rate
18:0 →	9-18:1	3.3
9-18:1 →	6,9-18:2	0.2
6,9-18:2 →	8,11-20:2	3.0
8,11-20:2 →	5,8,11-20:3	0.8
5,8,11-20:3 →	7,10,13-22:3	1.3

[a] Rates expressed as nanomoles of product produced $min^{-1} \cdot$ mg of rat liver microsomal protein^{-1} (Bernert and Sprecher, 1975).

(Table IV), subsequent to the 6-desaturase, are similar to those of analogous reactions in the linoleate pathway (Table III). These findings suggest that if an (n-9) and an (n-6) metabolite, beyond the rate limiting 6-desaturase, were fed to rats, that tissue lipids might then contain either substantial amounts of the (n-9) metabolite that is fed or 5,8,11-20:3. The results in Fig. 9 were obtained in a competitive feeding experiment in which a constant level of 6,9,12-18:3 was fed to rats raised on a fat free diet. The same animals also

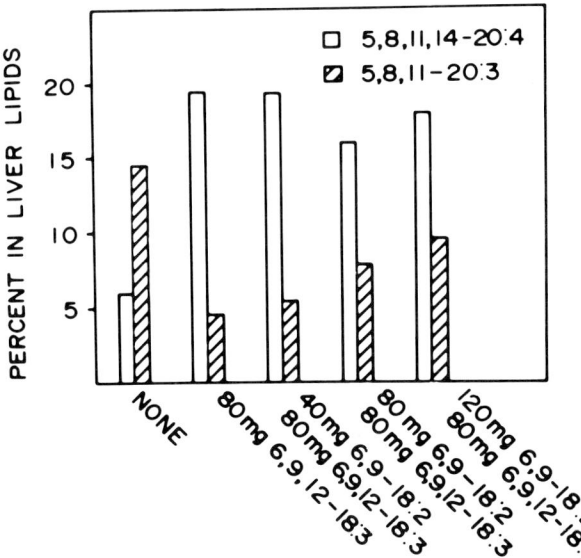

Fig. 9. The percent of 5,8,11-eicosatrienoic acid and arachidonic acid in total liver lipids after feeding rats a constant level of 6,9,12-octadecatrienoic acid and increasing levels of 6,9-octadecadienoic acid. From Sprecher (1974b).

received various amounts of 6,9-18:2. Liver lipids of animals raised on the fat free diet had the expected high level of 5,8,11-20:3 and low levels of 5,8,11,14-20:4. Liver lipids of animals fed only 6,9,12-18:3 had an increased level of arachidonate accompanied by a depressed amount of 5,8,11-20:3. Unless 6,9,12-18:3 acts by some type of feedback inhibition on the 6-desaturase it should not influence the rate for the conversion of oleate to 6,9-18:2 by a 6-desaturase. With all three other groups of animals this type of dietary fat alteration resulted in a depressed level of 5,8,11-20:3 in liver lipids which was accompanied by an increase in arachidonate. This alteration was observed even when the animals received 1.5 times as much 6,9-18:2 as 6,9,12-18:3. In addition they had a high level of 5,8,11-20:3 in tissue lipids at the start of the feeding experiment (Sprecher, 1974b). Almost identical compositional changes were found when similar competitive feeding experiments were carried out with 8,11-20:2 and 8,11,14-20:3 as well as with 5,8,11-20:3 and 5,8,11,14-20:4. These results show that the level of arachidonate is relatively independent of whether arachidonate or an (n-6) precursor for arachidonate is fed. In addition supplementation of the diet with an (n-9) metabolite did not markedly influence the amount of arachidonate incorporated into total liver lipids. The synthesis of arachidonate from its precursors and its incorporation into phospholipids must thus be a tightly coupled metabolic process which is not influenced by including (n-9) acids in the diet. Thus, even when competitive interactions are observed in enzymatic studies the role of this type of interaction *in vivo* must be documented before it can be assumed to be a major regulatory mechanism in dictating unsaturated fatty acid availability for phospholipid biosynthesis.

Tissue lipids never contain significant amounts of (n-7) fatty acids derived from palmitoleate. They do not accumulate even when rats are fed members of this metabolic sequence (Klenk, 1965; Sprecher, 1971). The virtual absence of these acids in tissue lipids may in part be explained by the very slow rates of the desaturation reactions required for converting palmitoleate to 4,7,10,13-20:4 (Budny and Sprecher, 1971; Bernert and Sprecher, 1975).

Recently there has been a renewed interest in the physiological function of the (n-3) fatty acids produced from dietary linolenate. Greenland Eskimos eat substantially larger quantities of fish than does the population of most Western countries. The prolonged bleeding time observed in Eskimos correlates in a positive manner with the elevated level of 5,8,11,14,17-20:5 in platelet lipids (Dyerberg et al., 1978; Dyeberg and Bang, 1979). When adult men were fed a high mackerel diet their bleeding times were also prolonged and shown to be related to the high level of 5,8,11,14,17-20:5 in plasma lipids (Siess et al., 1980). These findings suggest that platelet aggregation could be inhibited by replacing part of the arachidonate in platelet lipids with 5,8,11,14,17-20:5. When 5,8,11,14,17-20:5 was incubated directly with platelets it was converted primarily to a hydroxy fatty acid with only small amounts of thromboxane A₃ being produced (Needleman et al., 1979; Whitaker et al., 1979; Hamberg, 1980). When platelets were incubated together with arachidonate acid and 5,8,11,14,17-20:5 the synthesis of thromboxane A_2 was depressed (Needleman et al., 1979). The 5,8,11,14,17-20:5 appears to act as a competitive inhibitor of cyclooxygenase by depressing the synthesis of thromboxane A_2. This finding is consistent with competitive enzyme studies with the purified cyclooxygenase (Lands et al., 1971). Even if 5,8,11,14,17-20:5 is converted to small amounts of thromboxane A₃ this compound only weakly stimulates platelet aggregation (Needleman et al., 1979, 1980).

These findings have led to a school of thought advocating a higher dietary intake of (n-3) fatty acids to inhibit platelet aggregation. At present the control of (n-3) fatty acid biosynthesis is poorly understood even though the essential nature of these acids has long been a matter of debate (Tinoco et al., 1979). Before any dietary fat changes are recommended to mediate thrombosis it must be noted that 5,8,11,14,17-20:5 (Jakschik et al., 1980; Hammarström, 1981a), arachidonate (Murphy et al., 1979), 5,8,11-20:3 (Jakschik et al., 1980; Hammarström, 1981b) and 8,11,14-20:3 (Hammarström, 1981c) are all substrates for leukotriene biosynthesis.

In summary, it is now clear that unsaturated fatty acids serve as substrates for the synthesis of a variety of different prostaglandins, hydroxy fatty acids, and leukotrienes. The types and amounts of these compounds produced will in part depend on what type of dietary fat is included in the diet. In turn, the factors regulating the desaturation and chain elongation of fatty acids for subsequent incorporation into and release from phospholipids will contribute in defining what types and amounts of prostaglandins are produced to mediate and control physiological processes.

ACKNOWLEDGMENTS

These studies were supported in part by grants AM20387 and AM18844 from the United States Public Health Service.

REFERENCES

Adelman, R. C., Spolter, P. D., and Weinhouse, S. (1966). *J. Biol. Chem.* **241**, 5467–5472.

Albert, D. H., and Coniglio, J. G. (1977). *Biochim. Biophys. Acta* **489**, 390–396.

Ayala, S., Gaspar, G., Brenner, R. R., Peluffo, R. O., and Kunau, W. (1973). *J. Lipid Res.* **14**, 296–305.

Barden, R. E., and Cleland, W. W. (1969). *J. Biol. Chem.* **244**, 3677–3684.

Bernert, J. T., and Sprecher, H. (1975). *Biochim. Biophys. Acta* **398**, 354–363.

Bernert, J. T., and Sprecher, H. (1977). *J. Biol. Chem.* **252**, 6736–6744.

Bernert, J. T., and Sprecher, H. (1978). *Biochim. Biophys. Acta* **531**, 44–55.

Bernert, J. T., and Sprecher, H. (1979a). *J. Biol. Chem.* **254**, 11584–11590.

Bernert, J. T., and Sprecher, H. (1979b). *Biochim. Biophys. Acta* **573**, 436–442.

Bieri, J. G., and Prival, E. L. (1965). *Comp. Biochem. Physiol.* **15**, 275–282.

Bloomfield, D. K., and Bloch, K. (1958). *Biochim. Biophys. Acta* **30**, 220–221.

Bourre, J. M., Pollet, S., Chaix, G., Daudu, O., and Baumann, N. (1973). *Biochimie* **55**, 1473–1479.

Brenner, R. R., and Peluffo, R. O. (1966). *J. Biol. Chem.* **241**, 5213–5219.

Brenner, R. R., Peluffo, R. O., Mercuri, O., and Restelli, M. M. (1968). *Am. J. Physiol.* **215**, 63–70.

Brett, D., Howling, D., Morris, L. J., and James, A. T. (1971). *Arch. Biochem. Biophys.* **143**, 535–547.

Bridges, R. B., and Coniglio, J. G. (1970a). *J. Biol. Chem.* **245**, 46–49.

Bridges, R. B., and Coniglio, J. G. (1970b). *Biochim. Biophys. Acta* **218**, 24–35.

Budny, J., and Sprecher, H. (1971). *Biochim. Biophys. Acta* **239**, 190–207.

Carpenter, M. (1971). *Biochim. Biophys. Acta* **231**, 52–79.

Castuma, J. C., Catala, A., and Brenner, R. R. (1972). *J. Lipid Res.* **13**, 783–789.

Catala, A., Nervi, A. M., and Brenner, R. R. (1975). *J. Biol. Chem.* **250**, 7481–7484.

Cinti, D., and Montgomery, M. (1976). *Life Sci.* **18**, 1223–1228.

Comai, K., Farber, S. J., and Paulsrud, J. R. (1975). *Lipids* **10**, 555–561.

Cook, H. W., and Spence, M. W. (1973). *J. Biol. Chem.* **248**, 1786–1792.

Cook, H. W., and Spence, M. W. (1974). *Biochim. Biophys. Acta* **369**, 129–141.

Danon, A., Heimberg, M., and Oates, J. A. (1975). *Biochim. Biophys. Acta* **388**, 318–330.

de Alaniz, M. J. T., and Brenner, R. R. (1976). *Mol. Cell. Biochem.* **12**, 81–87.

de Alaniz, M. J. T., de Gomez Dumm, I. N. T., and Brenner, R. R. (1976). *Mol. Cell. Biochem.* **12**, 3–8.

DeTomas, M. E., Peluffo, R. O., and Mercuri, O. (1973). *Biochim. Biophys. Acta* **306**, 149–155.

Dhopeshwarkar, G. A., and Subramanian, C. (1975). *Lipids* **10**, 238–241.

Dhopeshwarkar, G. A., and Subramanian, C. (1976a). *J. Neurochem.* **26**, 1175–1179.

Dhopeshwarkar, G. A., and Subramanian, C. (1976b). *Lipids* **11**, 67–71.

Do, U. H., and Sprecher, H. (1975). *Arch. Biochem. Biophys.* **171**, 597–603.

Dommes, V., Baumgart, C., and Kunau, W. H. (1981). *J. Biol. Chem.* **256**, 8259–8262.

Donaldson, W. E., Wit-Peeters, E. M., and Scholte, H. P. (1970). *Biochim. Biophys. Acta* **202**, 35–42.

Dunbar, L. M., and Bailey, J. M. (1975). *J. Biol. Chem.* **250**, 1152–1153.

Dyerberg, J., and Bang, H. O. (1979). *Lancet* **2**, 433–435.

Dyerberg, J., Bang, H. O., Stofferson, E., Moncada, S., and Vane, J. R. (1978). *Lancet* **2**, 117–119.

Enoch, H. G., Catala, A., and Strittmatter, P. (1976). *J. Biol. Chem.* **251**, 5095–5103.

Fass, F. H., and Carter, W. J. (1980). *Lipids* **15**, 953–961.

Flick, P. K., Chen, J., and Vagelos, P. R. (1977). *J. Biol. Chem.* **252**, 4242–4249.

Gellhorn, A., and Benjamin, W. (1964). *Biochim. Biophys. Acta* **84**, 167–175.

Gellhorn, A., and Benjamin, W. (1965). *Ann. N.Y. Acad. Sci.* **131**, 344–356.

Gellhorn, A., and Benjamin, W. (1969). *Science* **146**, 1166–1168.

Goldberg, I., Shechter, I., and Bloch, K. (1973). *Science* **182**, 497–499.

Gurr, M. I., Robinson, M. P., James, A. T., Morris, L. J., and Howling, D. (1972). *Biochim. Biophys. Acta* **280**, 415–421.

Hamberg, M. (1980). *Biochim. Biophys. Acta* **618**, 389–398.

Hamberg, M., and Samuelsson, B. (1974). *Proc. Natl. Acad. Sci. U.S.A.* **71**, 3400–3404.

Hamberg, M., Svensson, J., and Samuelsson, B. (1975). *Proc. Natl. Acad. Sci. U.S.A.* **72**, 2994–2998.

Hammarström, S. (1981a). *Biochim. Biophys. Acta* **663**, 575–577.

Hammarström, S. (1981b). *J. Biol. Chem.* **256**, 2275–2279.

Hammarström, S. (1981c). *J. Biol. Chem.* **256**, 7712–7714.

Hiwatashi, A., Ichikawa, Y., and Yamano, T. (1975). *Biochim. Biophys. Acta* **388**, 397–401.

Holloway, C. T., and Holloway, P. W. (1974). *Lipids* 9, 196–200.

Holloway, P. W. (1971). *Biochemistry* 10, 1556–1560.

Holloway, P. W., and Wakil, S. J. (1970). *J. Biol. Chem.* **245**, 1862–1865.

Holman, R. T. (1964). *Fed. Proc., Fed. Am. Soc. Exp. Biol.* **23**, 1062–1067.

Houtsmuller, U. M. T. (1981). *Prog. Lipid Res.* **20**, 889–896.

Inkpen, C. A., Harris, R. A., and Quackenbusch, F. W. (1969). *J. Lipid Res.* **10**, 277–282.

Jakschik, B. A., Sams, A. R., Sprecher, H., and Needleman, P. (1980). *Prostaglandins* **20**, 401–410.

Jeffcoat, R., and James, A. T. (1977). *Lipids* **12**, 469–474.

Jeffcoat, R., and James, A. T. (1978). *FEBS Lett.* **85**, 114–118.

Jeffcoat, R., Brawn, P. R., and James, A. T. (1976). *Biochim. Biophys. Acta* **431**, 33–44.

Jeffcoat, R., Dunton, A. P., and James, A. T. (1978). *Biochim. Biophys. Acta* **528**, 28–35.

Jeffcoat, R., Roberts, A., and James, A. T. (1979). *Eur. J. Biochem.* **101**, 447–453.

Johnson, A. R., Fogerty, A. C., Pearson, J. A., Shenstone, F. S., and Bersten, J. A. (1969). *Lipids* **4**, 265–269.

Jones, P. D., Holloway, P. W., Peluffo, R. O., and Wakil, S. J. (1969). *J. Biol. Chem.* **244**, 744–745.

Joshi, V. C., and Aranda, L. P. (1979). *J. Biol. Chem.* **254**, 11779–11782.

Kameda, K., Valicenti, A. J., and Holman, R. T. (1980). *Biochim. Biophys. Acta* **618**, 13–17.

Kawashima, Y., Suzuki, Y., and Haskimoto, Y. (1977). *Lipids* **12**, 434–437.

Keyes, S. R., and Cinti, D. L. (1980). *J. Biol. Chem.* **255**, 11357–11364.

Keyes, S. R., Alfano, J. A., Jansson, I., and Cinti, D. L. (1979). *J. Biol. Chem.* **254**, 7778–7784.

Klenk, E. (1965). *Adv. Lipid Res.* **3**, 1–23.

Kloeze, J. (1969). *Biochim. Biophys. Acta* **187**, 285–292.

Kunau, W. H., and Bartnik, F. (1974). *Eur. J. Biochem.* **48**, 311–318.

Kunau, W. H., and Dommes, P. (1978). *Eur. J. Biochem.* **91**, 533–544.

Lamb, R. G., and Fallon, H. J. (1972). *J. Biol. Chem.* **247**, 1281–1287.

Lands, W. E. M., Lee, R., and Smith, W. (1971). *Ann. N.Y. Acad. Sci.* **180**, 107–122.

Lee, T. C., Baker, R. C., Stephens, N., and Snyder, F. (1977). *Biochim. Biophys. Acta* **489**, 25–31.

Leikin, A. I., Nervi, A. M., and Brenner, R. R. (1979). *Lipids* **14**, 1021–1026.

Lemarchal, P., and Bornens, M. (1968). *Bull. Soc. Chim. Biol.* **50**, 195–216.

Ludwig, S. A., and Sprecher, H. (1979). *Arch. Biochem. Biophys.* **197**, 333–341.

Maeda, M., Doi, O., and Akamatsu, Y. (1978). *Biochim. Biophys. Acta* **530**, 153–164.

Mahfouz, M. M., and Holman, R. T. (1980). *Lipids* **15**, 63–65.

Mahfouz, M. M., Valicenti, A. J., and Holman, R. T. (1980). *Biochim. Biophys. Acta* **618**, 1–12.

Marcel, Y. L., Christiansen, K., and Holman, R. T. (1968). *Biochim. Biophys. Acta* **164**, 24–35.

Marsch, J. B., and James, A. T. (1962). *Biochim. Biophys. Acta* **60**, 320–328.

Mead, J. F. (1971). *Prog. Chem. Fats Other Lipids* 9, 161–192.

Mercuri, O., Puluffo, R. O., and Brenner, R. R. (1966). *Biochim. Biophys. Acta* **116**, 409–411.

Mishkin, S., and Turcotte, R. (1974). *Biochem. Biophys. Res. Commun.* **57**, 918–926.

Mohrhauer, H., and Holman, R. T. (1963). *J. Lipid Res.* **4**, 151–159.

Montgomery, M., and Cinti, D. (1977). *Mol. Pharmacol.* **13**, 60–69.

Morris, L. J., Harris, R. V., Kelly, W., and James, A. J. (1968). *Biochem. J.* **109**, 673–678.

Murad, S., and Kishimoto, Y. (1978). *Arch. Biochem. Biophys.* **185**, 300–306.

Murphy, R. C., Hammarström, S., and Samuelsson, B. (1979). *Proc. Natl. Acad. Sci. U.S.A.* **76**, 4275–4279.

Needleman, P., Raz, A., Minkes, M. S., Ferrendelli, J. A., and Sprecher, H. (1979). *Proc. Natl. Acad. Sci. U.S.A.* **76**, 944–948.

Needleman, P., Whitaker, M. O., Wyche, A. Watters, H., Sprecher, H., and Raz, A. (1980). *Prostaglandins* **19**, 165–181.

Nugteren, D. H. (1965). *Biochim. Biophys. Acta* **106**, 280–290.

Ockner, R. K., Manning, J. A., Poppenhausen, R. B., and Ho, W. K. L. (1972). *Science* **177**, 56–58.

Oelz, O., Seyberth, H. W., Knapp, H. R., Sweetan, B. J., and Oates, H. R. (1976). *Biochim. Biophys. Acta* **931**, 268–277.

Okayasu, T., Ono, T., Shinojima, K., and Imai, Y. (1977). *Lipids* **12**, 267–271.

Okayasu, T., Nagao, M., Ishibashi, T., and Imai, Y. (1981). *Arch. Biochem. Biophys.* **206**, 21–28.

Oshino, N. (1972). *Arch. Biochem. Biophys.* **149**, 378–387.

Oshino, N., and Omura, T. (1973). *Arch. Biochem. Biophys.* **157**, 395–404.

Oshino, N., and Sato, R. (1972). *Arch. Biochem. Biophys.* **149**, 369–377.

Oshino, N., Imai, Y., and Sato, R. (1966). *Biochim. Biophys. Acta* **128**, 13–28.

Paulsrud, J. R., Stewart, S. E., Grass, G., and Holman, R. T. (1970). *Lipids* **5**, 611–616.

Paultauf, F., Prough, R. A., Masters, B. S. S., and Johnston, J. M. (1974). *J. Biol. Chem.* **249**, 2661–2662.

Podack, E. R., Saathoff, G., and Seubert, W. (1974). *Eur. J. Biochem.* **50**, 237–243.

Poisson, J. P., Blond, J. B., and Lemarchal, P. (1979). *Diabete Metab.* **5**, 43–46.

Pollard, M. R., Gunstone, F. D., James, A. T., and Morris, L. J. (1980a). *Lipids* **15**, 306–314.

Pollard, M. R., Gunstone, F. D., Morris, L. J., and James, A. T. (1980b). *Lipids* **15**, 690–693.

Prasad, M. K., and Joshi, V. C. (1979a). *J. Biol. Chem.* **254**, 997–999.

Prasad, M. R., and Joshi, V. C. (1979b). *J. Biol. Chem.* **254**, 6362–6369.

Pugh, E. L., and Kates, M. (1977). *J. Biol. Chem.* **252**, 68–73.

Reddy, V. R., Kupfer, D., Caspi, E. (1977). *J. Biol. Chem.* **252**, 2797–2801.

Robinson, N. C., and Tanford, C. (1975). *Biochemistry* **14**, 369–378.

Rüstow, B., Kunze, D., Hodi, J., and Egger, E. (1979). *FEBS Lett.* **108**, 469–472.

Schroepfer, G. J., and Bloch, K. (1965). *J. Biol. Chem.* **240**, 54–63.

Seubert, W., and Podack, E. R. (1973). *Mol. Cell. Biochem.* **1**, 29–40.

Shimakata, T., Mihara, K., and Sato, R. (1972). *J. Biochem.* (Tokyo) **72**, 1163–1174.

Siess, W., Scherer, B., Bohlig, B., Roth, P., Kurzman, I., and Weber, P. C. (1980). *Lancet* **1**, 441–444.

Sprecher, H. (1967). *Biochim. Biophys. Acta* **144**, 296–304.

Sprecher, H. (1968). *Lipids* **3**, 14–20.

Sprecher, H. (1971). *Biochim. Biophys. Acta* **231**, 122–130.

Sprecher, H. (1974a). *Biochim. Biophys. Acta* **360**, 113–123.

Sprecher, H. (1974b). *Biochim. Biophys. Acta* **369**, 34–44.

Sprecher, H., and James, A. T. (1979). *In* "Geometrical and Positional Fatty Acid Isomers" (E. A. Emken and H. J. Dutton, eds.). pp. 303–338. Am. Oil Chem. Soc., Champaign, Illinois.

Sprecher, H., and Lee, C. (1975). *Biochim. Biophys. Acta* **388**, 113–124.

Stoffel, W. (1961). *Biochem. Biophys. Res. Commun.* **6**, 270–275.

Stoffel, W. (1963). *Hoppe-Seyler's Z. Physiol. Chem.* **333**, 71–88.

Stoffel, W., Ecker, W., Assad, R., and Sprecher, H. (1970). *Hoppe-Seyler's Z. Physiol Chem.* **351**, 1545–1554.

Stone, K. J., Hart, W. M., Kirtland, S. J., Kernoff, P. B. A., and McNicol, G. P. (1979). *Lipids* **14**, 174–180.

Strittmatter, P., Spatz, L., Corcoran, D., Rogers, M. J., Setlow, B., and Redline, R. (1974). *Proc. Natl. Acad. Sci. U.S.A.* **71**, 4565–4569.

Tinoco, J., Babcock, R., Hincenbergs, I., Medwadowski, B., Miljanick, P., and Williams, M. A. (1979). *Lipids* **14**, 166–173.

Ullman, D., and Sprecher, H. (1971). *Biochim. Biophys. Acta* **248**, 186–197.

Vahouny, G.V., Hodges, V. A., and Tradwell, C. R. (1979). *J. Lipid Res.* **20**, 154–161.

Van Golde, L. M. G., and Van Den Bergh, S. G. (1977). *Lipid Metab. Mamm.* **1**, 35–149.

Verdino, B., Blank, M. L., Privett, O. S., and Lundberg, W. O. (1964). *J. Nutr.* **83**, 234–238.

Volpe, J. J., and Vagelos, P. R. (1976). *Physiol. Rev.* **56**, 339–417.

Walker, B. L. (1970). *J. Nutr.* **100**, 355–360.

Whitaker, M. O., Wyche, A., Sprecher, H., and Needleman, P. (1979). *Proc. Natl. Acad. Sci. U.S.A.* **76**, 5919–5923.

Willis, J. (1981). *Nutr. Rev.* **39**, 289–301.

17

Results Obtained with Feeding Low Erucic Acid Rapeseed Oils and Other Vegetable Oils to Rats and Other Species

J. K. G. KRAMER AND F. D. SAUER

413

High and Low Erucic Acid Rapeseed Oils
Copyright © 1983 by Academic Press Canada
All rights of reproduction in any form reserved.
ISBN 0-12-425080-7

I. INTRODUCTION

The success of plant breeders to develop rapeseed cultivars practically devoid of erucic acid stands out as an accomplishment to the potential of genetic manipulation in plants to remove chemical constituents (see Chapter 6). The oil from these new cultivars of rapeseed has an entirely different fatty acid composition compared to the older cultivars of rapeseed, and hence by right should be considered as a new vegetable oil. For example, a typical Canadian and European high erucic acid rapeseed (HEAR) oil, like mustard oil, is rich in erucic (22:1 n-9) and gandoic (20:1 n-9) acids, while the new low erucic acid rapeseed (LEAR) oil is rich in oleic acid (18:1 n-9) (Table I). The fatty acid composition of LEAR oil resembles that of peanut and olive oil except for linolenic acid (18:3 n-3), which is found in soybean oil at a similar level.

As defined by the Codex Alimentarius Commission (1979), LEAR oils contain less than 5% erucic acid. The LEAR oils from Canada are much lower than 5%, and are presently 0.4–2% (see Chapter 7). In 1965, personnel of the Research and Development Laboratories of Canada Packers Ltd., Toronto, Canada, proposed that the new low erucic acid cultivars of rapeseed be named "canbra" in order to clearly differentiate them from the older high erucic acid cultivars. The name "canbra" includes the rapeseed cultivars low in erucic but still high in glucosinolates. (The properties of glucosinolates are discussed in Chapters 1, 4, and 6.) In 1974 a new cultivar of rapeseed was released which was low both in erucic acid and glucosinolates (Stefansson and Kondra, 1975). This type of rapeseed, sometimes referred to as "double low" rapeseed, has now largely replaced the "single low" (low in erucic acid only) rapeseed and today constitutes 80% of the rapeseed planted in Canada (Prairie Grain Variety Survey, 1981). In 1978, the Western Canadian Oilseed Crushers Association of Canada suggested the name "canola" to identify the double low rapeseed and that the oil derived from the seed be called "canola oil." This generic term has been accepted in February of 1981 for usage on labels of retail products by the Health Protection Branch and by the Department of Consumer and Corporate Affairs in Canada.

In this chapter, the name LEAR oil will be used rather than canbra or

TABLE I

Fatty Acid Composition of Commonly Used Vegetable Oils

Fatty acids	Mustard[a]	HEAR (European)[b]	HEAR (Canadian)[c]	LEAR[d]	Peanut[e]	Olive[f]	Soybean[f]	Corn[c]	Sunflower[g]
16:0	1.1	3.4	2.9	5.7	10.7	11.6	12.4	13.1	6.4
16:1		0.1	0.3	0.1	0.3	1.2		0.2	0.1
18:0	0.7	0.9	1.4	2.1	3.4	2.5	3.7	2.0	3.8
18:1	21.0	11.4	33.0	57.7	49.0	75.5	25.4	26.5	23.8
18:2	16.0	14.4	15.4	24.6	28.0	7.3	50.6	56.1	64.6
18:3	6.5	7.9	6.2	7.9		0.7	7.9	0.7	0.3
20:0	0.8	0.8		0.2	2.0				0.2
20:1	7.5	7.1	12.2	1.0	1.6	0.4		0.4	0.2
22:0	2.5	0.6	0.5	0.2	3.0	0.1			0.2
22:1	42.0	50.6	25.5	0.2	<0.1				0.6
24:0			0.2	0.1	1.4				
24:1	1.5	1.3	0.2	0.1					

[a] Gopalan et al. (1974).
[b] Rocquelin et al. (1973).
[c] Kramer et al. (1979b).
[d] Kramer et al. (1979a).
[e] Rocquelin and Fouillet (1979).
[f] Hulan et al. (1977a).
[g] Abdellatif and Vles (1973).

canola since the scientific investigations were done with both oils. In any case, fully refined oils derived from either canbra or canola seed are for all practical purposes indistinguishable and appear to give the same results in nutritional experiments.

II. NUTRITIONAL AND PATHOLOGICAL PROPERTIES

A. Growth Performance

In the 1960s the new cultivars of rapeseed, with their very low content of erucic acid in the oil, became available for nutritional studies. It was immediately apparent that this new vegetable oil gave as good a growth rate in rats as other vegetable oils, such as olive or peanut oil (Craig and Beare, 1968; Rocquelin and Cluzan, 1968). Therefore, the undesirable growth depressing effect of the HEAR oils was removed by the elimination of erucic acid from the oil. Many other investigators have subsequently confirmed the finding that rats fed LEAR oils give the same growth rate in rats as corn oil (Hulan et al., 1977b; Hung et al., 1977; Kramer et al., 1973), linseed oil (Vles et al., 1978), olive oil (Vles et al., 1978), peanut oil (Engfeldt and Brunius, 1975b), poppyseed oil (Beare-Rogers et al., 1979), soybean oil (Ilsemann et al., 1976; Kramer et al., 1973, 1979a), and sunflower oil (Vles et al., 1976, 1978; Ziemlanski, 1977).

It was shown over two decades ago that optimum weight gain and feed efficiency were obtained when albino rats were fed fat mixtures that contained 30% saturated fatty acids (Murray et al., 1958). Since most vegetable oils, in particular rapeseed oils, are low in saturated fatty acids, it would seem reasonable to suppose that the growth rate of rats fed vegetable oils could be improved by adding saturated fatty acids to the diet. That this is, in fact, the case was shown by Beare et al. (1963) who improved the growth rate of rats fed HEAR oil by the addition of palm oil to raise the level of saturated fatty acids from 4% to 30%. Similar results were obtained with Sprague–Dawley rats fed a LEAR oil whose saturated fatty acid content was increased from 7 to 17% by the addition of cocoa butter (Farnworth et al., 1982a; Kramer et al., 1981). The addition of cocoa butter over a 16 week period resulted in a 14% increase in body weight ($P < 0.01$) and a 9% increase in feed efficiency (weight gain/feed consumption) over the same period of time. Preliminary evidence indicates that the increased weight gain, when a LEAR oil is supplemented with cocoa butter, represents an actual increase in carcass weight and is not simply an increase in body fat (Farnworth and Kramer, 1981). It would seem that further research is needed in order to explain this requirement for saturated fatty acids when rats are fed diets rich in vegetable oils that are high in mono- and polyunsaturated fatty acids. However, it is well recognized that saturated fatty acids are a "risk factor" for coronary heart disease in man, and most health regulatory agencies are unanimous in recommending a substantial decrease in the intake of saturated fatty acids (Vergroesen and Gottenbos, 1975). Therefore, the characteristic low level of saturated fatty acids in LEAR oils is probably beneficial.

B. Digestibility

The apparent digestibility of LEAR oil in rats was thoroughly investigated by Rocquelin and Leclerc (1969). A LEAR oil containing 1.9% erucic acid was digested as well as peanut oil. In fact, the digestibility of the LEAR oil was greater than that of peanut oil, which was 98 and 94%, respectively. The high digestibility of LEAR oils was confirmed recently. Male rats fed a LEAR oil with 0.6% erucic acid had a digestibility of 95% after 4 weeks on diet, 96% after 8 weeks and 98% after 12 weeks, while the values for soybean oil were 96, 95, and 98% for the same time periods (Farnworth et al., 1982a). These studies show that the new LEAR oils are digested as other vegetable oils.

Erucic (22:1 n-9) and gandoic (20:1 n-9) acids proved to be slightly more digestible when present in small amounts, such as in LEAR oils, than when present at higher concentrations such as in HEAR oils. A digestibility of 91% was recorded for erucic acid in LEAR oil that contained only 1.9% of this acid, whereas a HEAR oil with 45% erucic acid showed a digestibility of 73% for this acid (Rocquelin and Leclerc, 1969).

C. Myocardial Lipidosis

Concern was expressed in the early 1970s (Campbell, 1970) when it became clear that erucic acid in rapeseed oils was the cause of fat accumulation in rat hearts (Abdellatif and Vles, 1970). However, the fat accumulation was evident only in young rats fed a diet containing in excess of 1.5% (Beare-Rogers et al., 1971) or 2% (Engfeldt and Brunius, 1975a) 22:1 in the diet. LEAR oils (defined as rapeseed oils with <5% 22:1) did not cause myocardial lipidosis in young rats even when the oil was fed as the only fat in the diet at 20% by weight, because the concentration of 22:1 was less than 1% of the diet. Many research groups have demonstrated that the rat heart is essentially free of fat accumulation when fed LEAR oils (Beare-Rogers et al., 1971; Engfeldt and Brunius, 1975a; Kramer et al., 1973, 1979b; Rocquelin et al., 1973).

Health and Welfare Canada considered it prudent in 1970 to recommend that rapeseed cultivars that contained high levels of 22:1 in the oil be replaced as soon as practical with new cultivars of rapeseed essentially devoid of this acid (Campbell, 1970). These new cultivars of rapeseed had been successfully developed during the 1960s (Stefansson et al., 1961; Downey, 1964), and their production in 1970 was encouraged because these new rapeseed oils did not give rise to myocardial lipidosis. Health and Welfare Canada followed up their recommendation with a regulation in 1973 which limited the docosenoic fatty acid content in the fat of food products to 5% (News release 1973-76, Health and Welfare Canada, June 29, 1973). The European health regulatory agencies followed Canada's recommendation. Plant breeders in Europe developed new winter rapeseed cultivars that were also low in 22:1 (Morice,1979; Röbbelen and Thies, 1980). As of the 1st of July 1979, the level of erucic acid was limited to less than 5% of the fat components of food products by the Council of European Communities (see Chapter 11).

The average content of 22:1 in the Canadian rapeseed crop for human consumption has been below 2% since 1978 (see Chapter 7). With such a LEAR oil, myocardial lipidosis would not be present even if this vegetable oil were consumed as the only source of fat in a diet containing 20% by weight (or 40 calorie %) fat. At this concentration of 22:1, if LEAR oil were the only source of fat in a diet which contained 40 calorie % fat, the 22:1 content would represent 0.26% of the diet.

D. Myocardial Degenerative Changes

It was unexpected to find in 1968 that the oil from the newly developed rapeseed varieties practically devoid of 22:1 still caused a high incidence of myocarditis in male rats (Rocquelin and Cluzan, 1968). This problem, like the adverse effects that were found with HEAR oils, had been ascribed to the high content of 22:1 (Roine et al., 1960). Problems such as growth retardation, lower digestibility, myocardial lipidosis, susceptibility to cold stress, and impaired mitochondrial function were eliminated with the removal of 22:1 from rapeseed. However, the development of myocardial degenerative changes on prolonged feeding of these new rapeseed oils remained, despite the fact that the LEAR oil contained only 1.9% 22:1 (Rocquelin and Cluzan, 1968). These results with male rats were confirmed by many investigators (Abdellatif and Vles, 1973; Beare-Rogers et al., 1974a; Clandinin and Yamashiro, 1980; Ilsemann et al., 1976; Kramer et al., 1973; McCutcheon et al., 1976; Rocquelin et al., 1973; Vogtmann et al., 1975; Ziemlanski, 1977). The result that showed that the LEAR oils still caused heart lesions in male rats was of concern to the scientific community.

Representatives from some 19 countries met at an International Conference on Rapeseed Products in Ste. Adèle, Quebec, in 1970 and decided, according to an official conference statement, to switch to the low erucic acid rapeseed cultivars as soon as practical (Migicovsky, 1970). This decision prompted the dramatic changeover to LEAR oils in Canada and Europe. With evidence showing that the heart damage in male rats was not alleviated with the removal of 22:1, the wisdom of the decision to switch to the LEAR oils without placing any limitation on its use was questioned (Lancet, 1974). The possibility was raised by several investigators that cardiotoxins other than erucic acid may be present in rapeseed oils (Beare-Rogers, 1977; Beare-Rogers et al., 1974a; Rocquelin et al., 1973) and therefore pose a more serious problem than first anticipated.

Several countries, notably Canada, France, Germany, Netherlands, Norway, Poland, and Sweden, placed high priority on research to evaluate the cardiopathogenicity of LEAR oils and determine the cause of this problem. In Canada research was undertaken at Health and Welfare, Ottawa, at Agriculture Canada, Ottawa, and at several Canadian Universities. The French decided in 1974 to organize a cooperative research effort which involved several French and foreign investigators (Flanzy, 1979a). Their accumulated research results on LEAR oil were published in 1979 (Flanzy, 1979b). In Sweden, the Medical Research Council decided in 1971 to initiate a series of projects to investigate the biochemical, morphological, and physiological effects of high and low levels of erucic acid in the diet of the rat. The results were published in 1975 (Engfeldt, 1975). Research on the nutritional and pathological properties of LEAR oils in Germany was carried out at the Bundesanstalt für Fettforschung in Münster, while in the Netherlands work was done at the Unilever Research Laboratories at Vlaardingen. In Norway research was undertaken to determine if cetoleic acid (cis 22:1 n-11) which is present in marine oils, is a problem, since this acid has properties similar to erucic acid in rapeseed oils. In Poland, rapeseed research was carried out in the Institute of Food and Nutrition, Warsaw.

A decade of intense research was aimed at elucidating the nature of the cardiopathological factor in LEAR oils. The results are presented in the remaining portion of this chapter. In 1973 it was pointed out that male rats fed control diets developed myocardial lesions that were indistinguishable from the myocardial lesions found in rats fed LEAR oils (Kramer et al., 1973). Prior to this it had been assumed that male rats fed control diets were either devoid of myocardial lesions or else showed only very mild myocarditis. With the realization that myocarditis was present in a large percentage of male rats that were fed control diets, it became necessary to use a large number of rats for each experiment and to statistically analyze the incidence and severity of the lesions in order to make meaningful comparisons (Hulan et al., 1977d). Alternatively, some research workers used few rats but made many more histological sections from each heart for careful examination

(Engfeldt and Brunius, 1975a,b; Vles *et al.*, 1976; McCutcheon *et al.*, 1976). In either case, sophisticated statistical analyses were necessary in order to interpret the data (see Chapter 19).

III. RESULTS WITH RATS

A. Types of Rats and Length of Time on Experimental Diets

A summary of most of the data generated during the last decade is presented in the tables that follow. Research workers have used dietary protocols sufficiently different to necessitate giving detailed information as to the amount of fat used in the diet, the strains of rats used, the duration of feeding, and the pathological grading system employed. For details regarding the diet composition and pathological methods systems see Chapters 13 and 12, respectively. The other parameters are summarized in Table II. The authors are listed in alphabetical order with detailed information summarized for each study. In subsequent tables, therefore, only the references are cited.

1. EFFECT OF SEX

As indicated in Table III, female rats of all strains are less susceptible to the development of myocardial lesions than male rats fed control or rapeseed oil containing diets. In general, male rats of the albino strain have a higher incidence and severity of myocardial lesions when fed isocaloric amounts of LEAR oils than when fed other vegetable oils. The incidence and severity of heart lesions are similar in female rats fed control or LEAR oils.

Androgens may be involved in the development of heart lesions since castration lowered the incidence of lesions in male rats (Hulan *et al.*, 1977c). This is further supported by a recent study that shows that the male rat is rather susceptible to the development of heart lesions caused by 22:1 during the period of sexual maturation at about 14 weeks of age (Svaar, 1982). However, both these claims need to be confirmed because further questions remain. In the study by Hulan *et al.* (1977c), castration also resulted in a decrease in growth of male rats which may have contributed to a decrease in the incidence of heart lesions (Kramer *et al.*, 1979a). This appears to be supported by Vles (1979) who found no decreased growth in castrated male rats and no significant reduction in heart lesions. Similarly, the study by Svaar (1982) does not explain why heart lesions develop in rats fed diets that do not contain 22:1.

2. EFFECT OF STRAIN

The common laboratory rat strains have been maintained as closed populations with limited numbers of parents for many generations. Differences and similarities may therefore exist among and within these strains. In fact,

TABLE II

Experimental Protocol Used by Different Authors

Reference	% fat in diet	Strain of rat	Time on diet (weeks)	Cardiopathological grading system
Abdellatif and Vles, 1973	30	W	24	Mild, moderate
Ackman, 1974	20	SD	16	Severity index not given
Ackman and Loew, 1977	25	SD, W	2–20	n.a.
Astorg and Cluzan, 1976	15	W	16	Necrosis, macrophage infiltration
Astorg and Levillain, 1979	15	W	16	Histiocytes, granuloma, and necrosis
Beare-Rogers and Nera, 1977	20	SD, W	16	Severity scale 1 to 4 in increasing order
Beare-Rogers et al., 1974a	20	SD	16 or 26	n.a.
Beare-Rogers et al., 1974b	20	SD	16	Severity scale 1 to 4 in increasing order
Beare-Rogers et al., 1979	20	W	26	Severity scale 1 to 4 in increasing order
Bijster et al., 1979a	15	W	n.a.	No. of lesions/rat, sum of Feret diameters
Charlton et al., 1975	20	SD	16	No. of lesions/rat
Clandinin and Yamashiro, 1980, 1982	20	SD	16 or 28	n.a.
Cluzan et al., 1979	15	W	26	Doubtful, necrotic
Engfeldt and Brunius, 1975b	20	SD	6–23	n.a.
Farnworth et al., 1982c	20	SD	16	1, 2, 3, 4 and >4 lesions/heart (3 sections)
Hulan et al., 1976b, 1977a	20	SD	16	1–2, 3–5, 6–10 and >10 lesions/heart (3 sections)
Hulan et al., 1977b	20	SD, W, Sh	16	1, 2, 3, 4–10 and >10 lesions/heart (3 sections)
Hulan et al., 1977c	20	SD	16	1, 2, 3–7 and >10 lesions/heart (3 sections)
Hung et al., 1977	20	SD	16	n.a.
Ilsemann et al., 1976	20	W	12	n.a.
Kramer et al., 1973	20	SD	16	n.a.
Kramer et al., 1975b, 1979a,b	20	SD	16	1–2, 3–5, 6–10 and >10 lesions/heart (3 sections)
Kramer et al., 1981, 1982	20	SD	16	1, 2, 3, 4 and >4 lesions/heart (3 sections)
McCutcheon et al., 1976	20	W	25	Microvascular alteration, fresh and old focal necrosis

(Continued)

TABLE II (CONTINUED)

Reference	% fat in diet	Strain of rat	Time on diet (weeks)	Cardiopathological grading system
Nolen, 1981	20	SD	16	n.a.
Procter et al., 1974	20	SD	16	Severity scale 1 to 3 in increasing order
Rocquelin and Cluzan, 1968	15	W	26	Disputable, moderate, marked
Rocquelin et al., 1973	15	W	8.6	Mild, necrosis, fibrosis
Rocquelin et al., 1974	15	W	26	Necrosis, macrophage infiltration, fibrosis
Rocquelin et al., 1981	15	W	12	Number of lesions per rat
Rose et al., 1981	20	SD	16	Myocarditis index = mean no. of foci/section of heart
Slinger, 1977	20	W, W	18	Microvascular alteration, fresh and old focal necrosis
Svaar and Langmark, 1975	20	SD	16	n.a.
Svaar and Langmark, 1980	21	SD	30	1–3 small, >3 small or 1–3 large, >3 large lesions
Umemura et al., 1978	20	W, W	18	Microvascular alteration, fresh and old focal necrosis
Vles, 1974	30	W	12–24	1 minor lesion, 2 or more lesions
Vles, 1979	20	SD, W	16	No. of lesions/rat, sum of Feret diameters
Vles et al., 1976	20	W	24 or 26	No. of lesions/rat, sum of Feret diameters
Vles et al., 1978	20	SD, W	27 or 53	No. of lesions/rat, sum of Feret diameters
Vles et al., 1979	15	W	26	No. of lesions/rat, sum of Feret diameters
Vogtmann et al., 1975	15	SD	10	Lesion, multiple extensive lesions
Yamashiro and Clandinin, 1980	20	SD	16 or 28	n.a.

TABLE III

Effect of Sex on Incidence and Severity of Myocardial Lesions in Rats Fed High Fat Diets

Dietary oil	% 22:1	n^a	Incidence (%)[b] Male	Female	Severity difference (%)[c] 1	2	3	Ref.[d]
Peanut		20	20	10	10	—	—	1
LEAR, canbra	1.9	19	100	20	—	22	58	
Lard		20	10	0	n.a.			2
Corn		20	40	10				
LEAR (cv. Oro)	1.6	20	70	20				
LEAR (cv. Span)	4.3	20	70	0				
Soybean		20	60	20	30	10		3
LEAR (cv. Zephyr)	0.8	20	50	30	0	20		
LEAR (cv. Span)	3.7	20	60	10	—	50		
Corn		52	27*	15*	−4	4	12	4
LEAR (cv. Zephyr)	0.9	51	64**	12*	12	24	16	
Sunflower		20	—	—	−0.7	−10		5
LEAR (cv. Lesira)	7.0	20	—	—	8.9	78		
Soybean		30	27	13	n.a.			6

[a] The sum of rats used for the two groups examined.

[b] Number of rats affected/examined x 100. In this and subsequent tables the absence of asterisks indicates no statistical analysis was done. Statistically analyzed results indicated, not significant (one asterisk each), significant at $P < 0.05$ (different no. of asterisks).

[c] Severity ratings are those used by the investigators in increasing order of severity. Each severity rating is expressed as the percent difference and is obtained as follows. The number of male rats affected in severity category one is expressed as a percent of total number of male rats examined. The number of female rats affected in severity category one is expressed as a percent of the total number of female rats examined. The difference (expressed as percent) between male and female rats within category one is then recorded. This process is repeated for severity ratings 2,3 etc. A summation of all severity ratings (expressed as percent) always equals the difference in lesion incidence (expressed as percent), i.e., between males and females. A dash (—) signifies no rats affected for either group, i.e. male or female. All groups in the following tables are compared in the same manner.

[d] References (for abbreviated experimental protocol, see Table II): 1, Rocquelin and Cluzan (1968); 2. Kramer et al. (1973); 3. Vogtmann et al. (1975); 4. Hulan et al. (1977c); 5. Vles (1979); 6. Nolen (1981).

the genetic development of substrains with specific pathological disorders attest to this variation, i.e., diabetes mellitus, spontaneous hypertension, and rheumatoid arthritis. Differences have been observed between strains of mice in incidence and severity of mineralized heart lesions (Rings and Wagner, 1972).

Detailed cardiopathological investigations in male rats fed high levels of

LEAR and other vegetable oils have also provided evidence of differences in myocardial lesion response between different strains and substrains. The Chester Beatty (hooded) male rat has a much lower incidence and severity of myocardial lesions than the Sprague–Dawley male rat (Table IV). In addition, there appear to be no specific cardiopathological lesions attributable to any vegetable oil (LEAR or HEAR oil included) in the Chester Beatty male rat (Kramer et al., 1979b). Recently, Clandinin and Yamashiro (1980) obtained similar results.

A comparison between Sprague–Dawley and Wistar rats in myocardial lesion response to the feeding of high fat diets has shown very little difference in the incidence of lesions (Table IV, see also Vles et al., 1978). However, there is convincing evidence to indicate that the severity of myocardial necrosis in the Wistar male rat is significantly lower than in the

TABLE IV

Myocardial Lesions in Different Strains of Male Rats

Dietary oil	% 22:1	Incidence (%)[a]		Severity difference (%)[b]					Ref.[c]
				1	2	3	4	5	
		Sprague–Dawley (n)	Chester Beatty (n)						
Corn		20* (50)	20* (10)	−6	4	2	—	—	1
LEAR (cv. Zephyr)	0.9	68* (47)	0** (8)	26	17	6	15	4	
Corn		33* (24)	17* (24)	8	8	—	—		2
LEAR (cv. Zephyr)	0.8	83* (24)	25**(24)	4	21	17	17		
Soybean		0* (10)	0* (10)	n.a.					3
LEAR (n.a.)	0.9	60* (10)	30**(10)						
		Sprague–Dawley (n)	Wistar (n)						
Corn		20* (50)	17* (24)	6	0	2	−4	—	1
LEAR (cv. Zephyr)	0.9	68* (47)	64* (25)	−6	5	−2	11	−4	
Lard/corn (3/1)		15 (20)	0 (20)	5	—	10	—		4
LEAR (cv. Tower)	1.4	65 (20)	60 (20)	−30	10	20	5		

[a] Lesion incidence within a diet significantly different at $P < 0.05$ is indicated by different number of asterisks.

[b] Severity difference as described in footnote c of Table III.

[c] References (for abbreviated experimental protocol, see also Table II): 1, Hulan et al. (1977b); 2, Kramer et al. (1979b); 3, Clandinin and Yamashiro (1980); 4, Beare-Rogers and Nera (1977).

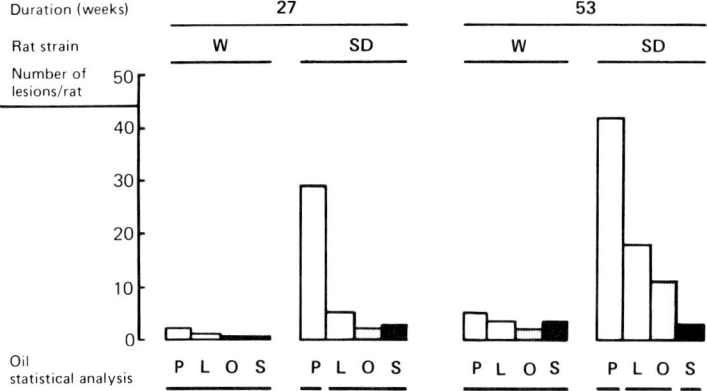

Fig. 1. The cardiopathological effects of feeding diets containing LEAR cv. Primor (P), lin-seed (L), olive (O), or sunflower (S) oils at 20% by weight for 27 and 53 weeks to male Wistar (W) and Sprague–Dawley (SD) rats. Lesion severity is indicated as number of lesions per rat (*n* = 18). Data were analyzed statistically and significant differences are represented by interruptions in the horizontal bar at bottom of figure. From Vles (1978); published with permission of *Rev. Franc. Corps Gras* and authors.

Sprague–Dawley male rat (Fig. 1). By counting the number of lesions per heart or measuring the size of the lesion, Vles *et al.* (1978) observed differences between Wistar and Sprague–Dawley rats as well as differences in lesion response between different vegetable oils. It is quite evident from these results that the Sprague–Dawley rat is by far the most susceptible strain of rat to the development of myocardial necrosis when fed high fat diets. It should be noted that there may also be substrain differences. Hulan *et al.* (1977b) used Sprague–Dawley rats obtained from two different sources and observed a significant difference ($P < 0.05$) in lesion response.

3. EFFECT OF DURATION OF EXPERIMENT

The duration of most experiments reported to date is 16 to 26 weeks. The results of Table V show the reason for this. Almost no myocardial lesions were observed when weanling rats were fed experimental diets for 4 weeks; lesions began to appear after 8 weeks and were quite evident at 16 weeks. Further feeding up to 1 year continues to increase the incidence and severity of heart lesions even more, but at a somewhat slower rate as shown in Fig. 1. If we assume the lifespan of the rat to be 36 months, then a maximum number of lesions have developed when rats have reached about 10 percent of their life-span.

4. EFFECT OF FAT LEVEL IN THE DIET

The National Research Council recommended level of fat in the diet of the rat is 10 calorie % containing a sufficient amount (2 calorie %) of essential

TABLE V

Myocardial Lesions in Male Rats Fed Fats and Oils for Different Length of Time

			Incidence (%)					
	%		Duration (weeks)					
Dietary oil	22:1	n	1	4	8	16	26	Ref.[a]
Lard		30	0	0		10		1
Corn		30	0	0		40		
LEAR (cv. Oro)	1.6	30	0	20		70		
LEAR (cv. Span)	4.3	30	0	10		70		
Peanut		15			0		13	2
LEAR (cv. Primor)	0.1	30			7		25	
LEAR, canbra	0.4	30			21		38	
Lard/Corn (3/1)		76				18	8	3
LEAR (cv. Oro)	0.6	39				45	21	
LEAR (cv. Span)	2.7	40				55	65	
Soybean		29	0		8	46		4
LEAR, canbra	n.a.	29	17		25	64		

[a] References (for abbreviated experimental protocol, see also Table II): 1, Kramer et al. (1973); 2, Rocquelin et al. (1974); 3, Beare-Rogers et al. (1974a); 4, Slinger (1977).

fatty acids. All the diets that were used to test the cardiopathogenicity of fats and oils exceeded this margin by a considerable amount, in order to duplicate the average Canadian diet. In the Canadian diet 40 calorie % is believed to be derived from fat. It is for this reason that the oils were tested at 30–70 calorie %, representing about 15–35% by weight in the diet. As shown by the results of Table VI as the fat content in the diet is increased from 5 to 20% by weight, the lesion incidence in rats fed corn or peanut oils increased three- to fourfold. A similar but less dramatic trend was observed in male Wistar rats fed 5, 10, and 15% by weight B. napus cv. Primor or peanut oil (Table VI). Research workers observed a significant difference (P < 0.03) in heart lesions between the 10 and 15% level of fat in the diet (Vles et al., 1979). Furthermore, a slow or gradual adaptation to high fat diets, which was achieved by increasing the fat content of the diet from 6 to 10 to 15 or 20% after 3, 3, and 2 weeks, did not decrease the heart lesion incidence from that observed in rats fed a 20% fat diet throughout the entire 16 weeks (Clandinin and Yamashiro, 1982). It is of interest to note that rats fed a fat free diet developed fewer heart lesions than rats fed the same basal diet supplemented with 5% corn oil (Hulan et al., 1977c). This evidence suggests that myocardial lesions are related to the amount of fat in the diet.

TABLE VI

Effect of Level of Fat in the Diet on Myocardial Lesions in Male Rats

Dietary oil	% 22:1	n	% fat in diet				Severity difference (%)[a]				Ref.[b]
			5	10	15	20	1	2	3	4	
			Incidence (%)[c]								
Corn		200	6			23***	−12	−2	−1	−2	1
Corn		52	12			27*	−3	0	−12	0	2
Peanut[d]		52	12			39	−15	−8	−4		3
			Mean no. lesions/rat								
Peanut		36	2.7	1.2	3.8[e]						4
LEAR (cv. Primor)	0.3	36	3.3	2.7	11.4[e]						4
Peanut		36	1.1	0.3	1.3						5
LEAR (cv. Primor)	0.3	36	1.7	1.4	4.7						5

[a] Severity difference as described in footnote c of Table III.

[b] References (for abbreviated experimental protocol, see also Table II): 1, Hulan et al. (1977b); 2, Hulan et al. (1977c); 3, Svaar and Langmark (1980); 4, Vles et al. (1979); 5, Cluzan et al. (1979).

[c] Significantly different at $P < 0.05\%$ (*) and $P < 0.01\%$ (***).

[d] Peanut oil fed at 4.3 and 21% by weight of the diet.

[e] For the two diets combined, only the 15% level of fat was significantly different ($P < 0.03$) from the 10% level of fat.

5. EFFECT OF PROTEIN LEVELS IN THE DIET

The results show that the incidence of heart lesions in male rats is not altered by increasing the level of protein from 20 to 25% (Beare-Rogers et al., 1974), or by substituting different quality proteins, i.e., casein vs. a casein–soybean meal mixture (Vles et al., 1976). Some recent publications suggest that the methionine content in the rat diets may be deficient if casein is fed as the only source of protein and that methionine deficiency increases the incidence of heart lesions (Clandinin and Yamashiro, 1980, 1982). Different results were obtained by Farnworth et al. (1982c), who found that the addition of methionine to casein-containing diets improved the growth of rats but that the incidence of heart lesions was not affected.

B. Myocardial Lesion Incidence and Severity Reported by Different Investigators

The data in Table VII to XI summarize the available published data from different research centers on heart lesions in male rats produced by feeding diets rich in LEAR and control oils.

TABLE VII

Incidence and Severity of Myocardial Lesions Reported by Agriculture Canada

Dietary oil	% 22:1	Incidence[a] %	n	Lesion frequency 1	2	3	4	5	Ref.[b]
Lard		10	10	n.a.					1
Corn		40	10						
LEAR (cv. Oro)	1.6	70	10						
LEAR (cv. Span)	4.3	70	10						
LEAR (cv. Oro)	1.8	40	10	0.7					2
LEAR (cv. Oro)	1.4	100	10	7.2					
LEAR (cv. Span)	5.1	100	10	14.6					
Olive	0.4	20*	20	4	0	0	0		3
LEAR (cv. Oro)	1.6	38**	50	9	7	3	0		
LEAR (cv. Span)	4.8	46**	50	13	4	6	0		
Olive	0.4	20*	10	2	0	0	0		3
LEAR (cv. Span)	4.8	60**	10	3	0	3	0		
Safflower		35*	20	7	0	0	0		3
Corn		45*	20	4	3	1	1		
LEAR (cv. Tower)	0.8	75**	20	8	2	2	3		
LEAR (cv. Span)	4.8	85**	20	2	1	3	11		
Lard	0.2	30*	20	4	1	1	0		4
LEAR (cv. Span)	4.8	67**	9	1	4	0	1		
Lard	0.2	16*	45	6	1	0	0		4
LEAR (cv. Span)	4.8	76**	45	12	11	8	3		
Corn		20*	50	7	2	1	0	0	5
LEAR (cv. Zephyr)	0.9	68**	47	12	8	3	7	2	
Corn		27*	26	3	1	3	0		6
LEAR (cv. Zephyr)	0.9	64**	25	4	8	4	0		
Corn		38*	45	11	4	2	0		7
Olive	0.1	42*	45	10	6	3	0		
Soybean		42*	45	12	4	3	0		
LEAR (cv. Tower)	0.3	80**	44	15	9	6	5		
LEAR (cv. Tower)	0.3	76**	45	18	9	6	1		
LEAR (cv. Zephyr)	0.9	80**	45	12	15	2	7		
Olive	0.1	29*	45	12	1	0	0		7
Soybean		41**	44	14	3	1	0		
LEAR (cv. Tower)	0.3	62***	45	14	6	5	3		

TABLE VII *(CONTINUED)*

Dietary oil	% 22:1	Incidence[a] %	n	Lesion frequency 1	2	3	4	5	Ref.[b]
Corn		33*	24	6	2	0	0		8
LEAR (cv. Zephyr)	0.8	83**	24	7	5	4	4		
Soybean		48*	46	17	4	0	1		9
LEAR (cv. Tower)	0.2	74**	46	22	8	2	2		
Soybean		57*	44	10	7	7	1	0	10
LEAR (cv. Tower)	0.6	61*	44	10	10	3	3	1	
Soybean		29	150	22	11	3	4	3	11

[a] Within an experiment significant differences ($P < 0.05$) in lesion incidence are indicated by different number of asterisks (*,**, and ***). Absence of asterisks means no statistics were done.

[b] References (for abbreviated experimental protocol, see also Table II): 1, Kramer *et al.* (1973); 2, Charlton *et al.* (1975); 3, Kramer *et al.* (1975b); 4, Hulan *et al.* (1976b); 5, Hulan *et al.* (1977b); 6, Hulan *et al.* (1977c); 7, Hulan *et al.* (1977a); 8, Kramer *et al.* (1979b); 9, Kramer *et al.* (1979a); 10, Kramer *et al.* (1981, 1982); 11, Farnworth *et al.* (1982c).

Table VII presents a summary of data by Agriculture Canada scientists since 1973. Statistical analyses were performed for all but the first two studies (Kramer *et al.*, 1973; Charlton *et al.*, 1975). Statistical significance at the 5% level ($P < 0.05$) are indicated by asterisks. As shown by results in Table VII, LEAR oils usually result in a higher incidence of myocardial lesions than other oils when fed to male Sprague–Dawley rats except in one experiment in which soybean oil and LEAR oil showed no significant difference in lesion incidence (Farnworth *et al.*, 1982a; Kramer *et al.*, 1981). Similarly, the lesion severity, indicated by number of lesions per heart, was generally significantly higher in rats fed LEAR oils than in rats fed other oils.

The results presented in Table VIII summarize the findings by Health and Welfare Canada. Statistical analyses have been applied since 1977. It is evident from the data that LEAR oil fed rats developed a higher incidence (and severity where indicated) of heart lesions than rats fed the lard/corn oil mixture or another oil.

The results obtained by researchers from the University of Guelph are shown in Table IX. Although no statistical analyses were carried out, the incidence and severity of myocardial lesions in rats fed control oils, i.e., safflower, corn, soybean, and hydrogenated coconut oils, appear to be as high or higher than that found with LEAR oils, with the exception of one report (Hung *et al.*, 1977). Similar results were obtained from the University

TABLE VIII

Incidence and Severity of Myocardial Lesions Reported by Health and Welfare Canada

Dietary oil	% 22:1	Incidence %	Incidence n	Lesion grades 1	2	3	4	Ref.[a]
Lard/corn (3/1)		0	17	n.a.				1
LEAR, canbra	2.9	63	16					
Lard/corn (3/1)		18	38	n.a.				2
LEAR (cv. Oro)	0.6	33	39					
Lard/corn (3/1)		8	38	n.a.				2
LEAR (cv. Span)	2.7	60	40					
Lard/corn (3/1)		7	56	n.a.				1
Soybean		15	27					
LEAR (cv. Oro)	1.9	37	27					
LEAR, canbra	2.5	35	20					
LEAR (cv. Span)	2.7	58	19					
LEAR (cv. Span)	3.4	59	27					
Lard/corn (3/1)		11	36	n.a.				1
LEAR (cv. Oro)	2.1	42	36					
LEAR (cv. Span)	2.7	50	36					
Lard/corn (3/1)		20	20	n.a.				1
Olive		5	20					
LEAR (cv. Zephyr)	0.7	35	20					
Lard/corn (3/1)		15	20	2	0	1	0	3
Olive		20	20	2	1	0	1	
LEAR (cv. Span)	0.2	40	20	4	1	3	0	
Lard/corn (3/1)		20	20	4	0	0	0	4
LEAR (cv. Span)	4.6	85	20	3	5	3	6	
Lard/corn (3/1)		15	20	1	0	2	0	4
LEAR (cv. Tower)	1.4	65	20	3	5	4	1	
Lard/corn (3/1)		0	20	0	0	0	0	4
LEAR (cv. Tower)	1.4	60	20	9	3	0	0	
Lard/corn (3/1)		7	15	1	0			4
Olive		7	15	1	0			
Linseed		13	15	2	0			
LEAR (cv. Tower)	1.4	40	15	4	2			

TABLE VIII *(CONTINUED)*

Dietary oil	% 22:1	Incidence %	Incidence n	Lesion grades 1	Lesion grades 2	Lesion grades 3	Lesion grades 4	Ref.[a]
Lard/corn (3/1)		7	15	0	1	0	0	5
Sunflower		20	15	2	1	0	0	
Poppyseed		7	14	1	0	0	0	
LEAR (cv. Tower)	1.0	60	15	9	0	0	0	

[a] References (for abbreviated experimental protocol, see also Table II): 1, Beare-Rogers *et al.* (1974a)(16 weeks); 2, Beare-Rogers *et al.* (1974a)(16 and 26 weeks); 3, Beare-Rogers *et al.* (1974b); 4, Beare-Rogers and Nera (1977); 5, Beare-Rogers *et al.* (1979).

TABLE IX

Incidence and Severity of Myocardial Lesions Reported by Other Canadian Laboratories

Dietary oil	% 22:1	Incidence %	Incidence n	Lesion grades 1	Lesion grades 2	Lesion grades 3	Ref.[a]
University of Guelph							
Safflower		40[b]	15	1	1	5	1
Coconut (hydr.)		50	14	3	4	4	
Soybean		64	14	4	5	7	
LEAR (cv. Tower)	0.8	64	14	5	8	8	
Corn		50	16	1	5	6	2, 3
LEAR (cv. Tower)	0.9	71	17	4	11	11	
LEAR (1788)	3.6	41	17	n.a.			
LEAR (1788)	3.6	63	27	15	13	8	
Soybean		45	11	n.a.			2
LEAR, canbra	n.a.	64	11				
Corn		0	15	n.a.			4
LEAR (cv. Tower)	Trace	20	15				
University of Alberta							
Soybean	0.1	60	10	5	1		5
LEAR (cv. Zephyr)	0.8	50	10	2	3		
LEAR (cv. Span)	3.7	60	10	0	6		
Fisheries Lab., Halifax							
Lard/corn (3/1)		0	20	0			6
Peanut	0.1	30	20	0.35			
University of Toronto							
Soybean		40*	10	n.a.			7
LEAR (n.a.)	0.9	80**	10				

TABLE IX *(CONTINUED)*

Dietary oil	% 22:1	Incidence %	n	Lesion grades 1	2	3	Ref.[a]
University of Toronto *(Cont.)*							
Soybean		10*	10	n.a.			7[c]
LEAR (n.a.)	0.9	50**	10				
Soybean		0*	10	n.a.			8, 9[c]
LEAR (n.a.)	0.9	60**	10				
Soybean		5*	20	n.a.			10
LEAR (n.a.)	0.9	45**	20				
Soybean		0*	20	n.a.			10[c]
LEAR (n.a.)	0.9	35**	20				
University of Saskatoon							
Lard/Corn (3/1)	<0.1	19	16	n.a.			11
Lard/Corn (3/1)		33*	9	0.25			12
LEAR (cv. Tower)	0.1–5.8	81**	94	4.63			

[a] References (for abbreviated experimental protocol, see also Table II): 1, McCutcheon et al. (1976); 2, Slinger (1977); 3, Umemura et al. (1978); 4, Hung et al. (1977); 5, Vogtmann et al. (1975); 6, Ackman (1974); 7, Clandinin and Yamashiro (1980)(16 weeks); 8, Clandinin and Yamashiro (1980)(28 weeks); 9, Yamashiro and Clandinin (1980); 10, Clandinin and Yamashiro (1982); 11, Ackman and Loew (1977)(SD & W); 12, Rose et al. (1981).
[b] Incidence of heart lesions was obtained from B. L. Walker (personal communication).
[c] Diets supplemented with methionine.

of Alberta (Table IX). Studies by researchers at the Fisheries Laboratory, Halifax, and the University of Saskatchewan showed a low incidence of heart lesions in male rats fed fats and oils. Researchers from the University of Toronto also found lesions in male rats fed soybean oil and claimed that these could be reduced by supplementing the dietary protein with methionine. Later studies, however, were unable to confirm this methionine effect (Farnworth et al., 1982c).

Scientists from the Unilever Research Laboratories used an experimental protocol in which 20–28 sections per heart were examined except in their first two studies (Abdellatif and Vles, 1970, 1973). Their results (Table X) with both Wistar and Sprague–Dawley rats showed no significant difference in lesion incidence with rats fed sunflower, olive, peanut, linseed or LEAR (cv. Tower, cv. Primor) oils. There was, however, a significant difference in severity rating (number of lesions per heart and lesion diameter) in Sprague–Dawley but not Wistar rats fed LEAR and other oils; more severe lesions were found in the former strain (Fig. 1).

TABLE X

Incidence and Severity of Myocardial Lesions Reported from Unilever Research Laboratories

Dietary oil	% 22:1	Incidence[a] %	Incidence[a] n	Lesion grades[a] 1	Lesion grades[a] 2	Ref.[b]
Sunflower		0	12	0	0	1
LEAR/sunflower (5/1)	7.1	67	12	6	2	
Olive		38	24	9	0	2
Peanut	0.1	21	24	5	0	
LEAR/sunflower (2/1)	4.7	38	24	7	0	
Sunflower		33*	12	0.4*	5*	3
LEAR (cv. Tower)/sun-flower/palm (6/1/1)	0.75	42*	12	3.1*	37*	
Sunflower		44*	16	1.1*	17*	3
LEAR (cv. Primor)	0.3	50*	16	1.0*	16*	
Sunflower		28*	18	0.4*	6*	4
Olive		39*	18	0.6*	6*°	
Linseed		50*	18	1.0*	13*°	
LEAR (cv. Primor)	0.4	56*	18	1.8*	61°	
Sunflower		78*	18	2.6*	52*	5
Olive		78*	18	2.1*	46*	
Linseed		89*	18	4.9*	135*	
LEAR (cv. Primor)	0.4	83*	18	29.3**	1276**	
Sunflower		94*	18	3.2*	49*	6
Olive		56°	18	2.1*	45*	
Linseed		72*°	18	3.6*	67*	
LEAR (cv. Primor)	0.4	78*°	18	5.1*	118*	
Sunflower		83*	18	3.3	62	7
Olive		94*	17	10.8*	296*	
Linseed		100*	18	17.8*	583*°	
LEAR (cv. Primor)	0.4	100*	17	42.0°	1478°	
Peanut	<0.1	75*	12	3.8*	85*	8
LEAR (cv. Primor)	0.3	67*	12	11.4*	339*	
Sunflower		—	20	0.8	3	9
Soybean		—	20	1.8	8	
LEAR (cv. Lesira)	7.0	—	20	10.5	84	

[a] Numbers having different symbols (none, * or °) are significantly different ($P < 0.05$).

[b] References (for abbreviated experimental protocol, see also Table II): 1, Abdellatif and Vles (1973); 2, Vles (1974); 3, Vles et al. (1976); 4, Vles et al. (1978)(W, 27 weeks); 5, Vles et al. (1978)(SD, 27 weeks); 6, Vles et al. (1978)(W, 53 weeks); 7, Vles et al. (1978)(SD, 53 weeks); 8, Vles et al. (1979); 9, Vles (1979).

TABLE XI

Incidence and Severity of Myocardial Lesions Reported in Male Rats from France, Germany, Sweden, and Norway

Dietary oil	% 22:1	Incidence %	Incidence n	Severity grades 1	Severity grades 2	Severity grades 3	Ref.[a]
France							
Peanut		20	10	2	0	0	1
LEAR, canbra	1.9	100	9	0	2	7	
Peanut		0	16	0	0	0	2
LEAR, canbra	0.9	63	16	6	2	2	
Peanut	0.4	25	8	13	13	0	3
LEAR (cv. Primor)	0.1	56	16	25	31	0	
LEAR, canbra	0.4	88	16	38	50	0	
Peanut		0	6	0	0		4
Oleic		33	9	3	0	0	5
Oleic + linoleic		100	9	8	1	0	
Elaidic		100	9	9	0	0	
Elaidic + linoleic		100	9	8	1	0	
Peanut	<0.1	42	12	0	5		6
LEAR (cv. Primor)	0.3	50	12	0	6		
Sunflower		13	8	0.1			7
LEAR (cv. Primor)	0.3	63	8	0.9			
Germany							
Soybean		13	8	n.a.			8
LEAR (cv. Lesira)	4.3	50	8				
Soybean		61	18	1	10		9
LEAR (cv. Lesira)	n.a.	65	17	2	18		
Sweden							
Peanut	0.1	—[b]	15	n.a.			10
LEAR (cv. Oro)	0.3	—[b]	15				

TABLE XI *(CONTINUED)*

Dietary oil	% 22:1	Incidence		Severity grades			Ref.[a]
		%	n	1	2	3	
Norway							
Lard		8	12	n.a.			11
Lard/corn (3:1)		25	12				
Peanut		39	26	27	8	4	12

[a] References (for abbreviated experimental protocol, see also Table II): 1, Rocquelin and Cluzan (1968); 2, Rocquelin et al. (1973); 3, Rocquelin et al. (1974); 4, Astor and Cluzan (1976); 5, Astorg and Levillain (1979); 6, Cluzan et al. (1979); 7, Rocquelin et al. (1981); 8, Ilsemann et al. (1976); 9, Bijster et al. (1978); 10, Engfeldt and Brunius (1975b); 11, Svaar and Langmark (1975); 12, Svaar and Langmark (1980).
[b] Only small myocardial lesions were detected which were considered "normal."

Considerable variability was reported when peanut oil was tested against LEAR (canbra and cv. Primor) oil by research workers in France (Table XI). In some experiments LEAR oils produced a much higher lesion incidence than peanut oils, whereas in other experiments no real differences were apparent. Similar results were obtained by German researchers. On the other hand, Swedish workers (Table XI) reported no cell infiltration or myocardial scarring in male Sprague–Dawley rats fed either peanut oil or a LEAR (cv. Oro) oil even though the hearts were serially sectioned (Engfeldt and Brunius, 1975b). Small myocardial lesions, which were present, were considered to be a "normal" finding. In addition, the Swedish workers established that the morphological effects on the myocardium were similar between conventional and germ-free rats (Engfeldt and Gustafsson, 1975). They concluded that myocardial lesions in male rats were not influenced by the presence or absence of a normal intestinal flora. Norwegian workers consistently found myocardial lesions in male rats fed fats or oils.

A few studies have permitted a comparison of nonrapeseed vegetable oils. The lesion incidence obtained when feeding control oils and fats is summarized in Table XII. From the data it is immediately apparent that the results show considerable variability due in part to studies in which small numbers of rats were used. But in spite of this fact, it is clear that all vegetable oils, and not just LEAR oils, develop the same type although not the same incidence of myocardial lesions. It is apparent that the lowest incidence of heart lesions was obtained when rats were fed lard or a 3/1 lard/corn oil mixture. Feeding single oils such as soybean, sunflower, corn, peanut, olive, or safflower oils gives an incidence of heart lesions somewhere between that obtained with LEAR oils and the lard/corn oil mixture. In a number of experi-

TABLE XII

Incidence (%) of Myocardial Lesions in Male Rats Fed Nonrapeseed Oils and Fats[a]

Corn	Lard	Lard/corn (3/1)	Olive	Peanut	Soybean	Sunflower	Other oils
0 (17)	8 (30)	0 (2, 5, 6, 7)	0 (33)	0 (4, 26, 33)	0 (9)	0 (1)	7 (8)
17 (15)	10 (19)	7 (5, 7, 8)	5 (7)	17 (33)	10 (9)	13 (28)	13 (5)
23 (15)	16 (13)	11 (7)	7 (5)	20 (25)	13 (18)	20 (8)	33 (23)
27 (16)	30 (13)	13 (3)	17 (33)	25 (27)	15 (7)	28 (35)	35 (20)
33 (22)		15 (5, 7)	20 (20)	30 (2)	27 (24)	33 (34)	40 (23)
36 (15)		17 (7)	29 (14)	39 (31)	29 (12)	44 (34)	50 (23, 35)
38 (14)		20 (5, 7)	39 (35)	42 (10)	40 (9)	78 (35)	72 (35)
40 (19)		25 (3, 30)	42 (14)	67 (10, 33)	41 (14)	83 (35)	89 (35)
45 (20)		33 (38)	56 (35)	75 (36)	42 (14)	94 (35)	100 (35)
50 (29, 32)			67 (33)	100 (36)	45 (29)		
			78 (35)		48 (21)		
			94 (35)		57 (11)		
					60 (37)		
					64 (23)		

[a] References in parentheses: 1, Abdellatif and Vles (1973); 2, Ackman (1974); 3, Ackman and Loew (1977); 4, Astorg and Cluzan (1976); 5, Beare-Rogers and Nera (1977); 6, Beare-Rogers et al. (1972); 7, Beare-Rogers et al. (1974a); 8, Beare-Rogers et al. (1974b); 9, Clandinin and Yamashiro (1980); 10, Cluzan et al. (1979); 11, Farnworth et al. (1982a); 12, Farnworth et al. (1982c); 13, Hulan et al. (1976b); 14, Hulan et al. (1977a); 15, Hulan et al. (1977b); 16, Hulan et al. (1977c); 17, Hung et al. (1977); 18, Ilsemann et al. (1976); 19, Kramer et al. (1973); 20, Kramer et al. (1975b); 21, Kramer et al. (1979a); 22, Kramer et al. (1979b); 23, McCutcheon et al. (1976); 24, Nolen (1981); 25, Rocquelin and Cluzan (1968); 26, Rocquelin et al. (1973); 27, Rocquelin et al. (1974); 28, Rocquelin et al. (1981); 29, Slinger (1977); 30, Svaar and Langmark (1975); 31, Svaar and Langmark (1980); 32, Umemura et al. (1978); 33, Vles (1974); 34, Vles et al. (1976); 35, Vles et al. (1978); 36, Vles et al. (1979); 37, Vogtmann et al. (1975); 38, Rose et al. (1981).

ments, the lesion incidence obtained with soybean oil equals that obtained with LEAR oils (Vogtmann et al., 1975; McCutcheon et al., 1976; Slinger, 1977; Farnworth et al., 1982a). In Sprague–Dawley rats some reports noted a similarity of heart lesions between corn, and safflower oils (Kramer et al., 1975b) and corn, olive and soybean oils (Hulan et al., 1977a), while in other studies differences in lesion incidence and severity were observed between vegetable oils, e.g., olive and soybean (Hulan et al., 1977a) and sunflower, olive, and linseed (Vles et al., 1978). With Wistar rats, both similarities (sunflower, olive, and linseed, Vles et al., 1978; sunflower and poppyseed, Beare-Rogers et al., 1979) and differences were noted in heart lesions between different oils (peanut and olive oils, Vles, 1974; safflower, hydrogenated coconut and soybean oils, McCutcheon et al., 1976).

C. Effects of Modifying LEAR Oils on Lesion Incidence

1. EFFECT OF COMMERCIAL PROCESSING

The effects on myocardial lesions by commercial processing of LEAR oils was studied by scientists at Agriculture Canada, Health and Welfare Canada, and Dijon, France, and summarized in Table XIII. Charlton et al. (1975) and Beare-Rogers et al. (1974a) concluded that there are no differences in incidence of myocardial lesions in rats fed either the crude or fully refined LEAR oils. Rocquelin et al. (1974) observed a decreased incidence of heart lesions in rats fed fully refined LEAR (canbra) oil compared to rats fed the crude oil. These workers, however, did not attach any significance to this finding and concluded that LEAR oils did not change their lesion causing properties during refining.

TABLE XIII

Effect of Commercial Processing of LEAR Oils on Myocardial Lesions in Male Rats

| | | | Incidence (%) | | |
| | % | | Crude | Refined | |
LEAR	22:1	n	LEAR	LEAR	Ref.[a]
cv. Oro	1.6	30	80	70	1
cv. Span	4.8	20	90	100	1
cv. Span	2.6	37	56	58	2
Canbra	2.5	40	55	35	2
cv. Zephyr	0.5	40	55	75	2
Canbra	0.4	24	100	38	3
cv. Tower	0.6	88	64	61	4

[a] References (for abbreviated experimental protocol, see also Table II): 1, Charlton et al. (1975); 2, Beare-Rogers et al. (1974a); 3, Rocquelin et al. (1974); 4, Farnworth et al. (1982b).

A fully refined LEAR oil prepared commercially was also compared to a LEAR oil prepared from the same batch of seed by an extraction procedure in which excessive heat treatments, commonly employed in industrial oil extractions, were eliminated. The results show no difference in cardiopathogenic response in male Sprague–Dawley rats between these two LEAR oils (Farnworth et al., 1982b).

2. EFFECT OF MIXING RAPESEED WITH NONRAPESEED OILS

Abdellatif and Vles (1973) showed that when a HEAR oil (50.1% 22:1) was fed to Wistar rats in different proportions with sunflower oil, a nearly linear response of lesion incidence to the concentration of 22:1 in the diet was observed (Chapter 11; Fig. 1). In similar experiments with two LEAR oils, cv. Primor (0.3% 22:1) and cv. Lesira (3.4% 22:1), this response disappeared (Fig. 2). From these results Vles et al. (1976, 1978), concluded that heart changes observed in Wistar rats fed increasing levels of LEAR oils (cv. Primor or cv. Lesira) were indistinguishable in nature, incidence, and severity from those observed in rats fed sunflower oil or soybean oil.

3. EFFECT OF HYDROGENATION

A typical change in the fatty acid composition of mild and extensive hydrogenation of a LEAR oil is shown in Table XIV. Mild hydrogenation to an iodine value of about 90–100, results in a decreased concentration of 18:3 and 18:2 and in an increased concentration of 18:1 and a small increase in 18:0. After extensive hydrogenation of the LEAR oil the concentration of

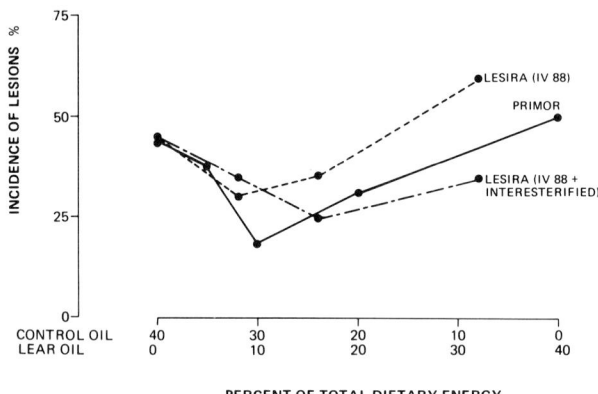

Fig. 2. Relationship between incidence of myocardial lesions and level of LEAR oil in the diet (Vles et al., 1976, 1978). The cultivars of LEAR oil used were Primor and Lesira. The latter was partially hydrogenated to iodine value (IV) 88 and interesterified.

TABLE XIV

Fatty Acid Composition of Unhydrogenated and Partially Hydrogenated LEAR Oil cv. Tower[a]

Fatty acids	Unhydro- genated	Hydrogenated[b]	
		Mild (I.V. 97.1)	Extensive (I.V. 76.6)
16:0	4.3	4.3	4.6
18:0	2.0	2.9	11.7
20:0	1.1	1.3	0.7
22:0	0.5	0.4	0.4
18:1	59.3	68.7	72.4
20:1	2.1	3.8	2.0
22:1	1.4	1.3	1.5
18:2	20.7	14.9	5.8
18:3	8.1	1.7	0.5
Total PUFA	28.8	16.6	6.3
cis-PUFA	25.2	7.3	0.3
trans-FA	0.8	13.6	29.9

[a] From Beare-Rogers and Nera (1977). Published with permission of *Lipids* and authors.
[b] IV = iodine value; PUFA = polyunsaturated fatty acids; cis-PUFA determined by lipoxygenase.

18:0 increased markedly and the concentration of polyunsaturated fatty acids (PUFA) decreased from about 30 to 6%, while the cis-PUFA concentration almost completely disappeared. The trans-fatty acid concentration increased to 30% after extensive hydrogenation.

The summary of data on the effect of hydrogenation of LEAR oils is given in Table XV. The reports from the different laboratories were not analyzed statistically; however, it is evident that extensive hydrogenation decreases both the incidence and severity of myocardial lesions. A much smaller and probably a nonsignificant effect was observed on mild hydrogenation. It therefore appears that by decreasing the concentration of 18:3 and increasing the concentration of 18:0 in LEAR oils, it was possible to reduce the lesion incidence in spite of the fact that extensive hydrogenation of vegetable oils also lowers the concentration of essential fatty acids. It is equally apparent that the increase of trans-fatty acids had no apparent adverse effect on the rat heart.

4. EFFECT OF RANDOMIZATION

The suggestion has been made that the position of fatty acids on the triglyceride molecule of rapeseed oils may affect the incidence of myocardial

TABLE XV

Effect of Hydrogenation or Myocardial Lesions in Male Rats

	%		Incidence (%)			Lesion severity			
LEAR	22:1	n	Ori-ginal	Mild[a]	Exten-sive[b]	Ori-ginal	Mild	Exten-sive	Ref.[c]
cv. Span	0.2	80	55	43	—	0.83	0.50	—	1
cv. Zephyr	0.5	120	63	58	33	1.00	0.90	0.35	1
cv. Zephyr	0.5	39	35	—	21	n.a.			2
cv. Tower	0.9	51	71	65	41	high	high	low	3
cv. Tower	1.4	118	63	62	33	high	high	low	4
Canbra	2.9	46	63	—	43	n.a.			2
"1788"	3.6	34	41	—	65	high	—	low	3
cv. Lesira	4.3	16	50	50	—	high	low	—	5
cv. Span	4.6	30	85	—	30	high	—	low	4

[a] Mild hydrogenation: iodine value 90 to 100; about 14% trans.

[b] Extensive hydrogenation: iodine value 70–80; about 30–50% trans.

[c] References (for abbreviated experimental protocol, see also Table II): 1, Procter et al. (1974); 2, Beare-Rogers et al. (1974a); 3, Slinger (1977); 4, Beare-Rogers and Nera (1977); 5, Ilsemann et al. (1976).

lesions (Kramer et al., 1975b). This was tested experimentally by chemically randomizing the fatty acids of oils. The procedure involves heating an oil in the presence of a catalyst ($NaOCH_3$) under anhydrous conditions. The results of a successful randomization are shown in Table XVI in which the fatty acid composition of the 2-monoglyceride was compared before and after randomization. The 2-monoglyceride was obtained from the trigly-

TABLE XVI

Fatty Acid Composition of LEAR (cv. Tower) and HEAR Oils before and after Randomization

	LEAR			HEAR		
Fatty acids	1, 2, 3	2 not randomized	2 randomized	1, 2, 3	2 not randomized	2 randomized
16:0	2.5	0.6	7.3	3.8	0.7	4.2
18:0	2.3	0.3	2.2	1.7	0.4	2.0
18:1	55.0	45.7	56.7	26.3	40.3	26.2
18:2	28.3	37.8	24.9	16.6	38.9	16.4
18:3	9.7	14.4	8.0	6.7	16.5	6.8
20:1	1.2	0.6	0.7	10.0	2.4	10.3
22:1	Trace	Trace	Trace	34.2	1.0	33.3

[a] Adapted from Hung et al. (1977) with permission from Lipids and authors.

TABLE XVII

Effect of Randomization on the Incidence of Myocardial Lesions in Male Rats

Dietary oil	% 22:1	n	Incidence (%) Original	Incidence (%) Randomized	Severity	Ref.[a]
HEAR	50.6	10	100	80	Similar	1
LEAR (cv. Tower)	Trace	30	20	13	n.a.	2
HEAR	34.2	30	40	20	n.a.	2

[a] References (for abbreviated experimental protocol, see also Table II): 1, Rocquelin et al. (1974); 2, Hung et al. (1977).

ceride following hydrolysis with pancreatic lipase. It is evident that the fatty acid composition of the 2-monoglyceride before randomization was characteristically different from the composition of the total oil, i.e., it contained less saturated and monounsaturated fatty acids, and was enriched in 18:2 and 18:3. After randomization these differences disappeared and the 2-monoglyceride had a fatty acid composition similar to the original oil.

Rocquelin et al. (1974) concluded that randomization of a HEAR oil (50.6% 22:1) decreased neither the incidence nor the severity of myocardial lesions (Table XVII). Hung et al. (1977) randomized a LEAR and a HEAR oil and their results showed a slight reduction of myocardial lesions. These authors did not appear to attach any significance to the changes in lesion incidence obtained by randomization. Of interest is their observation that LEAR and HEAR oils gave a very low incidence of heart lesions in male Sprague–Dawley rats not usually obtained by other research workers. It is probably safe to conclude that randomization of rapeseed oils has no effect on the incidence or severity of myocarditis in male rats.

D. Experiments with Highly Purified Triglyceride Fractions Isolated from LEAR Oils

1. EXPERIMENTS WITH LEAR CULTIVAR SPAN

There was an obvious possibility that the lesion causing effects of LEAR oils might reside in their nontriglyceride fraction, since there are a few nontriglyceride compounds that are specific to the *Brassica* family, e.g., brassicasterol and the sulfur containing products from glucosinolate hydrolysis during processing. Therefore, in 1973 Agriculture Canada undertook to fractionate large amounts of LEAR oil and purify to the highest degree possible the triglycerides of this oil.

The following procedure was undertaken: LEAR oil (cv. Span) from the 1971 crop (953 kg) was fractionated by molecular distillation under contract

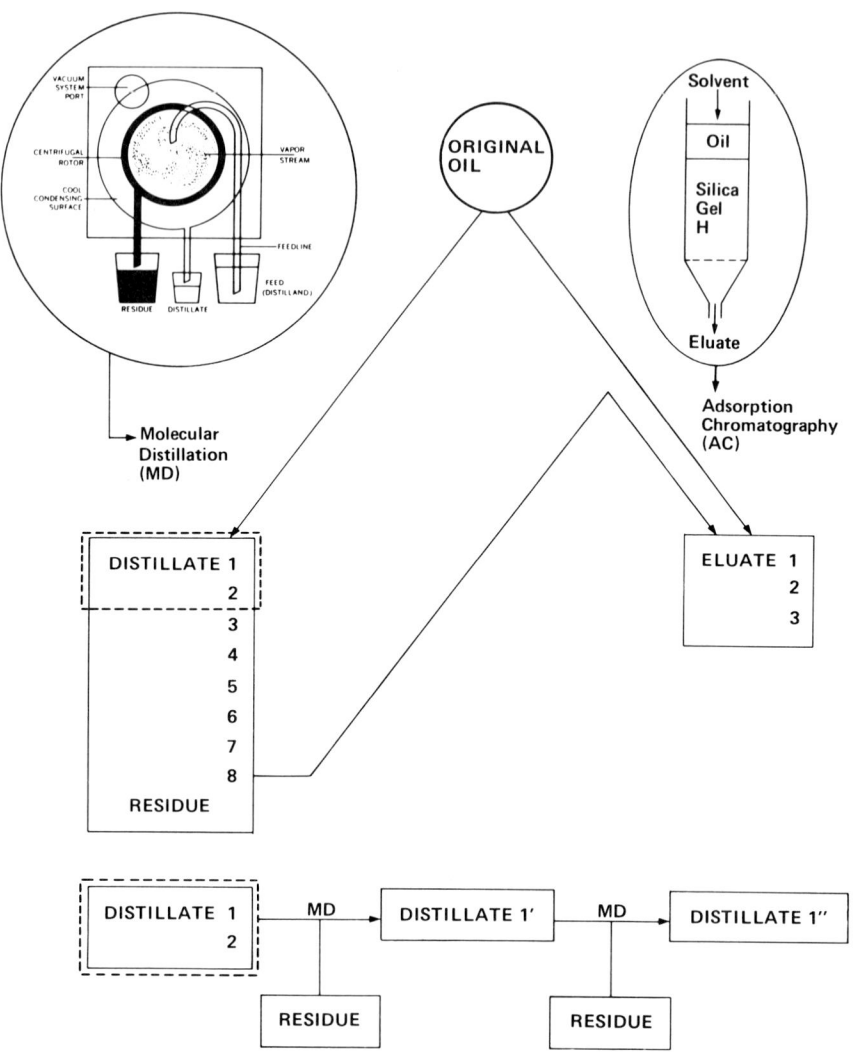

Fig. 3. A schematic of molecular distillation and column chromatography used to prepare pure triglycerides and concentrate non-triglyceride components from vegetable oils (Kramer *et al.*, 1975a, 1979a).

with Distillation Products Industries, Rochester, New York. A schematic of molecular distillation is shown in Fig. 3. The oil was distilled under vacuum (<10 μm at 250°–270°C) to yield six fractions on successive cycles through the still. Fractions 7 and 8 were obtained by continuous recycling of the residue. Figure 4 shows a thin-layer chromotograph of fractions 1 to 8.

The early distillates (MD1 and MD2) contained the major amounts of the

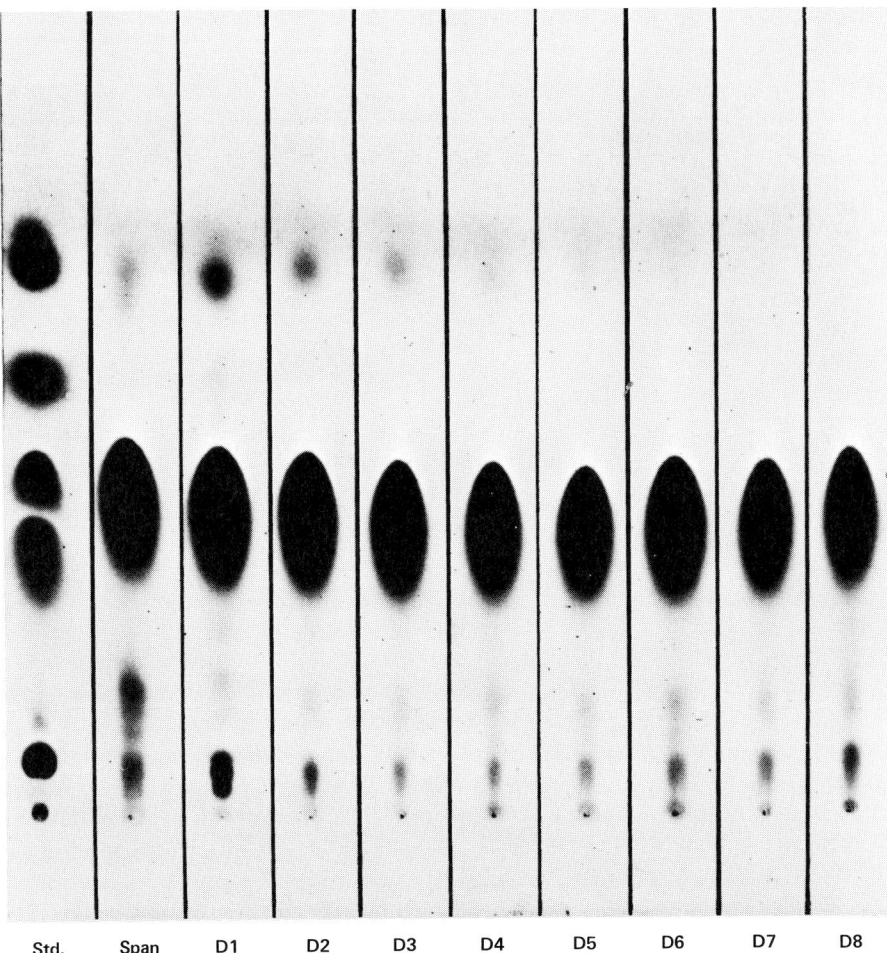

| Std. | Span | D1 | D2 | D3 | D4 | D5 | D6 | D7 | D8 |

Fig. 4. Thin-layer chromatogram of distillates (D) 1 through 8 obtained by molecular distillation of LEAR (cv. Span). Standard mixture (Std.) shows spots corresponding to cholesterol ester (top), methyl ester, triglyceride, free fatty acid, and cholesterol (bottom). Diglycerides migrate just below sterols; distillates 3 to 8 are free of sterols. Developing solvent: hexane/diethyl ether/acetic acid (85/15/1) (Kramer et al., 1975a). Published with permission from *Lipids*.

easily distilled components which include the more volatile triglycerides as well as free and esterified sterols, hydrocarbons, methyl esters, and aliphatic and terpenol alcohols. Distillate MD8 was relatively free of these contaminants. The sterol content of MD8 was 1/120 of the original sterol content, and based on the brassicasterol concentration, a 300-fold purification was realized (Table XVIII). The easily distilled components in molecular distillates 1 and 2 were further concentrated by repeated distillation to a final

TABLE XVIII

Fatty Acid Composition and Sterol Content of Fractions Obtained from Molecular
Distillation of LEAR cv. Span[a]

Fatty acids	LEAR (cv. Span)	Molecular distillates (MD)				
		MD1″	MD5	MD6	MD7	MD8
16:0	4.8	9.3	4.5	3.3	2.2	1.8
18:0	2.1	1.8	2.0	2.0	2.0	1.7
18:1	58.6	59.4	60.9	58.4	53.2	50.0
18:2	19.5	20.7	19.9	20.4	19.9	19.0
18:3	5.2	5.3	5.3	5.5	6.0	6.8
20:1	3.3	1.2	3.0	4.2	5.6	5.8
22:1	4.8	1.0	3.1	4.5	8.3	11.6
Sterol content (μg/g of oil)						
Total sterols	2,500	36,800	100	300	30	30
Brassicasterol	168	2,981	1.3	2.4	0.5	0.5

[a] From Kramer et al. (1975a).

sterol content of 36,800 μg per gram of oil (Table XVIII). In addition to free
and esterified sterols and triglycerides, molecular distillate 1″ contained hydrocarbons, methyl esters, alcohols, free fatty acids, mono- and diglycerides, and any other volatile compounds including possible isothiocyanates and cyanohydroxy by-products from glucosinolate hydrolysis during
processing. Figure 5 shows a thin-layer chromatogram of the light distillates.

The fatty acid composition and the sterol content of the fractions obtained
by molecular distillation are shown in Table XVIII. The early distillates contained higher levels of 16:0 and less 20:1 and 22:1, while the converse was
found in the later fractions. The sterol content, as mentioned earlier, decreased with each successive distillation. The relative concentration of brassicasterol decreased in the later distillates and was enriched in the light distillate MD1″.

A further 200 kg of the same LEAR oil (cv. Span) from the 1971 crop was
purified on a large scale by silica gel column chromatography (Fig. 3) under
contract with Applied Science Laboratories, State College, Pennsylvania.
This fractionation gave three fractions (see Fig. 6). Fraction AC1 was enriched in nonpolar components, e.g., hydrocarbons, sterol esters, and
methyl esters. Fraction AC2 contained highly purified triglycerides. Fraction
AC3 contained in addition to triglycerides, polar compounds, e.g., free fatty
acids, sterols, alcohols, and mono- and diglycerides.

The fatty acid composition and the sterol content of the three fractions

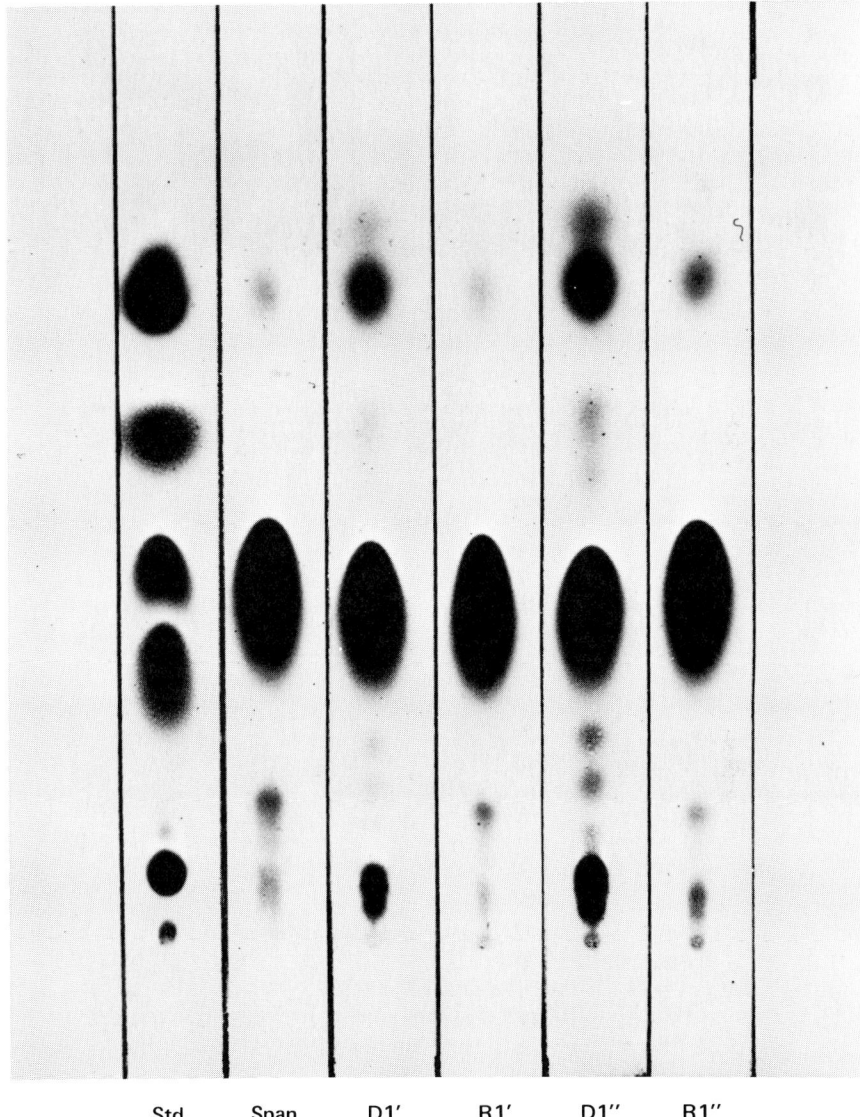

Std.   Span   D1'   R1'   D1''   R1''

Fig. 5. Thin-layer chromatogram of distillates (D) and residues (R) obtained during repeated molecular distillation of the light distillates (MD1 and MD2) of LEAR (cv. Span). For standard mixture and developing solvent see Fig. 4 (Kramer *et al.*, 1975a). Published with permission from *Lipids*.

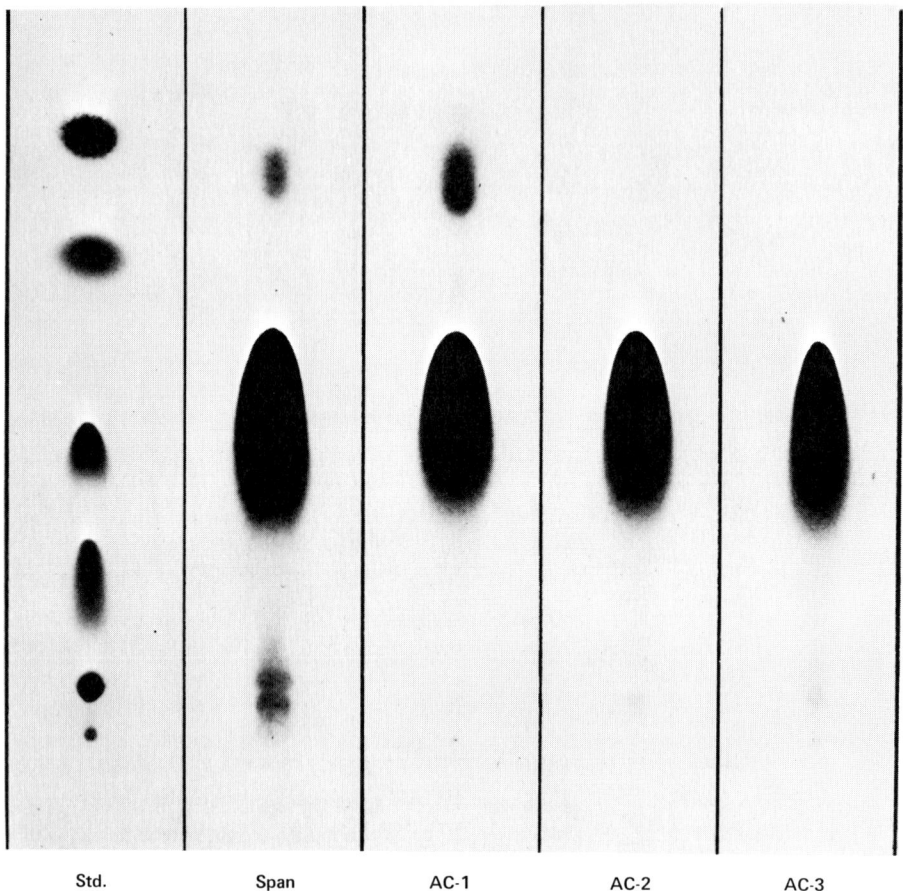

| Std. | Span | AC-1 | AC-2 | AC-3 |

Fig. 6.   Thin-layer chromatogram of LEAR (cv. Span) fractions 1, 2, and 3 obtained by elution through columns packed with silica gel H. For standard mixture and developing solvent see Fig. 4 (Kramer *et al.*, 1975a). Published with permission from *Lipids*.

obtained by column chromatography are shown in Table XIX. In addition to separating lipid classes, adsorption chromatography also slightly separates triglycerides. For example, the triglycerides first eluted from the column are enriched in saturated (18:0) and monounsaturated (18:1, 20:1, and 22:1) fatty acids whereas the later fractions are enriched in polyunsaturated fatty acids (18:2 and 18:3). The sterols in fraction AC1 are sterol esters while fraction AC3 contains free sterol. There were no detectable levels of sterols left in fraction AC2.

Several fractions obtained by molecular distillation were fed to male Sprague–Dawley rats at 20% by weight in the diet (Sauer, 1974; Kramer

TABLE XIX

Fatty Acid Composition and Sterol Content of Fractions Obtained by Adsorption Chromatography (AC) of LEAR cv. Span[a]

Fatty acids	LEAR (cv. Span)	Fractions		
		AC1	AC2	AC3
16:0	4.8	5.7	6.1	6.1
18:0	2.1	3.3	3.1	2.5
18:1	58.6	49.3	49.6	42.5
18:2	19.5	17.5	19.9	24.5
18:3	5.2	5.8	6.9	13.6
20:1	3.3	6.3	5.2	4.0
22:1	4.8	8.8	6.2	4.2
Sterol content (μg/g of oil)				
Total sterols	2500	6400	n.d.[b]	60
Brassicasterol	168	282	n.d.	3.1

[a] From Kramer et al. (1975a).
[b] n.d., not detectable.

TABLE XX

Myocardial Lesions in Male Sprague–Dawley Rats Fed for 16 Weeks LEAR (cv. Span) and Fractions Obtained from Molecular Distillation[a]

Dietary oil	Incidence		Lesion frequency[b]			
	%	n	1–2	3–5	6–10	>10
Stock diet, chow	30	20	4	2	0	0
Olive	20	20	4	0	0	0
LEAR (cv. Span)	46	50	13	4	6	0
MD5	62	50	17	9	4	1
MD6	48	50	8	10	3	3
MD8	80	50	9	11	9	11
MD1″	76	25	8	5	0	6

χ^2 analysis[c]	Incidence (df)	Severity (df)
Controls (chow and olive)	0.53 (1)	2.38 (1)
Among Span oil and its fractions	18.93*** (4)	35.18*** (12)
Controls vs. all Span oils	18.04*** (1)	9.12* (3)
Span, MD5 and MD6 vs. MD8 and MD1″	15.70*** (1)	16.24*** (3)
Controls vs. Span, MD5, and MD6	9.67*** (1)	5.87 (3)

[a] Adapted from Kramer et al. (1975b) with permission from Lipids.
[b] Number of rats with 1–2, 3–5, 6–10, and >10 lesions per three sections of heart.
[c] χ^2 significant at the 5%(*) and 0.1%(***) levels.

et al., 1975b). The histological results are presented in Table XX. The statistical analyses, as expected, showed that the control diets of either rat chow or semisynthetic diet containing 20% by weight olive oil gave a significantly lower incidence and severity of heart lesions than Span oil or its fractions. The results further indicate a significant difference in the incidence and severity among the original Span oil and its fractions. Of primary importance, however, was the observation that none of the fractions had a lower incidence or severity of myocardial lesions than that observed with the nonfractionated LEAR oil cv. Span. This clearly shows that it is not possible to remove a cardiotoxic compound from LEAR oil by means of fractional

TABLE XXI

Myocardial Lesions in Male Sprague–Dawley Rats Fed for 16 Weeks LEAR (cv. Span) and Fractions from Adsorption Chromatography[a]

Dietary oil	Incidence		Lesion frequency[b]			
	%	n	1–2	3–5	6–10	>10
Experiment A						
Stock diet, chow	55	20	5	4	1	1
Safflower	35	20	7	0	0	0
Corn	45	20	4	3	1	1
LEAR (cv. Span)	85	20	2	1	3	11
AC1	95	20	3	6	3	7
AC2	80	20	10	1	2	3
AC3	80	20	7	2	4	3
Experiment B						
Stock diet, chow	10	10	0	0	1	0
Olive	20	10	2	0	0	0
LEAR (cv. Span)	60	10	3	0	3	0
AC2	80	10	2	3	0	3

χ^2 analysis[c]	Incidence (df)		Severity (df)	
Experiment A				
Controls (chow, safflower, corn)	1.63	(2)	9.03	(6)
All Span oils	2.75	(3)	21.11*	(9)
Controls vs. all Span oils	25.61***	(1)	13.06**	(3)
Experiment B				
Controls (chow and olive)	0.40	(1)	3.82	(2)
All Span oils	0.97	(1)	12.39**	(3)
Controls vs. Span oils	13.21***	(1)	2.97	(3)

[a] Adapted from Kramer *et al.* (1975b) with permission from *Lipids*.
[b] Number of rats with 1–2, 3–5, 6–10, and >10 lesions per three sections of heart.
[c] χ^2 significant at the 5%(*), 1%(**), and 0.1%(***) levels.

distillation. A statistical analysis of the results shows that while none of the fractions gave a lower lesion incidence than the original LEAR oil, there was an indication that MD8 and MD1" gave a higher incidence than did the original oil or MD5 and MD6. The reason for this is not known but one may speculate that MD1", which highly concentrates all the volatile compounds of the original oil, may have, in some unknown manner, increased the lesion incidence, while in MD8 the increase in lesion incidence may be related to an enrichment of 20:1 and 22:1.

Scientists at Health and Welfare Canada tested several of the molecular distillates (MD5, MD6, and MD7) from the same distillation of Span rapeseed oil and obtained similar results (Beare-Rogers et al., 1974b).

The results presented in Table XXI support the same conclusion that the triglycerides are in fact the cardiotoxic principle. In this experiment, a very highly purified triglyceride fraction was obtained by column chromatography (Fig. 6, Table XIX), and no reduction in lesion incidence or severity was found when diets containing this fraction (AC2) was fed to male Sprague–Dawley rats. A repetition of this experiment gave identical results (Experiment B in Table XXI).

2.   EXPERIMENTS WITH LEAR CULTIVAR TOWER AND SOYBEAN OIL

Results similar to those described above were obtained when a new cultivar of LEAR (*B. napus* cv. Tower), which is practically devoid of 22:1 (and contains low levels of glucosinolates in the seed), was fractionated by the same procedures (Fig. 3). The results in Table XXII show the fatty acid com-

TABLE XXII

Fatty Acid Composition and Sterol Content of Soybean and LEAR (cv. Tower) Oils Fractionated by Molecular Distillation and Adsorption Chromatography[a]

Fatty acids	Soybean	Soybean MD8	LEAR	LEAR MD8	LEAR AC2a	LEAR AC2b
16:0	10.5	8.8	5.7	4.9	4.3	4.2
18:0	3.2	3.8	2.1	2.4	2.3	2.3
18:1	24.9	26.9	57.7	59.9	58.6	58.6
18:2	51.5	51.0	24.6	22.7	23.0	22.9
18:3	8.4	7.9	7.9	7.2	6.9	6.9
20:1	0.6	0.5	1.0	1.3	2.0	2.1
22:1	—	—	0.2	0.2	0.3	0.3
Sterol content (μg/g of oil)						
Total sterols	2800	200	7000	350	10	20
Brassicasterols	—	—	672	13	0.4	0.8

[a] From Kramer et al. (1979a).

position of soybean oil and Tower rapeseed oil before and after extensive molecular distillation. Molecular distillate 8 of Tower (but not from soybean oil) was further fractionated by silica gel chromatography exactly as described above to give two highly purified triglyceride fractions designated AC2a and AC2b. As shown in Table XXII, none of the fractionation procedures altered the fatty acid composition of the oils significantly. As before, the sterol content was greatly decreased. Based on the brassicasterol concentration, LEAR oil (cv. Tower) was purified at least 800-fold by both fractionation procedures.

As shown in Table XXIII, lesion formation from soybean molecular distillate 8 (MD8) was not significantly different from the original soybean oil in lesion incidence or severity. This supports the hypothesis that the lesion incidence of Span MD8 (Table XX) was the result of 22:1 enrichment and not a chemical or physical alteration brought about by the molecular distillation process. The Tower oil and its fractions gave a significantly higher lesion incidence than soybean or soybean MD8. Among the Tower oils, fraction AC2a gave a significantly lower lesion incidence than the remainder (Table XXIII). This was not considered of physiological significance because the rats fed AC2a consumed less feed and gained less weight than the rats fed AC2b from LEAR oil (cv. Tower) (Kramer et al., 1979a). It is recognized that

TABLE XXIII

Myocardial Lesions in Male Sprague–Dawley Rats Fed Pure Triglyceride Fractions of Soybean Oil and LEAR (cv. Tower) Oil for 16 Weeks[a]

Dietary oil	Incidence		Lesion frequency[b]			
	%	n	1–2	3–5	6–10	>10
Soybean	48	46	17	4	0	1
Soybean MD8	54	46	16	7	1	1
LEAR (cv. Tower)	74	46	22	8	2	2
MD8	72	46	15	11	2	5
AC2a	41	46	12	7	0	0
AC2b	63	46	16	12	1	0

χ^2 analysis[c]	Incidence (df)
Soybean vs. Tower	6.7** (1)
Oils vs. MD8 fractions	0.1 (1)
Among LEAR	12.9** (3)
Among LEAR except AC2a	1.4 (2)
LEAR-Tower AC2a vs. all other LEAR	11.5*** (1)
LEAR-Tower AC2a vs. AC2b	4.4* (1)

[a] From Kramer et al. (1979a). Published with permission from *Lipids*.
[b] Number of rats with 1–2, 3–5, 6–10, and >10 lesions per three sections of heart.
[c] χ^2 significant at the 5%(*), 1%(**), and 0.1(***) levels.

fast growing rats have a higher incidence of myocardial necrosis than their slower growing counterparts (Kramer *et al.*, 1979a, 1979b). Although the reason for the lower feed consumption of the diet containing AC2a is not clear, since AC2a and AC2b had the same fatty acid composition, this is the most likely explanation for the lower lesion incidence.

From the results it is clear that extensive purification of a LEAR oil, i.e., in excess of a 800-fold purification based on the brassicasterol content, did not lower the lesion causing properties of the oil. The conclusion seems reasonable that the myocardial lesions that result when rats are fed LEAR oils at a high level (i.e., 20% by weight) are a direct result of the triglycerides of these oils and not due to trace contaminants.

3. EXPERIMENTS WITH VARIOUS RAPESEED CONSTITUENTS

A different approach was used by the research team from the Federal Center for Lipid Research at Münster, Germany (Bijster *et al.*, 1979a). These workers fractionated entire rapeseed, both high and low in 22:1, by the

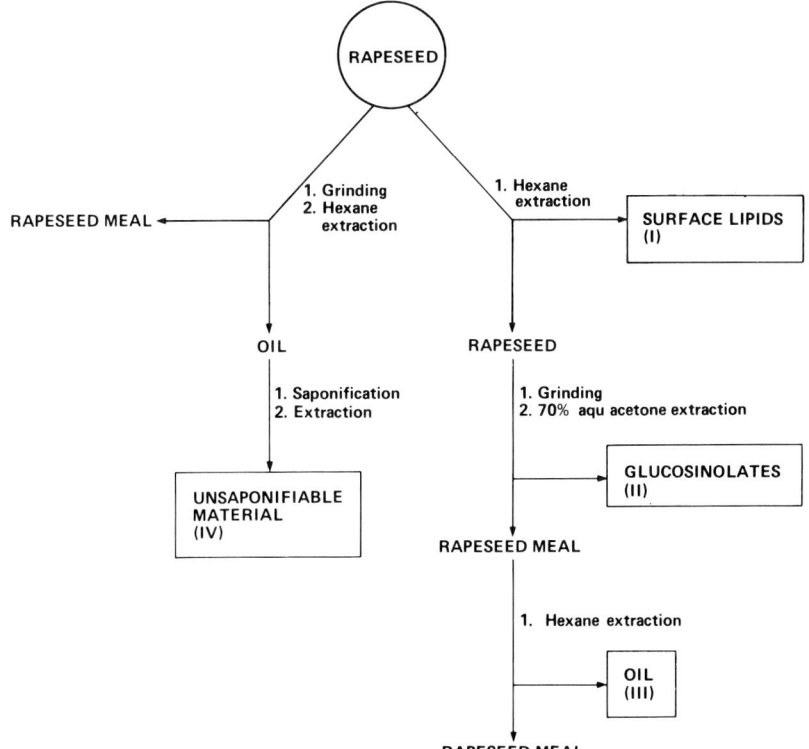

Fig. 7. Diagrammatic presentation of fractions obtained from entire rapeseed (Bijster *et al.*, 1979a).

452   J. K. G. Kramer and F. D. Sauer

protocol shown in Fig. 7, and obtained surface lipids (I), glucosinolates (II), unsaponifiable material (IV), and residual oil (III). The fractions were mixed into soybean oil at five times the concentration normally present in an equivalent amount of rapeseed oil. The residual rapeseed oil was mixed with soybean oil in the ratio of 1/4. These fractions were fed to male Wistar rats at 30 calorie % (15% by weight) as part of the semisynthetic diet. The authors observed no increase in lesion incidence or severity (as measured by the average number of lesions per heart or the average sum of Feret diameters) with the fractions from high or low erucic acid rapeseed or the rapeseed–soybean oil mixture. They concluded that, "with the exception of erucic acid, the seeds investigated did not contain any constituents which exhibit marked antinutritional effects" (Bijster et al., 1979a).

4. EXPERIMENTS WITH WEED SEED CONTAMINANTS

A research team at the University of Saskatchewan investigated the effect of substances extracted from weed seeds which are commonly found in rapeseed samples (Rose et al., 1981). Oil was extracted from three different rapeseed screenings as well as from stinkweed (Thlaspi arvense), a major contaminant of rapeseed, and added to LEAR oil (cv. Tower) at 5, 10, and 15% of the LEAR oil. The results shown in Table XXIV indicate that the contaminants have no effect on the cardiopathogenicity of the LEAR oil. The authors conclude that focal myocardial lesions in male albino rats fed LEAR

TABLE XXIV

Myocardial Lesions in Male Rats Fed Different Levels of Weed Seed Oil Contaminants in a LEAR Oil (cv. Tower) at 20% by Weight of the Diet for 16 Weeks[a]

Treatment	Incidence %	Incidence n	Myocarditis index[b]
Oil from LEAR screenings[c]			
Sample 1	78	23	4.66
Sample 2	83	24	4.50
Sample 3	75	24	3.67
Stinkweed oil	87	23	5.74
Level of contaminant in LEAR oil[c]			
5%	87	31	5.00
10%	87	31	4.34
15%	69	32	4.56

[a] Adapted from Rose et al. (1981) with permission from J. Nutr. and authors.
[b] Myocarditis index: number of foci of inflammation in all sections divided by number of sections examined.
[c] Differences in lesion incidence and myocarditis index not significant (P > 0.05).

oil are due to an imbalance in the fatty acid composition of the LEAR oil, and there was no evidence to indicate that weed seeds contained cardiotoxins extractable with hexane.

E. The Relationship of Dietary Fatty Acids to Heart Lesions

A statistical analysis of much of the published data on myocardial necrosis in male rats fed fats and oils for at least 16 weeks indicated that the heart lesions were negatively correlated to dietary saturated fatty acids (16:0 and 18:0) and linoleic acid (18:2), and positively correlated to dietary linolenic (18:3), oleic (18:1), eicosenoic (20:1), and docosenoic (22:1) acids (Table XXV). Most of the variation in lesion incidence between experiments could be explained by the concentration of 16:0 and 18:3. A plot of the observed incidence of heart lesion versus the lesion incidence predicted based on the concentration of 18:3 and 16:0 in the dietary oil of 23 experiments with over 2000 rats is shown in Fig. 8. Figure 8 shows a continuum of points representing a broad spectrum of fats, oils, and fat–oil mixtures. The more satu-

TABLE XXV

Summary of Regression Analysis of Aggregate Data to Assess Impact of Levels of Fatty Acids on Observed Incidence of Myocardial Lesions[a]

	Fatty acids							% variation explained[b]
	16:0	18:3	18:0	18:1	18:2	20:1	22:1	
Regression coefficients[c] from stepwise regression								
Step 1[d]								34.1
Step 2	−0.026							68.9
Step 3[e]	−0.020	0.017						73.3
Final[f]	−0.014	0.018	−0.002	0.002	0.001	−0.003	0.014	74.3
Correlations between incidence and fatty acid levels								
Overall	−0.64	0.54	−0.47	0.27	−0.08	0.33	0.26	
Partial[g]	−0.73	0.61	−0.56	0.40	−0.23	0.45	0.42	

[a] From Trenholm et al. (1979) with permission from Can. Inst. Food Sci. Technol. J.

[b] The "variation due to regression" divided by "overall variation" expressed as a percentage, that is, the square of the multiple correlation coefficient × 100.

[c] Expressed as incidence of lesions/% by weight of fatty acid in test oil. Incidence defined as no. of rats affected to no. of rats examined.

[d] Step 1 involved fitting constants for differences between experiments.

[e] Constant for final equation, averaged over experiments, is 0.55.

[f] Constant for final equation, averaged over experiments, is 0.35.

[g] Correlation after taking experimental differences into account.

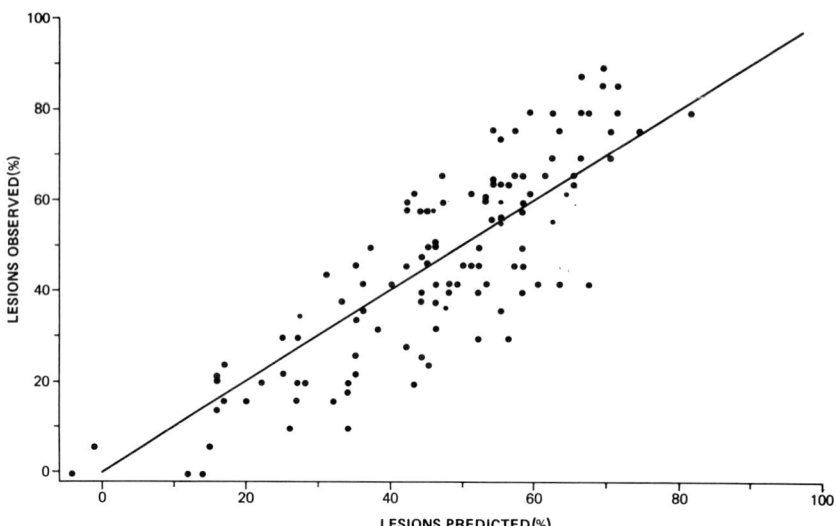

Fig. 8.   Observed versus predicted incidence of myocardial lesions in male rats. The observed incidence of heart lesions was taken from published data for which regression coefficients were calculated and used to determine the predicted incidence of lesions (Trenholm *et al.*, 1979). Published with permission of *Can. Inst. Food Sci. Technol. J.*

rated fats are clustered in the low incidence region whereas the LEAR oils are found in the higher incidence region.

A number of experiments have been carried out in which nonrapeseed oils have been altered in an attempt to make them more like rapeseed oil in fatty acid composition. As shown in Table XXVI the addition of free erucic acid to soybean oil (5.7%) and lard (5.4%) did not increase the incidence of myocardial lesions in male rats. On the other hand, similar quantities of 22:1 added to olive oil (3 and 4.4%) resulted in a significant increase in myocardial lesions. The reason for this may be found in the fatty acid composition of olive and soybean oils (Table I). It is apparent that olive oil resembles LEAR oil in fatty acid composition except for the lack of 22:1 and greatly reduced levels of 18:3 and 20:1. With the addition of 22:1 the fatty acid composition of olive oil more closely resembles that of LEAR oil. Soybean oil differs greatly from olive oil in that it has a lower level of 18:1 and a sixfold higher level of 18:2. The combination of low 18:1 together with the protective effect of the essential fatty acid (18:2) apparently nullifies the cardiopathogenic effects of 22:1.

Erucic acid was also fed with lard. Erucic acid was added in two ways: either as the free acid or biologically interesterified into the lard triglycerides. In order to prepare the latter, pigs were fed HEAR oils for 16 weeks, slaughtered, and lard obtained from the pigs by the usual rendering process

TABLE XXVI

The Effect of Adding Different Levels of Erucic Acid to Various Fats or Oils

Dietary oil	Final % 22:1 in oil	n	Incidence (%) Before	After	Severity difference[a] 1	2	3	4	Ref.[b]
LEAR (cv. Tower)	0.8[c]	90	62*	67*	−4	−11	7	4	1
Olive	3.0[d]	30	7	20	−13	0			3
Linseed	3.0[d]	30	13	27	−13	0			3
Olive	4.5[c]	90	29*	60**	−9	−13	−4	−4	1
Lard	5.6[c]	129	20*	20*	2	−2	0	0	2
LEAR (cv. Tower)	5.6[c]	90	62*	56*	4	0	2	0	1
Soybean	5.7[c]	89	41*	42*	−4	4	2	−4	1
Safflower	29[e]	59	40	93	−17	−19	−27		4
Sunflower	42[f]	24	0	100	−42	−25	−25	−8	5
Lard/corn (3/1)	29[d]	24	0	50	n.a.				6

[a] Severity difference as described in footnote c of Table III.

[b] References (for abbreviated experimental protocol, see also Table II: 1. Hulan et al. (1977a); 2. Hulan et al. (1976b); 3. Beare-Rogers and Nera (1977); 4. McCutcheon et al. (1976); 5. Abdellatif and Vles (1973); 6. Beare-Rogers et al. (1972).

[c] Free erucic acid was added.

[d] HEAR oil was added.

[e] Nasturtium oil (*Tropaeolum majus*) was added.

[f] Trierucin was added (84% 22:1).

(Hulan et al., 1976b). The fatty acid composition of this lard, here referred to as rendered pig fat (RPF), is shown in Table XXVII. As shown in Table XXVIII, neither lard plus free erucic acid (5.6%) nor RPF (5.6% 22:1) gave a significant increase in the lesion incidence above that obtained from lard alone, and the lesion incidence was significantly lower than that obtained with LEAR oil cv. Span (4.5% 22:1). As suggested by the lesion plot in Fig. 8,

TABLE XXVII

Fatty Acid Composition of Lard, Lard plus Erucic Acid, Rendered Pig Fat (RPF) and LEAR cv. Span[a]

Fatty acid	Lard	Lard + 22:1	RPF	LEAR (cv. Span)
16:0	25.6	22.4	12.6	4.8
18:0	14.6	14.7	4.8	2.1
18:1	43.2	37.9	39.1	58.6
18:2	9.6	10.4	18.5	19.5
18:3	0.6	0.7	6.1	5.2
20:1	1.2	1.4	8.3	3.3
22:1	0.2	5.6	5.6	4.8

[a] From Hulan et al. (1976b), with permission from *Lipids*.

TABLE XXVIII

The Effect of Feeding Lard Containing 22:1 to Sprague–Dawley Rats for 16 Weeks[a]

	Incidence		Lesion frequency[b]			
Dietary oil	%	n	1–2	3–5	6–10	>10
Lard	16	45	6	1	0	0
Lard + 22:1	22	45	8	1	1	0
Rendered Pig Fat (RPF)	31	45	14	0	0	0
LEAR (cv. Span)	76	45	12	11	8	3
χ^2 analysis[c]			Incidence	(df)	Severity	(df)
Lard vs. Lard + 22:1			0.66	(1)	1.13	(2)
Span vs. RPF			19.87***	(1)	22.06***	(3)
Lard plus Lard + 22:1 vs. Span plus RPF			24.50***	(1)	5.47	(3)
Span vs. Lard plus Lard + 22:1			43.65***	(1)	11.60**	(3)
RPF vs. Lard plus Lard + 22:1			2.46	(1)	3.96	(2)

[a] From Hulan et al. (1976b). Published with permission from Lipids.

[b] Number of rats with 1–2, 3–5, 6–10, and >10 lesions per three sections of heart.

[c] χ^2 significant at the 1%(**) and 0.1%(***) level.

this result is directly attributable to the twofold increase in 16:0 and 18:0 and the 33% decrease in 18:1 when compared to the LEAR oil. Thus, the protective effect of lard against the lesion causing properties of 22:1 appears to be related to the higher levels of saturated fatty acids.

It should be noted, however, that the protective effect of 18:2 and the saturated fatty acids only appear to be effective with moderate levels of erucic acid, but do not apply with the addition of extreme levels of erucic

TABLE XXIX

Incidence and Severity of Myocardial Lesions in Male Wistar Rats Fed Different Levels of 18:3 in Dietary Oils Rich in 22:1[a]

	Fatty acids		Incidence[b]		Types of lesions[c] (% of maximum)		
Dietary oil	18:3	22:1	%	n	1	2	3
Mixture 1[d]	8.3	29.9	100	15	40	58	49
Mixture 2[e]	—	28.7	93	14	21	21	38

[a] From McCutcheon et al. (1976).

[b] Incidence of heart lesions obtained from B. L. Walker (private communication).

[c] For lesion grades see Table II.

[d] Mixture of nasturtium oil (40%), olive oil (30%), safflower oil (18%), and linseed oil (12%).

[e] Mixture of nasturtium oil (45%), safflower oil (45%), and olive oil (10%).

acid. For example, when erucic acid was added at 29 or 42% of the fat mixture (Table XXVI) there was a sharp increase in myocardial lesions even when the fat mixture was rich in 18:2 (Abdellatif and Vles, 1973; McCutcheon et al., 1976) or saturated fatty acids (Beare-Rogers et al., 1972). Thus, whereas 22:1 at levels of 5% or less did not appear to cause an increase in myocardial lesions incidence when added to an oil rich in 18:2 (i.e., at least 20%), when high levels of 22:1 were used (i.e., 29 and 42%), the protective effect of 18:2 was no longer sufficient to protect the heart from damage by 22:1.

Again, by reference to Fig. 8, it is noted that linolenic acid increases lesion incidence. This has also been shown directly by experiments where 18:3 was removed from synthetic fat mixtures high in 22:1 (Table XXIX). With the removal of 18:3 the lesion incidence did not change but the severity of the lesions decreased markedly (McCutcheon et al., 1976). In agreement with this, Vles et al. (1978) was able to show that the severity of myocardial necrosis in Sprague–Dawley rats fed linseed oil rich in 18:3 for 53 weeks was significantly increased over that obtained by feeding sunflower oil, but still below that obtained with a LEAR oil cv. Primor (Fig. 1).

Linolenic acid (18:3) is known to be chain elongated and desaturated to

TABLE XXX

Modification of LEAR and Soybean Oils to Alter Their Saturated Fatty Acid Content[a]

Fatty acids	LEAR (cv. Tower)	LEAR + cocoa butter[b]	LEAR + triolein[c]	Soybean	Soybean + cocoa butter[b]	Soybean + triolein[d]	Soybean + triolein + 22:1[e]
16:0	4.4	8.2	3.4	12.1	17.9	5.8	5.8
18:0	1.5	7.5	1.2	3.5	9.2	2.6	2.7
Total saturates	7.2	16.6	5.4	16.2	27.6	8.7	9.1
18:1	57.5	55.8	66.9	24.6	26.9	56.1	55.3
20:1	1.9	1.2	1.2	0.3	0.2	0.1	1.0
22:1	0.6	0.4	0.4	0.1	Trace	Trace	0.4
Total monoenes	60.5	57.6	68.7	25.1	27.4	56.3	56.7
18:2	22.0	17.8	17.9	51.9	40.1	28.3	27.5
18:3	10.3	7.9	7.8	6.7	4.9	6.7	6.7
Total polyenes	32.3	25.7	25.7	58.6	45.0	35.0	34.2

[a] From Farnworth et al. (1982a) and Kramer et al. (1982).
[b] Cocoa butter added at 20% of the oil.
[c] Triolein added at 20% of the oil.
[d] Triolein added at 48% and linseed oil at 4% of the oil.
[e] Triolein added at 47.5%, linseed oil at 4% and long chain monoenes (20:1 and 22:1) at 1% of the oil.

22:6 n-3 which is incorporated into phospholipids, specifically phosphati-dylethanolamine. Supporting evidence was obtained by several investiga-tors who showed that LEAR and linseed oil (Beare-Rogers and Nera, 1977) as well as soybean oil (Kramer, 1980) fed rats have an increased concentration of 22:6 n-3 in cardiac lipids, in particular the cardiac phospholipid, phos-phatidylethanolamine (see Chapter 18 for more detail). The suggestion has been made that this increased concentration of 22:6 n-3 in cardiac phos-pholipids may lead to altered physical properties of membranes (Kramer, 1980).

Based on the statistical results of Table XXV and Fig. 8, a soybean oil and a LEAR oil (cv. Tower) were modified by adding cocoa butter or pure triolein (Table XXX). Cocoa butter was used to increase the total saturated fatty acids of soybean oil from 16% to 28% and of LEAR oil (cv. Tower) from 7% to 17%. Triolein was added in a proportion equal to that of cocoa butter to assure that the decrease in cardiotoxicity of the oil was not simply due to

TABLE XXXI

Effect of Feeding Modified LEAR and Soybean Oils on Myocardial Lesions in Male Sprague–Dawley Rats[a]

	Incidence			Lesion frequency[c]				
Dietary oil	% (Obs.)	n	% (Pred.)[b]	1	2	3	4	>4
LEAR oil (cv. Tower)	61	44	64	10	10	3	3	1
LEAR + cocoa butter	36	44	47	8	3	2	2	1
LEAR + triolein	55	44	62	11	9	3	1	0
Soybean	57	44	46	10	7	7	1	0
Soybean + cocoa butter	34	44	27	11	1	1	1	1
Soybean + triolein	59	44	55	16	4	1	4	1
Soybean + triolein + 22:1	55	44	55	16	3	2	2	1

χ^2 analyses	Incidence $(df)^d$	
All diets	13.0*	(6)
Diets with cocoa butter vs original oils	10.4**	(1)
Among diets with cocoa butter	0.05	(1)
Among diets without cocoa butter	0.6	(4)

[a] From Kramer et al. (1982). Published with permission from Lipids.

[b] The predicted incidence of heart lesions was calculated using the following equation: $Z_i = Y - 0.013(X_{1i} - \ddot{X}_1) + 0.016(X_{2i} - \ddot{X}_2)$, where Z_i is the predicted incidence of heart lesions, $\ddot{Y}$ the average observed incidence of heart lesions for all diets (0.51), -0.013, and 0.016 the correlation coefficients of 16:0 + 18:0 and 18:3, respectively, X_1 and X_2 the dietary concentration of 16:0 + 18:0 and 18:3 in the ith diet, and $\ddot{X}$ the overall mean concentration of the specific fatty acid(s) from all diets.

[c] Number of rats with 1, 2, 3, 4, or >4 lesions per three section of heart.

[d] χ^2 significant at the 5% (*) and 1% (**) levels.

dilution of a hypothetical cardiotoxin in the oil. As shown in Table XXXI, the heart lesion incidence in male rats was significantly decreased by the addition of cocoa butter to soybean and LEAR oil. Adding pure triolein to soybean oil or LEAR oil did not increase the incidence of myocardial lesions over that of the original oils. The indication that the saturated fatty acids contained in cocoa butter were able to decrease the incidence of myocardial necrosis in both the groups fed soybean and LEAR oils suggests that this lesion in the rat is related to a relative deficiency of saturated fatty acids in the diet. The results of this study provide evidence that the heart lesions are caused by the triglyceride fraction, and the lesion incidence can be either increased or decreased by manipulating the dietary fatty acids in accordance with the correlation shown in Table XXV (Farnworth *et al.*, 1982a, Kramer *et al.*, 1982).

The calculated incidence of heart lesions, as predicted by the regression coefficients (Table XXV) and the fatty acid composition of the dietary oils, is in good agreement with the observed incidence of heart lesions. The predicted incidence of heart lesions for each dietary oil in this experiment is included in Table XXXI (Kramer *et al.*, 1982).

Similar regression analyses with dietary fatty acids are being used by others. Tinsley *et al.* (1981) recently reported that the incidence of spontaneous mammary tumors in C3H mice appears to be positively correlated to linoleic (18:2) and palmitic (16:0) acids and negatively correlated to stearic (18:0), erucic (22:1), and myristic (14:0) acids, while oleic (18:1) and linolenic (18:3) acids show little correlation. This clearly shows that a fatty acid that increases the lesion incidence in one case may actually decrease the lesion incidence in another. As seen by the two types of conditions listed above, palmitic (16:0) and erucic (22:1) acids have both beneficial and harmful effects. It is possible that more evidence of lesion response to specific dietary fatty acids may be reported in the future.

IV.   RESULTS WITH PIGS

A.   Breeds of Pigs and Length of Time on Experimental Diets

1.   BREEDS OF PIGS

Several breeds of domestic pigs have been used by investigators over the past 20 years to test the cardiopathogenicity of rapeseed oils. Roine *et al.* (1960) and Friend *et al.* (1975a,b, 1976) used Yorkshire boars and gilts. Aherne *et al.* (1975, 1976) used Crossbred barrows and gilts. Svaar *et al.* (1980) used female piglets of the Norwegian Landrace breed. In a German–Dutch study (Petersen *et al.*, 1979; Vles, 1978) German Landrace barrows were used.

2.   AGE OF PIGS AND DURATION OF EXPERIMENT

The initial age of pigs when placed on the experimental diet and the duration of feeding were different for each reported experiment and therefore this information is given in each of the tables. In general, the starting age was between 4 and 13 weeks, and the experiments lasted 16, 23, 24, 27, or 52 weeks. The 52 week period represents approximately 8–10% of the life-span of the pig, which is the same percentage of the life-span generally used for the rat.

B.   Histopathological Results

The Finnish workers, Roine et al. in 1960, fed 11 Yorkshire pigs of both sexes diets containing soybean oil and HEAR oil at 28 calorie % for 60 days. No histological difference was seen between animals fed HEAR or soybean oil. A few pigs on both diets showed histological evidence of thyroid hyperfunction, interstitial myocarditis, and inflammatory reaction in the gastric mucosa. No lesions were found in the other tissues.

Agriculture Canada undertook an extensive program from 1972 to 1975 to test HEAR, LEAR, and various control oils with Yorkshire pigs. In the first study (Friend et al. 1975a) 192 Yorkshire boars and gilts (equally divided as to sex) were fed a control diet with no added oil and diets containing 10 and 20% by weight soybean oil or LEAR oil (cv. Span). Pigs were killed at the start and after 1, 4, and 16 weeks of feeding the diets ad libitum. The results are presented in Table XXXII. Minute focal interstitial infiltrations of mono-

TABLE XXXII

Number of Yorkshire Boars and Gilts Showing Heart Lesions When Fed Diets Containing 10 and 20% of Soybean Oil or LEAR (cv. Span) Oil Containing 4.3% 22:1[a]

Weeks on diet	Sex[b]	No fat added	Soybean		LEAR (cv. Span)	
			10%	20%	10%	20%
0	Boars	2[c]	—	—	—	—
	Gilts	2	—	—	—	—
1	Boars	1	3	4	3	1
	Gilts	3	2	2	3	5
4	Boars	1	3	2	1	1
	Gilts	2	1	3	4	0
16	Boars	1	0	0	2	1
	Gilts	0	2	0	0	1

[a] From Friend et al. (1975a). Published with permission from Can. J. Anim. Sci.
[b] All pigs were allotted to dietary treatments at 8 or 9 weeks of age.
[c] Each value was obtained from an examination of six pigs and represents the number exhibiting small focal infiltrations of mononuclear cells in the heart.

TABLE XXXIII

Incidence and Frequency of Cardiac Lesions in Yorkshire Boars Fed Diets with No Added Fat or Diets Containing 20% by Weight Vegetable Oils for 16 or 24 Weeks[a,b]

Dietary oil	22:1 %	Incidence %	Incidence n	Lesion frequency[c] 1–2	3–5	6–10	>10
Experiment A[d]							
Control, no fat	0	4	24			n.a.	
Soybean	0	4	24				
HEAR	22.3	13	24				
Experiment B[e]							
Control, no fat	0	60*	15	7	0	2	0
Corn	0	73*	15	7	3	1	0
LEAR (cv. Midas, Zephyr mixture)	0.9	60*	15	4	3	2	0
HEAR	24.1	87*	15	8	3	0	2

[a] Adapted with permission from *Can. J. Anim. Sci.*

[b] Initial age of boars was 10–13 weeks.

[c] Number of boars with scores of 1–2, 3–5, 6–10, and >10 lesions in a total of five cardiac sections examined per animal.

[d] Friend *et al.* (1975b).

[e] Friend *et al.* (1976).

nuclear cells were present in the myocardium of some pigs in each dietary treatment and also in the initial controls. Foci of overt myocardial necrosis usually seen in male rats were not observed in this pig study. It is evident from the cardiopathological results presented in Table XXXII that there are no differences in heart lesions between sexes, between the levels of fat in the diet (0, 10, or 20%), or with duration of feeding (0, 1, 4, and 16 weeks).

In the second study reported by Agriculture Canada (Friend *et al.*, 1975b), 72 Yorkshire boars were fed a control diet with no added oil and diets containing 20% by weight either soybean oil or HEAR oil for 16 weeks (Table XXXIII). The incidence of heart lesions was low in all groups and not significantly different between diets.

The results of the third study by Agriculture Canada scientists (Friend *et al.*, 1976) are shown in Table XXXIII. No significant differences were observed in the incidence and severity of heart lesions in Yorkshire boars fed a control diet with no added oil, or one of corn, LEAR, or HEAR oil at 20% by weight added to the diet.

Similar results were obtained by research workers at the University of Alberta. They reported feeding 112 Crossbred pigs (barrows and gilts, equally divided) a control diet with no added oil or diets containing 15% by weight of either soybean, LEAR, or HEAR oils from 20 to 90 kg live weight (Aherne *et al.*, 1975). These authors concluded from their results (Ta-

TABLE XXXIV

Incidence of Myocardial Lesions in Crossbred Gilts and Barrows[a]

| Dietary oils | % 22:1 | Affected/examined[b] | | |
		Ad libitum	80% of ad libitum	Free choice of four diets
Control, no fat	0.0	1/12	—	
Soybean	0.0	2/12	1/16	0/16
LEAR (cv. Span)	3.8	0/12	2/16	
HEAR	14.5	0/12	0/16	

[a] From Aherne et al. (1975).

[b] The pigs were equally divided as to sex and fed the vegetable oils at 15% by weight of the diet from 25 to 90 kg body weight or about 10 to 21 weeks of age.

ble XXXIV) that no marked differences in cardiomyopathy were evident between any of the diets fed either *ad libitum*, restricted at a level of approximately 80% of the calorie intake of the *ad libitum* group, or allowing continuous, free-choice access to each of the four diets.

In the second study reported by scientists from the University of Alberta (Aherne *et al.*, 1976), 80 Crossbred barrows and gilts (equally divided) were fed a control diet with no added oil, or three separate LEAR oils or a HEAR oil at 15% by weight of the diet for 4, 16, and 23 weeks (Table XXXV). No differences were observed in the incidence of heart lesions between diets or duration of feedings.

TABLE XXXV

Incidence of Myocardial Lesions in Crossbred Gilts and Barrows[a]

| Dietary oil | % 22:1 | Incidence[b] | |
		%	n[c]
Control, no fat	0	19	16
LEAR (cv. Tower)	0.3	0	16
LEAR (cv. Span)	1.2	0	16
LEAR (1788)	4.9	6	16
HEAR	34.2	6	16

[a] From Aherne et al. (1976).

[b] The pigs were equally divided as to sex and fed the vegetable oils at 15% by weight of the diet from 4 to 5 weeks of age.

[c] Four of the pigs on each diet were killed after 4 weeks on the diets, 8 after 16 weeks, and the last 4 after 23 weeks.

TABLE XXXVI

Myocardial Lesions in Female Norwegian Landrace Pigs[a]

Dietary fat	22:1	Incidence[b]			
		Duration (weeks)[c]			
		1	5	27	52
Lard	0	0	0	1	1
Hydrogenated soybean oil	0.5	1	0	1	1
HEAR	42.2	1	0	1	2
Raw capelin oil	18.0	0	0	0	—[d]
Hydrogenated capelin oil	17.8	0	0	1	1

[a] From Svaar and Langmark (1975) and Svaar et al. (1980). Published with permission of Acta Path. Microbiol. Scand. and authors.

[b] Number of pigs with positive histological findings in the heart out of two pigs examined at each time period. All lesions found were small focal groups of inflammatory cells (grade 1) except one heart (hydrogenated soybean oil, 27 weeks) which was scored grade 2.

[c] The initial age of the pigs varied from 6 to 11 weeks.

[d] These two pigs were killed at 37 weeks, no histological results were given.

The Norwegian workers (Svaar et al., 1980) reported a study using female Norwegian Landrace pigs. They fed diets containing 16% by weight lard, hydrogenated soybean oil, HEAR oil, a fish oil (raw capelin oil), or the same fish oil partially hydrogenated. The results are presented in Table XXXVI. The heart lesions which appeared in all groups were mild in comparison to those observed in a parallel rat study (Svaar and Langmark, 1980).

In a cooperative German–Dutch study (Vles, 1978; Peterson et al., 1979; Bijster et al., 1979b) 68 German Landrace pigs from an SPF-herd were fed for 17 weeks diets containing either no added oil, soybean oil, LEAR oil (cv. Lesira), or three mixtures of each of two rapeseed oils, a LEAR oil cv. Primor (5% 22:1) and a HEAR oil (48% 22:1). As seen in Table XXXVII the diets contained different levels of added fat (0, 4, and 8%) and different levels of erucic acid (0–2.2%). The histological results (Table XXXVII) of pigs fed these diets showed that there was no significant difference in the incidence and severity of heart lesions in pigs fed different levels of fat in the diet (0, 4, and 8% by weight), the kind of oil added (soybean, LEAR or mixtures) or the concentration of 22:1 in the diet (0–2.2%).

In summary, the results with pigs are consistent. Pigs fed diets without added oil, or diets containing different levels of control, LEAR or HEAR oils do not develop specific diet induced heart lesions. Any heart lesions that are present, are generally much less severe than those found in rats.

TABLE XXXVII

Myocardial Lesions in German Landrace Pigs Fed Experimental Diets for 17 Weeks[a]

Description	Control (no fat added)	Soybean oil		LEAR (cv. Lesira)			Mixture of HEAR and LEAR (cv. Primor)			
No. of pigs examined	7	6	7	6	7	7	7	7	7	7
Oil added (% of diet)	—	4	8	8	4	8	4	8	4	8
Erucic acid (% of diet)	—	—	—	0.2	0.5	0.7	0.9	1.5	1.4	2.2
No. of lesions/pig	25	21	46	26	39	39	32	45	30	39
Sum Feret diameters/pig[b]	28	18	45	24	36	41	34	45	33	38

[a] From Bijster et al. (1979). Adapted with permission of Fette, Seifen, Anstrichm. and authors.
[b] Measured in mm, magnification ×125.

V. RESULTS WITH DOGS

Research workers at the University of Saskatoon have conducted a preliminary study with beagle dogs to test the cardiopathogenicity of LEAR oils (D.L. Hamilton and B. Schiefer, private communication). Eighteen beagle dogs of mixed sex, 8 to 20 weeks of age, were divided into 2 groups of 9 each. After an adjustment period of 2 weeks, all dogs were fed the control diet (20% by weight lard/corn oil, 3/1) for 1 week. At this time the experimental group was fed the LEAR oil diet (20% by weight *B. napus* cv. Tower). Both groups were maintained on the experimental diets for 12 weeks.

The physical appearance and growth rates of the dogs were not affected by feeding of LEAR oil. Routine blood analyses showed no abnormalities. Standard serum analyses for lactic dehydrogenase, creatine phosphokinase, and serum glutamic oxalacetic transaminase activity, cholesterol and triglyceride concentrations, and Na, K, and Mg concentrations did not show significant differences between the control and LEAR oil fed groups.

Necropsy did not reveal any gross abnormalities. In the heart sections from either group, no lesions were observed that were morphologically similar to that reported in LEAR oil fed rats. A certain amount of hypercellularity around some of the myocardial vessels was seen in both groups. Such changes are often seen in young canine heart sections and have not been associated with any functional derangement.

In conclusion, this short-term dog study did not indicate that the feeding of LEAR oil had any significant detrimental effects on growth, hematology, serum chemistry, gross pathology, or myocardial histology.

VI. RESULTS WITH SWISS MICE

Research workers at the Unilever Research Laboratories in the Netherlands tested sunflower, olive, linseed, and LEAR (cv. Primor) oils using 3-week old SPF male Swiss mice (Vles *et al.*, 1978). The animals were housed individually in disposable cages. Four groups of 28 mice were formed on the basis of body weight (10.6 g) and fed diets containing 20% by weight of these oils for 21 weeks. At the end of the experiment organs were removed, weighed, and fixed in 10% buffered formalin for subsequent histopathology. The hearts were subjected to the same histological procedures described for rats (Vles *et al.*, 1976); 20 sections of each heart were examined.

The results of the pathological examination of the hearts of Swiss mice are shown in Table XXXVIII. The 28 mice that had received 20% by weight LEAR oil (cv. Primor) for 5 months showed no heart lesions. The mice fed

TABLE XXXVIII

Incidence and Severity of Myocardial Lesions in Swiss Mice[a]

Dietary oil	Incidence[b]		No. of lesions per mouse[b]	Average sum of Feret diameters[c,d]
	%	n		
Sunflower	4*°	26	0.08*	3.2*
Olive	19°	26	0.35°	11.3°
Linseed	8*°	24	0.08*°	2.8*°
LEAR (cv. Primor)	0*	28	0.0*	0.0*
P values	0.051		0.055	0.058

[a] From Vles et al. (1978). Adapted with permission from Arch. Toxicol. and authors.

[b] Application of χ^2 test was in fact not justified due to low frequency of lesion incidence. Values that differ in superscripts are significantly different.

[c] Kruskal–Wallis analysis of variance on ranked observations.

[d] In mm, magnification ×160.

sunflower and linseed oil had a low incidence of heart lesions not significantly different from mice fed LEAR oil. The incidence of heart lesions in mice fed olive oil was significantly higher than those fed LEAR oil. This the authors attributed to the low level of 18:2 in olive oil.

VII. RESULTS WITH MONKEYS

There is only one study reported to date in which monkeys were used to test LEAR oils (Kramer et al., 1978a, 1978b). The authors used jungle reared male and female cynomolgus monkeys (Macaca fascicularis) weighing between 2 and 5.6 kg, which were tested and proved to be free of tuberculosis and intestinal parasites. The authors randomly assigned the monkeys to two diets, and housed them individually in primate cages of conventional design. The design and duration of this experiment are given in Table XXXIX.

In this experiment it was found that the hearts from monkeys fed high fat diets do not develop the myocardial necrosis and fibrosis that were found in male rats. As described in Table XXXIX, two types of lesions were observed. The (±) or trace lesion was a myocardial lesion consisting of focal interstitial collections of inflammatory cells, occasional focal scars, or groups of swollen interstitial fibroblasts. The authors concluded that these heart lesions were not diet related. The (+) myocardial lesion consisted of multiple small foci of fibroblasts, mononuclear cells, and Antischkow's myocytes. As shown in Table XXXIX this lesion was but rarely observed and then only in male monkeys fed the soybean oil diet.

TABLE XXXIX

Myocardial Lesions in Cynomolgus Monkeys (*Macaca fascicularis*) Fed Soybean Oil or
LEAR Oil at 20% by Weight of the Diet for 1 to 24 Weeks[a]

Sex	Time on diet (weeks)	n	Number of monkeys affected[b]	
			Soybean	LEAR (cv. Tower)
Male	1	2	2±	0
	2	2	—	1±
	8	2	—	0
	16	3	—	2±
	24	5	2±, 2+	4±
Female	1	2	2±	2±
	2	2	—	1±
	8	2	—	2±
	16	1	—	0
	24	5	4±	2±

[a] From Kramer *et al.* (1978b). Adapted with permission from *Can. J. Anim. Sci.*

[b] Lesions of the type designated (±) are trace lesions consisting of focal interstitial collections of inflammatory cells, occasional focal scars or groups of swollen interstitial fibroblasts. Lesions of the type designated (+) are represented by degenerative cardiac muscle cell alterations around which mononuclear, predominantly histiocytic, cellular infiltrative and/or fibroblastic proliferation were present.

Qureshi (1979) stated that chronic interstitial myocarditis is frequently seen in monkey hearts, is of unknown cause, and does not appear to have clinical importance. He found myocardial histological lesions in the squirrel, rhesus, cynomolgus, and assam species of monkeys, and postulated that this type of myocarditis may be precipitated by stress.

The following conclusions can be made from the results of the study with cynomolgus monkeys. First, there were no significant differences in heart lesion incidence between the LEAR oil and the control diet containing soybean oil. Second, lesions in the hearts of monkeys were generally focal interstitial infiltrations of mononuclear cells and only very few lesions of greater severity were observed.

Kramer *et al.* (1978a) showed that platelet counts and clotting times were the same in monkeys fed diets containing either soybean or LEAR oils. The prothrombin times, however, were shorter in the monkeys fed the LEAR oils than in the monkeys fed the soybean oil, and in addition it was observed that the female monkeys had shorter prothrombin times than the males.

The authors also measured the activity of serum glutamic oxalacetic transaminase (SGOT), creatine phosphokinase (CPK), and lactic dehydrogenase (LDH) to determine if there was evidence of myocardial necrosis. They found that the enzyme activities were all in the normal range given by Loew *et al.* (1978) for cynomolgus monkeys. The authors did find, however, that for SGOT and LDH the activity was higher in the male monkeys fed LEAR oil than in the male monkeys fed soybean oil, and that the male monkeys fed either diet had higher activities than the female. The activity of CPK was not different in any of the monkeys. The authors concluded that the serum enzyme changes were not indicative of myocardial damage since all values fell

TABLE XL

ECG Data from Monkeys Fed Soybean or LEAR (cv. Tower) Oils for 24 Weeks[a]

Sex	Time on diet[b] (weeks)	Heart rate (beats/min)		PR interval (msec)		QRS Interval (msec)		QT Interval (msec)	
		Soy	LEAR	Soy	LEAR	Soy	LEAR	Soy	LEAR
Male	-2	250[c]	226	60	62	34	42	164	160
	-1	254	230	58	60	38	36	162	166
	1	252	226	60	56	40	38	158	170
	4	246	222	68	80	44	40	132	160
	8	244	232	66	62	38	34	132	138
	16	254	242	60	66	36	40	160	180
	24	258	244	58	58	38	40	154	170
Female	-2	242	244	58	66	40	42	174	176
	-1	244	240	60	58	40	32	172	162
	1	236	236	64	54	40	38	186	174
	4	228	230	72	88	32	40	124	164
	8	248	234	54	60	32	28	138	132
	16	246	240	58	68	38	36	170	166
	24	240	246	56	62	42	38	172	174

Analysis of variance				
Source of variation	df	Mean square[d]		
Diet (D)	1	40.2	0.41	0.01	1.6
Sex (S)	1	1.2	0.01	0.07	1.1
D × S	1	26.6	0.07	0.03	0.9
Error	16	9.7	0.16	0.08	0.4
Time (T)	6	5.3**	0.87**	0.11*	4.1**
D × T	6	0.6	0.22**	0.06	0.8**
S × T	6	1.1	0.07	0.06	0.3
D × S × T	6	1.5	0.04	0.06	0.2
Error	96	1.1	0.06	0.05	0.2

[a] From Kramer *et al.* (1978a). Published with permission from *Can. J. Anim. Sci.*
[b] Negative values denote weeks prior to start of experiment.
[c] All values are means of five animals.
[d] Significance is indicated at the 5%(*) and 1%(**) levels.

within the normal range. Some studies have been done with cynomolgus monkeys that were fed HEAR (Loew *et al.*, 1978) or mustard (Gopalan *et al.*, 1974) oils which contained 25 and 44% erucic acid, respectively. Loew *et al.* (1978) found that the monkeys fed HEAR oil had similar LDH and CPK activities, but the SGOT activity was significantly greater in the group fed HEAR oil. On the other hand, Gopalan *et al.* (1974) fed mustard oil to macaque monkeys and found that the SGOT and serum glutamic pyruvic transaminase values were virtually the same as in the monkeys fed peanut or hydrogenated peanut oils.

Kramer *et al.* (1978a) did electrocardiogram studies in cynomolgus monkeys fed soybean and LEAR oils. There were no significant differences observed between the two diets or between male and female monkeys. The heart rate and duration of waves in the electrocardiograms are given in Table XL. These results with nonhuman primates are in agreement with electrocardiogram patterns obtained with rats fed LEAR and HEAR oils (Berglund, 1975; Hulan *et al.*, 1976a; Hunsaker *et al.*, 1977). This indicates that the conduction system of the heart was not affected by the feeding of rapeseed oil.

VIII. CONCLUSION

This chapter deals with the new cultivars of rapeseed, i.e., the LEAR oils and does not cover the HEAR oils. The results obtained for HEAR oils are summarized in Chapter 11.

1. All the commonly used vegetable oils when fed at 40 calorie % (or 20% by weight) of the diet increased the incidence of myocardial lesions in male albino rats (e.g., Sprague–Dawley, Wistar, and Sherman). The lowest incidence of heart lesions is found when, instead of a single oil, a 3/1 mixture of lard and corn oil was fed (11%). The mean lesion incidence observed for the different oils were soybean, 36%; sunflower, 49%; corn, 30%; peanut, 37%; olive, 38%; and safflower, 37%.

2. LEAR oils (cv. Oro, Span, Zephyr, Tower, Primor, and Lesira) consistently produce a higher incidence of lesions than other oils. The results from some of the different laboratories are Agriculture Canada, 68 vs. 34%; Health and Welfare Canada, 50 vs. 11%; University of Guelph, 54 vs. 41%; Unilever Research Laboratory, 64 vs. 60%; and Dijon, France, 69 vs. 42%. Vles *et al.* (1976) from Unilever was the first to point out that lesion size can vary. Thus, rats fed HEAR oils have larger myocardial lesions than rats fed LEAR oils. No difference in lesion size was observed between rats fed LEAR oils or other vegetable oils.

3. Female rats have a lower incidence and severity of myocardial lesions than male rats when fed vegetable oils at 20% by weight of the diet. The incidence of heart lesions in male and female rats fed LEAR oils is 69 vs.

14%, and for control oils including lard the incidence is 30 vs. 12%. The suggestion has been made (Kramer et al., 1979a) that the lower sensitivity of female rats to myocardial lesions may be related to the fact that the female rat grows more slowly than the male and reaches a lower adult weight.

4. Among rats there is a strain difference in sensitivity. The male hooded Chester Beatty rat gives a lower incidence and severity of heart lesions, and the male Wistar rat gives a lower severity of heart lesions than the male Sprague–Dawley rat when fed the same diet.

5. Commercial processing does not appear to have any effect on the lesion causing properties of LEAR oils. Results from several laboratories indicate that as many lesions are observed with fully refined oils as with the crude oils.

6. Chemical modification of LEAR oils, such as mild hydrogenation (iodine value 97), did not discernibly affect the lesion causing properties of these oils. On the other hand, extensive hydrogenation (iodine value 77) invariable lowered the lesion causing properties of LEAR oils, usually to the same level as that observed with other oils. Randomization, i.e., the rearrangement of the fatty acids on the triglyceride molecule, appears to have no effect on the lesion causing properties of LEAR oils. This suggests that the positional isomers have no influence on the incidence of lesions.

7. A 1000-fold purification of LEAR oils, as indicated by the decreased concentration of brassicasterol, was achieved by molecular distillation and adsorption chromatography, or a combination thereof. The resultant triglycerides, which were free of detectable impurities, when fed to male Sprague–Dawley rats gave as high an incidence of heart lesions as did the original LEAR oil. From these results it was concluded that extensive purification of LEAR oils by completely separate procedures, i.e., molecular distillation and adsorption chromatography, failed to remove a nontriglyceride lesion causing agent from the oil. It is apparent that myocardial lesions in male rats are caused by the triglycerides per se and may relate to the particular fatty acid composition present in LEAR oils.

8. Comparison of LEAR oils to other commonly used vegetable oils shows that palmitic (16:0) and stearic (18:0) acids are lower than that of soybean and peanut oils, oleic acid (18:1) is as high as in olive oil, linolenic acid (18:3) is as high as in soybean oil, and levels of linoleic acid (18:2) are intermediate compared to the other oils. Statistical analyses of published data show that of the fatty acids in vegetable oils, 18:1, 18:3, 20:1, and 22:1 are lesion promoting, while 16:0, 18:0, and 18:2 have a definite protective effect against the development of myocardial lesions (Hulan et al., 1976b; McCutcheon et al., 1976; Vles et al., 1978; Trenholm et al., 1979; Kramer et al., 1981, 1982; Farnworth et al., 1982a). It should, however, be kept in mind that some of the fatty acids that induce myocardial lesions in rat heart may be nutritionally valuable in other species, e.g., linolenic acid (Holman,

1981), and this should not be taken as a suggestion to the plant breeders that linolenic acid should be removed or palmitic acid raised in rapeseed oil.

9. Results indicated that species other than the rat, i.e., pigs, dogs, mice, and nonhuman primates, do not respond to dietary LEAR oils like the rat. The heart lesions found in these species are not generally fat related and seem to have a different etiology.

ACKNOWLEDGMENT

This work is Contribution No. 1095 from the Animal Research Centre, Ottawa, Ontario, Canada.

REFERENCES

Abdellatif, A. M. M., and Vles, R. O. (1970). *Nutr. Metab.* **12**, 285–295.

Abdellatif, A. M. M., and Vles, R. O. (1973). *Nutr. Metab.* **15**, 219–231.

Ackman, R. G. (1974). *Lipids* **9,** 1032–1035.

Ackman, R. G., and Loew, F. M. (1977). *Fette, Seifen, Anstrichm.* **79**, 15–24, 58–69.

Aherne, F. X., Bowland, J. P., Christian, R. G., Vogtmann, H., and Hardin, R. T. (1975). *Can. J. Anim. Sci.* **55**, 77–85.

Aherne, F. X., Bowland, J. P., Christian, R. G., and Hardin, R. T. (1976). *Can. J. Anim. Sci.* **55**, 275–284.

Astorg, P.-O., and Cluzan, R. (1976). *Ann. Nutr. Aliment.* **30**, 581–602.

Astorg, P.-O., and Levillain, R. (1979). *Ann. Nutr. Aliment.* **33**, 643–658.

Beare, J. L., Campbell, J. A., Youngs, C. G., and Craig, B. M. (1963). *Can. J. Biochem. Physiol.* **41**, 605–612.

Beare-Rogers, J. L. (1977). *Prog. Chem. Fats Other Lipids* **15**, 29–56.

Beare-Rogers, J. L., and Nera, E. A. (1977). *Lipids* **12**, 769–774.

Beare-Rogers, J. L., Nera, E. A., and Heggtveit, H. A. (1971). *Can. Inst. Food Technol. J.* **4**, 120–124.

Beare-Rogers, J. L., Nera, E. A., and Craig, B. M. (1972). *Lipids* **7,** 548–552.

Beare-Rogers, J. L., Nera, E. A., and Heggtveit, H. A. (1974a). *Nutr. Metab.* **17,** 213–222.

Beare-Rogers, J. L., Nera, E. A., and Heggtveit, H. A. (1974b). *Proc. Int. Rapskongr., 4th, 1974,* pp. 685–691.

Beare-Rogers, J. L., Gray, L., Nera, E. A., and Levin, O. L. (1979). *Nutr. Metab.* **23**, 335–346.

Berglund, F. (1975). *Acta Med. Scand., Suppl.* **585**, 47–49.

Bijster, G. M., Hudalla, B., Kaiser, H., Mangold, H. K., and Vles, R. O. (1979a). *Proc. Int. Rapeseed Conf., 5th, 1978,* pp. 141–143.

Bijster, G. M., Timmer, W. G., and Vles, R. O. (1979b). *Fette, Seifen, Anstrichm.* **81**, 192–194.

Campbell, J. A. (1970). *Proc. Int. Conf. Sci., Technol. Market. Rapeseed Rapeseed Prod., 1970,* pp. 467–469.

Charlton, K. M., Corner, A. H., Davey, K., Kramer, J. K. G., Mahadevan, S., and Sauer, F. D. (1975). *Can. J. Comp. Med.* **39**, 261–269.

Clandinin, M. T., and Yamashiro, S. (1980). *J. Nutr.* **110**, 1197–1203.

Clandinin, M. T., and Yamashiro, S. (1982). *J. Nutr.* **112**, 825–828.

Cluzan, R., Suschetet, M., Rocquelin, G., and Levillain, R. (1979). *Ann. Biol. Anim. Biochim., Biophys.* **19**, 497–500.

Codex Alimentarius Commission (1979). "F.A.O. Report of the 13th session of the Codex Committee on Fats and Oils, December 4–8, 1978." Codex Aliment. Comm., Rome.

Craig, B. M., and Beare, J. L. (1968). *Can. Inst. Food Technol. J.* **1**, 64–67.

Downey, R. K. (1964). *Can. J. Plant Sci.* **44**, 295.

Engfeldt, B. (1975). *Acta Med. Scand. Suppl.* **585**.

Engfeldt, B., and Brunius, E. (1975a). *Acta Med. Scand., Suppl.* **585**, 15–26.

Engfeldt, B., and Brunius, E. (1975b). *Acta Med. Scand., Suppl.* **585**, 27–40.

Engfeldt, B., and Gustafsson, B. (1975). *Acta Med. Scand., Suppl.* **585**, 41–46.

Farnworth, E. R., and Kramer, J. K. G. (1981). *Can. Fed. Biol. Soc.* **24**, 114 (abst. 55).

Farnworth, E. R., Kramer, J. K. G., Thompson, B. K., and Corner, A. H. (1982a). *J. Nutr.* **112**, 231–240.

Farnworth, E. R., Kramer, J. K. G., Jones, J. D., Thompson, B. K., and Corner, A. H. (1982b). *Can. Inst. Food Sci. Technol. J.* (in press).

Farnworth, E. R., Kramer, J. K. G., Corner, A. H., and Thompson, B. K. (1982c). *J. Am. Oil Chem. Soc.* **59**, 299A (abstract 243).

Flanzy, J. (1979a). *Ann. Biol. Anim. Biochim., Biophys.* **19**, 467–470.

Flanzy, J. (1979b). *Ann. Biol. Anim. Biochim., Biophys.* **19**, No. 2B.

Friend, D. W., Corner, A. H., Kramer, J. K. G., Charlton, K. M., Gilka, F., and Sauer, F. D. (1975a). *Can. J. Anim. Sci.* **55**, 49–59.

Friend, D. W., Gilka, F., and Corner, A. H. (1975b). *Can. J. Anim. Sci.* **55**, 571–578.

Friend, D. W., Kramer, J. K. G., and Corner, A. H. (1976). *Can. J. Anim. Sci.* **56**, 361–364.

Gopalan, C., Krishnamurthi, D., Shenolikar, J. S., and Krishnamachari, K. A. V. R. (1974). *Nutr. Metab.* **16**, 352–365.

Holman, R. T. (1981). *Chem. Ind. (London)* Oct. 17, 704–709.

Hulan, H. W., Hunsaker, W. G., Kramer, J. K. G., and Mahadevan, S. (1976a). *Can. J. Physiol. Pharmacol.* **54**, 1–6.

Hulan, H. W., Kramer, J. K. G., Mahadevan, S., Sauer, F. D. and Corner, A. H. (1976b). *Lipids* **11**, 9–15.

Hulan, H. W., Kramer, J. K. G., and Corner, A. H. (1977a). *Lipids* **12**, 951–956.

Hulan, H. W., Kramer, J. K. G., and Corner, A. H. (1977b). *Can. J. Physiol. Pharmacol.* **55**, 258–264.

Hulan, H. W., Kramer, J. K. G., Corner, A. H., and Thompson, B. (1977c). *Can. J. Physiol. Pharmacol.* **55**, 265–271.

Hulan, H. W., Thompson, B., Kramer, J. K. G., Sauer, F. D., and Corner, A. H. (1977d). *Can. Inst. Food Sci. Technol. J.* **10**, 23–26.

Hung, S., Umemura, T., Yamashiro, S., Slinger, S. J., and Holub, B. J. (1977). *Lipids* **12**, 215–221.

Hunsaker, W. G., Hulan, H. W., Kramer, J. K. G., and Corner, A. H. (1977). *Can. J. Physiol. Pharmacol.* **55**, 1116–1121.

Ilsemann, K., Reichwald, I., and Mukherjee, K. D. (1976). *Fette, Seifen, Anstrichm.* **78**, 181–187.

Kramer, J. K. G. (1980). *Lipids* **15**, 651–660.

Kramer, J. K. G., Mahadevan, S., Hunt, J. R., Sauer, F. D., Corner, A. H., and Charlton, K. M. (1973). *J. Nutr.* **103**, 1696–1708.

Kramer, J. K. G., Hulan, H. W., Mahadevan, S., and Sauer, F. D. (1975a). *Lipids* **10**, 505–510.

Kramer, J. K. G., Hulan, H. W., Mahadevan, S., Sauer, F. D., and Corner, A. H. (1975b). *Lipids* **10**, 511–516.

Kramer, J. K. G., Hulan, H. W., Procter, B. G., Dussault, P., and Chappel, C. I. (1978a). *Can. J. Anim. Sci.* **58**, 245–256.

Kramer, J. K. G., Hulan, H. W., Procter, B. G., Rona, G., and Mandavia, M. G. (1978b). *Can. J. Anim. Sci.* **58**, 257–270.

Kramer, J. K. G., Hulan, H. W., Corner, A. H., Thompson, B. K., Holfeld, N., and Mills, J. H. L. (1979a). *Lipids* **14**, 773–780.

Kramer, J. K. G., Hulan, H. W., Trenholm, H. L., and Corner, A. H. (1979b). *J. Nutr.* **109**, 202–213.

Kramer, J. K. G., Farnworth, E. R., Thompson, B. K., and Corner, A. H. (1981). *Prog. Lipid. Res.* **20**, 491–499.

Kramer, J. K. G., Farnworth, E. R., Thompson, B. K., Corner, A. H., and Trenholm, H. L. (1982). *Lipids* **17**, 372–382.

Lancet (1974). **2**, 1359–1360.

Loew, F. M., Schiefer, B., Laxdal, V. A., Prasad, K., Forsyth, G. W., Ackman, R. G., Olfert, E. D., and Bell, J. M. (1978). *Nutr. Metab.* **22**, 201–217.

McCutcheon, J. S., Umemura, T., Bhatnagar, M. K., and Walker, B. L. (1976). *Lipids* **11**, 545–552.

Migicovsky, B.B. (1970). *Proc. Int. Conf. Sci., Technol. Market. Rapeseed Rapeseed Prod. 1970*, pp. 556–560.

Morice, J. (1979). *Ann. Biol. Anim., Biochim., Biophys.* **19**, 471–477.

Murray, T. K., Beare, J. L., Campbell, J. A., and Hopkins, C.Y. (1958). *Can. J. Biochem. Physiol.* **36**, 653–657.

Nolen, G. A. (1981). *J. Am. Oil Chem. Soc.* **58**, 31–37.

Petersen, U., Oslage, H.-J., and Seher, A. (1979). *Fette, Seifen, Anstrichm.* **81**, 177–181.

Prairie Grain Variety Survey (1981). Canadian Co-operative Wheat Producers Ltd., Regina, Saskatchewan.

Procter, B. G., Dussault, P., Rona, G., and Chappel, C. I. (1974). Bio-Research Laboratories Ltd., Project 4178 and 5879 for the Rapeseed Association of Canada.

Qureshi, S. R. (1979). *Vet. Pathol.* **16,** 486–487.

Rings, R. W., and Wagner, J.E. (1972). *Lab. Anim. Sci.* **22**, 344–352.

Röbbelen, G., and Thies, W. (1980). *In* "Brassica Crops and Wild Allies" (S. Tsunoda, K. Hinata, and C. Gomez-Campo, eds), pp 253–283. Jpn. Sci. Soc. Press, Tokyo.

Rocquelin, G., and Cluzan, R. (1968). *Ann. Biol. Anim., Biochim., Biophys.* **8**, 395–406.

Rocquelin, G., and Fouillet, X. (1979). *Ann. Biol. Anim., Biochim., Biophys.* **19**, 479–481.

Rocquelin, G., and Leclerc, J. (1969). *Ann. Biol. Anim., Biochim., Biophys.* **9**, 413–426.

Rocquelin, G., Sergiel, J.-P., Astorg, P. O., and Cluzan, R. (1973). *Ann. Biol. Anim. Biochim., Biophys.* **13**, 587–609.

Rocquelin, G., Sergiel, J. P., Astorg, P. O., Nitou, G., Vodovar, N., Cluzan, R., and Levillain, R. (1974). *Proc. Int. Rapskongr. 4th, 1974*, pp. 669–683.

Rocquelin, G., Juaneda, P., and Cluzan, R. (1981). *Ann. Nutr. Metab.* **25**, 350–361.

Roine, P., Uksila, E., Teir, H., and Rapola, J. (1960). *Z. Ernährungswiss.* **1**, 118–124.

Rose, S. P., Bell, J. M., Wilkie, I. W., and Schiefer, H.B. (1981). *J. Nutr.* **111**, 355–364.

Sauer, F. D. (1974). *Proc. Int. Rapskongr. 4th, 1974*, pp. 725–731.

Slinger, S. J. (1977). *J. Am. Oil Chem. Soc.* **54**, 94A–99A.

Stefansson, B. R., and Kondra, Z. P. (1975). *Can. J. Plant Sci.* **55**, 343–344.

Stefansson, B. R., Hougen, F. W., and Downey, R. K. (1961). *Can. J. Plant Sci.* **41**, 218–219.

Svaar, H. (1982). *In* "Nutritional Evaluation of Long-Chain Fatty Acids in Fish Oil" (S. M. Barlow and M. E. Stansby, eds.), pp. 163–184. Academic Press, New York (in press).

Svaar, H., and Langmark, F. T. (1975). *INSERM Sér. Action Thématique* pp. 329–335.

Svaar, H., and Langmark, F. T. (1980). *Acta Pathol. Microbiol. Scand., Sect. A.* **88**, 179–187.

Svaar, H., Langmark, F. T., Lambertsen, G., and Opstvedt, J. (1980). *Acta Pathol. Microbiol. Scand., Sect. A.* **88**, 41–48.

Tinsley, I. J., Schmitz, J. A., and Pierce, D. A. (1981). *Cancer Res.* **41**, 1460–1465.

Trenholm, H. L., Thompson, B. K., and Kramer, J. K. G. (1979). *Can. Inst. Food Sci. Tech. J.* **12**, 189–193.

Umemura, T., Slinger, S. J., Bhatnagar, M. K., and Yamashiro, S. (1978). *Res. Vet. Sci.* **25**, 318–322.

Vergroesen, A. J., and Gottenbos, J. J. (1975). *In* "The Role of Fat in Human Nutrition" (A. J. Vergroesen, ed.), pp. 1–41. Academic Press, New York.

Vles, R. O. (1974). *Proc. Int. Rapskongr. 4th, 1974*, pp. 17–30.

Vles, R. O. (1978). *Rev. Fr. Corps Gras* **25**, 289–295.

Vles, R. O. (1979). *Proc. Nouv. Huile Colza, Cent. Nat. Coord. Etud. Rech. Nutr. Aliment., 1978*, pp. 149–156.

Vles, R. O., Bijster, G. M., Kleinekoort, J. S. W., Timmer, W. G., and Zaalberg, J. (1976). *Fette, Seifen, Anstrichm.* **78**, 128–131.

Vles, R. O., Bijster, G. M., and Timmer, W. G. (1978). *Arch. Toxicol., Suppl.* **1,** 23–32.

Vles, R. O., Timmer, W. G., and Zaalberg, J. (1979). *Ann. Biol. Anim. Biochim. Biophys.* **19**, 501–508.

Vogtmann, H., Christian, R., Hardin, R. T., and Clandinin, D. R. (1975). *Int. J. Vitam. Nutr. Res.* **45**, 221–229.

Yamashiro, S., and Clandinin, M. T. (1980). *Exp. Mol. Pathol.* **33**, 55–64.

Ziemlanski, S. (1977). *Bibl. "Nutr. Dieta"* **25**, 134–157.

18

Cardiac Lipid Changes in Rats, Pigs, and Monkeys Fed High Fat Diets

J. K. G. KRAMER AND F. D. SAUER

High and Low Erucic Acid Rapeseed Oils
Copyright © 1983 by Academic Press Canada
All rights of reproduction in any form reserved.
ISBN 0-12-425080-7

I. INTRODUCTION

Much attention was focused in the past decade on determining changes in heart lipids due to dietary fatty acids because of the apparent relationship of dietary rapeseed oil and two cardiopathological conditions, lipidosis and necrosis. The initial interest centered on the effect of erucic acid, a docosenoic fatty acid present at concentration of 25–50% in the original high erucic acid rapeseed (HEAR) oils, and which was shown to result in a specific accumulation of triglycerides (TG) in the heart of experimental animals (Abdellatif and Vles, 1970). However, these studies became less critical with the development of the new low erucic acid rapeseed (LEAR) oils.

The second cardiopathological condition observed in rats fed diets containing rapeseed oil, namely myocardial necrosis (Roine et al., 1960), continues to encourage studies on how heart lipid changes might affect function. The reason for this research effort was the fact that erucic acid did not appear to be the sole cause of this particular problem. Both LEAR oils (Rocquelin and Cluzan, 1968) as well as other vegetable oils (Kramer et al., 1973) appeared to cause the same myocardial necrosis. Furthermore, there was no evidence to indicate that minor components in vegetable oils were responsible, since the pure triglycerides from the vegetable oils were equally effective in producing this heart lesion (Kramer et al., 1975b, 1979a). These results suggest that the common C_{16} and C_{18} fatty acids found in most vegetable oils may also be involved in cardiac necrosis of the rat.

When one is designing experiments to test the nutritional and toxicological properties of vegetable oils, each fatty acid should be examined by itself. However, such studies would be too costly and furthermore complicate the issue since some fatty acids would be poorly absorbed and others would promote essential fatty acid deficiency. Therefore, it seems appropriate to consider cardiac lipid changes in animals fed diets which contain different vegetable oils or fats that in fact possess a characteristic fatty acid composition (Table I). A knowledge of the differences in cardiac lipids with different dietary fats and with different animal species may help us to understand the causes of this cardiopathological condition.

The protocol in most of these published experimental studies involved feeding postweaned animals diets that contained high levels of fats, oils, or fat–oil mixtures. Diets with vegetable oils contain more than adequate amounts of essential fatty acids with a linoleic acid content from 7% in olive oil to over 70% in sunflower oil (Table I). The fatty acid composition of the animal body, including the heart, generally will reflect the dietary fatty acids (Carroll 1965; Holub and Kuksis, 1978). This in part may be due to the fact that the high levels of fat in the diet depress significantly the *de novo* synthe-

TABLE I

Fatty Acid Composition of Selected Vegetable Oils, Fats, and Fat–Oil Mixtures

Fatty acid	Corn oil	HEAR oil	LEAR oil	Linseed oil	Mustard oil	Olive oil	Peanut oil	Soybean oil	Sunflower oil	Fish oil[a]	Lard	Lard/corn oil (3/1)
14:0	0.2	0.1	0.1	—	—	—	—	0.1	0.1	7.2	1.5	1.1
16:0	13.1	2.9	5.7	5.9	2.6	11.0	10.7	10.5	6.6	12.1	25.6	21.5
16:1	0.2	0.3	0.1	0.2	0.3	0.8	0.3	0.2	0.1	8.7	2.6	2.3
18:0	2.0	1.4	2.1	3.8	1.0	3.0	3.4	3.2	4.3	1.5	14.6	10.6
18:1	26.5	33.0	57.7	19.2	17.3	75.3	49.0	24.9	18.3	13.2	43.2	40.6
18:2	56.1	15.4	24.6	16.8	14.1	7.8	28.0	51.5	68.7	1.7	9.6	20.9
18:3	0.7	6.2	7.9	53.4	7.2	0.7	—	8.4	0.4	1.3	0.6	0.6
20:0	—	—	0.2	—	1.0	0.5	2.0	1.0	0.4	0.2	0.1	0.3
20:1	0.4	12.2	1.0	0.4	11.0	0.6	1.6	0.6	0.3	13.9	1.2	0.9
22:0	—	0.5	0.2	—	0.4	—	3.0	—	0.7	—	0.1	<0.1
22:1	—	25.5	0.2	—	44.2	—	<0.1	—	—	14.6	0.2	<0.1
24:0	—	0.2	0.1	—	—	—	1.4	—	—	—	trace	<0.1
24:1	—	0.2	0.1	—	—	—	—	—	—	—	0.1	<0.1
Ref.[b]	1	1	2	3	4	3	5	2	3	6	7	8

[a] In addition 4.1% 18:4; 8.2% 20:5; 7.2% 22:6.

[b] References: 1, Kramer et al. (1979b); 2, Kramer et al. (1979a); 3, Vles et al. (1978); 4, Sen and Sen Gupta (1980); 5, Rocquelin and Fouillet (1979); 6, Opstvedt et al. (1979); 7, Hulan et al. (1976); 8, Ackman and Loew (1977).

sis of fatty acids in the experimental animals (Romsos and Leveille, 1974; Iritani and Fukuda, 1980).

II. FATTY ACID COMPOSITION OF VEGETABLE OILS AND FATS

Animal fats and fully refined vegetable oils contain 95–97% TG which show characteristic differences in their fatty acid composition (Table I). Some vegetable oils are rich in 18:2 (corn and sunflower oils) while others are rich in 18:1 (olive and peanut oils), but these oils have in common a low level of 18:3. Other vegetable oils contain about 10% 18:3 together with high (soybean oil) or intermediate (LEAR, HEAR and mustard oils) levels of 18:2 and high levels of monounsaturated fatty acids greater than C_{18} (HEAR and mustard oils). In fish oils, approximately 20% of the fatty acids belong to the linolenic acid family that includes 20:5 *n*-3, 22:5 *n*-3, and 22:6 *n*-3. Linseed oil is unique in its high content of 18:3.

The content of saturated fatty acids in vegetable oils ranges from a low of 5% in rapeseed and mustard oils to about 15% in corn and soybean oils. Peanut oil contains about 20% saturated fatty acids, appreciable amounts of which are longer than C_{18}. In contrast to vegetable oils, animal fats and fish oils are much richer in saturated fatty acids, i.e., chicken fat 33%, lard 40%, beef and mutton tallow 48%, and fish oils 17–33% (Sheppard *et al.*, 1978).

III. CHANGES IN CARDIAC LIPIDS OF RATS FED DIFFERENT OILS AND FATS

A. Cardiac Lipid Composition of Rats

The average lipid content of the rat heart is about 3% of the fresh weight. The phospholipids comprise 60–70% of the total lipids, the remainder consists of triglycerides (~25%), cholesterol (~6%), cholesteryl ester (~2%) and minor amounts of diglycerides and free fatty acids (Simon and Rouser, 1969; Kramer, 1980). A detailed compositional analysis of the rat heart phospholipids was published by Simon and Rouser (1969) and shown in Table II. The structure of the phospholipids is also given in Table II.

Cardiac TG usually occur as droplets of various sizes within the cells and do not appear to be membrane enclosed (Rouser *et al.*, 1968). They are readily mobilized for energy production or serve as phospholipid precursors. On the other hand, the phospholipids occur almost exclusively in the membranous structures of the cell (McMurray, 1973). Cholesterol also occurs in membranes, has a condensing effect, and controls the movement of the hydrocarbon chains of phospholipids (Papahadjopoulos, 1973).

TABLE II

Structure and Concentration of Phospholipid in Rat Heart

CH_2—$OOCR^1$ CH_2—OPO_2O—CH_2 $C_{13}H_{27}C$=C—CH—OH

R^2COO—CH HC—OH HC—$OOCR^3$ HC—$NHCOR$

CH_2—OPO_2O—X CH_2 CH_2—$OOCR^4$ CH_2—OPO_2O—X

I   II   III

Name (abbreviation)	Compound—X	Concentration[a] (% lipid phosphorus)
Phosphatidylcholine (PC)	I—$CH_2CH_2N(CH_3)_3$	36
Phosphatidylethanolamine(PE)	I—$CH_2CH_2NH_2$	30
Diphosphatidylglycerol (DPG)	I–II	11
Phosphatidylinositol (PI)	I—$C_6H_{11}O_5$	3.7
Phosphatidylserine (PS)	I—$CH(NH_2)COOH$	3.2
Sphingomyelin (SP)	III—$CH_2CH_2N(CH_3)_3$	3.1
Phosphatidylglycerol (PG)	I—$CH_2CH(OH)CH_2OH$	1.0
Lysophosphatidylcholine (LPC)	Replace R^2CO by H in PC	0.5
Phosphatidic acid (PA)	I—H	0.2

[a] Simon and Rouser (1969).

B. Changes in Cardiac Neutral Lipids of Rats

1. TRIGLYCERIDES

a. Quantitative Changes.   The level of cardiac TG in rats remains constant even when the animals are changed from a diet low in fat to one containing 20% fat, provided that the level of erucic acid does not exceed 1% of the diet (Kramer and Hulan, 1978b). The concentration of TG in rat hearts was observed to range from 2 to 3 mg/g wet tissue under these experimental conditions. LEAR oils, which by definition contain less than 5% erucic acid (Codex Alimentarius Commission, 1979), can therefore be fed as the sole fat in the diet at 20% by weight without affecting the level of cardiac TG in the heart.

When rats are fed a diet rich in erucic acid, a rapid accumulation of cardiac TG occurs within the first week. The concentration of heart TG then declines, but even after 16 weeks the level still remains about twice that found in the controls (Fig. 1). Similar results were obtained by other investigators (Table III). The amount of cardiac TG which accumulate during peak lipidosis is proportional to the erucic acid content in the diet (Table III).

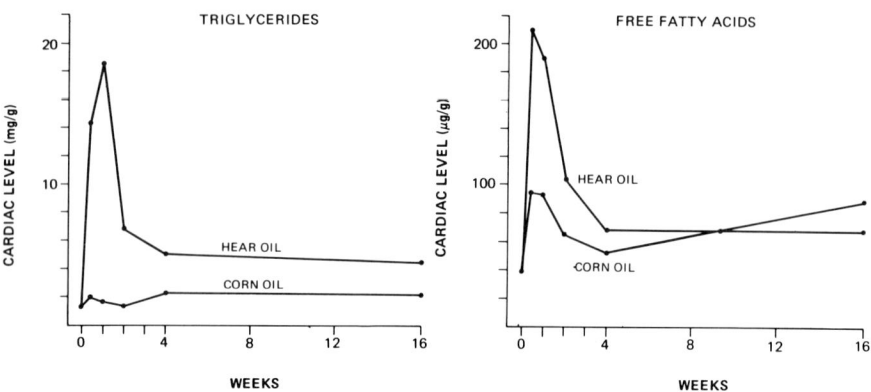

Fig. 1.   Concentration of cardiac triglycerides and free fatty acids in rats fed HEAR oil or corn oil at 20% by weight of the diet for up to 16 weeks. Data for 0 to 3 days were taken from Kramer and Hulan (1978b) and the remaining values from Kramer et al. (1979b).

There appear to be no strain differences among rats in the development of myocardial lipidosis. The most widely used strains of rats, namely, the Sprague–Dawley and the Wistar strains have not been compared in controlled experiments. However, the Sprague–Dawley rats have been compared to the Chester Beatty (hooded) rats (Kramer et al., 1979b). These two strains of rats differ in their major route of fat absorption (Johnston, 1968), and in their response to myocardial necrosis (Hulan et al., 1977b), but the accumulation of the cardiac TG remains similar throughout the experimental period (Table IV).

The influence of other fatty acids in the dietary fat on the accumulation of the cardiac TG was also investigated. It is evident from the results presented in Table V, that neither 18:2, 18:3, nor saturated fatty acids affect the erucic acid induced accumulation of cardiac TG in the rat. The digestibility of erucic acid was not significantly different between each of these diets in the three studies reported in Table V. It may therefore be concluded that erucic acid, once absorbed, will accumulate in the heart in the form of TG quite independent of the concentration of other fatty acids.

Cetoleic acid, a positional isomer of erucic acid, also causes an accumulation of cardiac TG in the rat. It may be concluded from the results of these studies (Table VI) that the accumulation of heart TG is similar if the concentration of cetoleic and erucic acids in the diet is the same. In only one report does there appear to be a difference between the two positional isomers. Astorg and Cluzan (1977) reported less myocardial TG accumulation with cetoleic than erucic acid. The authors attributed this decrease in TG accumulation to an increase of the 18:2 content in the dietary fat from 1.33 to

TABLE III

Cardiac Triglyceride Levels in Rats Fed Diets Containing Different Levels of Erucic Acid for Different Periods of Time

Diet (% 22:1 in oil)	Heart triglycerides[a] (mg/g wet weight)				Ref.[b]
	Days on diet				
	3	7	35	70	
Corn oil	3.4	3.0	2.3	3.4	1
HEAR oil (31-34%)	23.7*	33.1*	6.0*	6.7*	2
	3	7			
Peanut oil	6.4	3.1			3
HEAR oil mixture (14.2%)	10.5	7.0			
HEAR oil mixture (27.2%)	17.1	15.9			
	5	60			
Sunflower oil	2.3	2.6			4
HEAR oil mixture (23.7%)	7.1*	3.3			
HEAR oil (47.4%)	11.7*	5.0*			
	4				
Lard	4.0				5
HEAR oil mixture (11.6%)	5.5				
HEAR oil mixture (36.6%)	27.1*				
	3	10			
Sunflower oil	34[c]	—			6
HEAR oil (44%)	300	—			
Trierucin (60%)	716	509			

[a] Values significantly different from controls $P < 0.05$ (*) whenever analyzed.

[b] References: 1, Hung et al. (1977) (weanling Sprague–Dawley rats); 2, Hung and Holub (1977) (weanling Sprague–Dawley rats); 3, Astorg and Cluzan (1977) (weanling Wistar rats); 4, Bellenand et al. (1980) (180g Sprague–Dawley rats); 5, Opstvedt et al. (1979) (weanling Wistar rats); 6, Stam et al. (1980) (200–250 g Wistar rats).

[c] Values expressed as μmoles triglycerides/g of protein.

TABLE IV

Cardiac Triglycerides in Two Strains of Male
Rats Fed a HEAR Oil at 20% by Weight and
Containing 25.5% Erucic Acid[a]

Time on diet (weeks)	Heart triglycerides (mg/g wet tissue)	
	SD[b]	CB[c]
1	18.6	17.7
2	6.8	9.6
4	5.0	6.4
16	4.5	3.7

[a] From Kramer et al. (1979b).
[b] SD, Sprague–Dawley.
[c] CB, Chester Beatty.

9.2%. This suggestion is not supported by the data in Table V which indicate that the level of myocardial TG is independent of the 18:2 content of the dietary oil.

   b. *Fatty Acid Changes.*   The fatty acid composition of cardiac TG in rats fed diets rich in fat for a week is shown in Table VII. The cardiac TG have a higher content of saturated fatty acids (14:0, 16:0 and 18:0), and palmitoleic

TABLE V

Effect of Other Fatty Acids on the Triglyceride Accumulation in the Rat Heart as a Result of Feeding Erucic Acid Containing Diets

Dietary fatty acids					% fat in diet	Time on diet (days)	Heart triglycerides		Ref.[b]
Saturates	18:1	18:2	18:3	22:1			mg/g	P < 0.05[a]	
2.2	47.2	—	—	48.7	15	4	32.7		1
2.2	46.0	2.2	—	47.3			30.8		
2.9	43.6	4.4	—	47.2			28.8	N.S.	
3.6	37.4	10.8	—	46.6			37.2		
4.6	26.0	22.2	—	45.3			44.6		
11.7	26.9	13.0	—	47.2	15	8	21.8	N.S.	2
6.1	8.9	7.8	28.1	48.6			21.8		
8.2	18.6	40.0	4.8	23.7	15	5	7.1	N.S.	3
47.7	10.6	8.7	4.6	23.7			6.6		

[a] N.S., not significant.
[b] References: 1, Astorg and Compoint (1978); 2, Rocquelin (1979); 3. Bellenand et al. (1980).

TABLE VI

Comparison of Erucic and Cetoleic Acid in Accumulating Cardiac
Triglycerides in Rat Hearts

Diet[a] (% 22:1 in oil)	Heart triglycerides (mg/g wet weight)		Ref.[b]
	3 days	7 days	
Fish oil (14.8%)	7.1	5.6	1
Fish oil (13.1%)	5.0	7.7	
HEAR oil (14.2%)	10.5	7.0	
Fish oil (29.7%)	12.9	10.6	1
Fish oil (25.6%)	6.1	7.1	
HEAR oil (27.2%)	17.1	15.9	
	4 days		
Fish oil (10.9%)	9.8		2
HEAR oil (11.6%)	5.5		

[a] All fish oils were partially hydrogenated containing 30–50% *trans*
fatty acids (Astorg and Cluzan, 1977).
[b] References: 1, Astorg and Cluzan (1977); 2, Opstvedt *et al.* (1979).

acid (16:1), and a significantly reduced level of linoleic (18:2) and linolenic
(18:3) acids than the dietary oils. The concentration of oleic acid (18:1) in
the dietary oil and the heart TG is about the same. When rats are fed diets
containing docosenoic acids, the heart TG during maximum lipidosis generally have the same composition as the diet. The concentration of monounsaturated fatty acids 18:1, 20:1, and 22:1 is virtually unchanged compared
to the dietary oils. This is so regardless of the origin of the docosenoic acid,
i.e., whether derived from HEAR oil, mustard oil, fish oil, or from synthetic
triglycerides.

A detailed analysis of the positional distribution of the dietary and heart
TG revealed that position 2 of TG undergoes the greatest change (Myher *et
al.*, 1979). As seen in Table VIII, the concentration of molecular species of
heart TG with 18:2 and 18:3 in position 2 is significantly lower than that
found in the dietary oil. The opposite is found for molecular species with
saturated and monounsaturated fatty acids in position 2. Of particular interest is the relatively high concentration of erucic acid in position 2 of cardiac
TG, which in rapeseed oil (Brockerhoff an Yurkowski, 1966) and mustard oil
(Myher *et al.*, 1979), is in positions 1 and 3.

The fatty acid composition of heart TG in rats fed diets containing docose-

TABLE VII

Fatty Acid Composition of the Dietary Oils and the Heart Triglycerides of Rats Fed the Oils for about 1 Week

Fatty acid	Corn oil Diet	Heart TG[a]	LEAR oil Diet	Heart TG	Olive oil Diet	Heart TG	Peanut oil Diet	Heart TG	Soybean oil Diet	Heart TG	Sunflower oil Diet	Heart TG	HEAR oil Diet	Heart TG	Mustard oil Diet	Heart TG	Trierucin Diet	Heart TG	Fish oil Diet	Heart TG
14:0	0.2	1.6		1.2		1.7		1.9		1.6		1.8	0.1	0.4	0.03	0.2			6.6	3.7
16:0	13.1	23.8	6.1	12.1	11.6	22.2	10.5	22.6	12.4	21.4	6.1	24.4	2.9	6.2	2.6	5.2	2.2	7.2	13.2	15.7
18:0	2.0	7.8	2.0	4.4	2.5	6.0	3.0	8.2	3.7	7.4	4.4	7.6	1.4	3.8	0.9	1.9	1.9	3.1	3.3	5.1
16:1	0.2	2.1	0.1	1.0	1.2	1.5	0.3	4.3	0.1	1.4	0.4	4.0	0.3	0.9	0.1	0.9		0.9	8.4	3.8
18:1	26.5	28.4	56.5	54.1	75.5	60.1	49.3	40.7	25.4	28.8	23.4	29.8	33.0	31.6	23.0	28.1	26.0	23.5	14.8	25.6
20:1	0.4	1.0	1.5	2.6	0.4	0.7	1.3	2.2		0.7		0.4	12.2	13.6	11.2	12.6	1.0	2.7	14.0	11.5
22:1		0.1	0.3	1.3	0.1	0.2	0.4	1.2		0.3			25.5	27.4	31.6	34.4	45.3	45.7	14.0	18.0
18:2	56.1	30.5	26.0	17.7	7.3	5.9	29.2	9.5	50.6	32.1	64.4	30.7	15.4	7.3	15.3	11.1	22.2	10.1	4.2	11.2
18:3	0.7	0.1	7.1	2.1	0.7	0.1		_b	7.9	2.2	0.5	_b	6.2	0.8	12.6	3.7			0.1	
Days on diet	7		7		7		7		7		5		7		7		4		4	
Ref.[c]	1		2		2		3		2		4		1		5		6		7	

[a] TG, triglycerides.

[b] 18:3 and 20:1 are combined.

[c] References: 1, Kramer et al. (1979b); 2, Kramer and Hulan (1978b); 3, Rocquelin et al. (1973); 4, Bellenand et al. (1980); 5, Myher et al. (1979); 6, Astorg and Compoint (1978); 7, Opstvedt et al. (1979).

TABLE VIII

Molecular Species of Triglycerides of Mustard Oil and Heart Triglycerides of Rats Fed the Mustard Oil in Their Diet[a]

Triglyceride position 2	Mustard oil (mole %)	Heart triglyceride (mole %)
16:0 and 18:0	3.27	11.14
18:1	38.42	47.00
18:2	32.26	19.39
18:3	27.74	4.50
20:1	0.20	5.13
22:1	0.50	10.38

[a] Myher et al. (1979).

noic fatty acids changes following the period of maximum accumulation within the first week. The relative concentration of erucic acid declines while the level of 18:1 increases proportionally (Table IX). This may be the result of chain shortening (Craig and Beare, 1967; Clouet and Bezard, 1978).

2. FREE FATTY ACIDS

Fairbairn in 1945 reported that the concentration of free fatty acids in normal tissues is quite low unless there is autolysis of the lipids during the

TABLE IX

Changes in the Fatty Acid Composition of Heart Triglycerides of Rats Fed a HEAR Oil over a 16-Week Period

Fatty acid	Time on diet[a]						HEAR oil
	0	3 days	1 week	2 weeks	4 weeks	16 weeks	
14:0	3.0	0.5	0.4	0.9	0.9	0.7	0.1
16:0	32.4	7.6	6.2	10.2	8.6	10.3	2.9
16:1	2.1	0.8	0.9	1.1	0.8	2.1	0.3
18:0	16.7	3.3	3.8	4.7	5.6	4.3	1.4
18:1	33.3	34.5	31.6	34.9	35.4	42.8	33.0
18:2	12.0	8.8	7.3	8.8	9.1	11.6	15.4
18:3		1.1	0.8	1.2	0.9	1.4	6.2
20:1		12.8	13.6	12.0	11.6	10.2	12.2
22:1		25.6	27.4	20.4	22.2	13.7	25.5
mg/g	1.3	14.4	18.6	6.8	5.0	4.5	

[a] Values for time 0 and 3 days were taken from Kramer and Hulan (1978b); all other values were taken from Kramer et al. (1979b).

TABLE X

Level of Cardiac Free Fatty Acids in Rats Fed HEAR Oil or Other Oils as Determined Using Conventional Homogenization Techniques or a New Quick Freezing Technique

Age of rat (weeks)	Diet (% 22:1)	Fat in diet %	Time on diet (days)	Heart free fatty acids (μg/g)	Ref.[a]
Conventional homogenization technique					
8	Control	—	0	1900	1
	HEAR oil (45%)	25	3	4500	
3	Control (72% 18:1)	20	7	2400	2
	Synthetic oil (73%)	20	7	7100	
n.a	Peanut oil	15	5	1500	3
	HEAR oil (48%)	15	5	2500	
6–7	Corn oil	20	3	1200	4
	HEAR oil (22.3%)	20	3	2300	
3	Peanut oil	15	60	508	5
	HEAR oil (50%)	15	60	728	
4–5	Peanut oil	15	2	321	6
	Mustard oil (47.1%)	15	2	970	
Quick freezing technique					
3	Control	—	0	39	7
	Soybean oil	20	3	95	
	Olive oil	20	3	45	
	LEAR oil (0.3%)	20	3	73	
	HEAR oil (22.3%)	20	3	210	

[a] References: 1, Houtsmuller et al. (1970); 2. Beare-Rogers et al. (1972); 3. Rocquelin (1972); 4. Dow-Walsh et al. (1975); 5. Joffrain et al. (1975); 6. Bhatia et al. (1979); 7. Kramer and Hulan (1978b).

homogenization process. The use of faulty isolation techniques is indicated by the results of numerous investigators who have reported free fatty acid levels in the heart tissue in excess of 1000 μg/g wet weight (Table X). As shown by Kramer and Hulan (1978a), when proper extraction techniques are employed, normal free fatty acid levels of <100 μg/g wet weight are found in cardiac tissues. These authors resorted to a quick freezing of the heart tissue between blocks of dry ice, and then pulverizing the heart tissue at dry ice temperature. This was followed by solvent extraction at 0°C. The low free fatty acid values in heart tissue reported by Kramer and Hulan (1978a) have been confirmed by Opstvedt et al. (1979).

An increase in cardiac free fatty acids has been observed by all investiga-

TABLE XI

Comparison between the Fatty Acid Composition of Free Fatty Acids in the Hearts of Rats Fed HEAR Oil for 7 Days and That of Dietary Oil[a]

Fatty acids	Dietary HEAR oil	Heart free fatty acids
14:0	0.1	1.0
16:0	4.0	8.1
16:1	0.3	1.3
18:0	1.7	4.6
18:1	36.2	18.6
18:2	15.1	6.2
18:3	5.9	1.4
20:1	12.3	7.5
20:4	—	0.8
22:1	22.3	42.5
24:1	0.2	3.8

[a] From Kramer and Hulan (1978a).

tors in rats fed HEAR oil during the period of myocardial lipidosis regardless of the extraction technique employed. The level of cardiac free fatty acid increases within the first week to a maximum, then declines rapidly, and after 4 weeks approaches the level found in control rats (Fig. 1). During maximum lipidosis, 22:1 and 24:1 were present in relatively greater proportion in the cardiac free fatty acids than in the original oil (Table XI). In fact, the long chain monoenoic fatty acids 20:1, 22:1, and 24:1 constitute over 50% of the cardiac free fatty acids during this period. These results suggest that these fatty acids are metabolized slower than shorter chain fatty acids.

3. DIGLYCERIDES

Diglycerides are known intermediates in the biosynthesis of triglycerides and phospholipids (Hill and Lands, 1970; Holub and Kuksis, 1978). The concentration of diglycerides is very low in heart tissue. A value of 72 μg/g of wet heart tissue was reported in 3-week-old rats which increased to 260 and 340 μg/g in rats fed HEAR oil for 3 and 7 days, respectively (Kramer and Hulan, 1978a). The increase in cardiac diglycerides in rats fed HEAR oils may accompany the increase in TG synthesis in these animals (Hung and Holub, 1977). The fatty acid composition of cardiac diglycerides was similar to that of cadiac TG (Kramer and Hulan, 1978a).

4. CHOLESTEROL

Cholesterol is a membrane component, which varies in concentration from one type of membrane to another, but within any given type of membrane the cholesterol concentration is remarkably similar (McMurray, 1973; Thompson, 1980). The concentration of cholesterol in mammalian tissues, except for liver, does not appear to change with diet (Okano et al., 1980; Lewis et al., 1981), not even when increasing levels of cholesterol are fed (Beher et al., 1963; Raicht et al., 1975). Therefore, it was not surprising to find that the cholesterol content in cardiac lipids, which is 2–3 mg/g wet heart tissue (or 5–6% of total lipids), was not affected in rats fed diets rich in fats (Egwim and Kummerow, 1972) or vegetable oils including LEAR oil (Kramer, 1980) or HEAR oil (Houtsmuller et al., 1970; Krámer, 1973; Bhatia et al., 1979).

5. CHOLESTERYL ESTERS

The cholesteryl ester content in the cardiac lipids is much lower than the cholesterol content, i.e., ca. 1.5 mg/g dry weight (Okano et al., 1980), or 0.34 mg/g wet weight based on a water content of 77% in rat hearts (Seoane and Gorrill, 1975). Most vegetable oils have no effect on the cholesteryl ester content of the rat heat. Only when HEAR oil was fed did the cholesteryl ester content increase from 70 to 210 µg/g wet weight (Dow-Walsh et al., 1975; Kramer and Hulan, 1978a). These cholesteryl esters contain a high concentration of 20:1 and 22:1 (Kramer and Hulan, 1978a). With an 18:2 deficient diet Egwim and Kummerow (1972) reported an increase in fatty acids of the oleic acid family in the cholesteryl ester fraction.

C. Changes in Cardiac Phospholipids of Rats

1. QUANTITATIVE CHANGES

The phospholipid content of rat heart muscle is approximately 22 mg/g fresh weight (Simon and Rouser, 1969; Okano et al., 1980). The large variability in the phospholipid content of wet hearts reported by different investigators is probably due to incomplete phosphorus recovery. Nevertheless, the results in Table XII clearly show that different fats and oils in the diet do not influence total phospholipid concentration of the rat heart.

Analyses of the individual phospholipids is given in Table XIII. For reference Table XIII includes the cardiac phospholipid composition from normal rats fed stock diets (Simon and Rouser, 1969; Okano et al., 1980). It is evident from these results that the level and type of dietary fat does not influence the concentration of cardiac phosphatidylcholine (PC) and diphosphatidylglycerol (DPG). Several investigators have reported a decrease in cardiac phosphatidylethanolamine (PE) in rats fed diets rich in docosenoic acid.

The plasmalogen content is about 12 and 2% in cardiac PE and PC of the rat, respectively (Okano et al., 1980). Different dietary vegetable oils appear

TABLE XII

Phospholipid Content in Hearts of Rats Fed Different Vegetable Oils, Fat–Oil Mixtures, and Fish Oils[a]

Diet (% 22:1)	% fat in diet	Phospholipids (mg/g wet wt) (Time on diet, days)		Ref.[b]
		(8 days)		
Triolein—peanut oil	15	15.3		1
Triolein—linseed oil	15	13.2		
Trierucin—peanut oil (47.2%)	15	15.5		
Trierucin—linseed oil (48.6%)	15	13.4		
		(5 days)	(60 days)	
Coconut oil	15	11.3	13.7	2
Sunflower oil	15	11.9	13.7	
HEAR oil (47.4%)	15	11.3	14.1	
		(10 days)	(120 days)	
Peanut oil	15	23.0	21.4	3
Mustard oil (47.1%)	15	18.7	23.4	
		(4 days)		
Peanut oil	15	15.4		4
Trierucin—corn oil (45.3%)	15	17.7		
		(3 days)	(7 days)	
Corn oil	20	11.4	11.7	5
LEAR oil (trace)	20	11.0	12.1	
HEAR oil (34.2%)	20	9.7	12.1	
		(7 days)	(28 days)	
Lard/corn oil (3/1)	20	20.7[a]	21.6[a]	6
Fish oil (24.1%)	20	18.4	21.8	
HEAR oil (32.9%)	20	17.0	21.4	
		(7 days)		
Stock diet (chow)	—	23.3		7
Corn oil	20	23.5		
Mustard oil (31.6%)	20	22.2		

(Continued)

TABLE XII (CONTINUED)

Diet (% 22:1)	% fat in diet	Phospholipids (mg/g wet wt) (Time on diet, days)		Ref.[b]
		(3 days)	(7 days)	
Peanut oil	15	16.2	15.4	8
Fish oil (14.8%)	15	10.9	13.5	
Fish oil (29.7%)	15	15.4	15.0	
HEAR oil—peanut oil (14.2%)	15	18.1	14.4	
HEAR oil—peanut oil (27.2%)	15	12.4	13.2	
		(7 days)		
Stock diet (chow)	—	12.2[a]		9
Corn oil	20	12.5		
HEAR oil (22.3%)	20	11.6		

[a] Amount lipid phosphorous per heart was converted to mg phospholipid per heart by assuming an average phosphorus content of 4.18% in the phospholipids, i.e.,(4.06% PC + 4.30% PE/2 = 4.18%).

[b] References: 1, Rocquelin (1979); 2, Bellenand et al. (1980); 3, Bhatia et al. (1979); 4, Astorg and Compoint (1978); 5, Hung et al. (1977); 6, Beare-Rogers et al. (1971); 7, Myher et al. (1979); 8, Astorg and Cluzan (1977); 9, Dow-Walsh et al. (1975).

TABLE XIII

Distribution of Cardiac Phospholipids in Rats Fed Stock Diets or Diets Containing Different Vegetable Oils or Fish Oils

Diet (% 22:1)	% fat in diet	Time on diet (weeks)	Heart phospholipids[a] (mg/g wet weight)						Ref.[b]
			PC	PE	DPG	SP	PS+ PI	LPC	
Peanut oil	15	8.6	4.7	2.1	1.2	0.7			1
HEAR oil (50%)	15	8.6	3.8	1.5	2.5	0.4			
Olive oil	n.a.	n.a.	(38.3	36.9	13.6	4.8	4.2	1.7)[d]	2
HEAR oil (n.a.)[c]			(36.2	31.1*	13.0	13.5**	3.2	2.6)	
Peanut oil	15	20	8.7	6.7	3.9	0.6			3
LEAR oil (3.7%)	15	20	10.1	6.7	5.6*	0.7			
HEAR oil (46.2%)	15	20	8.7	5.8*	4.4	1.0**			
Control	—	12	5.8	4.9	0.9				4
Cod liver oil	10	12	4.2	3.2	1.0				
Corn oil	20	16	10.2	6.2	2.6	0.9	0.7	0.4	5
Olive oil	20	16	10.7	6.5	2.7	1.4	0.8	0.4	
Soybean oil	20	16	10.4	6.4	2.4	1.2	0.7	0.5	
LEAR oil (0.3%)	20	16	10.0	6.2	2.4	1.2	0.9	0.4	
HEAR oil (25.5%)	20	16	9.4	6.7	2.5	1.1			6

TABLE XIII *(CONTINUED)*

Diet (% 22:1)	% fat in diet	Time on diet (weeks)	Heart phospholipids[a] (mg/g wet weight)						Ref.[b]
			PC	PE	DPG	SP	PS+ PI	LPC	
Lard/corn oil (3/1)	20	26	(39.2	36.0	13.1	3.4	3.0	5.0)[d]	7
Poppyseed oil	20	26	(37.7	33.9	13.6	3.3	2.8	8.6)	
Sunflower oil	20	26	(37.8	34.4	13.1	3.4	3.0	8.1)	
LEAR oil (1.0%)	20	26	(37.5	33.3	13.0	3.2	2.7	10.1)	
Peanut oil	15	1.4	(44.9	45.4	19.5)[e]				8
HEAR oil (42.5%)	15	1.4	(47.2	42.6	18.7)				
Mustard oil (44.2%)	15	1.4	(52.3*	38.2	19.7)				
Peanut oil	15	21.4	(45.3	46.3	16.7)[e]				8
HEAR oil (42.5%)	15	21.4	(46.0	41.9	18.3)				
Mustard oil (44.2%)	15	21.4	(48.1	40.9*	18.1)				
Erucic acid[f]	5	2.9	8.8	8.4	3.2	0.9			9
Sunflower oil	15	12	15.5	8.6	3.3	0.5	0.4	0.1	10
LEAR oil (0.3%)	15	12	11.3	6.7	4.3	1.8	1.8	0.3	
HEAR oil (42.9%)	15	12	13.2	6.4	4.5	1.4	0.9	0.3	
Stock diet[g]	—	n.a.	(36.8	30.8	10.5	2.7	6.8 Trace)[d]		11
Stock diet[h]	—	n.a.	7.9	6.2	2.1	0.5	1.5 Trace		
Stock diet[g]	—	n.a.	(41.7	37.1	6.7	3.2	6.8	1.5)[d]	12
Stock diet[h]	—	n.a.	10.5	8.8	3.1	0.7	1.8	0.3	

[a] PC, phosphatidylcholine; PE, phosphatidylethanolamine; DPG, diphosphatidylglycerol; SP, sphingomyelin; PS, phosphatidylserine; PI, phosphatidylinositol; and LPC, lysophosphatidycholine. Values marked with an asterisk are significantly different from controls at the 5% (*) and 1% (**) level.

[b] References: 1, Joffrain et al. (1975); 2, Beare-Rogers (1975); 3, Dewailly et al. (1977); 4, Gudbjarnason et al. (1978); 5, Kramer (1980); 6, Kramer et al. (1979b); 7, Beare-Rogers et al. (1979); 8, Ray et al. (1979); 9, Yasuda et al. (1980); 10, Rocquelin et al. (1981); 11, Simon and Rouser (1969); 12, Okano et al. (1980).

[c] The erucic acid content in the HEAR oil was probably about 30%.

[d] Values are expressed as % phosphorous of total phospholipids.

[e] Values are expressed as μg lipid/mg protein.

[f] Values for the control diet were not given but said to be similar.

[g,h] Results from the same group of rats in which the content of the cardiac phospholipids is expressed as % phosphorous of total phospholipids[g] and as mg/g of wet weight[h].

to have no effect on the concentration of the plasmalogen lipids in either phospholipid (Kramer, 1980).

The isolation of cardiac sphingomyelin (SP) is difficult and cannot be achieved without the use of two-dimensional thin-layer chromatography (Rouser et al., 1970) or the use of selective enzymatic treatments to remove the common contaminants PC and lysophosphatidylcholine (LPC) (Myher et al., 1981). Therefore, with good techniques a low content of SP is found in cardiac tissue (0.5–1 mg/g wet weight or 3% based on phosphorus content). The high values that have been reported probably contain other phospholipids such as PC and LPC. An increase in cardiac SP concentration has been reported for rats fed HEAR oils (Beare-Rogers, 1975; Dewailley et al., 1977; Rocquelin et al., 1981). This observaion, however, was not confirmed by other investigators who found no change in cardiac SP content in rats fed either HEAR oil (Joffrain et al., 1975; Kramer et al., 1979b) or erucic acid (Yasuda et al., 1980). In view of the difficulty in accurately determining the SP content in lipid mixtures, more carefully controlled work needs to be done to determine if dietary docosenoic acids have any effect on this lipid fraction.

The concentration of LPC is usually very low in heart lipids. Any significant amount of this phospholipid class is probably only an indication of lipid autolysis during the homogenization process (Kramer and Hulan, 1978a; Kramer, 1980).

In general, the results show that the proportion of the different phospholipid classes remains remarkably constant whatever the dietary fat. This appears to agree with the observations by Simon and Rouser (1969) that even among different vertebrate species the phospholipid class distribution in the heart is essentially the same.

2. FATTY ACID CHANGES

Dietary fatty acids will influence the cardiac phospholipid composition of the rat but to a lesser extent than the cardiac neutral lipids (Carroll, 1965). Dietary fatty acids may be incorporated as saturated and monounsaturated fatty acids, or they may be incorporated and converted to more polyunsaturated fatty acids (PUFA), i.e., linoleic acid to 20:4 n-6, 22:4 n-6, and 22:5 n-6 and linolenic acid to 20:5 n-3, 22:5 n-3, and 22:6 n-3 (see Chapter 16). Dietary saturated and monounsaturated fatty acids are incorporated to a small extent into cardiac phospholipids (Carroll, 1965). On the other hand, dietary 18:2 n-6 and 18:3 n-3 cause a significant increase in the pentaenoic and hexaenoic acids which is greater in the heart than in any other organ of the rat (Rieckehoff et al., 1949; Widmer and Holman, 1950). With this background it may be useful to discuss changes in the composition of the different cardiac phospholipids with diet.

a. Phosphatidylcholine (PC) and Phosphatidylethanolamine (PE). The fatty acid profile of the two major cardiac phospholipids in the rat, PC and PE, are shown in Tables XIV and XV, respectively. These two phospholipids have a characteristic composition. PC contains higher levels of saturated fatty acids and arachidonic (20:4 *n*-6) acid, and lower levels of C_{22} PUFA and plasmalogen lipids than PE. The concentration of monounsaturated, linoleic (18:2 *n*-6) and linolenic (18:3 *n*-3) acids is similar in the two cardiac phospholipid classes.

This fatty acid profile is maintained over a wide range of different dietary fatty acids. When the dietary concentration of 18:3 is low, the relative concentration of saturated fatty acids, C_{22} PUFA, 20:4 and plasmalogenic lipids appears to be unaffected, except for differences in the level of monounsaturated fatty acids and 18:2 which occur in proportion to the dietary level. Dietary saturated fatty acids do not appear to alter the fatty acid profile to a great extent. The C_{22} PUFA from both the linoleic (*n*-6) and linolenic (*n*-3) acid family are found, the former is more abundant with diets rich in 18:2, while the latter is more abundant with diets rich in 18:1. However, the sum total of the C_{22} PUFAs is the same for each PC and PE on all these diets.

When diets are fed which contain 18:3, the fatty acid profile remains basically the same as with oils low in 18:3 except for the relative proportion of the C_{22} PUFAs. Dietary 18:3 markedly reduced the concentration of the C_{22} *n*-6 PUFA which were substituted by C_{22} *n*-3 PUFA. However, only when the dietary concentration of 18:3 is very high, as in linseed oil (53% 18:3), is the formation of C_{22} *n*-6 PUFA totally blocked and the level of 20:4 is also reduced. These results indicate that a dietary concentration of about 10% 18:3 does not inhibit desaturation and chain elongation of 18:2 *n*-6 to 20:4 *n*-6, *in vivo*, but only the subsequent conversion of 20:4 *n*-6 to 22:4 *n*-6 and 22:5 *n*-6. This is in contrast to *in vitro* experiments with rat liver preparations where 18:3 already inhibits the conversion of 18:2 *n*-6 to 20:2 *n*-6 (Mohrhauer *et al.*, 1967).

The inclusion of HEAR oil in the diet resulted in a marked alteration of the fatty acid profile of both PC (Table XIV) and PE (Table XV). The concentration of saturated and monounsaturated fatty acids increased, while the concentration of 20:4 and C_{22} PUFA decreased. It should be noted, however, that the increase in monoenoic acids in cardiac PC and PE from rats fed HEAR oil could not be accounted for by an accumulation of 20:1 and 22:1. The concentration of 20:1 plus 22:1 in PC and PE was only 5 and 6.5%, respectively.

Rats fed unhydrogenated fish oils which are rich in cetoleic (22:1 *n*-11) and gadoleic (20:1 *n*-11) acids and those belonging to the linolenic acid family (*n*-3) (Table I) also show a marked alteration of the fatty acid profile of cardiac PC and PE. The concentration of 20:4 is decreased and the level of C_{22}

TABLE XIV

Fatty Acid Profile of Cardiac Phosphatidylcholine in Rats Fed Diets Rich in Different Fats and Oils for at Least 4 Months

Type of dietary fat or oil	Di-methyl ace-tal	Satu-rates	Mono-un-satu-rates (22:1)	Polyunsaturates 18:2 n-6	18:3 n-3	20:4 n-6	22:4 22:5 n-6	22:5 22:6 n-3
Low 18:3, high 18:2								
Poppyseed oil[a]	—	43.7	4.9	9.2	—	31.9	3.2	4.2
Sunflower oil[a]	—	45.3	4.7	9.5	—	32.7	3.0	3.0
Corn oil[b]	0.8	40.5	7.2	13.3	—	31.0	4.5	1.8
Low 18:3, high 18:1								
Olive oil[b]	0.8	40.4	13.4	3.9	—	34.3	1.8	3.9
Peanut oil[c]	—	45.7	12.4	7.3	—	32.0	1.1	1.6
Low 18:3, high saturates								
Lard/corn oil (3/1)[a]	—	46.4	7.2	4.7	—	33.0	2.7	4.0
18:3, plus 18:2								
Soybean oil[b]	0.8	41.5	7.2	10.5	0.3	31.0	1.0	7.1
LEAR oil[b]	0.6	39.5	12.7 (0.1)	10.5	0.3	27.9	0.6	7.2
Linseed oil[d]	—	49.5	11.4	6.1	2.3	15.8	—	14.9
18:3, plus 22:1								
HEAR oil[e]	—	49.3	20.2 (1.8)	10.8	0.3	17.4	0.2	1.1
Fish oil[f]	—	44.3	12.3 (na)	12.9	—	13.4	—	16.1[g]
18:3, plus saturates or 18:1[h]								
LEAR	0.8	39.8	14.0	6.9	0.3	30.3	0.4	7.0
LEAR + cocoa butter	0.9	41.9	11.7	5.7	0.2	31.1	0.5	7.2
LEAR + triolein	1.1	39.3	15.0	6.0	0.3	29.5	0.6	7.7
Soybean	1.0	41.9	7.6	10.5	0.2	32.0	1.0	4.8
Soybean + cocoa butter	0.8	42.9	7.4	8.4	0.2	34.2	1.0	4.8
Soybean + triolein	1.0	39.3	12.8	8.8	0.1	29.6	0.5	7.5

[a] Beare-Rogers et al. (1979).
[b] Kramer (1980).
[c] Dewailly et al. (1977).
[d] Landes and Miller (1975).
[e] Kramer et al. (1979b).
[f] Gudbjarnason et al. (1978).
[g] Includes 20:5 n-3.
[h] Kramer et al. (1982).

TABLE XV

Fatty Acid Profile of Cardiac Phosphatidylethanolamine in Rats Fed Diets Rich in Different Fats and Oils for at Least 4 Months

Type of dietary fat or oil	Di-methyl ace-tal	Satu-rates	Mono-un-satu-rates (22:1)	18:2 n-6	18:3 n-3	20:4 n-6	22:4 22:5 n-6	22:5 22:6 n-3
Low 18:3, high 18:2								
Poppyseed oil[a]	—	37.4	4.4	8.5	—	18.9	13.5	11.1
Sunflower oil[a]	—	37.7	4.6	8.8	—	20.3	14.5	9.3
Corn oil[b]	6.8	31.4	7.1	9.4	—	20.4	16.5	7.4
Low 18:3, high 18:1								
Olive oil[b]	6.2	32.4	10.0	2.3	—	26.5	6.3	15.5
Peanut oil[c]	—	33.4	11.6	5.1	—	29.8	3.8	16.3
Low 18:3, high saturates								
Lard/corn oil (3/1)[a]	—	39.0	6.3	3.6	—	19.9	13.2	11.5
18:3, plus 18:2								
Soybean oil[b]	7.8	32.3	5.9	6.7	0.3	17.9	2.6	25.9
LEAR oil[b]	5.4	32.1	9.8 (0.1)	6.0	0.4	19.7	1.4	25.0
Linseed oil[d]	1.1	30.2	6.3	21.5	2.9	11.0	—	26.9[g]
18:3, plus 22:1								
HEAR oil[e]	—	43.8	28.0 (2.3)	7.9	0.5	13.0	0.7	5.5
Fish oil[f]	—	33.2	9.3 (na)	9.1	—	6.9	—	36.6[g]
18:3, plus saturates or 18:1[h]								
LEAR	5.1	32.8	12.0	4.2	0.3	19.4	1.2	24.4
LEAR + cocoa butter	7.1	34.0	10.4	3.6	0.3	18.8	1.2	24.3
LEAR + triolein	4.3	31.9	14.4	3.3	0.2	18.7	1.4	25.1
Soybean	7.1	33.7	7.4	8.5	0.3	18.9	3.3	20.5
Soybean + cocoa butter	8.9	34.9	7.3	6.2	0.2	19.9	3.6	19.4
Soybean + triolein	5.0	32.4	11.7	5.2	0.3	18.8	1.4	25.3

[a] Beare-Rogers et al. (1979).
[b] Kramer (1980).
[c] Dewailly et al. (1977).
[d] Landes and Miller (1975).
[e] Kramer et al. (1979b).
[f] Gudbjarnason et al. (1978).
[g] Includes 20:5 n-3.
[h] Kramer et al. (1982).

PUFAs is increased. The C_{22} PUFAs were only those derived from the linolenic acid family (n-3). The influence of dietary saturated fatty acids on the
relative concentration of saturates in these two major phospholipids was
generally not apparent from the results from various laboratories. In a recent
study, however, in which the addition of saturated fatty acids (in the form of
cocoa butter) was controlled, it became evident that the addition of saturated fatty acids in the diet significantly increased the level of saturates in
both PC (Table XIX) and PE (Table XV) (Kramer et al., 1982). For detailed fatty
acid composition of the dietary oils see Table XXX in Chapter 17.

b. Diphosphatidylglycerol (DPG). DPG is found mostly in mitochondria, particularly the inner membrane (White, 1973). The heart contains
about 10% DPG in the total cardiac phospholipids because of the large
number of mitochondria. Linoleic acid (18:2 n-6) is the major fatty acid in
DPG, comprising about 80% of this phospholipid class (White, 1973). The

TABLE XVI

Fatty Acid Profile of Cardiac Diphosphatidylglycerol in Rats Fed Diets Rich in Different Fats and Oils for at Least 4 Months

Type of dietary fat or oil	Satu- rates	Mono- unsatu- rates (22:1)	Polyunsaturates				
			18:2 n-6	18:3 n-3	20:4 n-6	22:4 22:5 n-6	22:5 22:6 n-3
Low 18:3, high 18:2							
Sunflower oil[a]	9.1	8.4	79.6	—	—	—	—
Corn oil[b]	3.9	4.8	85.0	—	1.9	1.2	0.5
Low 18:3, high 18:1							
Olive oil[b]	4.9	22.9	51.7	—	7.5	2.7	5.4
Peanut oil[c]	8.5	11.6	73.2	—	3.6	0.3	2.8
18:3, plus 18:2							
Soybean oil[b]	2.7	4.4	85.8	1.3	1.8	0.2	1.8
LEAR oil[b]	2.1	8.3 (trace)	81.7	1.6	1.6	0.1	1.8
18:3, plus 22:1							
HEAR oil[d]	7.5	20.1 (3.0)	67.5	1.1	1.3	0.4	0.6
Fish oil[e]	11.0	12.2 (n.a.)	67.5	—	1.1	—	3.4

[a] Rocquelin et al. (1981).
[b] Kramer (1980).
[c] Dewailly et al. (1977).
[d] Kramer et al. (1979b).
[e] Gudbjarnason et al. (1978).

fatty acid composition of cardiac DPG is altered significantly by feeding a diet high in monounsaturated fatty acids and low in 18:2, e.g., olive oil (Table XVI). A dietary oil which contains 20–30% 18:2 in the presence of high monounsaturated fatty acids (e.g., HEAR oil, peanut oil, and LEAR oil) also causes a decrease in 18:2 of cardiac DPG, but the reduction is less than that with olive oil. The lower concentration of 18:2 in cardiac DPG is accompanied by a higher concentration of 20:4 n-6, 22:5 n-6, and 22:6 n-3 in an apparent attempt by the heart tissue to maintain an average number of two double bonds per fatty acid in DPG. Erucic acid is incorporated into cardiac DPG, the relative concentration is greater than that found in either PC and PE (Blomstrand and Svensson, 1974).

c. *Sphingomyelin (SP).* The fatty acid composition of SP consists almost entirely of long chain saturated and monounsaturated fatty acids from C_{14} to C_{26} including several odd chain fatty acids (White, 1973). When rats are fed dietary oils high in monounsaturated fatty acids there is an increase in nervonic acid (24:1 n-9) in cardiac SP (Table XVII). Nervonic acid is derived by

TABLE XVII

Fatty Acid Composition of Cardiac Sphingomyelin in Rats Fed Diets Rich in Oils for at Least 4 Months

Fatty acid	Sunflower oil[a]	Corn oil[b]	Olive oil[b]	Peanut oil[c]	Soybean oil[b]	LEAR oil[b]	HEAR oil[d]
16:0	14	13.4	13.1	13.7	12.5	11.4	17.2
18:0	24	11.5	12.4	12.6	12.6	12.5	18.5
20:0	10	24.9	19.7	17.3	15.9	21.0	10.8
21:0	—	1.3	1.6	—	2.0	0.9	0.9
22:0	30	16.1	15.7	38.6	26.4	21.6	14.7
23:0	—	4.5	4.6	—	5.2	2.5	1.0
24:0	14	13.9	10.7	10.3	12.1	8.6	3.9
Total saturates	92	88.8	81.2	92.5	89.7	82.8	75.8
18:1	4	1.6	2.9	3.1	1.4	2.1	—
20:1	—	—	—	—	—	—	—
22:1	—	0.2	0.5	—	0.4	0.5	5.5
24:1	—	5.9	11.2	3.5	5.8	12.3	12.6
Total monoenes	4	7.7	14.6	7.5	7.6	14.9	18.1

[a] Rocquelin et al. (1981).
[b] Kramer (1980).
[c] Dewailly et al. (1977).
[d] Kramer et al. (1979b).

chain elongation from members of the oleic acid family (n-9), i.e., oleic (18:1 n-9), gandoic (20:1 n-9), and erucic acids (22:1 n-9) (Fulco and Mead, 1961; Kishimoto and Radin, 1963; Bourre et al., 1976). It is of interest to note that none of the intermediates in the chain elongation process builds up to any great extent except erucic acid when it is fed at a high level in the diet. In fact, the highest concentration of erucic acid in the cardiac phospholipids of rats fed HEAR oil is found in the SP fraction (Dewailly et al., 1977; Kramer et al., 1979b).

D. Changes in Cardiac Mitochondrial Lipids of Rats

Mitochondrial lipids are more than 90% phospholipids with minor amounts of neutral lipids (~6%) and cholesterol (~1%) (Rouser et al., 1968). The major phospholipids are PC, PE, and DPG in a ratio of ca. 2:2:1. DPG occurs almost exclusively in mitochondria, particularly in the inner mitochondrial membrane (Rouser et al., 1968). In general the relative distribution of mitochondrial phospholipids shows little variability between different organs, or species (Rouser et al., 1968), although the concentration of lipids in mitochondria may differ between species. For example, human liver mitochondria contain twice the lipid of rat liver mitochondria, but the proportion of the different lipid classes is the same (Benga et al., 1978).

Some studies in which HEAR oils were tested in rats suggest that the proportion of the mitochondrial phospholipids may be altered when feeding this oil. There was a reported tendency for PC to increase and PE to decrease while the relative concentration of DPG remained unchanged (Blomstrand and Svensson, 1974; Sen and Sen Gupta, 1980; Innis and Clandinin, 1981). These reported changes in cardiac mitochondrial phospholipids due to dietary oils should be confirmed in view of the fact that previous reports which showed that the relative proportion of phospholipids varied in accordance with diet could, in fact, be traced to insufficient purification of the mitochondrial lipid fraction (Kuksis, 1978) or to improper procedures of lipid handling (Rouser et al., 1968). As Rouser et al. (1968) pointed out, PE may easily be decreased simply by evaporating the lipid extract to constant weight.

The fatty acid composition of each of the mitochondrial phospholipids shows a characteristic profile which differs between species. For instance, liver mitochondria in man, pig, and rat differ in the fatty acid profile of PC, PE and DPG (Table XVIII). In general the rat contains the highest concentration of 20:4 and C_{22} PUFA, while in man the level of 18:2 predominates in all three lipid classes. The gastric mitochondrial lipids of pigs, rabbits, and

TABLE XVIII

Fatty Acid Profile of Mitochondrial Phosphatidylcholine (PC), Phosphatidylethanolamine (PE), and Diphosphatidylglycerol (DPG) from Different Species, Different Tissues, and from Rats Fed Different Diets

Description	Lipid class	Description	Saturates	Mono-unsaturates	18:2 n-6	20:4 n-6	C_{22} PUFA
Liver	PC	Man[a]	36.9	24.1	34.0	5.5	—
		Pig[b]	48.2	10.0	34.1	3.1	0.4
		Rat[a]	34.8	25.7	19.8	17.5	2.9
	PE	Man[a]	39.0	20.9	35.0	3.6	—
		Pig[b]	48.1	4.1	20.7	20.5	3.6
		Rat[a]	43.1	18.0	16.9	17.0	5.2
	DPG	Man[a]	18.9	11.1	70.2	—	—
		Pig[b]	6.5	8.2	77.1	0.8	0.1
		Rat[a]	11.0	8.5	77.3	3.2	1.1
Gastric[c]	PC	Pig	41.2	24.3	22.8	6.3	—
		Rabbit	37.2	24.5	14.1	5.2	6.0
		Frog	40.2	27.3	20.1	5.8	—
	PE	Pig	23.5	28.9	24.1	22.3	—
		Rabbit	28.1	21.7	24.6	8.1	6.0
		Frog	27.6	22.4	19.7	21.9	—
	DPG	Pig	27.0	23.6	39.1	7.2	—
		Rabbit	32.5	17.3	26.1	2.9	10.0
		Frog	20.1	28.4	41.1	trace	—
Rat Liver	PC	Stock diet[d]	49.0	16.9	12.4	17.7	2.9
		Safflower oil[e]	47.1	11.0	11.4	21.6	4.8
		HCO[e,f]	48.2	11.0	7.7	7.6[g]	2.4
	PE	Stock diet[d]	54.2	15.2	5.4	22.0	3.2
		Safflower oil[e]	44.7	6.3	11.6	20.5	11.9
		HCO[e,f]	47.6	15.1	7.5	10.1[h]	2.7
	DPG	Stock diet[d]	10.8	27.5	58.8	1.8	—
		No fat[i]	8.1	59.6	16.3	2.8	2.9
		5% corn oil[i]	4.7	18.8	66.7	1.4	0.4
		25% HCO[i]	6.8	60.1	16.5	2.0	2.7
		HCO + corn oil[i,j]	4.5	12.2	72.9	1.0	0.5

[a] Benga et al. (1978).
[b] Parkes and Thompson (1970).
[c] Sen and Ray (1980).
[d] Colbeau et al. (1971).
[e] Divakaran and Venkataraman (1977).
[f] HCO, hydrogenated coconut oil.
[g] In addition 20.4% 20:3 n-9.
[h] In addition 16.9% 20:3 n-9.
[i] Williams et al. (1972).
[j] 20% HCO + 5% corn oil.

TABLE XIX

Fatty Acid Profile of Mitochondrial Phosphatidylcholine (PC), Phosphatidylethanolamine (PE), and Diphosphatidylglycerol (DPG) in Hearts of Rats Fed Different Vegetable Oils

Lipid class	Diet (% 22:1)	Satu-rates	Mono-unsatu-rates (22:1)	18:2 n-6	18:3 n-3	20:4 n-6	22:4 22:5 n-6	22:5 22:6 n-3
PC	Stock diet[a]	46.7	10.2	14.0	—	21.5	—	3.8
	Peanut oil[b]	48.2	7.8	8.6	—	32.7	0.8	1.8
	LEAR oil (3.7%)[b]	47.7	12.3 (0.1)	11.4	0.3	23.9	1.9	2.3
	HEAR oil (7%)[c]	43.1	10.3 (–)	16.7	—	28.1	—	1.4
	HEAR oil (13%)[c]	42.7	11.1 (0.8)	17.2	0.1	28.4	—	0.2
	HEAR oil (46%)[b]	47.7	12.9 (0.5)	16.0	0.3	20.8	0.7	1.1
PE	Stock diet[a]	46.8	8.6	8.9	—	24.1	0.8	5.5
	Peanut oil[b]	42.2	11.5	6.7	—	23.8	2.7	13.1
	LEAR oil (3.7%)[b]	45.0	17.7 (0.2)	6.0	0.5	15.1	3.3	11.0
	HEAR oil (7%)[c]	43.4	12.8 (0.5)	14.0	—	22.8	—	6.4
	HEAR oil (13%)[c]	38.8	12.9 (1.0)	9.6	—	23.4	—	14.8
	HEAR oil (46%)[b]	41.8	21.4 (0.7)	8.9	0.1	17.2	2.5	7.1
DPG	Stock diet[a]	13.0	12.2	65.6	—	0.5	—	—
	Peanut oil[b]	6.8	13.1	77.3	—	2.8	Trace	Trace
	LEAR oil (3.7%)[b]	3.2	12.2 (0.1)	81.1	1.2	1.0	0.2	0.7
	HEAR oil (7%)[c]	1.7	8.0 (0.6)	87.4	0.2	1.1	—	0.9
	HEAR oil (13%)[c]	3.2	8.8 (1.3)	85.2	0.3	1.3	—	0.5
	HEAR oil (46%)[b]	4.4	16.2 (5.6)	76.2	1.8	1.1	0.1	0.2

[a] Gloster and Harris (1970).
[b] Dewailly et al. (1978).
[c] Blomstrand and Svensson (1974).

frogs are compared in Table XVIII. The fatty acid profile again indicates differences between species in all the three major phospholipid classes. Not only is the fatty acid pofile of mitochondrial phospholipids different between species from the same organ, but also between organs from the same species (Table XVIII).

The fatty acid profile of mitochondrial phospholipids is also altered by dietary fatty acids (Table XVIII). A vegetable oil rich in 18:2, such as corn or safflower oil, will increase the 18:2 and decrease the saturated and monounsaturated fatty acid content. On the other hand, an essential fatty acid deficient diet, i.e., one that contains hydrogenated coconut oil, results usually in a higher content of monounsaturated fatty acids and a lower content of 18:2 and 20:4.

The changes in heart mitochondrial phospholipids when rats are fed peanut oil, LEAR oil, or HEAR oils are shown in Table XIX. The pattern of fatty acids in each of the three phospholipids is very similar to that observed for the total cardiac phospholipids shown in Tables XIV to XVI as pointed out by Blomstrand and Svensson (1974) who found no significant difference in the relative distribution of phospholipids between mitochondrial and total heart phospholipids. Some of the phospholipid fatty acid patterns that have been reported are at variance with the generally accepted values. This could be the result of insufficient purification of mitochondrial lipids or improper procedures of lipid extraction (Rouser et al., 1968; Kuksis, 1978).

IV. CHANGES IN CARDIAC LIPIDS OF PIGS AND MONKEYS FED DIFFERENT OILS AND FATS AND HOW THESE CHANGES COMPARE TO THOSE OBSERVED IN RATS

A. Total Cardiac Lipids

It is apparent from the results presented in Table XX that the amount of total cardiac fat in pigs is the same irrespective of the level or type of fat in the diet. For example, a basal diet with no added fat results in a similar cardiac lipid content as a diet that contains corn, soybean, LEAR, HEAR, or fish oils. The level of fat in the heart may be different between breeds of pigs, but within an experiment the level of cardiac fat remains constant with diet and time on experiment (Table XX). In this respect, the pig differs from the rat which shows a marked increase in cardiac TG with dietary 22:1 (Fig. 1). The docosenoic acid content in the cardiac lipids of the pig remains relatively low compared to the cardiac lipid of the rat fed a diet containing a similar content of docosenoic acid (see Fig. 2, Chapter 14).

The cardiac lipid content of monkey (Macaca fascicularis) appears to be the same irrespective of the kind of dietary fat or oil, even with diets rich in docosenoic acid (Table XXI). There is one exception in a 6 month study reported by Ackman (1980) in which the lard/corn oil control diet gave an unusually low cardiac lipid value. However, this result needs to be confirmed, since in an earlier study by the same author there was no difference in the cardiac lipid content (Ackman and Loew, 1977). Except for this one report, it appears that the cardiac lipids in monkeys respond like those of pigs (Table XX), and respond differently from those of rats (Table III).

The docosenoic acid content in the cardiac lipids of monkeys fed this acid reached about 75% of the dietary concentration of 22:1, and remained at this level throughout a 2.5 year feeding trial (Table XXI). In this respect, all three species showed a different response to similar concentrations of dietary 22:1. The pig accumulates a low content of 22:1 in the cardiac lipids which is ca. 20% of the dietary concentration of 22:1. This level is attained within the first week on diet and remains unchanged for up to 1 year (Table

TABLE XX

Lipid and Docosenoic Acid Content of Hearts of Pigs Fed Experimental Diets for a Short Term (1–2 Weeks) or for a Longer Period of Time (8–52 Weeks)

Diets	Dietary fat % added	Dietary fat % 22:1	Breed of pig[a]	Lipid weight (mg/g) 1–2 weeks	Lipid weight (mg/g) >8 weeks	Heart lipids (% 22:1) 1–2 weeks	Heart lipids (% 22:1) >8 weeks	Ref.[b]
No added fat	0	< 0.1	DL	43	56[c]	0.2	< 0.1	1
Soybean oil	8	< 0.1		51	66	0.1	< 0.1	
LEAR oil	8	1.7		37	75	0.5	0.3	
HEAR oil	8	7.5		44	53	0.8	0.7	
HEAR oil	8	14.8		48	56	1.0	1.3	
HEAR oil	8	22.4		38	55	1.6	2.1	
Initial control	0	0	Y	16	—	Trace	—	2
Corn oil	20	0		17	19[d]	Trace	Trace	
HEAR oil	20	22.3		16	19	2.6	2.2	
Lard	21	0	NL	25		1.0[e]		3
HEAR oil	21	11.6		25		1.7		
Fish oil	21	14.2		26		2.7		
HEAR oil	21	36.6		28		2.7		
Lard	16	0	NL	48[f]	48[f,g]	0	0	4
Fish oil	16	18.0				3.8[e]	1.7[e]	
HEAR oil	16	42.2				5.4	6.4	

[a] DL, Deutsche Landrasse; Y, Yorkshire; NL, Norwegian Landrace.

[b] Reference: 1, Seher et al. (1979); 2, Kramer et al. (1975a); 3, Opstvedt et al. (1979); 4, Svaar et al. (1980).

[c] Pigs fed for 17 weeks.

[d] Pigs fed for 8 weeks.

[e] The 22:1 content of heart triglycerides.

[f] Average cardiac lipid content from all diets. No significant differences between diets or time.

[g] Pigs fed for 52 weeks.

TABLE XXI

Cardiac Lipids of Monkeys (*Macaca fascicularis*) Fed Different Fats and Oils for up to 2-1/2 Years

Diets	Dietary fat		Time on diet (months)	Heart lipids[a]						Ref[b]
				Total		TG		PL		
	% fat	% 22:1		mg/g	% 22:1	mg/g	% 22:1	mg/g	% 22:1	
Lard/corn oil (3/1)	25	<0.1	6	69	0.3	n.a.	0.6	—	—	1
HEAR oil	25	24.6	6	97	15.5		23.7			
H₂ fish oil[c]	25	22.9	6	86	10.8		13.6			
Lard/corn oil (3/1)	25	<0.1	6	11	0.1	2	1.0	—	—	2
	25	ʺ	12	25	0.1	13	3.3	10	3.0	
	25	ʺ	18	43	0	21	0.2	15	1.8	
	25	ʺ	24	38	0	22	0.3	14	1.8	
	25	ʺ	30	38	0	23	0.4	15	1.7	
H₂ fish oil[c]	25	20.8	6	53	14.8	28	14.1	—	—	2
	25	ʺ	12	32	14.2	17	14.5	13	—	
	25	ʺ	18	54	14.4	36	15.2	14	10.7	
	25	ʺ	24	51	16.6	34	15.9	17	11.2	
	25	ʺ	30	43	16.4	21	16.4	13	8.3	
Soybean oil	20	0.1	0.3	68	0.1					3
Soybean oil	20	0.1	6	55	0.2					
LEAR oil	20	0.2	0.3	63	0.2					
LEAR oil	20	0.2	0.5	85	0.3					
LEAR oil	20	0.2	2	71	0.2					
LEAR oil	20	0.2	4	65	0.5					
LEAR oil	20	0.2	6	49	0.5					

[a] TG, triglyceride; PL, phospholipid.

[b] References: 1, Ackman and Loew (1977); 2, Ackman (1980); 3, Kramer et al. (1978).

[c] H₂, partially hydrogenated.

XX). The rat, on the other hand, shows an initial accumulation of 22:1 to about 100% of the dietary level which is followed by a decline to a concentration of ca 20% of dietary levels. The concentration of 22:1 in the cardiac lipids of monkeys fed diets rich in 22:1 plateaus just like in the pig, but at a considerably higher level (Table XXI). These results are summarized in Fig. 2 of Chapter 14.

B. Cardiac Lipid Classes

The cardiac triglycerides (Opstvedt et al., 1979) and phospholipid classes (Kramer and Hulan, 1977) of pigs do not appear to be affected by dietary fats even with oils which contain 22:1. The same is true for monkeys (Table XXI). This clearly differentiates the pig and the monkey from the rat, for only in the rat do the cardiac triglycerides increase in response to dietary 22:1.

These three test animals also differ in their cardiac lipid composition. For example, in the test animals fed control fats or oils, the cardiac triglycerides comprise about 10% of the total cardiac lipids of the pig (Opstvedt et al., 1979), 10–20% of the rat (Astorg and Cluzan, 1977; Bellenand et al., 1980; Hung and Holub, 1977; Kramer and Hulan, 1978b; Myher et al., 1979; Rocquelin, 1979), and 50% of the monkey (Ackman, 1980). The proportion of the cardiac phospholipids is also different between species. For example, in the pig the concentration of SP is greater than PE, whereas in the rat the inverse is found (Table XXII).

The concentration of long chain PUFA in cardiac tissue also differs among rats, pigs ,and monkeys (Table XXIII). The rat shows the highest levels of C_{22} PUFA whereas the pig contains only trace amounts of these PUFA. The level of 20:4 n-6 in cardiac PC and PE is also higher in the rat than in the pig.

TABLE XXII

Comparison of the Relative Concentration (%) of Cardiac Phospholipids of Rats and Pigs Fed Diets Containing 20% Vegetable Oils

Lipid class	Rat[a]		Pig[b]	
	Corn oil	LEAR oil	Corn oil	HEAR oil
Phosphatidylcholine	48.5	47.3	44.8	39.3
Phosphatidylethanolamine	29.5	29.5	10.7	15.9
Diphosphatidylglycerol	12.5	11.5	8.8	10.3
Sphingomyelin	4.5	5.7	24.3	24.8
Phosphatidylserine and phosphatidylinositol	3.1	4.1	5.1	6.3

[a] Kramer (1980).
[b] Kramer and Hulan (1977).

TABLE XXIII

Concentration of Long Chain Polyunsaturated Fatty Acids (PUFA) in the Total Lipids and Phosphatidylcholine (PC) and Phosphatidylethanolamine (PE) of Rat, Pig, and Monkey Hearts

Species	Diet	20:4 n-6			C_{22} PUFA		
		Total	PC	PE	Total	PC	PE
Rat[a]	Corn oil	9.1	31.0	20.4	5.1	6.3	23.9
Monkey[b]	Soybean oil	7.2	—	—	2.4	—	—
Pig[c]	Corn oil	10.5	4.5	13.3	<1	<1	<1
Rat[a]	LEAR oil	10.3	27.9	19.7	4.8	7.8	26.4
Monkey[b]	LEAR oil	7.7	—	—	3.1	—	—
Pig[c]	HEAR oil	7.2	3.9	7.2	<1	<1	<1

[a] Kramer et al. (1979b); and Kramer (1980).
[b] Kramer et al. (1978); composition of PC and PE not available.
[c] Kramer et al. (1975a); and Kramer and Hulan (1977).

These long chain PUFAs occur almost exclusively in position 2 of cardiac PC and PE in both the rat (Dewailly et al., 1977) and the pig (Kramer and Hulan, 1977). In addition, position 2 of cardiac PC and PE of the pig contains most of the 18:1 and 18:2 while position 1 is highly enriched in 18:0 and 16:0 (Kramer and Hulan, 1977). In the rat, position 1 is also high in saturated fatty acids (16:0 and 18:0) but 18:1 and 18:2 are found in both positions of cardiac PC and PE (Dewailly et al., 1977). In general, the docosenoic acid is found preferentially in position 2 of PC and PE of both rat (Dewailly et al., 1977) and pig heart lipids (Kramer and Hulan, 1977).

TABLE XXIV

Fatty Acid Composition of Cardiac Sphingomyelin in Pigs Fed Corn Oil or HEAR Oil[a,b]

Fatty acids	Corn oil	HEAR oil[a]
16:0	16.5	18.5
18:0	16.9	13.8
20:0	21.5	20.5
22:0	12.1	10.1
22:1	—	2.0**
23:0	3.5	1.0*
24:0	10.7	11.8
24:1	4.5	9.0**

[a] From Kramer and Hulan (1977).
[b] Significantly different from corn oil fed pigs at $P < 0.05$ (*) and $P < 0.01$ (**).

C. Fatty Acid Compositions

The fatty acid composition of cardiac SP is similar in rats (Table XVII) and pigs (Table XXIV) fed corn oil. Diet induced changes in cardiac SP however were more pronounced in the rat than in the pig as evidenced by changes in the concentration of 20:0, 22:1, 24:0, and 24:1 (compare Tables XVII and XXIV). The greatest changes in cardiac SP of the rat were due to dietary monounsaturated fatty acids (18:1, 20:1, and 22:1), which resulted in a decrease of 24:0 and an increase of 22:1 and 24:1 (Table XVII). On the other hand, in pig heart SP, 24:0 remained unaltered, 22:1 was incorporated to a lesser extent, and 24:1 increased when the same HEAR oil was fed.

V. CAN THE MYOCARDIAL DISORDERS ASCRIBED TO THE FEEDING OF RAPESEED OIL BE CORRELATED TO CARDIAC LIPID CHANGES?

Attempts have been made repeatedly to detect specific lipid changes in tissues which show evidence of functional disorders. Several examples may be cited. A high level of 20:3 n-9 in the serum phospholipids (Holman, 1977) and reduced levels, or reduced production, of prostaglandins (Galli, 1980) are associated with dermal lesions, a sign of essential fatty acid deficiency. A high content of blood lipids, in particular, cholesterol, may be an indicator of the development of artherosclerotic vascular lesions (Vergoesen and Gottenbos, 1975). An accumulation of various sphingolipids is associated with some forms of mental retardation which appear to be due to single enzymatic defects, e.g., in Niemann–Pick disease, sphingomyelin builds up because of a deficiency of sphingomyelinase (Thompson, 1980). In the area of rapeseed oil research three problems have been ascribed to the consumption of this oil: myocardial lipidosis, impaired oxidative phosphorylation, and myocardial necrosis. Whenever studies are undertaken to correlate these disorders to lipid changes, one should consider the following criteria to avoid reaching erroneous conclusions: (1) there must be convincing evidence that a malfunction exists; (2) the correlation must be based on reliable lipid analyses; and (3) critical experiments must be designed to test the hypothesis in question.

A. Myocardial Lipidosis

Excellent progress has been made with the problem of myocardial lipidosis. The problem was well documented. The fatty acid and triglyceride analyses were fairly simple. Finally, a variety of dietary fats were chosen with different concentrations of the causative agent (docosenoic acid) in the presence of different dietary fatty acids. For greater detail see Chapter 11, Section IV.

B. Oxidative Phosphorylation

It has been demonstrated convincingly that significant changes in dietary fatty acids, such as may be found in EFA deficiency, do not cause an impairment of function of mitochondrial metabolic activity (Ito and Johnson, 1964; Stancliff et al., 1969). In a series of carefully controlled experiments Williams et al. (1972) showed that liver mitochondria from rats fed diets deficient in essential fatty acids were normal with respect to the efficiency of oxidative phosphorylation, respiratory control, and passive and energy dependent ion and metabolite transport. The only changes observed for EFA depleted mitochondria was an increased oscillation period for swelling and decreased freedom of motion of added spin labels. What has been suggested is that EFA depleted mitochondria may be more fragile during and after isolation (Smith and DeLuca, 1964; Ito and Johnson, 1964; Houtsmuller et al., 1969). This may be the result of an altered fatty acid composition or from increased phospholipase A_2 activity (Waite and van Golde, 1968), but in any case the physiological significance of this, if any, is as yet unknown.

A state of EFA deficiency probably causes the most severe change in tissue lipids that can be induced in laboratory animals by dietary means. Although there are reports to the contrary (Divakaran and Venkataraman, 1977; Haeffner and Privett, 1975), it seems fairly certain that mitochondria from EFA deficient rats are fully coupled and have normal respiratory control.

In view of the fact that even in EFA deficiency mitochondria function normally, it seems improbable that the less severe lipid alterations associated with the feeding of HEAR oils could functionally impair mitochondria. Nevertheless, reports continue to appear which suggest that heart mitochondria can be uncoupled by dietary erucic acid (Clandinin, 1978, 1979; Hsu and Kummerow, 1977; Innis and Clandinin, 1981; Sen and Sen Gupta, 1980). Evidence has been presented that shows if heart mitochondria are carefully isolated from rats fed diets with erucic acid (as HEAR oil) these are fully active, tightly coupled, and with no loss of respiratory control (Dow-Walsh et al., 1975; Cheng and Pande, 1975; Beare-Rogers and Gordon, 1976). In agreement with the results of liver mitochondria from EFA deficient rats, Dow-Walsh et al. (1975) showed that heart mitochondria from HEAR oil fed rats showed evidence of increased fragility of mitochondrial membranes.

It would appear that criterion number 1, i.e., "convincing evidence that a malfunction exists," in this case has not been met and that any attempts to correlate tissue lipid changes to impaired oxidative phosphorylation are probably futile.

C. Myocardial Necrosis

The problem of myocardial necrosis in rats has received the greatest attention in attempts to correlate myocardial necrosis to myocardial lipid changes. Myocardial necrosis has been observed in male rats irrespective of the dietary oil or fat, provided a sufficient number of heart sections (Vles et al., 1976 and 1978) or a sufficient number of rats (Hulan et al., 1977d) are examined. Dietary fats high in saturated fatty acids (Hulan et al., 1976; Kramer et al., 1981; Farnworth et al., 1982) and linoleic acid (Vles et al., 1978) are associated with a low incidence of heart lesions, whereas diets which contain linolenic acid (McCutcheon et al., 1976; Hulan et al., 1977a; Vles et al., 1978) and erucic acid (Abdellatif and Vles, 1973) are associated with a high incidence of heart lesions. A statistical analysis of much of the heart lesion data clearly showed this correlation of dietary fatty acids to heart lesions (see Chapter 17, Table XXV).

1. CORRELATION OF CARDIAC SPHINGOMYELIN AND FREE FATTY
    ACIDS TO MYOCARDIAL NECROSIS

Among the cardiac lipid classes which have been proposed to account for myocardial necrosis are free fatty acids (Houtsmuller et al., 1970; Vles, 1975) and sphingomyelin (Beare-Rogers, 1975). This increase in cardiac free fatty acids (Table X) has been attributed to improper extraction (Section III,B,2) while the large increase reported in cardiac sphingomyelin could not be reproduced and was probably due to improper isolation (Section III,C,1.). Therefore, it seems highly unlikely that a change in cardiac lipid classes in the rat will account for myocardial necrosis in this species, since these remain essentially unchanged.

2. CORRELATION OF 22:6 n-3 LEVELS TO MYOCARDIAL NECROSIS

The fatty acid changes in the rat heart are more numerous and may provide an answer to the cause of myocardial necrosis in male rats. For instance, the level of 22:6 n-3 has been suggested as an indicator of myocardial necrosis. This would certainly be supported by the fact that the 22:6 n-3 content in the rat heart is greater than that in the pig or monkey heart (Table XXIII), and myocardial necrosis appears to occur most frequently in the rat. However, this argument also has its weaknesses. Male and female rats (Hulan et al., 1977c) and male rats of the Sprague–Dawley and Chester Beatty (Hooded) strain (Kramer et al., 1979b) have a similar content of 22:6 n-3 in the cardiac lipids even though they do not show the same incidence of myocardial lesions.

The C_{22} n-3 PUFAs, which include 22:6 n-3, and 22:5 n-3, arise from linolenic acid (18:3 n-3) by chain elongation and desaturation. It is of interest that 18:3 n-3 when fed in moderate concentrations (~10%), as found in soybean oil or LEAR oil, does not interfere with arachidonic acid (20:4 n-6) biosynthesis but will inhibit the subsequent chain elongation and desaturation of 20:4 n-6 to 22:4 n-6 and 22:5 n-6 (Kramer, 1980). What is important is that when the 18:3 content in the dietary fat is kept constant the 22:6 n-3 concentration remains unaltered if the content of saturated fatty acids is raised (Bellenand et al., 1980; Kramer et al., 1982) and is only slightly reduced when the 18:2 n-6 content is increased (Beare-Rogers et al., 1979; Bellenand et al., 1980). However, the increase in saturated fatty acids (Farnworth et al., 1982; Kramer et al., 1982) and 18:2 (Beare-Rogers et al., 1979) decreased significantly the incidence of heart lesions in male rats. These results suggest that the 22:6 n-3 content in the heart lipids may not be a reliable indicator of heart lesions. Nevertheless, 22:6 n-3 is probably involved in heart lesions, since no dietary oils have been observed to give a high incidence of heart lesions in male rats when the content of 22:6 n-3 is low in heart lipids.

The possibility that 22:6 n-3 may contribute to myocardial necrosis has received support from the studies by Gudbjarnason and co-workers. For example, 22:6 n-3 increases in cardiac PE with the age of the rat from 20% at 2 months of age to 29% at 18 months of age (Gudbjarnason, 1980), and so does the incidence of heart lesions (Kaunitz and Johnson, 1973). Furthermore, rats that were stressed with norepinephrine (Gudbjarnason et al., 1978; Emillsson and Gudbjarnason, 1981) or isoproterenol (Gudbjarnason and Hallgrimsson, 1976) had an increase in the 22:6 n-3 content of the cardiac phospholipids. These agents also increased the development of myocardial necrosis and mortality rate (Gudbjarnason and Hallgrimsson, 1976). The very long chain PUFA, 22:6 n-3, does appear to be involved in myocardial necrosis, but the process is not understood. It is of interest to note that the 22:6 n-3 content in the cardiac phospholipids of man also increases with age (Gudbjarnason, 1980). More research is needed to evaluate interspecies differences and the effect of dietary fatty acids on the content and function of the C_{22} PUFA in the heart.

3. CORRELATION OF 22:1 LEVELS TO MYOCARDIAL NECROSIS

Another fatty acid which is correlated to heart lesions is docosenoic acid (22:1). Apart from the accumulation of 22:1 in the cardiac triglycerides and free fatty acids, 22:1 is also incorporated into the cardiac phospholipids in decreasing order of SP > DPG > PE ~ PC (Section III,C,2.). Generally, the 22:1 is incorporated into position 2 of cardiac PE and PC, a position normally occupied by 18:2 n-6 and 20:4 n-6. In this position 22:1 may possibly interfere with prostaglandin formation, although this appears unlikely from skin analyses that showed that a depression of prostaglandin formation is correlated to the 18:1 content of the dietary oil (Hulan and Kramer, 1977). The significance of 22:1 in cardiac SP and DPG is not understood. These phospholipids are mainly membrane constituents (Rouser et al., 1968), and, as such, the presence of a significant amount of a very long chain monounsaturated fatty acid, four carbons longer than the common C_{18} monounsaturated fatty acid, may cause a change in the physical property of the membrane (de Kruyff et al., 1974). A change in the physical properties of the mammalian membrane with changes in the fatty acid composition is generally not seen (Williams et al., 1972), partly because cholesterol regulates the fluidity of the membrane (Thompson, 1980).

More work is necessary to explain the function of lipids in the process of myocardial necrosis. In addition to improving the lipid analyses and the separation techniques of subcellular particles, new approaches will need to be used to study the intact organ. The use of ^{31}P NMR, freeze-fracture electron microscopy, and X-ray diffraction techniques (Hui et al., 1981; Matthews et al., 1981) may be necessary to elucidate the role of lipids in myocardial degeneration.

ACKNOWLEDGMENT

This work is Contribution No. 1095 from the Animal Research Centre, Ottawa, Ontario, Canada.

REFERENCES

Abdellatif, A. M. M., and Vles, R. O. (1970). Nutr. Metab. **12**, 289–295.
Abdellatif, A. M. M., and Vles, R. O. (1973). Nutr. Metab. **15**, 219–231.
Ackman, R. G. (1980). "A Report to Fisheries and Oceans Canada," PRD Contract No. 08SC-01532-9-0244. Fisheries and Oceans Canada, Ottawa.
Ackman, R. G., and Loew, F. M. (1977). Fette, Seifen, Anstrichm. **79**, 15–24 and 58–69.
Astorg, P.-O., and Cluzan, R. (1977). Ann. Nutr. Aliment. **31**, 43–68.
Astorg, P.-O., and Compoint, G. (1978). Ann. Biol. Anim., Biochim. Biophys. **18**, 1117–1128.
Beare-Rogers, J. L. (1975). In "Modification of Lipid Metabolism" (E. G. Perkins and L. A. Witting, eds.), pp. 43–57. Academic Press, New York.
Beare-Rogers, J. L. and Gordon, E. (1976). Lipids **11**, 287–290.
Beare-Rogers, J. L., Nera, E. A., and Heggtveit, H. A. (1971). Can. Inst. Food Technol. J. **4**, 120–124.
Beare-Rogers, J. L., Nera, E. A., and Craig, B. M. (1972). Lipids **7**, 548–552.
Beare-Rogers, J. L., Gray, L., Nera, E. A., and Levin, O. L. (1979). Nutr. Metab. **23**, 335–346.
Beher, W. T., Baker, G. D., and Penney, D. G. (1963). J. Nutr. **79**, 523–530.
Bellenand, J. F., Baloutch, G., Ong, N., and Lecerf, J. (1980). Lipids **15**, 938–943.
Benga, G., Hodarnau, A., Böhm, B., Borza, V., Tilinca, R., Dancea, S., Petrescu, I., and Ferdinand, W. (1978). Eur. J. Biochem. **84**, 625–633.

Bhatia, I. S., Sharma, A. K., Gupta, P. P., and Ahuja, S. P. (1979). *Indian J. Med. Res.* **69**, 271–283.

Blomstrand, R., and Svensson, L. (1974). *Lipids* **9**, 771–780.

Bourre, J.-M., Daudu, O., and Baumann, N. (1976). *Biochim. Biophys. Acta* **424**, 1–7.

Brockerhoff, H., and Yurkowski, M. (1966). *J. Lipid Res.* **7**, 62–64.

Carroll, K. K. (1965). *J. Am. Oil Chem. Soc.* **42**, 516–528.

Cheng, C.-K., and Pande, S. V. (1975). *Lipids* **10**, 335–339.

Clandinin, M. T. (1978). *J. Nutr.* **108**, 273–281.

Clandinin, M. T. (1979). *FEBS Lett.* **102**, 173–176.

Clouet, P., and Bezard, J. (1978). *FEBS Lett.* **93**, 165–168.

Codex Alimentarius Commission (1979). "F. A. O. Report of the 13th session of the Codex Committee on Fats and Oils, December 4–8, 1978." Codex Aliment. Comm., Rome.

Colbeau, A., Nachbaur, J., and Vignais, P. M. (1971). *Biochim. Biophys. Acta* **249**, 462–492.

Craig, B. M., and Beare, J. L. (1967). *Can. J. Biochem.* **45**, 1075–1079.

de Kruyff, B., van Dijck, P. W. M., Demel, R. A., Schuijff, A., Brants, F., and van Deenen, L. L. M. (1974). *Biochim. Biophys. Acta* **356**, 1–7.

Dewailly, P., Nouvelot, A., Sezille, G., Fruchart, J. C., and Jaillard, J. (1978). *Lipids* **13**, 301–304.

Dewailly, P., Sezille, G., Nouvelot, A., Fruchart, J. C., and Jaillard, J. (1977). *Lipids* **12**, 301–306.

Divakaran, P., and Venkataraman, A. (1977). *J. Nutr.* **107**, 1621–1631.

Dow-Walsh, D. S., Mahadevan, S., Kramer, J. K. G., and Sauer, F. D. (1975). *Biochim. Biophys. Acta* **396**, 125–132.

Egwim, P. O., and Kummerow, F. A. (1972). *J. Lipid Res.* **13**, 500–510.

Emilsson, A., and Gudbjarnason, S. (1981). *Biochim. Biophys. Acta* **664**, 82–88.

Fairbairn, D. (1945). *J. Biol. Chem.* **157**, 645–650.

Farnworth, E. R., Kramer, J. K. G., Thompson, B. K., and Corner, A. H. (1982). *J. Nutr.* **112**, 231–240.

Fulco, A. J., and Mead, J. F. (1961). *J. Biol. Chem.* **236**, 2416–2420.

Galli, C. (1980). *Adv. Nutr. Res.* **3**, 95–126.

Gloster, J., and Harris, P. (1970). *Cardiovasc. Res.* **4**, 1–5.

Gudbjarnason, S. (1980). *Nutr. Metab.* **24** *Suppl. 1*, 142–146.

Gudbjarnason, S., and Hallgrimsson, J. (1976). *Acta Med. Scand., Suppl.* **587**, 17–27.

Gudbjarnason, S., Doell, B., and Oskarsdottir, G. (1978). *Acta Biol. Med. Ger.* **37**, 777–784.

Haeffner, E. W., and Privett, O. S. (1975). *Lipids* **10**, 75–81.

Hill, E. E., and Lands, W. E. M. (1970). *In* "Lipid Metabolism" (S. J. Wakil, ed.), pp. 185–277. Academic Press, New York.

Holman, R. T. (1977). *In* "Polyunsaturated Fatty Acids" (W.-H. Kunau and R. T. Holman, eds.), pp. 163–182.

Holub, B. J., and Kuksis, A. (1978). *Adv. Lipid Res.* **16**, 1–125.

Houtsmuller, U. M. T., Struijk, C.B., and Van der Beek, A. (1970). *Biochim. Biophys. Acta* **218**, 564–566.

Houtsmuller, U. M. T., Van der Beek, A., and Zaalberg, J. (1969). *Lipids* **4**, 571–574.

Hsu, C. M. L., and Kummerow, F. A. (1977). *Lipids* **12**, 486–494.

Hui, S. W., Stewart, T. P., Yeagle, P. L., and Albert, A. D. (1981). *Arch. Biochem. Biophys.* **207**, 227–240.

Hulan, H. W., and Kramer, J. K. G. (1977). *Lipids* **12**, 604–609.

Hulan, H. W., Kramer, J. K. G., Mahadevan, S., Sauer, F. D., and Corner, A. H. (1976). *Lipids* **11**, 9–15.

Hulan, H. W., Kramer, J. K. G., and Corner, A. H. (1977a). *Lipids* **12**, 951–956.

Hulan, H. W., Kramer, J. K. G., and Corner, A. H. (1977b). *Can. J. Physiol. Pharmacol.* **55**, 258–264.

Hulan, H. W., Kramer, J. K. G., Corner, A. H., and Thompson, B. (1977c). *Can. J. Physiol. Pharmacol.* **55**, 265–271.

Hulan, H. W., Thompson, B., Kramer, J. K. G., Sauer, F. D., and Corner, A. H. (1977d). *Can. Inst. Food Sci. Technol. J.* **10**, 23–26.

Hung, S., and Holub, B. J. (1977). *Nutr. Rep. Internat.* **15**, 71–79.

Hung, S., Umemura, T., Yamashiro, S., Slinger, S. J., and Holub, B. J. (1977). *Lipids* **12**, 215–221.

Innis, S. M., and Clandinin, M. T. (1981). *Biochem. J.* **193**, 155–167.

Iritani, N., and Fukuda, E. (1980). *J. Nutr.* **110**, 1138–1143.

Ito, T., and Johnson, R. M. (1964). *J. Biol. Chem.* **239**, 3201–3208.

Joffrain, C., Millot, S., Boucrot, P., Escousse, A., and Rocquelin, G. (1975). *C.R. Hebd. Seances Acad. Sci., Ser. D* **280**, 2489–2492.

Johnston, J. M. (1968). *In* "Handbook of Physiology" (C. F. Code, ed.), Sect. 6, Vol. III, pp. 1353–1375. Am. Physiol. Soc., Washington, D.C.

Kaunitz, H., and Johnson, R.E. (1973). *Lipids* **8**, 329–336.

Kishimoto, Y., and Radin, N. S. (1963). *J. Lipid Res.* **4**, 444–447.

Kramer, J. K. G. (1980). *Lipids* **15**, 651–660.

Kramer, J. K. G., and Hulan, H. W. (1977). *Lipids* **12**, 159–164.

Kramer, J. K. G., and Hulan, H. W. (1978a). *J. Lipid Res.* **19**, 103–106.

Kramer, J. K. G., and Hulan, H. W. (1978b). *Lipids* **13**, 438–445.

Kramer, J. K. G., Mahadevan, S., Hunt, J. R., Sauer, F. D., Corner, A. H., and Charlton, K. M. (1973). *J. Nutr.* **103**, 1696–1708.

Kramer, J. K. G., Friend, D. W., and Hulan, H. W. (1975a). *Nutr. Metab.* **19**, 279–290.

Kramer, J. K. G., Hulan, H. W., Mahadevan, S., Sauer, F. D., and Corner, A. H. (1975b). *Lipids* **10**, 511–516.

Kramer, J. K. G., Hulan, H. W., Procter, B. G., Rona, G., and Mandavia, M. G. (1978). *Can. J. Anim. Sci.* **58**, 257–270.

Kramer, J. K. G., Hulan, H. W., Corner, A. H., Thompson, B. K., Holfeld, N., and Mills, J. H. L. (1979a). *Lipids* **14**, 773–780.

Kramer, J. K. G., Hulan, H. W., Trenholm, H. L., and Corner, A. H. (1979b). *J. Nutr.* **109**, 202–213.

Kramer, J. K. G., Farnworth, E. R., Thompson, B. K., and Corner, A. H. (1981). *Prog. Lipid Res.* **20**, 491–499.

Kramer, J. K. G., Farnworth, E. R., Thompson, B. K., Corner, A. H., and Trenholm, H. L. (1982). *Lipids* **17**, 372–382.

Krámer, M. (1973). *Nahrung* **17**, 643–651.

Kuksis, A. (1978). *Handb. Lipid Res.* **1**, 381–442.

Landes, D. R., and Miller, J. (1975). *J. Agric. Food Chem.* **23**, 551–555.

Lewis, D. S., Masoro, E. J., and Yu, B. P. (1981). *J. Lipid Res.* **22**, 1094–1101.

Matthews, P. M., Bland, J. L., Gadian, D. G., and Rodda, G. K. (1981). *Fed. Proc., Fed. Am. Soc. Exp. Biol.* **40**, 1625, Abstr. No. 497.

McCutcheon, J. S., Umemura, T., Bhatnagar, M. K., and Walker, B. L. (1976). *Lipids* **11**, 545–552.

McMurray, W. C. (1973). *BBA Libr.* **3**, 205–251.

Mohrhauer, H., Christiansen, K., Gan, M. V., Deubig, M., and Holman, R. T. (1967). *J. Biol. Chem.* **242**, 1507–1514.

Myher, J. J., Kuksis, A., Vasdev, S. C., and Kako, K. J. (1979). *Can. J. Biochem.* **57**, 1315–1327.

Myher, J. J., Kuksis, A., Breckenridge, W. C., and Little, J. A. (1981). *Can. J. Biochem.* **59**, 626–636.

Okano, G., Matsuzaka, H., and Shimojo, T. (1980). *Biochim. Biophys. Acta* **619**, 167–175.

Opstvedt, J., Svaar, H., Hansen, P., Pettersen, J., Langmark, F. T., Barlow, S. M., and Duthie, I. F. (1979). *Lipids* **14**, 356–371.

Papahadjopoulos, D. (1973). *BBA Libr.* **3**, 143–169.

Parkes, J. G., and Thompson, W. (1970). *Biochim. Biophys. Acta* **196**, 162–169.

Raicht, R. F., Cohen, B. I., Shefer, S., and Mosbach, E. H. (1975). *Biochim. Biophys. Acta* **388**, 374–384.

Ray, S., Sen Gupta, K. P., and Chatterjee, G. C. (1979). *Indian J. Exp. Biol.* **17**, 918–921.

Rieckehoff, I. G., Holman, R. T., and Burr, G. O. (1949). *Arch. Biochem.* **20**, 331–340.

Rocquelin, G. (1972). *C.R. Hebd. Seances Acad. Sci., Ser. D* **274**, 592–595.

Rocquelin, G. (1979). *Nutr. Metab.* **23**, 98–108.

Rocquelin, G., and Cluzan, R. (1968). *Ann. Biol. Anim., Biochim., Biophys.* **8**, 395–406.

Rocquelin, G., and Fouillet, X. (1979). *Ann. Biol. Anim., Biochim., Biophys.* **19**, 479–481.

Rocquelin, G., Juaneda, P., and Cluzan, R. (1981). *Ann. Nutr. Metab.* **25**, 350–361.

Rocquelin, G., Sergiel, J.-P., Astorg, P. O., and Cluzan, R. (1973). *Ann. Biol. Anim., Biochim., Biophys.* **13**, 587–609.

Roine, P., Uksila, E., Teir, H., and Rapola, J. (1960). *Z. Ernährungswiss.* **1**, 118–124.

Romsos, D. R., and Leveille, G. A. (1974). *Adv. Lipid Res.* **12**, 97–146.

Rouser, G., Fleischer, S., and Yamamoto, A. (1970). *Lipids* **5**, 494–496.

Rouser, G., Nelson, G. J., Fleischer, S., and Simon, G. (1968). *In* "Biological Membranes" (D. Chapman, ed.), pp. 5–69. Academic Press, New York.

Seher, A., Arens, M., Krohn, M., and Petersen, U. (1979). *Fette, Seifen, Anstrichm.* **81**, 181–187.

Sen. A., and Sen Gupta, K. P. (1980). *Indian J. Exp. Biol.* **18**, 1012–1015.

Sen, P. C., and Ray, T. K. (1980). *Biochim. Biophys. Acta* **618**, 300–307.

Seoane, J. R., and Gorrill, A. D. L. (1975). *Can. J. Anim. Sci.* **55**, 749–757.

Sheppard, A. J., Iverson, J. L., and Weihrauch, J. L. (1978). *Handb. Lipid Res.* **1**, 341–379.

Simon, G., and Rouser, G. (1969). *Lipids* **4**, 607–614.

Smith, J. A., and DeLuca, H. F. (1964). *J. Cell Biol.* **21**, 15–26.

Stam, H., Geelhoed-Mieras, T., and Hülsmann, W. C. (1980). *Lipids* **15**, 242–250.

Stancliff, R. C., Williams, M. A., Utsumi, K., and Packer, L. (1969). *Arch. Biochem. Biophys.* **131**, 629–642.

Svaar, H., Langmark, F. T., Lambertsen, G., and Opstvedt, J. (1980). *Acta Pathol. Microbiol. Scand., Sect. A* **88**, 41–48.

Thompson, G. A. (1980). "The Regulation of Membrane Lipid Metabolism." CRC Press, Boca Raton, Florida.

Vergroesen, A. J., and Gottenbos, J. J. (1975). *In* "The Role of Fat in Human Nutrition" (A. J. Vergroesen, ed.), pp. 1–41. Academic Press, New York.

Vles, R. O. (1975). *In* "The Role of Fat in Human Nutrition" (A. J. Vergroesen, ed.), pp. 433–477. Academic Press, New York.

Vles, R. O., Bijster, G. M., Kleinekoort, J. S. W., Timmer, W. G., and Zaalberg, J. (1976). *Fette, Seifen, Anstrichm.* **78**, 128–131.

Vles, R. O., Bijster, G. M., and Timmer, W. G. (1978). *Arch. Toxicol., Suppl.* **1**, 23–32.

Waite, M., and van Golde, L. M. (1968). *Lipids* **3**, 449–452.

Widmer, C., and Holman, R. T. (1950). *Arch. Biochem.* **25**, 1–12.

Williams, M. A., Stancliff, R. C., Packer, L., and Keith, A. D. (1972). *Biochim. Biophys. Acta* **267**, 444–456.

White, D. A. (1973). *BBA Libr.* **3**, 441–482.

Yasuda, S., Kitagawa, Y., Sugimoto, E., and Kito, M. (1980). *J. Biochem. (Tokyo)* **87**, 1511–1517.

19

The Use of Statistics in Assessing the Results from Experiments with Vegetable Oils Fed to Test Animals

B. K. THOMPSON

I. INTRODUCTION

The trend in the use of statistical methodology in rapeseed research parallels that in many other scientific disciplines. The early reports included little,

High and Low Erucic Acid Rapeseed Oils
ISBN 0-12-425080-7

if any, formal statistical methodology; most publications presented at most means and their standard deviations. However, more recent work shows a considerably greater sophistication in the application of statistical techniques. Researchers not only carry out analyses of variance but they also partition the sums of squares into component parts (Astorg and Cluzan, 1976) or apply various multiple range procedures such as Neuman–Keuls' (Vles et al., 1978), Duncan's (Vogtmann et al., 1975), or Tukey's (Nolen, 1981) to the means. When the distribution of the data precludes the use of methods based on the normal distribution, the researcher turns to a number of nonparametric methods: for example, the Mann–Whitney U test and Friedman two-way analysis of variance (Cluzan et al., 1979), the Wilcoxon rank sum test (Svaar and Langmark, 1980), or the Kruskal and Wallis one-way analysis of variance (Vles et al., 1978). Contingency table data are analyzed not just by the standard χ^2 approach (Astorg and Cluzan, 1976) but also by Fisher's exact test (Beare-Rogers et al., 1979) or by the log–linear approach (Kramer et al., 1982; Clandinin and Yamashiro, 1980).

It is the intention here to review the statistical methodology associated with rapeseed research. To do so, it will be necessary to consider a variety of problems encountered in analyzing the different types of data collected by researchers. Having been personally involved with the Agriculture Canada research program for the past 8 years, I have certainly encountered my share of such problems. One of the laments of an applied statistician is that there is all too often no appropriate place to air his views on statistical problems relating to the scientific area of personal interest. The biological journals label his work "too theoretical" while the statistical journals consider his work "not original." Furthermore, publication in statistical journals would not reach the desired readership. Hence, I welcome the opportunity not only to discuss the statistical techniques that have been applied to data arising from rapeseed studies but also to address some of the problems inherent in their application.

It should be emphasized that the lists of examples cited here for the various statistical approaches are not to be taken as exhaustive. In particular, research groups will often adopt specific methods and use them repeatedly in their studies. A citation to Astorg and Cluzan (1976), for example, might equally apply to Astorg and Levillain (1979) or to Cluzan et al. (1979). It is also quite possible that I may have misconstrued the purpose or the mechanics of a statistical application in some studies. For this, I apologize. While the application of statistical methodology may have improved dramatically over time in rapeseed research, the same cannot always be said for the descriptions of the statistical methods in the various publications.

II. BODY WEIGHT AND GROWTH DATA

Like reports on nutritional studies in general, most publications dealing with rapeseed oil present growth data as means of final body weights or weight gains over the experimental period. Usually standard errors or some other measures of variation are given and little else. There are a few exceptions. Vles et al. (1978), for example, applied Neuman–Keuls multiple range test to their means while Astorg and Cluzan (1976) partitioned the treatment sums of squares into component parts. Whatever the statistics used, these data relate to two weight measurements only, the first and last of the experiment.

An obvious shortcoming of this approach is that measurements taken in the middle of the experiment are ignored. This may be appropriate when the experimental period is defined by commercial or other considerations but generally the duration of a rapeseed study is arbitrary. Growth over the experiment is certainly of interest but so are the growth patterns during the interim. Some authors such as Astorg and Cluzan (1976) and Hung et al. (1977) show gains over specific periods of their experiment; others undoubtedly analyze similar data but, perhaps because of space restrictions, they omit these results from the final publications. The mean body weights at each date are sometimes plotted (e.g., Abdellatif and Vles, 1973; Beare-Rogers et al., 1979). In a few instances, the authors also include standard errors of the means in the plots (e.g., Hulan et al., 1977b; Svaar and Langmark, 1980). Although these plots provide an effective means of presenting data, they lack the rigor of a formal statistical analysis. It is all too easy to read into a plot, especially one without standard errors indicated, inferences that are not warranted when variation or serial correlation is taken into consideration. A simple example serves to illustrate the effectiveness of incorporating all measurements into the analysis.

Friend et al. (1981) placed six young pigs on each of three diets for a period of 21 days. The experiment consisted of two replicates of three pens, each pen containing three pigs on a single diet. Body weights were measured every 3 days so that each pig was weighed eight times over the course of the experiment. The means for the diets are plotted in Fig. 1. Three different analyses of variance obtained from these data are shown in Table I. The first is based on final weights only and suggests no significant differences among diets. The second represents an analysis of the gains over the experiment and indicates that diet differences were significant at about the 0.4% level. This marked improvement in the power of the statistical test has been achieved simply by including the initial weights as part of the analysis. It should be noted that an improvement of this magnitude is unlikely in experiments in which all test subjects have similar body weights at the outset, a

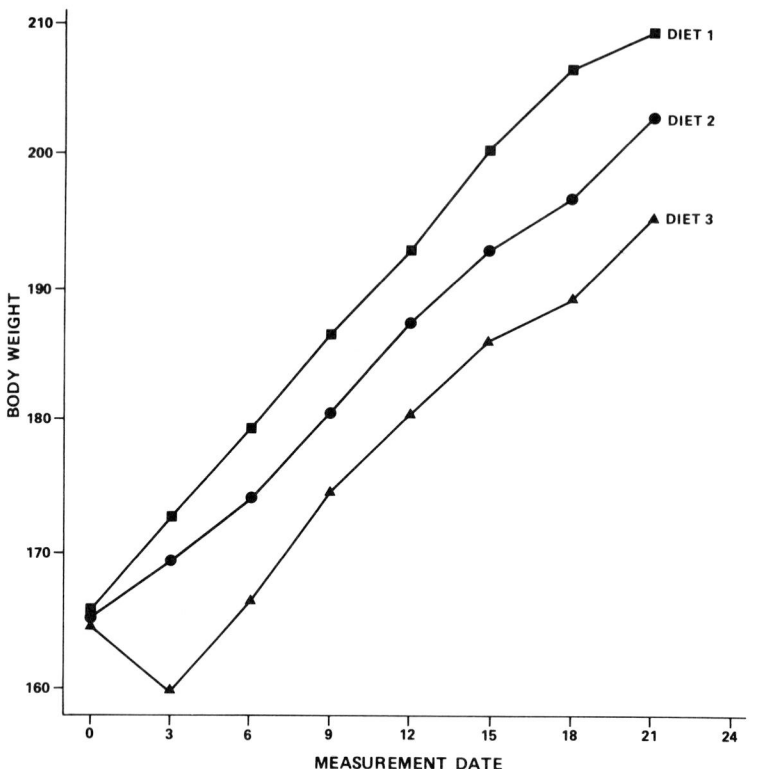

Fig. 1. The average weights, measured at 3-day intervals, of young pigs on each of three diets during a 21-day nutritional study.

situation typical of many rapeseed studies involving rats, or in which animals are grouped by body weights as, for example, in the Vles et al. (1978) study. The third analysis incorporates all the data in the experiment, handling repeated measurements from the same plot (animal) as a split-plot design (see Snedecor and Cochran, 1967, Sect. 12.12, for an example). The lower part of the analysis of variance table is relevant in the present context as it represents an analysis of the growth patterns. The large time mean square reflects the "common" growth pattern of all pigs, irrespective of diet. The significant diet × time interaction indicates that the growth patterns on the three diets differ. The significance level here is about 0.001%, a considerable increase even over the gain analysis.

By using orthogonal polynomials (see Snedecor and Cochran, 1967, Sect. 12.6), the third analysis of variance can be readily extended to enable examination of the actual shapes of the growth curves. In the present example, the linear component accounts for almost all of the time (99.1%) and the diet

TABLE I

Three Different Ways of Analyzing the Data Plotted in Fig. 1

Source	d.f.	(1) Final weight (mean square)	(2) Average daily gain (mean square)
Replicates	1	1089	0.341
Diets	2	287	0.561
Error[a]	14	92	0.067
		(3) Over all data (mean square)	
Among animals			
Replicates	1	9571	
Diets	2	1786	
Error A[a]	14	542	
Within animal			
Time	7	3522	
Time × diet	14	38	
Error B[a]	105	9	

[a] Distinction between random variation within and among pigs ignored as both were of the same magnitude throughout.

× time interaction (90.3%) sums of squares. Hence, the interpretation of the results rests mainly with differences in linear trends over time. It can be seen from Fig. 1 that the differences in diets arise from the initial effect of the diet; after the first 3 days, the growth "curves" parallel each other. In fact, there are no significant differences in weight gains among diets, if the gains are measured from day 3.

For several reasons, notably the lack of randomization, some statisticians question the use of a split-plot analysis with repeated measurements over time. Although this type of analysis is probably satisfactory for most applications, one could turn to multivariate profile analysis (see Timm, 1975) to obtain more theoretical rigor.

Measurements taken in the middle of the experiment are also useful in a different context, the screening of data for aberrant values. Although most "outlier" methods have been developed for the univariate situation, the use of highly correlated measurements such as weights at adjacent measuring dates can be very effective. Thompson (1974), for example, worked with two large poultry data sets, each including a few body weights apparently in error by 500 g (in relation to an overall weight of 2000 g). He found that about 80% of these observations could be identified using earlier weight measurements, compared to about 20% if standard univariate screening techniques were applied.

TABLE II

Percentage Arachidonic (20:4) and Erucic (22:1) Acids in the Cardiac Lipids of Male Albino Rats[a]

Fatty acid	Dietary oil					SEM[b]
	Lard	Corn	LEAR cv. Oro	LEAR cv. Span	HEAR	
20:4	7.8	15.4	9.5	11.3	4.1	1.0
22:1	0.0	0.0	0.9	2.9	23.2	0.4

[a] Extracted from Kramer et al. (1973).
[b] Standard error of mean.

III. SOME PROBLEMS IN ANALYZING RELATIVE MEASUREMENTS

In studies pertaining to rapeseed oil, various types of data are presented as relative measurements. Some measurements, such as organ weights relative to body weight, represent only a small part of the whole. Others, such as percent fatty acid composition of lipids, constitute almost the whole, that is, the percentages sum to nearly 100%. These types of data present particular problems in statistical analysis, several of which will be discussed below.

One common problem with relative measurements is heterogeneity of variance. As an illustration, some of the data presented by Kramer et al. (1973) are given in Table II. For the levels of erucic acid (22:1), the sample variance associated with the lard and corn diets is clearly 0 while that associated with high erucic acid rapeseed (HEAR) oil is considerably higher than the others, yet the standard error is based on a pooled estimate across diets. This standard error is almost meaningless for comparisons among diets. A transformation such as the arcsin (Snedecor and Cochran, 1967, Sect. 11.16) would help to stabilize the variance in less extreme cases but here, where most differences are obvious, it might be simplest to deal with specific comparisons individually, estimating variances from appropriate subsets of the data when necessary. Some authors (e.g., Hung et al., 1977; Beare-Rogers et al., 1979) have tackled the problem by presenting standard errors for each diet separately. Although these estimates are certainly an improvement over a single pooled estimate where there is considerable heterogeneity, they may be rather unreliable if sample size is small. Furthermore, it will become increasingly difficult to obtain suitable estimates when complex experimental designs are used.

Another problem introduced by relative measurements, especially those summing to nearly 100%, is the correlation between the different variables.

Figure 2 gives an excellent example: the percentage composition of lipids in the livers of 36 chickens (E. R. Farnworth, unpublished). The plot to the left shows the relationship between the areas under the peaks for triglycerides and phospholipids obtained from an Iatrascan; the plot to the right the relationship between the relative percentages. Because the lipids were generally about 95% triglycerides and phospholipids, the percentages indicate a very close relationship between the two lipid classes, even though there was little evidence of a relationship in the original data. There is no point carrying out analyses of variance for both sets of percentages in such circumstances since the results will be similar.

These correlations among constituents can produce patterns in the relative measurements which are misleading. For example, the level of one major constituent, totally unaffected by some treatment, will appear to fall if the level of another is increased. Early research with HEAR provides just such a situation. It was observed that the levels of arachidonic acid (20:4) seemed to drop in the cardiac lipids of rats fed HEAR. The results in Table II show this pattern. However, the "trend" is an artifact of the data. Lipidosis is reflected almost completely by increases in triglycerides and these contain little 20:4. The relative amounts of 20:4 are falling but absolute amounts remain the same. This particular issue has since been resolved as results are usually presented separately for triglycerides and phospholipids. However, the potential for similar difficulties is there whenever data are expressed in relative terms.

For example, if the relative measurements per se of fatty acids in cardiac lipids are not of interest, it is possible to avoid the problem by expressing the measurements in absolute units or as weights relative to heart weights (see,

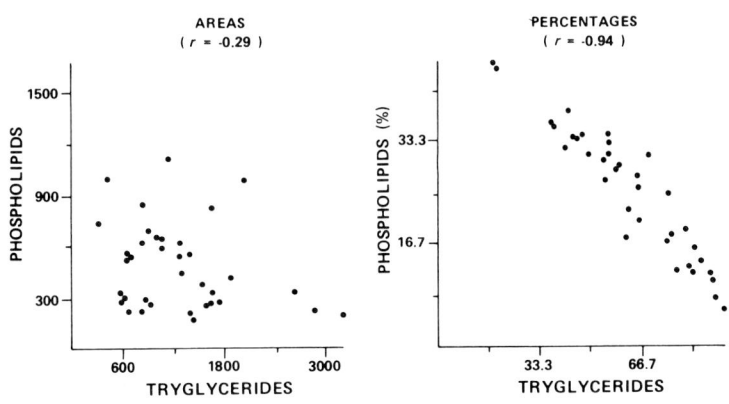

Fig. 2. The relationship between levels of phospholipids and triglycerides in the livers of 36 chickens; actual area under the curves given by the Iatrascan to the left, percentage composition to the right.

e.g., Rocquelin, 1979). The latter is admittedly a relative measurement but the fatty acid levels constitute only a small part of the heart weight. This approach should lead to a satisfactory representation if the percentages are an accurate reflection of the composition of the lipids. However, the usual method of calculating these weights, that is, by multiplying the percentages by total lipid weight, can still introduce misleading results if one of the major constituents is prone to measurement error. Consider, for example, the areas plotted in Fig. 2. The extreme values for the triglycerides do not correspond to extreme values for the phospholipids, a surprising result in that one of the major sources of variation to be expected here is the amount of solution tested and this will tend to produce a positive correlation. Hence, the extremes may represent errors in measurement of one constituent, say the triglycerides. If so, there will be errors in the weight estimates not only of the triglycerides but also of the phospholipids because the relative amount of the latter will be underestimated (overestimated) if the area of the former is too large (small). The difficulty is that each major constituent serves essentially as an internal standard for the other, the contribution of the minor constituents being small. A marked, but spurious, negative correlation between the weights of the triglycerides and phospholipids can be introduced by the error in measuring the triglycerides. Whereas an error in measuring an internal standard in the usual situation, where it is added to the solution, will have similar consequences, at least the data from the standard are not of interest per se so that the negative correlation is of little consequence.

IV. RANDOM VARIATION IN THE MEASUREMENT OF DIETARY COMPONENTS

One type of data that is seldom, if ever, subjected to statistical appraisal is the report of dietary components, such as the fatty acid levels in the diets. These data are generally presented as obtained from chemical analysis with no recognition of any possible variability in the measurements. That such variability exists is without question. Firestone and Horwitz (1979), for example, after conducting a collaborative trial of the IUPAC gas chromatographic determination of fatty acid levels, reported typical coefficients of variation ranging from 3% (for levels of about 50%) to 15% (for levels of about 2%). The sampling of the dietary material prior to chemical analysis could be expected to introduce further variation in the measurements. Yet, measures of the precision of the data are almost never given. Admittedly, there are many applications where the relatively small variability is inconsequential. It is certain, for instance, that high and low erucic acid rapeseed oils will not be confused by imprecision in the measurement of the fatty acid levels. The model derived by Trenholm et al. (1979) indicated that a difference of 1% in the levels of the saturated fatty acids, a rather sizable discrepancy in view of precision, would lead to a trivial shift of about 0.025 in the incidence of myocardial lesions.

There are, however, circumstances where the variability is of considerable importance. One pertinent example is that of regulatory controls. The consolidated regulations of Canada (Statute Revision Committee, 1978) state that "no person shall sell cooking oil, margarine, . . . , if the product contains more than five percent C_{22} Monoenoic Fatty Acids . . . ." A review of Western Canadian rapeseed production by Daun (see Chapter 7) suggests that the producers should have little difficulty in meeting this regulation as erucic acid levels are considerably lower than 5%. But, suppose that the regulations were modified to set a limit of 2%, a value much closer to present erucic acid levels.

Figure 3 shows the probability of accepting samples containing different concentrations of erucic acid. Two levels of variation are represented, both plausible in view of the Firestone and Horwitz (1979) results. The two curves, referred to as operational characteristic curves in quality control the-

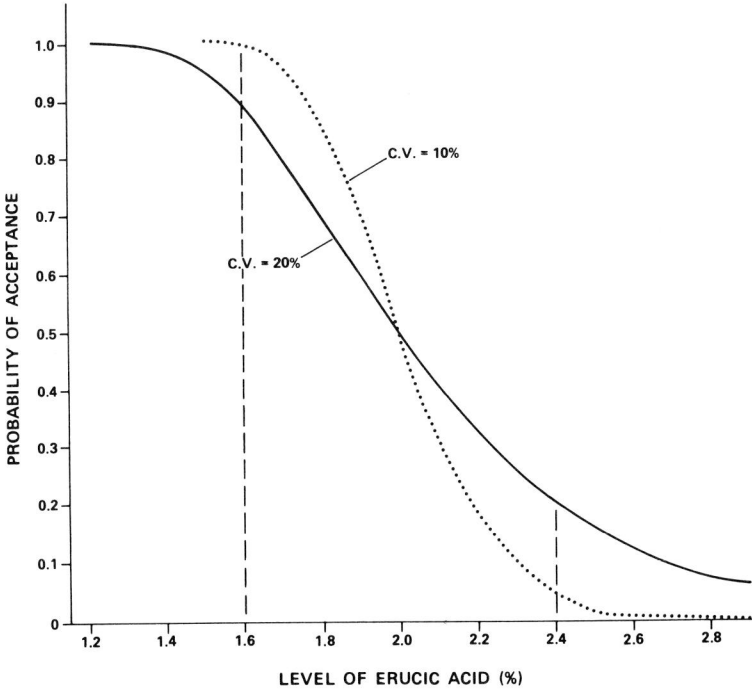

Fig. 3. The probability of accepting samples of various actual levels of erucic acid when a measured value of 2% or more is to be rejected; two levels of measurement precision are considered, with coefficients of variation of 10 and 20%.

ory, have been derived assuming that deviations of the measured concentration from the actual concentration arising from imprecision are distributed normally and that the analytical techniques are unbiased so that the mean deviation is zero. Hence, because of imprecision, the actual concentration will be underestimated half the time and overestimated the other half so that, for example, 50% of samples containing 2.0% erucic acid will be rejected even though the hypothetical legislation deems them to be acceptable. Two types of error are involved here: the risk of a good sample being rejected and the risk of a bad sample being accepted. In quality control applications, these risks are labeled producer's (α risk) and consumer's (β risk), respectively. Ideally these risks would have zero probability, that is, the "curve" would be a perpendicular line at 2% concentration in Fig. 3. As the variation increases, so do the error rates. With a coefficient of variation of 20%, it can be seen that samples with erucic acid levels as low as 1.6% will be judged unacceptable about a tenth of the time, while samples with levels of 2.4% will be judged acceptable about a fifth of the time. Consequently, even the small amount of variation associated with imprecision in the gas chromatographic methods could have considerable impact on producer and consumer alike.

It should be emphasized that the problems discussed above will be magnified by other potential sources of even greater variation. In particular, if the regulation were aimed at the seed rather than the oil, all the inherent variation due to sampling of the seed and the processing of the sample would come into play. The point to be made is that the levels of dietary components are not measured without error; hence, such information should not be treated as if it were.

V. THE ANALYSIS OF INCIDENCE DATA

Almost all papers reporting cardiac pathology in rapeseed studies present incidence data, that is, the proportions of test animals found with myocardial lesions. A variety of approaches is used to indicate the severity of these lesions. Some authors give the frequency of the numbers of lesions identified in the hearts (e.g., Kramer et al., 1975, 1979; Beare-Rogers et al., 1979) or the average numbers of lesions per heart (e.g., Vles et al., 1978; Cluzan et al., 1979). Others grade the lesions in some way (e.g., Beare-Rogers and Nera, 1977; Svaar and Langmark, 1980). McCutcheon et al. (1976) went further and devised a rating scale using the grades of lesions. Vles et al. (1978) obtained a measure of severity reflecting the amount of lesions in the heart by summing the Feret diameters of the lesions found in a large number of sections of each heart.

In view of the importance of the cardiac pathology in rapeseed research, it is rather surprising that incidence data are so often presented without any formal statistical analyses. In fact, examples can be found in the earlier reports of all major laboratories where statistical methods were not used (Kramer et al., 1973; Abdellatif and Vles, 1973; Rocquelin et al., 1973; Beare-Rogers et al., 1974; McCutcheon et al., 1976). The practice continues in more recent work (e.g., Hung et al., 1977; Umemura et al., 1978; Nolen, 1981; Svaar and Langmark, 1980). However, various statistical procedures now routinely appear .with publications of many researchers. With a few exceptions such as Vogtmann et al. (1975) who used the analysis of variance, incidence data are processed using one of the statistical methods developed for contingency tables. Perhaps the most common approach in dealing with two-way tables is to apply the standard χ^2 test (e.g., Vles et al., 1978; Astorg and Cluzan, 1976). Several researchers (Kramer et al., 1975, 1982; Clandinin and Yamashiro, 1980) have analyzed their data using a method for multidimensional contingency tables described by Fienberg (1970, 1977). There is at least one instance (Beare-Rogers et al., 1979) of the application of the Fisher–Irwin test, often referred to as Fisher's exact test (see, e.g., Gart, 1971). Perhaps it should be added there are other statistical methods available for analyzing this type of data, but these methods seem not to have been applied in rapeseed research (see Snedecor and Cochran, 1967, Chapter 9, or for more detailed information, see Cox, 1970).

As Cochran (1954) notes in discussing two-way tables, the standard analysis is a test of the overall contingency table; it may be quite insensitive to specific comparisons or hypotheses relating to the table. Even when the overall χ^2 is not significant, subtables formed by eliminating or summing over rows and columns of the table may have a significant χ^2. Cochran goes on to discuss a number of ways of strengthening the test. In rapeseed studies, it is likely that the researcher will wish to compare the results from specific diets. The most direct approach is to form the appropriate subtables and calculate the corresponding χ^2. Kendall and Stuart (1961, Sect. 33.53) describe the mechanics of the data manipulation; an example in a practical context is found in Astorg and Cluzan (1976). A problem in this approach is that the χ^2 obtained by partitioning the table do not sum to the overall χ^2. Kimball (1954) has described a method to obtain χ^2 that do sum to the total but, as Kendall and Stuart (1961) observe, "the approximate partition is good enough for most practical purposes." It may be noted that, in many applications, the components do sum to the total using the method described by Fienberg (1970, 1977; see, e.g., Hulan et al., 1977a).

A more serious complication arises with the multiway contingency table, that is, data from experiments including more than one classification variable, e.g., diets, strains of animals, and laboratories. By simply summing over the levels of one classification to obtain the totals for another, one ignores the interaction between the two. Lancaster (1951) proposed a method of analyzing multiway tables by partitioning the χ^2 much as one does the sums of squares in an analysis of variance (see Kendall and Stuart, 1961, Sect. 35.59, for a description) but, though the computations are straightforward, the results are only approximate. Here, the log–linear model approach described by Fienberg (1977) is more appropriate as it is based on sounder theoretical grounds. Kramer *et al.* (1979) provide a practical example involving two factors: laboratories and dietary oils. However, if the necessary computing algorithms are not available, the Lancaster procedure will probably suffice. Bishop *et al.* (1975) observe that the Lancaster method should be viewed as an empirical method, where the statistics from it and the log–linear procedure are often about the same. It should be emphasized that these methods do not remove the problems of interpretation raised by an interaction between factors; they only enable the user to test for such an interaction.

The analysis of severity data introduces several additional problems. For those researchers representing severity as the frequencies of the numbers of lesions in individual hearts, the extension of the contingency table to c columns from the two columns of the incidence table is straightforward (see Hulan *et al.*, 1977a, for an example involving the log–linear approach). However, there are often too few observations in many cells of the table to justify analysis by the various contingency table procedures (e.g., Beare-Rogers *et al.*, 1979; Kramer *et al.*, 1979). Pooling columns of the table to obtain sufficient numbers throughout may lead to a table not much different than the original incidence table. Much of the power of the test is consequently sacrificed, a point discussed by Cochran (1954). It may be possible to avoid the problem by treating the count for each heart separately and applying a nonparametric test. Vles *et al.* (1978) and Cluzan *et al.* (1979) have applied the Kruskal–Wallis and Mann–Whitney tests, respectively, in such circumstances. However, these tests become unwieldy and the results questionable when there are large numbers of ties, a likely occurrence with cardiac pathology where one or two lesions per heart are so commonly observed. Vles *et al.* (1979) apparently encountered yet another problem in analyzing their quantitative data (Feret diameters), that of nonnormality. They resolved the difficulty by using a nonparametric test (Kruskal–Wallis). In this context, the test should work satisfactorily as the data arise from a continuous distribution, one of the assumptions associated with the test.

It may be noted that "incidence," "rats affected," or other equivalent expressions are really misnomers. When pathology fails to detect cardiac lesions, there are two possible explanations: (1) the heart is free of lesions or (2) the examination has failed to detect extant lesions. Only the former represents a truly "unaffected" individual. In studies where large numbers of sections have been examined from each heart, the hearts of most male albino rats have been found to have at least one lesion, regardless of the type of dietary oil. Vles *et al.* (1978), for example, after studying more than 20 sections per heart, did not analyze the incidence data from their Sprague–Dawley rats on test for 53 weeks because the incidence approached 100% for all diets. Hence, if almost every heart has lesions, the incidence data reflect the difficulty in finding extant lesions rather than the presence or absence of lesions. Incidence becomes an indication of the number and size of the lesions in each heart. Indeed, if only two or three sections per heart are examined so that lesions may be frequently overlooked, incidence serves as a rough index for the same parameter that the Feret diameters of Vles *et al.* (1978) measure more precisely.

The question may be asked, Is it more effective to examine many sections on each heart or to obtain a coarser measure on a much larger number of hearts? The answer depends on the variability in response among individuals and on the importance placed in the measurement from each heart. Certainly, if the response to a dietary oil varies considerably from one test subject to another, it is important that the sample size be large. However, if the sample size is limited by considerations other than the time to be spent on pathology, Feret diameters or similar measurements will unquestionably give a more accurate picture of the response of each individual.

VI. SAMPLE SIZE AND INCIDENCE DATA

The question of sample size as it relates to incidence data in rapeseed research has been discussed in detail by Hulan *et al.* (1977c). The intention here is not to repeat that discussion but rather to consider several issues not raised there. A brief comment relating to their concerns is included at the end of this section.

In working with contingency tables, one consequence of small sample size is that cell frequencies will approach zero. Because the χ^2 approximation relies on large sample theory, users are cautioned against applying the standard χ^2 approach when expected cell frequencies are small. Cochran (1954) suggests that for a 2×2 contingency table, the χ^2 (corrected for continuity) may be used when the total number of observations (N) exceeds 40 or when $20 < N < 40$ and the smallest expectation is greater than 5. Otherwise, he recommends the use of Fisher's exact test, that is, the test used by Beare-Rogers *et al.* (1979). For contingency tables with more than one degree of freedom, where most expectations are less than 5, he suggests that χ^2 be applied only if the minimum expectation exceeds 2. When expectations are generally greater than 5 (say 80% of the time), the restriction can be relaxed to allow expectations greater than 1.

TABLE III

A Contingency Table with Small Cell Frequencies[a]

(a) Entire table (6 individuals per dietary oil)

Lesions	Peanut oil	Herring oil 1 2 3 4	HEAR oil 1 2
Yes	0	2 4 3 2	5 6
No	6	4 2 3 4	1 0

	χ^2(6 df)	Prob.
Standard χ^2	16.6	0.011
Fienberg method	21.5	0.001
Fisher's exact	—	0.007

(b) Subtable (peanut vs. herring oil[b])

Lesions	Peanut oil	Herring oil
Yes	0	11
No	6	13

	χ^2(1 df)	Prob.
Standard χ^2	2.6	0.107
Fienberg method	6.3	0.012
Fisher's exact	—	0.061

[a] Extracted from Astorg and Cluzan (1976).
[b] Results for herring oil obtained by summing over the four types.

The data of Astorg and Cluzan (1976) provide a useful example involving small cell frequencies (Table III). Although the frequencies are small, the expectations are all near 3, that is, within the constraints set by Cochran, so the χ^2 approximation is likely to be satisfactory. Table III shows the results from the standard χ^2 corrected for continuity, the log–linear approach and Fisher's exact test. The probabilities associated with the three methods are not exactly the same, but there is little question as to inference—all methods indicate differences in response among diets. Consider now the subtable formed to compare the data from the peanut and herring oils. There are less than 40 observations and the expectations for cells relating to peanut oil are both less than 5 so the 2×2 table falls outside the Cochran constraints. Here the probabilities associated with the two tests relying on the χ^2 distribution differ considerably from that of Fisher's exact test. In fact, the inference that will be drawn from the table depends greatly on the choice of test. Astorg and Cluzan (1976) gave a χ^2 value of 4.34% for the table, a result that differs from the χ^2 of Table III because they did not apply the correction for continuity. Owing to the small numbers involved, the correction factor alone has changed the probability level of the test by about 7% (from 0.037 to 0.107).

The problem is that a contingency table has only a limited number of configurations, especially when sample size is small. In the present example, there are only seven possible outcomes for peanut oil and 25 for herring oil. If the marginal totals are assumed fixed, as in Fisher's exact test, there are only seven possible configurations for the entire table. Perhaps more revealing, however, is the fact that Fisher's exact test almost certainly will not give a significant result at the 5% level in this case. The only configuration that leads to a probability less than 5% is 6 of 6 and 5 of 24 individuals affected for peanut and herring oils, respectively, an outcome that is very unlikely from a biological point of view. If both margins are assumed to be fixed, the outcome obtained in Table III is the most extreme one plausible from biological considerations and yet it is not "significant."

Another source of confusion arising from small sample size is the interpretation of 0% incidence in a sample. As Hulan et al. (1977c) observe, there was considerable debate in early rapeseed research as to whether myocardial lesions appeared in rats fed the control oils. As incidence levels markedly higher than 0% are now routinely reported for control oils, it is of interest to speculate whether earlier reports of 0% might be explained by random variation arising from small samples. Beare-Rogers and Nera (1972) and Abdellatif and Vles (1973), for example, observed 0% incidence for control oils fed to 10 and 6 rats, respectively. Table IV shows the proportions of experiments that can be expected to yield 0% incidence for various sample sizes and probabilities of an individual being affected. It can be seen that about 25 and 10% of the experiments involving 6 and 10 individuals, respectively, will produce 0% incidence, even when the underlying probability is as high as 0.2. Apparently, the results from the two research groups are not at all surprising for moderate incidence levels. In later publications involving similar sample sizes, Beare-Rogers et al. (1974) and Vles (1974) present data showing 0% intermingled with higher incidence levels for the identical diets used in different experiments or time periods, a good practical illustration of the results in Table IV. The entries in the right-hand column of the table give the upper limits for the 95% confidence intervals when 0% incidence is observed, that is, such an outcome would occur less than once in 20 trials for any value larger than that shown. Hence, for example, any hypothesis postulating a probability less than 0.39 would not be rejected (at the 5% level) by an experiment that yielded 0 affected of 6 individuals tested. The lesson to be learned is that samples showing zero affected should not be deemed conclusive evidence that a diet is "safe" unless large numbers of individuals have been included in the sample. Even then the definition of "safe" should be considered carefully.

TABLE IV

Proportion[a] of Experiments Showing 0% Incidence

Sample size	Probability of being affected				Upper limit for 95% confidence interval[b]
	0.1	0.2	0.3	0.4	
6	0.53	0.26	0.12	0.05	0.39
10	0.35	0.11	0.03	0.006	0.26
15	0.21	0.04	0.005	0.000	0.18
20	0.12	0.01	0.001	0.000	0.14
45	0.01	0.000	0.000	0.000	0.06

[a] Calculations based on binomial distribution.
[b] When a sample with 0 affected is observed.

Let us consider for a moment the influence of sample size in comparisons among diets, the topic of interest to Hulan et al. (1977c). Concern is often expressed when observed differences in incidence between diets are found to be not significant where it is generally accepted that there are real differences. Vogtmann et al. (1975), for example, found no difference in incidence between a LEAR oil, a HEAR oil, and soybean oil; Cluzan et al. (1979) reported no difference in incidence between a LEAR oil and peanut oil. These researchers used 10 and 12 rats per diet, respectively. Hulan et al. (1977c) produced a table showing the approximate probabilities of detecting differences among diets for different sample sizes. Using the table and assuming a real difference in incidence between the HEAR and LEAR oils and the control oils of 0.5 and 0.2, respectively, it can be shown that a nonsignificant result will be expected about 50% of the time in the Vogtmann et al. (1975) case and about 75% of the time in the Cluzan et al. (1979) case. Even if the sample size is increased considerably, the problem remains. Kramer et al. (1982) report no significant difference in incidence between LEAR and soybean oil after testing 45 animals per diet, a result to be expected about 40% of the time if the real difference is about 0.2. These examples show clearly that a nonsignificant difference should not be considered "proof" that there are no real differences between diets. The implication is serious: unless sample size is increased dramatically, it will be extremely difficult to "prove" that a rapeseed oil is as "safe" as a control oil.

VII. SAMPLE SIZE IN THE CONTEXT OF ESTIMATION

In the last section, we considered the question of sample size as it relates to significance testing. A model discussed by Trenholm et al. (1979) provides an example of the influence of sample size in the context of estimation.

To summarize briefly, Trenholm et al. (1979) combined the results from 23 experiments (111 diets involving over 2200 test animals) at four independent laboratories and used the methods of linear regression to study the relationships between incidence of myocardial lesions in young growing male albino rats and the levels of dietary fatty acids. These experiments were chosen because of similar protocol, with high fat diets (20% by weight) fed to the rats for 16 or 24 weeks. The authors found that the levels of palmitic and linolenic acids were closely associated with the differences in incidence among diets within experiments.

The relevant analysis of variance from that study is shown in Table V. The residual term represents the variation in incidence within experiment, unexplained by the linear relationship with the fatty acid levels. The term can be separated into two distinct components: (1) systematic departure from the model, that is, lack of fit, and (2) random variation arising from binomial sampling. While the former reflects the limitations of the model in describing the biological phenomenon, the latter is simply a function of the underlying incidence levels and the sample size in each experiment. The larger the sample size, the smaller will be the second component. Even if the model were to describe the biology completely, this component would not disappear. It seems of interest then to determine what proportion of the residual variation in the study (Trenholm et al., 1979) can be attributed to binomial sampling, for this variation will remain unexplained whatever the accuracy of the model.

The calculation of an estimate of the binomial variation is straightforward. For a sample of n individuals on a diet with probability p of inducing lesions, the variance of the estimate of p, that is, the observed incidence, is given by

TABLE V

Analysis of Variance of the Incidence Data from 23 Experiments as Collected by Trenholm et al. (1979)

Source of variation	df[a]	SS[a]	MS[a]
Among experiments	22	1.87	0.085
Regression on 16:0 and 18:3	2	2.15	1.075
Residual	86	1.47	0.017
Binomial sampling		0.92	0.011
Lack of fit		0.55	0.006
Total	110	5.48	

[a] df, degrees of freedom; SS, sum of squares; MS, mean square.

$p(1-p)/n$. Although the incidence itself could be used as an estimate of p to substitute into the formula, an estimate provided by the model is preferable here because it is based on a much larger data base. Averaging over the diets in each experiment and then summing over the experiments, each average weighted by the appropriate degrees of freedom (one less than the number of diets), we obtain a pooled estimate of the component in the residual sums of squares due to binomial sampling. Following this approach, the component in the study (Trenholm *et al.*, 1979) is estimated to be 0.92.

Apparently then, about 65% of the residual sums of squares or 25% of the total sums of squares within experiment cannot be explained. Even though the authors took steps to escape the effects of small sample size by including only diets with at least nine individuals, a sizable proportion of the overall variation arises from this source. The sums of squares representing a poor fit by the model is only about 0.55 or 15% of the within experiment sums of squares. This suggests a remarkably good fit for a model that is in itself very simple mathematically and is intended to describe a complicated biological process. If the levels of the dietary fatty acids per se are not directly involved in the etiology, then whatever is must be highly correlated with them. It may be noted that the model has since been supported experimentally by Kramer *et al.* (1982) who found that the incidence of lesions could be changed by manipulating levels of fatty acids in soybean and rapeseed based diets.

One last comment on the model—it relates not to causality but rather to increases and decreases in incidence of myocardial lesions. It seems that with high fat diets lesions will appear, irrespective of the oil type. Furthermore, 34% of the variation observed by Trenholm *et al.* (1979) arises from differences among experiments [although analyses of covariance suggest that a considerable proportion of this variation (about 25%) can be attributed to differences in dietary fatty acid levels across experiments].

VIII. CONCLUSION

Some of the statistical methods used in analyzing data from rapeseed studies have been reviewed in this chapter; the problems in their application and interpretation have been discussed. It is impossible in so little space to do justice to all methods or even to consider any method in depth. Furthermore, little has been said of one of the essential components of any sound statistical analysis, the initial screening and plotting of one's data, not because these procedures are unimportant but rather because they are seldom discussed in scientific publications. It is hoped, however, that the reader will have a better appreciation of the wide range of methodology available and of the care required in the application of the techniques. Statistical methods, properly used, should be recognized for what they are: not as hurdles to be cleared in the path to publication but rather as valuable tools to assist in the design of effective experiments and the interpretation of the resulting data.

REFERENCES

Abdellatif, A. M. M., and Vles, R. O. (1973). *Nutr. Metab.* **15**, 219–231.
Astorg, P.-O., and Cluzan, R. (1976). *Ann. Nutr. Aliment.* **30**, 581–602.
Astorg, P.-O., and Levillain, R. (1979). *Ann. Nutr. Aliment.* **33**, 643–658.
Beare-Rogers, J. L., and Nera, E. A. (1972). *Lipids* **7**, 548–552.
Beare-Rogers, J. L., and Nera, E. A. (1977). *Lipids* **12**, 769–774.
Beare-Rogers, J. L., Nera, E. A., and Heggtveit, H. A. (1974). *Nutr. Metab.* **17**, 213–222.
Beare-Rogers, J. L., Gray, L., Nera, E. A., and Levin, O. L. (1979). *Nutr. Metab.* **23**, 335–346.
Bishop, Y. M. M., Fienberg, S. E., and Holland, P. W. (1975). "Discrete Multivariate Analysis: Theory and Practice." MIT Press, Cambridge, Massachusetts.
Clandinin, M. T., and Yamashiro, S. (1980). *J. Nutr.* **110**, 1197–1203.
Cluzan, R., Suschetet, M., Rocquelin, G., and Levillain, R. (1979). *Ann. Biol. Anim., Biochim. Biophys.* **19**, 497–500.
Cochran, W. G. (1954). *Biometrics* **10**, 417–451.
Cox, D. R. (1970). "The Analysis of Binary Data." Chapman & Hall, London.
Fienberg, S. E. (1970). *Ecology* **51**, 419–433.
Fienberg, S. E. (1977). "The Analysis of Cross-Classified Categorical Data". MIT Press, Cambridge, Massachusetts.
Firestone, D., and Horwitz, W. (1979). *J. Assoc. Off. Anal. Chem.* **62**, 709–721.
Friend, D. W., Elliot, J. I., Trenholm, H. L., Thompson, B. K., and Hartin, K. E. (1981). *Proc. 31st Annu. Meet. Can. Anim. Soc.* p. 22.
Gart, J. J. (1971). *Rev. Int. Statist. Inst.* **39**, 148–169.
Hulan, H. W., Kramer, J. K. G., and Corner, A. H. (1977a). *Lipids* **12**, 951–956.
Hulan, H. W., Kramer, J. K. G., Corner, A. H., and Thompson, B. (1977b). *Can. J. Physiol. Pharmacol.* **55**, 265–271.
Hulan, H. W., Thompson, B., Kramer, J. K. G., Sauer, F. D., and Corner, A. H. (1977c). *Can. Inst. Food Sci. Technol. J.* **10**, 23–26.
Hung, S., Umemura, T., Yamashiro, S., Slinger, S. J., and Holub, B. J. (1977). *Lipids* **12**, 215–221.
Kendall, M. G., and Stuart, A. (1961). "The Advanced Theory of Statistics, 2. Inference and Relationship." Griffin, London.
Kimball, A. W. (1954). *Biometrics* **10**, 452–458.
Kramer, J. K. G., Mahadevan, S., Hunt, J. R., Sauer, F. D., Corner, A. H., and Charlton, K. M. (1973). *J. Nutr.* **103**, 1696–1708.
Kramer, J. K. G., Hulan, H. W., Mahadevan, S., Sauer, F. D., and Corner, A. H. (1975). *Lipids* **10**, 511–516.
Kramer, J. K. G., Hulan, H. W., Corner, A. H., Thompson, B. K., Holfeld, N., and Mills, J. H. L. (1979). *Lipids* **14**, 773–780.
Kramer, J. K. G., Farnworth, E. R., Thompson, B. K., Corner, A. H., and Trenholm, H. L. (1982). *Lipids* **17**, 372–382.
Lancaster, H. O. (1951). *J. R. Statist. Soc., Ser. B* **13**, 242–249.
McCutcheon, J. S., Umemura, T., Bhatnagar, M. K., and Walker, B. L. (1976). *Lipids* **11**, 545–552.
Nolen, G. A. (1981). *J. Am. Oil Chem. Soc.* **58**, 31–37.

Rocquelin, G. (1979). *Nutr. Metab.* **23**, 98–108.

Rocquelin, G., Sergiel, J.-P.; Astorg, P. O., and Cluzan,R. (1973). *Ann. Biol. Anim., Biochim., Biophys.* **13**, 587–609.

Snedecor, G. W., and Cochran, W. G. (1967). "Statistical Methods," 6th ed. Iowa State Univ. Press, Ames.

Statute Revision Commission (1978). "Consolidated Regulations of Canada," Vol. VIII. Government of Canada.

Svaar, H., and Langmark, F. T. (1980). *Acta Pathol. Microbiol. Scand. Sect. A* **88**, 179–187.

Thompson, B. K. (1974). Ph.D. Thesis, University of Edinburgh.

Timm, N. H. (1975). "Multivariate Analysis with Applications in Education and Psychology." Brooks/Cole, Monterey, California.

Trenholm, H. L., Thompson, B. K., and Kramer, J. K. G. (1979). *Can. Inst. Food Sci. Technol. J.* **12**, 189–193.

Umemura, T., Slinger, S. J., Bhatnagar, M. K., and Yamashiro, S. (1978). *Res. Vet. Sci.* **25**, 318–322.

Vles, R. O. (1974). *Proc. Int. Rapskongr., 4th, 1974* pp. 17–30.

Vles, R. O., Bijster, G. M., and Timmer, W. G. (1978). *Arch. Toxicol., Suppl.* **1**, 23–32.

Vogtmann, H., Christian, R., Hardin, R. T., and Clandinin, D. R. (1975). *Int. J. Vitam. Nutr. Res.* **45**, 221–229.

20

Studies with High and Low Erucic Acid Rapeseed Oil in Man

B. E. McDONALD

I. INTRODUCTION

Although rapeseed oil has long been used for edible purposes in China and India and was the chief source of edible fat in Germany during World War II, it did not become a major food in the Canadian diet until the latter part of the 1960s. From the modest beginnings of a wartime crop, grown to produce oil for industrial purposes, it has emerged as the principal edible oil in the Canadian diet. In spite of long-term use in the Orient and widespread use in Europe, Japan, and Canada today, very little has been published on

535

the nutritional properties or the metabolism of rapeseed oil in the human. The purpose of this chapter is to describe the research carried out on the utilization and metabolism of high and low erucic acid rapeseed oils in humans. For purposes of convenience high erucic acid rapeseed oil will be referred to simply as HEAR oil while low erucic acid rapeseed oil will be referred to as LEAR oil.

II. DIGESTIBILITY OF HEAR OIL AND LEAR OIL BY HUMANS

A study at the University of Manitoba (Vaisey et al., 1973) confirmed earlier studies (Deuel et al., 1949; Holmes, 1918) that showed HEAR oil was well utilized by the adult human. Vaisey et al. (1973) also found that LEAR oil was readily digested and absorbed by the adult human.

The subjects used by Vaisey et al. (1973), three male and three female college students, were supplied with diets based on foods reported to be popular with college students and which permitted the introduction of appreciable amounts of the test fats. HEAR oil and LEAR oil were used as the cooking medium for foods, such as doughnuts, French-fried potatoes, and chicken, and as the fat in mayonnaise, salad dressing, and baked products. Table spreads were made from hydrogenated HEAR and LEAR oils. Total fat intake averaged 128 g per day on the HEAR oil diet and 120 g per day on the LEAR diet. In both instances, fat accounted for 38% of the total energy intake, with test oils making up 58% of the total dietary fat; an average intake of 72 g HEAR oil and 70 g LEAR oil per day. Total fat digestibility during a 4 day metabolic study at the end of an 8 day feeding period was nearly identical for the two dietary regimens. Mean apparent digestibility of dietary fat was 96.0% for the HEAR oil diet and 96.5% for the LEAR oil diet. The mean digestibility coefficient for erucic acid, which was regarded as an index of the digestibility of HEAR oil, was 99.5, a value that corresponded precisely with the mean digestibility of 99% reported by Deuel et al. (1949).

The design of the study by Deuel et al. (1949) was very similar to that of Vaisey et al. (1973). Eight subjects were given diets, based on customary foods, in which rapeseed oil provided about 88% of the dietary fat, or approximately 52 g per day. Average digestibility was 99.0%. HEAR oil and LEAR oil appear to be completely digested and absorbed by the adult human. This situation for man contrasts with that for several other species. Man and the dog handle HEAR oil well whereas digestibility is appreciably lower for the rat, pig, and guinea pig. This species difference does not appear to apply to LEAR oil. Rocquelin and Leclerc (1969) found the coefficient of digestibility for LEAR oil for the rat was 95.5% whereas the value for HEAR oil was 77.6%.

III. SERUM LIPID CHANGES ACCOMPANYING THE INGESTION OF HEAR AND LEAR OILS

A number of well-controlled studies have demonstrated an appreciable effect of dietary fat on serum lipid patterns, in particular cholesterol level. Type of dietary fat, in terms of chain length and saturation of the constituent fatty acids, can induce large changes in serum cholesterol level. It is generally accepted that increasing the polyunsaturated fatty acid content of the diet results in a decrease in serum cholesterol level whereas increasing the saturated fatty acid content results in an appreciable increase in serum cholesterol level (McGandy and Hegsted, 1975). Monounsaturated fatty acids do not appear to play a significant role in the regulation of serum cholesterol level; increasing or decreasing the monounsaturated fatty acid content of the diet has little effect on serum cholesterol (McGandy and Hegsted, 1975).

LEAR oil and HEAR oil are characterized by a low level of saturated fatty acids and a relatively high level of monounsaturated fatty acids. The principal difference is a replacement of the erucic acid (22:1) in HEAR oil by oleic acid (18:1) in LEAR oil. Both oils contain 30–35 percent polyunsaturated fatty acids (18:2 + 18:3). Although LEAR and HEAR oils are not particularly rich sources of polyunsaturated fatty acids, their substitution for the customary fat in a diet would be expected to result in a decrease in serum cholesterol, since a decrease in the intake of saturated fatty acids is twice as effective in lowering serum cholesterol levels as an equivalent increase in the intake of polyunsaturated fatty acids (Keys et al., 1957, 1965).

Malmros and Wigand (1957) reported a series of studies with human subjects fed various dietary fat sources at levels providing 40% of calories. In one of the studies replacement of hydrogenated coconut oil by HEAR oil, in a basic diet consisting of bread, cereals, vegetables, potatoes, rice, fruits, and sugar, resulted in an average decrease in serum cholesterol of 40 mg/100 ml in only 1 week. Continuation of the subjects on the HEAR oil regimen for an additional 1.5 weeks resulted, in some cases, in small increases in serum cholesterol level. Replacement of HEAR oil by corn oil produced a further decrease in serum cholesterol of approximately 20 mg/100 ml. Grande et al. (1962) also found HEAR oil effective in lowering serum cholesterol in physically healthy, middle-aged men. They compared three experimental fats: butterfat, HEAR oil, and a mixture of corn and olive oil. The fats were incorporated into diets that provided an average 2770 kcal and 128 g of fat, of which 95 g or approximately 30% of calories was experimental fat. The study was designed so that each subject consumed each of the diets for a 3 week period. Mean serum cholesterol levels were significantly lower after 3 weeks on the HEAR oil or mixture of corn and olive oil than after 3 weeks on the butterfat regimen; mean serum cholesterol levels were 187,

188, and 233 mg/100 ml, respectively. Total serum phospholipid followed a similar pattern; mean values were 213, 208, and 245 mg/100 ml on the HEAR oil, the mixture, and the butterfat regimens, respectively.

The response of serum lipid patterns in healthy young men to the ingestion of HEAR oil and LEAR oil was studied in a series of four metabolic studies in the Department of Foods and Nutrition at the University of Manitoba. The subjects, who were either students or employees of the University, resided in their own homes and maintained their usual activity patterns but ate all their meals in the Department. The experimental diets in these studies were designed so dietary fat could be carefully controlled and yet the diets consist of familiar foods (Bruce and McDonald, 1977). Fat supplied about 38–40% of the total energy in the diets, with the specific fat being studied supplying about 95% of this total. The diets consisted of customary foods but contained no meat. Textured soybean protein, egg albumen, and skim milk were the main protein sources. Soy protein was incorporated into en-

TABLE I

Typical Menu for Fat Controlled Diets

Breakfast	Dinner
Orange juice	Beef stew[a]
Cooked rolled oats[a]	Mashed potatoes[a]
Skim milk	Carrots
Brown sugar	Coleslaw with dressing[a]
Scrambled egg albumen[a]	Whole wheat bread
Whole wheat bread	Fruit cocktail
Butter or margarine[b]	Butter or margarine[b]
Coffee or tea[c]	Coffee or tea[c]
Lunch	Snacks
Spaghetti with meat balls and tomato sauce[a]	Ginger ale
Whole wheat bread	Raisin·oatmeal cookies[a]
Lettuce and tomato salad with piquant dressing[a]	Spicy carrot cake[a]
Apple	
Skim milk	
Butter or margarine[b]	
Coffee or tea[c]	

[a] Experimental fat added.
[b] Butter for days on mixed fat regimen and margarine prepared from specific fat source being studied for days on experimental fat.
[c] Allowed *ad libitum*; alcohol and other beverages prohibited.

trées such as meat balls, hamburgers, and beef stew. Egg albumen was used in preparing baked products and scrambled eggs. Composition of a typical menu is shown in Table I.

The first two metabolic studies compared the effects of HEAR and LEAR oils, when eaten as the sole source of added dietary fat, on serum lipid patterns (McDonald et al., 1974). Each study was divided into three phases: a 9 day preliminary period when dietary fat was supplied by a mixture of fats typical of Canadian diets; a 22 day experimental period when either LEAR oil and margarine or HEAR oil and margarine supplied the fat in the diet; and an 8 day postexperimental period when the mixed fat again was eaten. The fatty acid compositions of the diets are shown in Table II.

Seven male subjects were used in each of the 39 day metabolic studies. Twelve hour fasting blood samples were taken on days 1, 10, 18, 25, 32, and 39 of each study and samples of serum analyzed for cholesterol, lipid phosphorus, and the fatty acid composition of the phospholipids (McDonald et al., 1974).

The subjects were in good health and body weights remained essentially constant throughout both the LEAR oil and HEAR oil studies. No digestive upsets were reported during these studies even though the subjects consumed in excess of 120 g of these fats in the form of oil and margarine daily for 22 days. This observation contrasted with that of Trémolières et al. (1971) who reported diarrhea in subjects given a single dose (0.5 g/kg body weight; approximately 30 g each) of HEAR oil following an overnight fast. The differences between these studies may relate to the method of administration although Trémolières et al. (1971) found no digestive upset to a similar dose

TABLE II

Percentage Fatty Acid Composition of Diets

Fatty acid	Mixed fat[a]	LEAR	HEAR
14:0	3.3	—	—
16:0	19.5	6.2	4.0
16:1	1.8	Trace	Trace
18:0	12.5	3.8	3.0
18:1	36.9	56.2	23.1
18:2	19.5	20.1	11.4
18:3[b]	1.5	8.5	5.0
20:1	1.5	2.1	12.8
22:1	—	2.2	39.1

[a] Mixture comprised 39.3% butter, 21.5% corn oil, 7.1% edible tallow, 14.3% lard, 10.7% margarine (Parkay Brand, Kraft Foods Ltd., Montreal) and 7.1% shortening (Crisco, Proctor and Gamble, Toronto).
[b] 18:3 and 20:0 not resolved with columns used.

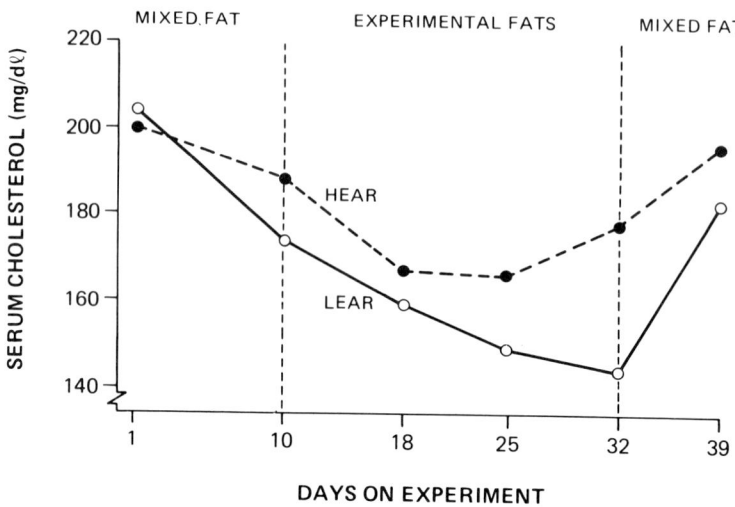

Fig. 1.   Mean serum cholesterol level of subjects fed HEAR oil (•) and LEAR oil (○).

of peanut oil. No digestive upsets in response to HEAR oil or LEAR oil have been observed in our laboratory even though in one study subjects consuming approximately 125 g of LEAR oil daily consumed 68 g in a single meal on two separate occasions (Lake, 1975).

Changes in serum lipids followed slightly different patterns during the LEAR and HEAR oil studies. Serum cholesterol level decreased appreciably during the preexperimental mixed fat period (day 1 vs. 10) in the LEAR oil group (Fig. 1). Mean serum cholesterol level also decreased during the pre-experimental period of the HEAR oil group but the change was less pronounced. These decreases may reflect the relatively low cholesterol level of the mixed fat diet. Serum cholesterol continued to decrease in a fairly consistent manner in response to the LEAR oil diet (day 10 vs. 32). Mean serum cholesterol level also decreased during the first week on the HEAR oil diet but then increased again during the third week. Increases in serum cholesterol levels between day 25 and day 32 were observed for 6 of the 7 subjects with the increase being particularly pronounced for one subject (48 mg/100 ml). Malmros and Wigand (1957) also reported an increase in serum cholesterol in some subjects following a marked decrease in response to the feeding of HEAR oil. Grande et al. (1962), however, did not report this tendency for serum cholesterol to increase again after 2 weeks on a HEAR oil regimen.

Keys et al. (1957) stated that the major change in serum cholesterol level occurs during the first week following a change in dietary fat. A further decrease may occur during the second week with little change thereafter.

TABLE III

Mean Serum Lipid Phosphorus Levels (mg/100 ml) in Subjects Given Diets That Contain LEAR and HEAR Oils

Group	Initial observation Day 1	Mixed fat Day 10	Test fat[a]			Mixed fat Day 39
			Day 18	Day 25	Day 32	
LEAR	11.7	10.1	9.2	7.9	6.7	10.5
HEAR	9.9	7.8	7.4	6.6	7.1	8.9

[a] Contains either LEAR or HEAR in the form of oil and margarine.

The fact that mean serum cholesterol levels increased sharply when the subjects were returned to the mixed fat diet (days 32 vs. 39; Fig. 1) strongly suggests that the decreases on the LEAR and HEAR oil diets were attributable to the dietary fat. In addition, serum lipid phosphorus followed a similar pattern to serum cholesterol (Table III). This relationship is consistent with the general observation that the pattern of change of serum phospholipids in response to a change in diet composition is similar to that of serum cholesterol (McGandy et al., 1970).

Fatty acid patterns of the serum phospholipids (precipitated with acetone) changed considerably in response to the feeding of LEAR oil and HEAR oil (Table IV). The decrease in palmitic acid in response to LEAR and HEAR oils (day 32 vs. day 10) reflected the low level of palmitic acid in the test fats. The decrease in palmitic acid was offset by an increase in oleic acid which also reflects the fatty acid composition of the diet and the suggestion that erucic acid is metabolized to oleic acid.

All changes in fatty acid level, with the exception of linoleic acid in the LEAR oil group, returned to pretest fat levels (day 10) with a return to the mixed fat diet for 8 days (day 32 to 39). As in other species (Walker, 1972), very little erucic acid was incorporated into serum phospholipids even though erucic acid comprised nearly 40% of the fat on the HEAR oil regimen. Fatty acid analyses of fat biopsies obtained from the middorsal abdomen of two of the subjects on the HEAR oil regimen on days 10 and 32 indicated that very little erucic acid was incorporated into adipose tissue. Erucic acid made up 0.4 and 2.0% of the total fatty acids of the two subjects on day 32. The low levels of erucic acid incorporated into phospholipids and triglycerides suggest that erucic acid is converted to other fatty acids, presumably oleic acid, in man as in other species.

The response of serum cholesterol to LEAR oil was also studied in two experiments at the University of Manitoba that were undertaken primarily to

TABLE IV

Percentage Fatty Acid Composition of Serum Phospholipids in Subjects Given Diets That Contain LEAR and HEAR Oils

	LEAR oil group			HEAR oil group		
Fatty acid	Mixed fat Day 10	LEAR oil Day 32	Mixed fat Day 39	Mixed fat Day 10	HEAR oil Day 32	Mixed fat Day 39
16:0	24.6	19.6	22.4	30.2	21.4	31.3
16:1	2.0	2.0	2.2	1.5	1.4	—
18:0	14.2	12.9	14.8	14.9	12.0	16.1
18:1	11.8	19.2	11.2	14.3	23.7	14.2
18:2	24.0	20.9	20.0	26.9	24.4	26.4
18:3[a]	—	—	—	—	1.0	Trace
20:1	2.6	1.6	2.0	Trace	3.5	1.8
20:3	2.1	2.4	2.8	2.0	1.6	2.2
20:4	7.2	8.8	7.5	7.5	6.2	6.4
22:1	—	—	—	—	1.5	—
Unknown[b]	2.9	2.7	3.1	Trace	Trace	Trace

[a] 18:3 and 20:0 not resolved with column used .
[b] Peak appears between 22:1 and 20:5 using a column packed with EGSS-Y (Applied Science Laboratories, State College, Pennsylvania).

compare energy metabolism in healthy young men given either LEAR oil or soybean oil (Bruce *et al.*, 1980). The diets and the general conduct of the studies were essentially the same as described earlier. The general design of the studies, however, differed from that described earlier in that all subjects were given both test fats; half were supplied with the LEAR oil first and the other half the soybean oil first. In study 1 the subjects consumed each fat source for 11 days and in study 2 for 8 days. A mixed fat diet was eaten for a 10 day preexperimental period in study 1 and an 8 day preexperimental and a 7 day postexperimental period in study 2. The subjects, four healthy young men in study 1, and eight in study 2, were either students or employees at the university.

Mean serum cholesterol levels decreased during the preexperimental mixed fat regimen in both studies (Figs. 2 and 3). A similar response has been observed in all studies in which this diet, which is comprised of a mixture of fats characteristic of those in the Canadian diet, was fed during the preexperimental period. The most likely explanation for the drop in serum cholesterol level, when the mixed fat diet replaced the customary diet of the subjects, is the low cholesterol content of the test diet. The mixed fat diet supplied approximately 145 mg of cholesterol daily, which is about 400 mg less than the estimated daily per capita intake in Canada. A 400 mg

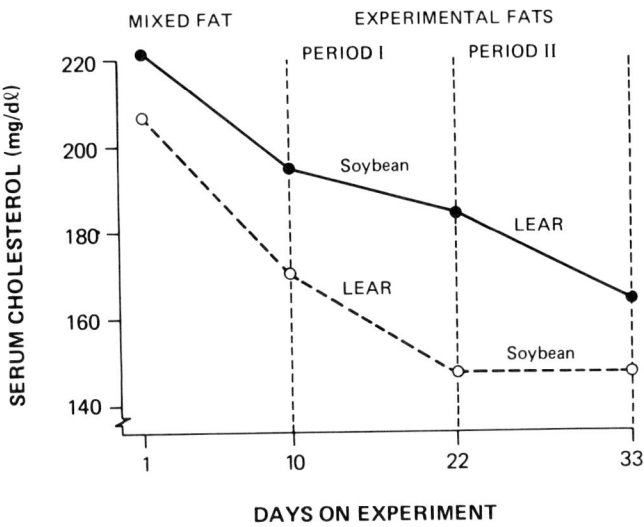

Fig. 2. Comparison of the effect of soybean oil and LEAR oil on the serum cholesterol levels of young men (study 1).

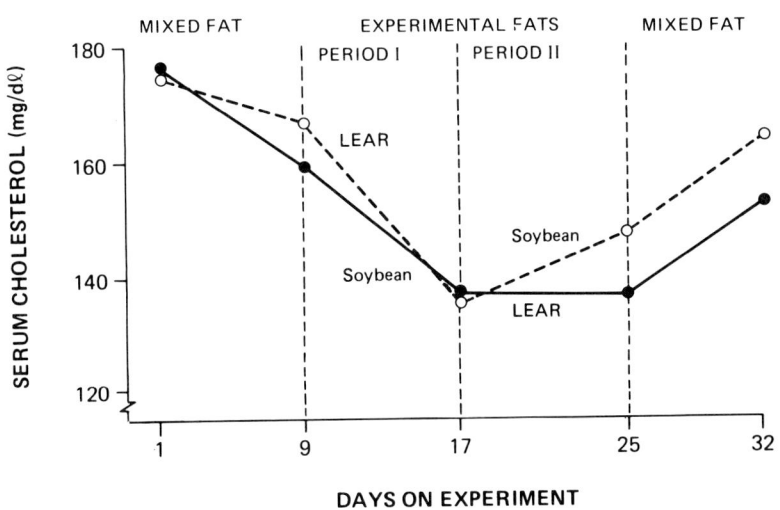

Fig. 3. Comparison of the effect of soybean oil and LEAR oil on the serum cholesterol levels of young men (study 2).

TABLE V

Percentage Fatty Acid Composition of Mixed Fat, Soybean
Oil, and LEAR Oil Diets[a]

Fatty acid	Mixed fat diet	Soybean oil diet	LEAR oil diet
14:0	1.7	0.6	0.4
16:0	21.5	11.0	6.0
18:0	12.9	4.5	3.8
18:1	40.3	43.9	59.0
18:2	20.7	34.8	19.2
18:3	1.5	3.8	7.6
20:1	Trace	Trace	2.3
22:0	—	—	0.5
22:1	—	—	0.5

[a] Gas chromatographic analysis of diets used in study 2; fatty acid composition of diets in study 1 was similar.

decrease in dietary cholesterol on a 3000 kcal diet would be expected to result in a 17 mg/100 ml decrease in mean serum cholesterol level (Mattson et al., 1972).

Substitution of LEAR oil or soybean oil for mixed fat resulted in a further decrease in mean serum cholesterol level in both studies. The decrease was especially marked in study 2 (Fig. 3). Although the response to LEAR oil appears to be slightly greater than to soybean oil, the changes in serum cholesterol level in response to the two test fats did not differ significantly ($P > 0.05$). This finding coincides with what would be predicted from the observed relationship between changes in the fatty acid composition of the diet and changes in serum cholesterol level (Keyes et al., 1965). Although the polyunsaturated fatty acid content of the LEAR oil diet was considerably lower than that of the soybean oil diet, the LEAR oil diet provided a much lower level of saturated fatty acids than the soybean oil diet (Table V).

IV. HEMATOLOGICAL PROFILES OF SUBJECTS GIVEN HEAR AND LEAR OIL DIETS

Most reports on the effect of diet on whole blood hematology deal with the effect of vitamins and minerals on these parameters. The fact that the basic diet used in the studies on LEAR and HEAR oils contained no meat meant that the test diets in these studies differed from the subjects' customary diets in form of iron, type of protein, and level of micronutrients, all of which have been found to affect whole blood hematology.

In the studies described earlier (Bruce et al., 1980), samples of blood were taken for hematological determinations (hemoglobin, hematocrit, red cell, reticulocyte, platelet, and leukocyte counts) at the same time fasting samples were obtained for serum lipid analysis. All hematological parameters, except blood platelet count for subjects who had eaten diets containing HEAR oil and margarine, remained within normal ranges for all subjects throughout the studies (McDonald et al., 1974). Changes were observed in hematological parameters other than platelet count but there was no evidence that the changes were functionally or physiologically meaningful or that they were related to the dietary treatment.

A pronounced drop was observed in the blood platelet count of subjects given HEAR oil and margarine as essentially the sole source of dietary fat for 22 days (Table VI). Blood platelet counts dropped to values that were considered to be below normal for five of the seven subjects on this regimen. Platelet counts in what was the low normal range (120,000–150,000/mm^3) also were observed in the LEAR oil group but no consistent pattern devel-

TABLE VI

Blood Platelet Counts of Normal Young Men Given HEAR and LEAR Oil Diets

Description	HEAR oil group	LEAR oil group
No. of subjects	7	7
Length of study (days)	39	39
Platelet counts[a]		
Initial, day 1	5N[b]	6N; 1LN
After preexperimental mixed fat, day 9	7N	7N
After 8 days on test fat, day 17	6LN; 1L	3N; 4LN
After 15 days on test fat, day 25	5LN; 2L	7N
After 22 days on test fat, day 32	2LN; 5L	6N; 1LN
After postexperimental mixed fat, day 39	5N; 2LN	4N; 3LN

[a] Platelet counts above 150,000/mm^3 were classed as normal (N); counts between 120,000 and 150,000/mm^3 as low-normal (LN); and counts below 120,000/mm^3 as low (L).

[b] Two blood samples were accidentally lost on day 1.

oped nor did any counts fall below the levels considered normal for humans. The fact that platelet counts returned to normal for five of the subjects supplied the HEAR oil diet and to the low normal range for the other two subjects when the mixed fat diet was reinstated strongly suggests that the decrease was related to the eating of HEAR oil. The reason and the significance of this response to the intake of a high level of HEAR oil in man is not clear although HEAR oil has been reported to induce hematological changes in other species (Abdellatif and Vles, 1970; Abdellatif et al., 1972).

In two other studies at the University of Manitoba, involving comparisons of LEAR oil with soybean oil, no changes that could be attributed to dietary fat source were observed in hematological parameters. Jacotot (1979), on the other hand, found a marked difference in the functional properties of blood platelets from subjects given LEAR oil or butter for only 5 days. Fourteen subjects, seven males and seven females, were supplied with each of the test diets for 5 days with an 8 day interval between, during which they consumed their customary diets. The test diets provided 2200 kilocalories, of which 810 kilocalories were supplied by dietary fat, i.e., either LEAR oil or butter. Blood clotting time was prolonged somewhat on the LEAR oil diet but there was no difference between the two treatments in the activity of the cephalin-kaolin sensitive plasma clotting factors. There also was no difference between the two fat sources in the number of blood platelets. However, the aggregation of blood platelets in response to ADP, both in terms of velocity and intensity, was much more pronounced following the eating of butter than LEAR oil. This difference in platelet function in response to LEAR oil and butter is probably related more to the relative saturation of the two fats than any other property per se. Renaud et al. (1980) found that platelet function could be improved by increasing the intake of polyunsaturated fatty acids at the expense of saturated fatty acids.

V. ENERGY METABOLISM IN YOUNG MEN GIVEN LEAR AND SOYBEAN OILS

Two studies designed primarily to assess energy metabolism in men given high levels of LEAR oil were undertaken at the University of Manitoba (Bruce et al., 1980) following reports by Trémolières et al. (1971) of a difference in respiratory quotient (RQ) in subjects given HEAR oil and peanut oil. They reported a lower RQ while at rest and a slightly higher RQ following light exercise (120 and 240 kpm/min) when subjects were given HEAR oil than when given peanut oil.

The design of the studies undertaken at the University of Manitoba and the description of the diets and subjects were described previously. Both studies were of a crossover design with half of the subjects (four in study 1 and eight in study 2) given the LEAR oil diet first and the other half the soybean oil diet first.

In study 1 the subjects were exercised on an electrically braked bicycle ergometer for 10 min at 950 kpm/min, 2.5 hr after a meal designed to contain over half (68 g) of the total fat intake for the day. Exercise was carried out on the last day on the preexperimental mixed fat regimen and on the first and last day on each of the test fat regimens. Each exercise period was preceded by a 20 min rest period and followed by a 20 min recovery period. In study 2 the subjects were only exercised at the end of each dietary period. Work load intensities (70% $V_{O_2\ max}$; 651–870 kpm/min) were commensurate with the subjects' individual levels of fitness. The exercise protocol consisted of a 5 min rest period, a 15 min exercise period, and a 10 min recovery period. Half of the subjects also were exercised at 60% $V_{O_2\ max}$ and the other half at 80% $V_{O_2\ max}$ during each of the dietary regimen.

Respiratory exchange measurements, namely, oxygen consumption (V_{O_2}) and the concentration of oxygen and carbon dioxide in expired gases, were monitored throughout rest, exercise, and recovery. Blood samples taken at the end of the rest, exercise, and recovery periods, were analyzed for glucose, lactate, and pyruvate. Blood samples were obtained during the $V_{O_2\ max}$ exercise sessions only in study 2.

There was no evidence in these studies with healthy young men of any differences in energy utilization due to dietary fat source (Tables VII and VIII). The mean increase in oxygen consumption in response to exercise was similar for all diets. Oxygen consumption tended to increase with an increase in work load (study 2, Table VII), which coincides with the linear relationship between oxygen consumption and increasing work load described by Åstrand and Rodahl (1970), but the increase was similar for all diets at any particular work load. Likewise, there was no differences in RQ due to dietary fat source. In study 1 respiratory parameters (V_{O_2} and RQ) and the concentration of blood metabolites were monitored after a single meal of the test fats as well as after a similar meal following 10 days on the LEAR oil and soybean oil regimens. There were no differences due to dietary fat source or the length of time subjects had consumed the test fats. These results differed from those reported by Trémolières et al. (1971) who observed a significantly ($P < 0.05$) lower resting RQ when subjects were given a single dose of HEAR oil than when given a similar dose of peanut oil. Trémolières et al. (1971) suggested that subjects fed HEAR oil containing 43% erucic acid oxidized fatty acids preferentially.

Dietary fat source also had no effect on the concentration of blood metabolites during rest, exercise, or recovery (Table VIII). Serum glucose remained within normal physiological limits during all phases. Although there was a small decrease in serum glucose due to exercise, the values did not differ significantly from those while at rest. Plasma lactate and pyruvate increased

TABLE VII

Mean Oxygen Consumption (V_{O_2}) and Respiratory Quotient (RQ) in Response to Exercise for Subjects Given Mixed Fat, Soybean Oil, and LEAR Oil Diets

Study	Work load[a] %$V_{O_2 max}$	Phase	$V_{O_2 max}$ (liters/min)			RQ		
			Mixed fat	Soybean oil	LEAR oil	Mixed fat	Soybean oil	LEAR oil
1[b]		Rest	0.35	0.30	0.39	0.78	0.80	0.79
		Exercise	2.12	2.07	2.00	0.91	0.94	0.92
		Recovery	0.49	0.47	0.57	0.92	0.88	0.90
2[c]	60	Rest	0.18	0.20	0.21	0.90	0.95	0.82
		Exercise	1.66	1.49	1.81	1.03	1.00	0.98
		Recovery	0.20	0.21	0.21	0.81	0.81	0.80
	70	Rest	0.21	0.25	0.23	1.04	0.97	0.96
		Exercise	1.69	1.82	1.88	1.04	1.01	1.04
		Recovery	0.25	0.27	0.28	0.88	0.87	0.89
	80	Rest	0.21	0.25	0.19	1.00	0.98	0.98
		Exercise	1.93	2.12	2.02	1.07	1.00	1.03
		Recovery	0.27	0.31	0.30	0.89	0.88	0.88

[a] 950 kpm/min for all subjects in study 1.
[b] Values are means for four subjects.
[c] Values are means for four subjects at 60 and 80% $V_{O_2 max}$ and means for eight subjects at 70% $V_{O_2 max}$.

TABLE VIII

Mean Serum Glucose, Plasma Lactate, and Plasma Pyruvate of Subjects Given Mixed Fat, Soybean Oil, and LEAR Oil Diets

Study	Phase	Glucose (mg/dl)			Lactate (mg/dl)			Pyruvate (mg/dl)		
		Mixed fat	Soybean oil	LEAR oil	Mixed fat	Soybean oil	LEAR oil	Mixed fat	Soybean oil	LEAR oil
1[a]	Rest	116	97	90	15.0	8.0	7.4	1.0	0.8	0.6
	Exercise[b]	84	83	74	52.3	43.9	36.9	2.2	1.6	1.5
	Recovery	78	82	78	19.4	19.4	14.9	1.5	1.2	1.1
2[c]	Rest	91	85	83	8.4	6.8	7.0	0.7	0.6	0.6
	Exercise[d]	82	80	78	43.5	45.3	44.1	1.5	1.6	1.6
	Recovery	85	81	77	30.2	27.8	27.1	1.5	1.7	1.5

[a] Means for four subjects.
[b] 950 kpm/min for 10 min.
[c] Means for eight subjects.
[d] 70% $V_{O_2 max}$ (651–870 kpm/min) for 15 min.

appreciably during exercise. There were no differences ($P > 0.05$), how-ever, among the dietary fat sources. Mean lactate/pyruvate ratios during rest, exercise, and recovery were the same for all diets. This suggests that the oxidation/reduction potential of the muscle was not altered by substituting LEAR or soybean oil for mixed fat in the diet. Similar results were reported by Trémolières *et al.* (1971) in response to HEAR oil and peanut oil. They concluded that dietary fat did not alter mitochondrial function. Lactate/py-ruvate ratio, however, does not necessarily reflect the state of oxidation/reduction of cytoplasmic NADH (Olson, 1963).

HEAR oil and peanut oil. They concluded that dietary fat did not alter mito-chondrial function. Lactate/pyruvate ratio, however, does not necessarily reflect the state of oxidation/reduction of cytoplasmic NADH (Olson, 1963).

REFERENCES

Abdellatif, A. M. M., and Vles, R. O. (1970). *Proc. Int. Conf. Sci., Technol., Market., Rapeseed, Rapeseed Prod., 1970* pp. 435–449.

Abdellatif, A. M. M., Starrenburg, A., and Vles, R. O. (1972). *Nutr. Metab.* **14**, 17–27.

Åstrand, P.-O., and Rodahl, K. (1970). "Textbook of Work Physiology," pp. 280–286. Mc-Graw-Hill, New York.

Bruce, V. M., and McDonald, B. E. (1977). *J. Can. Diet. Assoc.* **38**, 90–97.

Bruce, V. M., McDonald, B. E., Lake, R., and Parker, S. (1980). *Nutr. Rep. Int.* **22**, 503–511.

Deuel, H. J.; Jr., Johnson, R. M., Calbert, C. E., Gardner, J., and Thomas B. (1949). *J. Nutr.* **38**, 369–379.

Grande, F., Matsumoto, Y., Anderson, J. T., and Keys, A. (1962). *Circulation* **26**, 653–654.

Holmes, A. D. (1918). *U.S., Dep. Agric. Bull.* **687**.

Jacotot, B. (1979). *Proc. Int. Rapeseed Conf., 5th, 1978* Vol. 2, pp. 99–102.

Keys, A., Anderson, J. T. and Grande, F. (1957). *Lancet* **2**, 959–966.

Keys, A., Anderson, J. T., and Grande, F. (1965). *Metab. Clin. Exp.* **14**, 776–787.

Lake, R. E. (1975). M.Sc. Thesis, University of Manitoba, Winnipeg.

McDonald, B. E., Bruce, V. M., LeBlanc, E. L., and King, D. J. (1974). *Proc. Int. Rapskongr., 4th, 1974* pp. 693–700.

McGandy, R. B., and Hegsted, D. M. (1975). *In* "The Role of Fats in Human Nutrition" (A. J. Vergroesen, ed.), pp. 211–230. Academic Press, New York.

McGandy, R. B., Hegsted, D. M., and Myers, M. L. (1970). *Am. J. Clin. Nutr.* **23**, 1288–1298.

Malmros, H., and Wigand, G. (1957). *Lancet* **2**, 1–8.

Mattson, F. H., Erickson, B. A., and Kligman, A. M. (1972). *Am. J. Clin. Nutr.* **25**, 589–594.

Olson, R. E. (1963). *Ann. Intern. Med.* **59**, 960–963.

Renaud, S., Dumont, E., Godsey, F., Morazain, R., Thevenon, C., and Ortchanian, E. (1980). *Nutr. Metab.* **24** Suppl. 1, 90–104.

Rocquelin, G., and Leclerc, J. (1969). *Ann. Biol. Anim., Biochim., Biophys.* **9**, 413–426.

Trémolières, J., Lowy, R., Griffaton, G. and Carré, L. (1971). *Cah. Nutr. Diét.* **6**, 70–74.

Vaisey, M., Latta, M., Bruce, V. M., and McDonald, B. F. (1973). *Can. Inst. Food Sci. Techol. J.* **6**, 142–147.

Walker, B. L. (1972). *Nutr. Metab.* **14**, 8–16.

The Relevance to Humans of Myocardial Lesions Induced in Rats by Marine and Rapeseed Oils

H. C. GRICE AND H. A. HEGGTVEIT

I. HISTORICAL BACKGROUND AND REGULATORY CONCERN

During the 1940s and 1950s several investigations with laboratory animals indicated that the feeding of high erucic acid rapeseed (HEAR) oils to rats caused adverse effects. Growth retardation was observed by Boer et al.

551

High and Low Erucic Acid Rapeseed Oils
Copyright © 1983 by Academic Press Canada
All rights of reproduction in any form reserved.
ISBN 0-12-425080-7

(1947), Deuel et al. (1948), Thomasson and Boldingh (1955), and Roine et al. (1960). Erucic acid was incriminated by Thomasson who suggested that fatty acids with 20 or more carbon atoms exerted an unfavorable influence on the growth of young animals (Thomasson, 1955). An increase in the cholesterol content of the adrenal gland was described by Carroll in 1951. This same author attributed these changes to erucic acid in HEAR oil (Carroll, 1953). In 1956 concern over these findings stimulated officials of the Food and Drug Directorate of the Department of Health and Welfare Canada to issue a directive to the Saskatchewan Wheat pool to "cease and desist immediately all shipments and production of edible rapeseed oil for the Canadian consumer." It was apparent from this that the regulatory officials considered HEAR oil constituted a potential human health hazard. The federal officials further emphasized the concern by indicating that deviations from full compliance of their directives would be taken as a contravention of the Food and Drug Act.

In the same year the Food and Drug Directorate modified the restrictions on the use of rapeseed oil pending a submission showing the safety of HEAR oil for human use. A review of the published information on the nutritional properties of rapeseed oil presented in 1956 led the Canadian Committee on Fats and Oil to conclude there was no evidence to indicate that limited use of HEAR oil constituted a human health hazard. However, the committee indicated the need for further information on the nutritional properties of HEAR oil. It was also decided that research should be initiated to determine if the erucic acid content of HEAR oil could be lowered by selective plant breeding. Erucic acid in HEAR oil was further implicated as a component of concern in 1957 by Carroll and Nobel who suggested that erucic acid affected the reproduction of rats.

In 1958 further clarification relating to the regulatory concern for HEAR oil was offered in the following statement from officials of the Food and Drug Directorate to the Edible Oil Institute of Canada:

> While rapeseed oil has been employed in Europe as a constituent of margarine and in other foods, it has been shown to have certain undesirable characteristics which have been related largely to its erucic acid content. It may also be noted that no detailed information has been available regarding its possible effect on humans. As a result of this situation a comprehensive program was undertaken eighteen months ago by the Food and Drug Directorate to investigate the status of rapeseed oil with albino rats. These experiments have now indicated no harmful effects of rapeseed oil from a nutritional standpoint when fed at levels which would ordinarily be consumed by humans. We would, therefore, have no objection at this time to the use of rapeseed oil in moderate amounts in food in Canada.

In 1960 additional concern about the safety of HEAR oil was raised when Roine et al. (1960) reported on the occurrence of interstitial inflammatory changes in the myocardium of rats that had been fed high levels of HEAR oil.

In retrospect the studies of Roine *et al.* (1960) are interesting and noteworthy inasmuch as they provide the first indication of the particular susceptibility of the rat to myocardial effects of HEAR oil. These authors found that when rats were fed 15–70 calorie % HEAR oil they developed myocarditis and small necrotic foci after 13–53 days on the diet. They compared these results with rats to pigs that were fed high levels of HEAR oil or soybean oil in the diet. It is important to note that an interstitial inflammation was observed in the myocardium of all the rapeseed oil and soybean oil fed pigs. The authors described this as being fairly mild on the whole. In some cases a slight cloudy swelling of the muscle was observed. However, no difference could be observed between pigs fed different oils. This study was the first indication that the problem, at least in the pig, was not specifically related to HEAR oil but to the high level of fat in the diet. In spite of this the findings of the various toxic effects in rats and the association of the effects with erucic acid gave further impetus and urgency to the ongoing program to lower the erucic acid content of HEAR oils. The success of this program is described in detail in Chapter 5.

An additional concern relating to the safety of HEAR oil arose from the report of Abdellatif and Vles (1970) who described fatty accumulation in the heart, skeletal muscles, and adrenals of rats given 60 calorie % of HEAR oil in the diet. It was observed that the fatty infiltration of the heart muscle that developed after 3 days of feeding HEAR oil in the diet decreased even on continuous feeding of the oil and more rapidly still when feeding of the oil was discontinued. These findings suggested an adaptation to the high levels of fat in the diet and indicated that the increased deposition of fat in tissues was reversible.

In 1970 new varieties of rapeseed oil low in erucic acid (LEAR oil) were used in comparative studies in rats fed oils containing long chain fatty acids. In these studies there was no lipid accumulation in the hearts of rats fed the LEAR oil (Beare-Rogers *et al.*, 1971). The results of these studies led officials of the Canadian Food and Drug Directorate to the conclusions that "it is considered prudent as a sound public health measure to replace erucic acid containing rapeseed oil with LEAR oil as soon as practical" (Campbell, 1970).

In 1973 an Expert Committee assembled by Health Protection Branch of Health and Welfare Canada made the following recommendations concerning safe levels of intake of the long chain fatty acids for humans:

> 1. Rapeseed oil high in erucic acid should not be used as a source of fat in products intended for human or animal consumption. A similar recommendation is made with respect to marine oils rich in C_{20} (or greater) fatty acids. If such oils are to be used in human or animal nutrition, the long chain fatty acid content of the oil should be diluted down to acceptable levels.

2. We regard the development of low erucic acid rapeseed oil as an important step in the right direction. We would encourage the progressive elimination of all long chain fatty acids (greater than C_{20}) from rapeseed oil.

3. In the interim we would suggest that C_{22} monoenoic fatty acids should not constitute more than 5.0% by weight of the total fatty acids of fats and oils. This objective should be sought as soon as possible. Ideally, the low erucic acid rapeseed oils should be used as an admixture with other oils which will increase the content of palmitic acid and essential fatty acids.

4. We would also recommend that the % content of C_{22} monoenoic fatty acids be indicated on the label or package of food products intended for human consumption.

5. We recommend a greatly increased support program by government and other granting agencies for basic and applied research on the nutritional and physiological effects of fats intended for human and animal consumption.

In addition to these recommendations a number of studies were recommended:

1. To study the effect of long-term feeding of low erucic acid rapeseed oil in suitable animal species including primates.

2. Do epidemiological, clinical and pathological investigations in man.

3. In the interim we would suggest that C_{22} monoenoic fatty acids should not constitute more than 5.0% by weight of the total fatty acids of fats and oils. This objective should be sought as soon as possible. Ideally, the low erucic acid rapeseed oils should be used as an admixture with other oils which will increase the content of palmitic acid and essential fatty acids.

4. Efforts should be made to establish the morphological characteristics of myocardial alterations elicited by long chain fatty acids in rapeseed and marine oils. These studies should include histochemistry and electron microscopy in addition to light microscopic techniques. Proper identification of the myocardial changes would help to delineate background changes in animal experiments that hinder the establishment of "no response levels" of long chain fatty acids and the comparison of data obtained by various investigators.

6. We recommend that there be further development and testing of new strains of rapeseed with very low or zero levels of erucic acid and higher levels of palmitic and linoleic acids.

In the same year these recommendations were followed by a news release that placed restrictions in the content of C_{22} monoenoic fatty acids in processed edible fats and oils. In the news release the minister of Health and Welfare announced that, "the maximum content of C_{22} monoenoic fatty acids in processed edible fats will be restricted to 5% of the total fatty acids present as of December 1, 1973."

A concern for the then current shortage of edible fats and oils and the need to consider the impact of the policy on the availability to Canadians of oils of suitable quality was brought out in the news release. It appeared that, at the time, the 5% limit on long chain fatty acids could be met by the use of the new low erucic varieties of rapeseed developed in Canada. It was also pointed out that to meet the 5% maximum level it would be necessary to reduce the percentage of certain other sources of long chain fatty acids, such as marine oils, in processed products. Further evidence of the regulatory concern was indicated by the announcement that adherence to the program would be monitored and if necessary appropriate amendments would be made to the Food and Drug regulations.

The regulations to limit the erucic acid content of rapeseed oils are outlined in the Consolidated Foods and Drug Regulations of Canada under Division 9 section B.09.022 as follows: "No person shall sell cooking oil, margarine, salad oil, simulated dairy product, shortening or food that resembles margarine or shortening, if the product contains more than five per cent C_{22} monoenoic fatty acids calculated as a proportion of the total fatty acids contained in the product."

However, the fact that fatty infiltration was not observed when LEAR oils were fed (Beare-Rogers et al., 1971) did not completely allay the concerns of regulatory agencies since later studies indicated that myocardial necrosis was observed when LEAR oils were included at high levels in the diets of rats.

In 1968, Rocquelin and Cluzan reported increases in weight of the heart, liver, kidneys, spleen in 3-month-old rats fed either rapeseed oil with 44% of erucic acid, or rapeseed oil with 1.9% erucic acid. Myocardial lesions (myocarditis) were observed in 7-month-old male or female rats fed either rapeseed oil diet. The frequency of myocarditis was higher with males than with females. The authors suggested that common characteristics of the two rapeseed oils, such as a low content of saturated fatty acids, unbalanced ratio between saturated and monounsaturated fatty acids, or unsaponifiable matter of the oil, might account for these results.

These reports led to a growing international concern about the C_{22} monoenoic acids and their potential adverse human health effects. As a result a vast amount of research was undertaken into the pathogenesis, etiology, and mechanisms of the observed adverse effects. This research is critically reviewed elsewhere in the book.

II. THE LABORATORY RAT AS AN EXPERIMENTAL MODEL FOR SAFETY ASSESSMENT OF RAPESEED OILS

As in most toxicological research, the first investigations with rapeseed oil used the rat as the test animal. Toxic effects in the form of myocardial lesions were observed in the rat and its use was continued in subsequent studies. From the outset it was assumed there was some toxic factor associated with rapeseed oil that was responsible for the cardiac lesions. The question arose as to whether some nontriglyceride toxic factor in LEAR oils might be responsible for the heart lesions in rats and much research was devoted to attempts to define the toxic factors. However, there is convincing evidence that there are no toxic compounds in LEAR oils (Bijster *et al.*, 1979a; Kramer *et al.*, 1975, 1979) and that the heart lesions found in rats are not caused by toxins present in LEAR oils or any other oil (Kramer *et al.*, 1979). It is apparent that the problem with the oil was nutritional and metabolic. Cardiac lesions ensue because the rat cannot utilize high levels of vegetable oils in the diet. Because of this the rat is an inappropriate model for testing the nutritional properties and safety of these oils for humans.

There are several reasons why the rat is unsuitable in determining whether or not vegetable oils might pose a problem in the human diet. Most of the studies that have been undertaken have involved feeding the oils to rats at a concentration of 20% by weight in the diet. This concentration was chosen because the North American diet can contain up to 20% lipid. However, it is known that all the rat requires to meet all physiological activities is a concentration of 5% fat in the diet (National Academy of Sciences, 1972). In fact, feeding rats 20% vegetable oil in the diet reduces life span (Spindler *et al.*, 1978). This suggests that the rat is not physiologically capable of metabolically handling high concentrations of vegetable oil in the diet. An additional concern with the rat as an experimental model for these studies is the fact that the lesions in the heart are commonly seen in animals on control diets. These lesions do not differ in morphology from those seen in the rats fed 20% soybean oil, sunflower oil, corn oil, peanut oil, olive oil, safflower oil, or coconut oil in the diet (Chapter 17). The myocardial lesion is commonly seen in older rats that are used as controls and indeed a mild degree of this lesion is frequently not diagnosed so that the actual incidence may be considerably higher than the reported background incidence of 17–33% (Goodman *et al.*, 1979).

This suggests that the heart of the laboratory rat has a particular predisposition to the development of this particular type of myocardial lesion. It is not known if an infectious agent is involved, but it appears that vegetable oils function as provocative factors in increasing the background incidence of heart lesions in the rat. In other words, the vegetable oils may be capable of contributing to the unmasking of a latent lesion of the myocardium of laboratory rats.

It becomes more apparent that the rat is uniquely sensitive when it is compared with other species in which the incidence and severity of the heart lesions is much lower and certainly not related to the consumption of LEAR oils. The studies in primates are particularly interesting since they indicate that feeding of low erucic acid rapeseed or soybean oil did not cause heart lesions (Kramer *et al.*, 1978a, 1978b).

Other species tested, i.e., mice (Vles et al., 1978), swine (Friend et al., 1975a,b, 1976; Aherne et al., 1975, 1976; Bijster et al., 1979b; Svaar et al., 1980), and dogs (D. L. Hamilton and B. Schiefer, private communication), do not develop myocardial lesions in response to feeding of high levels of LEAR oil or any other vegetable oil in the diet. As a matter of fact, the necrotic myocardial lesions associated with the male albino rat are not generally found in these other species, and are never associated with LEAR oil or other vegetable oils.

An additional explanation for the absence of heart lesions attributable to vegetable oils in species such as swine and primates lies in their ability to metabolize fat differently than the rat. It is known that primates and pigs fed LEAR oil have less polyunsaturated fatty acids in the heart than does the rat. Furthermore, pigs and primates have greater capacity to oxidize erucic acid than does the rat and accordingly accumulate smaller amounts of erucic acid in the heart. While it is clear that the rat is not a suitable model to test the safety of vegetable oils, the results of studies from the other species that can metabolize and utilize vegetable oils are appropriate for assessing the safety of the oils in humans.

III. STUDIES IN HUMANS

In considering the safety of substances in the food supply, information on human consumption of the substance and any suggested relationship between consumption and adverse health effects should be assessed.

In making such an assessment, with respect to rapeseed oil, it is important to recognize there is a major difference between the common heart disease of humans and heart disease of the laboratory rat. The major problem of heart disease in man relates to atherosclerosis in the large coronary arteries with secondary changes in the heart muscle. In the rat, dietary intake of rapeseed oil affects primarily the myocardium with no evidence of changes in the coronary arteries.

Focal myocardial lesions are sometimes found incidentally in human hearts. These have some histological features that are similar to those seen in the rat. However, the etiology of these lesions is different. A low grade infectious myocarditis may be involved in some cases (Kline et al., 1963; Pomerantz and Davies, 1975). A number of drugs, poisons, and clinical conditions are known to cause diffuse or multifocal myocardial necrosis in humans but there is no hard evidence indicating dietary fat consumption as a factor in this respect (McKinney, 1974).

The available data in humans concerning a possible relationship between consumption of rapeseed oil and adverse health effects can be reviewed from the standpoint of accumulation of 22:1 in human myocardium as related to diet, reports of lipidosis in man, and the incidence of cardiomyopathies in man and their relationship to diet.

A study initiated by the Indian Council of Medical Research and reported in the Annual Report of the National Institute of Nutrition, Hyderabad, India (Anonymous, 1976, 1977), indicated that levels of erucic acid in the myocardium were related to the vegetable oils principally consumed in that particular district. In Calcutta, mustard oil high in erucic acid (40–44% 22:1) is the main edible oil; in Madras, peanut and sesame oil are the principal edible oils; and in Trivandrum, coconut is the primary edible oil. The lipid analyses of 50 hearts from each center were reported. The mustard oil consuming center of Calcutta showed significant amounts of erucic acid in the myocardium (5.6% with a range of 0.9–9.9%), whereas the other two regions showed no detectable amounts of erucic acid (Anonymous, 1977). Although the shortcomings of epidemiological studies of this nature are readily apparent, they do indicate that dietary erucic acid intake is reflected in the levels of this fatty acid in the myocardium. When the hearts from the Calcutta region were examined it was apparent that the presence of the 22:1 was not associated with heart damage. Over 100 hearts were examined for histological evidence of fibrosis and none was found that could be related to the consumption of mustard oil. A report presented at the International Symposium of Rapeseed and Mustard, November 22–24, 1976, at Mysore, India, indicated that in 38 hearts from Madras and 25 hearts from Trivandrum there was also no evidence of myocardial fibrosis that could be related to the consumption of peanut, sesame, or coconut oils.

An epidemiological study conducted 1974 in France (Chone, 1977) indicated that of 254,788 cases of death due to heart failure, 269 cases, or 0.11%, were identified as cardiomyopathies that were somewhat similar in histology to the observed cardiomyopathies in rats. Of the 269 cases, there was a significant association with alcohol consumption but not with dietary fat and vegetable oil. This study is of particular interest, since France, like India, is a major consumer of rapeseed oil which until 1974 was of the high erucic acid variety.

Data on the accumulation of 22:1 fatty acids in humans are also available from the work of Svaar who examined autopsy material from 54 hearts selected from Norwegian men, age 20 to 69, who had died suddenly from accidents (Svaar, 1982). These hearts were selected from a larger group on the basis of being without myocardial infarction, severe coronary stenosis, cardiac hypertrophy or valvular disease by macroscopical examination. No focal myocardial lesions were present. A mild to moderate lipidosis was found in 50% of the hearts but this was not correlated with the concentration of 22:1 which was present at less than 1% of the total lipids (Svaar,

1982). According to Svaar with the knowledge available at the present time there is no indication that the consumption of 22:1 fatty acids from rapeseed or marine oil sources cause harmful effects in the human heart.

The recent outbreak of oil-related poisoning from Spain in 1981 (Tabuenca, 1981) which caused illness in about 12,000 persons (Valenciano, 1981) and at least 200 deaths in 5 months (Torrey, 1981; Gilsanz, 1982) was initially blamed on rapeseed oil (Anonymous, 1981) and erucic acid (McMichael, 1981). The fact that this toxicity was ascribed to rapeseed oil is evidence of the widespread misunderstanding of the biochemical and nutritional properties of erucic acid among some scientists and the popular media. Once the outbreak was investigated in detail, it was quickly established that the rapeseed oil originally incriminated was, in fact, a mixture of industrial rapeseed oil, soybean oil, olive oil, and animal fats that had been purposely denatured with 2% aniline and was never intended to be used as an edible oil (Tabuenca, 1981; Gollob, 1981). When attempts were made to remove aniline by heating the oil, anilides would be formed with the unsaturated fatty acids which in turn would be toxic (Gollob, 1981; Gordon, 1981). Murphy and Vodyanoy (1982) suggest that these fatty acid anilides may be incorporated into normal cell membranes resulting in membrane destabilization and membrane destruction. Anilides are reported to have been found in the fatty tissues of the victims (Gollob, 1981). Although the etiology of the disease is in doubt, the disease appears to have two clinical phases, one toxic in character with pneumonia-like symptoms, and the second with similarities to autoimmune disease in which neuromuscular changes predominate (Gilsanz, 1982). In order to reproduce the toxic oil syndrome in laboratory animals, oleyl and linoleyl anilides have been prepared and fed mixed with pure olive oil. The rats showed lung lesions on this diet which resembles that of the toxic syndrome (Tena, 1982). Kemper et al. (1982) fed diets that contained olive oil, olive oil with aniline, or olive oil with anilides of oleic acid. Rats fed the aniline or anilides showed increased lung weight, and a trend to a decrease in thymus weight. The same aniline and anilide containing diets when fed to chicks decreased the size of the immunocompetent organs, i.e., the thymus and bursa of Fabricius. These authors concluded that the toxic oil syndrome may be related to long-term adverse effects on the immune system. Another suggestion that has been made is that the toxic oil syndrome may be related to the presence of superoxides and epoxides which may damage cell membranes through the presence of free radicals. At any rate it is clear that this oil-related outbreak in Spain is in no way specifically related to rapeseed oil or erucic acid (Sinclair, 1981), but rather the result of the fraudulent introduction into the food chain of a denatured oil intended for industrial purposes. In fact, an epidemic with a similar clinical picture was reported from Germany and Holland some years ago and was labeled "the margarine disease." These poisonings were also associated with the adulteration of edible oils by toxic chemicals (Ross, 1981).

IV. SAFETY ASSESSMENT CONSIDERATIONS

Information that is used in the safety assessment of substances for human use includes a knowledge of the chemistry of the substance, the results of studies in experimental animals, and evidence from epidemiological studies in humans. In Chapter 4 of this book it is made evident that LEAR oil is similar in its fatty acid composition to numerous other vegetable oils that are commonly consumed by humans. From this standpoint it is apparent that LEAR oil as normally consumed is, like other vegetable oils, a safe substance for humans.

For a number of years during the 1970s studies in experimental animals and in the rat in particular caused concern about the safety of rapeseed oil for human use. Initially it was theorized that the oil might contain a toxic factor or substance. However, it has been demonstrated that vegetable oils contain no cardiopathogenic nontriglyceride compounds responsible for myocardial lesions in rats. Futhermore, it is now apparent that concerns about safety were unfounded since it is established that the rat is not suitable as an experimental model for safety assessment studies on vegetable oils intended for human use. The reasons why the rat is not a suitable model may be summarized as follow:

1. The male rat readily develops areas of focal myocardial necrosis irrespective of diet. The lesion is commonly seen in rats on control oils.

2. The amount and type of vegetable oil in the diet can alter the incidence of this lesion.

3. No other animal tested (i.e., pig, dog, and monkey) shows specific heart lesions in response to the amount and type of vegetable oil in the diet.

4. The rat is much more sensitive to myocardial lipidosis than most other species and rapidly accumulates triglycerides and erucic acid in the myocardium when fed mustard oil or high erucic acid rapeseed oil.

5. Long chain polyunsaturated fatty acids of the linolenic acid family extensively accumulate in cardiac phospholipids of the rat but not in other species.

6. The ability to oxidize C_{20} and C_{22} fatty acids is much reduced in the rat as compared to other animals. Furthermore, in the rat these long chain fatty acids cause secondary inhibition of the tricarboxylic acid cycle oxidation. Pigs and primates do not respond in this manner.

7. Common heart disease in humans is different than the myocardial disease observed in rats.

In attempting to make extrapolations from rats to humans it is important to bear in mind that the heart disease of major concern in man is one in which the vascular system is primarily affected and alterations in the myocardium are secondary. The heart disease in rats that occurs spontaneously and is increased in incidence when rats are fed high levels of vegetable oils is confined to the heart muscle; no involvement of blood vessels similar to vascular disease in man has been reported to occur in rats fed rapeseed oils. Although lesions similar in morphology to those that occur in rats are seen in humans and other animals, such lesions are usually associated with or accompany another disease process rather than existing as a principal disease entity, as in the case with the rat.

Studies in humans include surveys for the presence of 22:1 in human hearts as related to diet, reports of myocardial lipidosis, and the incidence of myocardial lesions similar to those observed in experimental animals. The available evidence indicates that 22:1 may occur in human cardiac muscle in geographic areas where vegetable oils containing these fatty acids are consumed. However, there is no relationship in humans between myocardial lesions of the type observed in rats and the consumption of rapeseed oils.

V. CONCLUSIONS

In making a safety assessment of LEAR oil it is evident that initial concerns with the safety of rapeseed oil were based on studies conducted in the rat. It has been established that these concerns were unfounded since the rat is not a suitable model for safety assessment studies of vegetable oils. Studies in humans and other species, and from the knowledge of the substances, make it apparent that LEAR oil, as normally consumed, is like other vegetable oils, a safe substance for human consumption.

REFERENCES

Abdellatif, A. M. M., and Vles, R. O. (1970). *Nutr. Metab.* **12**, 289–295.

Aherne, F. X., Bowland, J. P., Christian, R. G., Vogtmann, H., and Hardin, R.T. (1975). *Can. J. Anim. Sci.* **55**, 77–85.

Aherne, F. X., Bowland, J. P., Christian, R. G., and Hardin, R. T. (1976). *Can. J. Anim. Sci.* **56**, 275–284.

Anonymous (1976). "Annual Report of the National Institute of Nutrition," pp. 31–32. Indian Council of Medical Research, Hyderabad, India.

Anonymous (1977). "Annual Report of the National Institute of Nutrition," pp. 49–55. Indian Council of Medical Research, Hyderabad, India.

Anonymous (1981). New Scientist **91**, 276.

Beare-Rogers, J. L., Nera, E. A., and Heggtveit, H. A. (1971). *Can. Inst. Food Technol. J.* **4**, 120–124.

Bijster, G. M., Hudalla, B., Kaiser, H., Mangold, H. K., and Vles, R. O. (1979a). *Proc. Int. Rapeseed Conf., 5th, 1978* Vol. 2, pp. 141–143.

Bijster, G. M., Timmer, W. G., and Vles, R. O. (1979b). *Fette, Seifen, Anstrichm.* **81**, 192–194.

Boer, J., Jansen, B. C. P., and Kentie, A. (1947). *J. Nutr.* **33**, 339–358.

Campbell, J. A. (1970). *Proc. Int. Conf. Sci. Technol., Market., Rapeseed, Rapeseed Prod., 1970* pp. 467–469.

Carroll, K. K. (1951). *Endocrinology* **48**, 101–110.

Carroll, K. K. (1953). *J. Biol. Chem.* **200**, 287–292.

Carroll, K. K., and Noble, R. L. (1957). *Can. J. Biochem. Physiol.* **35**, 1093–1105.

Chone, E. (1977). *Bull. CETIOM* **68**, 19–24.

Deuel, H. J., Greenberg, S. M., Straub, E. E., Jue, D., Gooding, C. M., and Brown, C. F. (1948). *J. Nutr.* **35**, 301–314.

Friend, D. W., Corner, A. H., Kramer, J. K. G., Charlton, K. M., Gilka, F., and Sauer, F. D. (1975a). *Can. J. Anim. Sci.* **55**, 49–59.

Friend, D. W., Gilka, F., and Corner, A. H. (1975b). *Can. J. Anim. Sci.* **55**, 571–578.

Friend, D. W., Kramer, J. K. G., and Corner, A. H. (1976). *Can. J. Anim. Sci.* **56**, 361–364.

Gilsanz, V. (1982). *Lancet* **1**, 335–336.

Gollob, D. (1981). *Lancet* **2**, 1102.

Goodman, D. G., Ward, J. M., Squire, R. A., Chu, K. C., and Linhart, M. S. (1979). *Toxicol. Appl. Pharmacol.* **48**, 237–248.

Gordon, R. S. (1981). *Lancet* **2**, 1171–1172.

Kemper, F. H., Luepke, N.-P., Renhof, M., and Weiss, U. (1982). *Lancet* **1**, 98–99.

Kline, I. K., Kline, T. S., and Saphir, O. (1963). *Am. Heart J.* **65**, 446–457.

Kramer, J. K. G., Hulan, H. W., Mahadevan, S., Sauer, F. D., and Corner, A. H. (1975). *Lipids* **10**, 511–516.

Kramer, J. K. G., Hulan, H. W., Procter, B. G., Dussault, P., and Chappel, C. I. (1978a). *Can. J. Anim. Sci.* **58**, 245–256.

Kramer, J. K. G., Hulan, H. W., Procter, B. G., Rona, G., and Mandavia, M. G. (1978b). *Can. J. Anim. Sci.* **58**, 257–270.

Kramer, J. K. G., Hulan, H. W., Corner, A. H., Thompson, B. K., Holfeld, N., and Mills, J. H. L. (1979). *Lipids* **14**, 773–780.

McKinney, B. (1974). "Pathology of the Cardiomyopathies." Butterworth, London.

McMichael, J. (1981). *Lancet* **2**, 1172.

Murphy, R. B., and Vodyanoy, V. (1982). *Lancet* **1**, 98.

National Academy of Sciences (1972). "Nutritional Requirements of Laboratory Animals," 2nd rev. ed. Nat. Acad. Sci., Washington, D. C.

Pomerantz, A., and Davies, M. J. (1975). "Pathology of the Heart." Blackwell, Oxford.

Rocquelin, G., and Cluzan, R. (1968). *Ann. Biol. Anim., Biochim. Biophys.* **8**, 395–406.

Roine, P. Uksila, E., Teir, H., and Rapola, J. (1960). *Z. Ernährungswiss.* **1**, 118–124.

Ross, G. (1981). *Br. Med. J.* **283**, 424–425.

Sinclair, H. (1981). *Lancet* **2**, 1293.

Spindler, A. A., Dupont, J., and Mathias, M. M. (1978). *Age* **1**, 85–92.

Svaar, H. (1982). *In* "Nutritional Evaluation of Long-Chain Fatty Acids in Fish Oil" (S. M. Barlow and M. E. Stansby, eds.), Academic Press, New York (in press).

Svaar, H., Langmark, F. T., Lambertsen, G., and Opstvedt, J. (1980). *Acta Path. Microbiol. Scand., Sect. A* **88**, 41–48.

Tabuenca, J. M. (1981). *Lancet* **2**, 567–568.

Tena, G. (1982). *Lancet* **1**, 98.

Thomasson, H. J. (1955). *J. Nutr.* **56**, 455–468.

Thomasson, H. J., and Boldingh, J. (1955). *J. Nutr.* **56**, 469–475.

Torrey, L. (1981). *New Scientist* **91**, 640.

Valenciano, L. (1981). *Morb. Mort. Weekly Rep.* **30**, 436–438.

Vles, R. O., Bijster, G. M., and Timmer, W. G. (1978). *Arch. Toxicol., Suppl.* **1**, 23–32.

22

Some Recent
Innovations in Canola
Processing Technology

A. D. RODEN

I. INTRODUCTION

In the very recent past, there have been advancements made in canola processing. These advancements are made possible due to the results of the breeding programs, new equipment design, and recent research on processing of the oil. The developments have improved the crude oil quality sufficiently to make substantial changes in the refining process possible. They have reduced processing costs both in crushing and refining of the oil, making canola oil more economical and competitive when compared to most other oils.

The three developments discussed in this chapter will be cold pressing of seed, chemical degumming, and physical refining. These innovations are

563

High and Low Erucic Acid Rapeseed Oils
Copyright © 1983 by Academic Press Canada
All rights of reproduction in any form reserved.
ISBN 0-12-425080-7

past the development stage and have been in operation at CSP Foods (Dundas, Ontario) for some time. They offer great economic advantages by reducing capital costs and processing costs substantially. They are presented here to help the canola processors reduce their costs and improve their product, which in the long term will increase canola's market potential.

II. COLD PRESSING OF CANOLA SEED

As outlined in the chapter on crushing and extraction, canola is currently flaked, cooked, pressed, and then solvent extracted. Krupp Industries und Stahlbau Werk (Harburt, West Germany) has designed a new screw press which eliminates the need for flaking and cooking of the seed.

The principle of operation is that extraction will be possible provided the cells containing the oil are ruptured and the cake is porous enough for extraction. The press mechanically breaks down the cells by compression and shear as in a normal press operation. However, the screw press is equipped with several compression and decompression stages along its length. By compressing, decompressing, and compressing again, the cake structure is sufficiently changed in each stage to create a cake with good porosity.

The advantages of the system are that it uses less energy and produces an oil of a significantly higher quality than previously. This quality difference is shown by a low content of free fatty acids, sulphur, chlorophyll, and phosphorus.

III. CHEMICAL DEGUMMING

For many years degummed canola oil has been traded with a typical phosphorus content of 180–200 ppm. This is equal to approximately 0.5 acetone insoluble content. The phosphorus content was this high due to the presence of nonhydratable phospholipids. For this reason, the crude oil, even when contacted with 2% water, would still contain a high concentration of phosphorus. The action of various acids on the ester linkages and/or calcium and magnesium salts of the phosphorus compounds has been known for some time. Unfortunately, the problems of contacting the phosphorus compounds with the acid, separation of the precipitated gums, and selecting the best acid for splitting the linkages have prevented its use on an industrial scale. For canola oil, these problems have now been solved.

The best acid tested for reactivity and economic viability is citric acid. It has the further advantage of removing trace metals by chelating them, after which they are removed in the separating centrifuge. This process further increases the quality of the oil stock.

There are two methods used for the addition of the acid. In one method of addition, the acid is added to oil that has been cooled to about 35°C and mixed slowly for at least 1 hour. In the second method, the acid is added to oil at 60°C and mixed vigorously for a shorter period of time. It is important to ensure that the oil is thoroughly mixed with the acid and that the precipitated gums are not broken up into particles which are too small to be separated in the centrifuge.

Other variables apart from the degree of mixing and temperature of addition influence the effectiveness of this process. These include the quantity of citric acid added, the concentration of citric acid, and the quantity of water present. In general the citric acid concentration should be 50% and the level of addition between 1000 and 3500 ppm. The water concentration should be greater than 0.75%. If correct procedures are followed, it is possible to chemically degum with a solid bowl centrifuge, although there are advantages to the use of a split bowl centrifuge.

Chemically degummed oil is currently sold under several different names; the most common include semirefined oil, acid degummed oil, and special quality degummed oil. This oil will generally contain less than 50 ppm phosphorus at the point of shipping and, in most other respects, be of similar quality to oil degummed by other currently used processes.

IV. PHYSICAL REFINING

Physical or steam refining has been employed for a number of years in the processing of palm and soya oils. The problems encountered with the application of the process to canola oil have been related to the nature and the quantity of the phospholipids present in the crude degummed oil. Therefore, almost as a direct result of the development of higher quality crude degummed stocks, physical refining is now possible.

The method employed is very similar to that used for processing palm oil. The basic technique is to pretreat the oil with an acid solution and then bleach the mixture with activated clay. This procedure requires strict control of the feed stock quality as well as frequent checks on the bleached oil quality.

Crude oil should contain less than 70 ppm phosphorus in order to be physically refined. The process works best when the feed stock contains less than 50 ppm phosphorus. It is possible, in some cases, to physically refine oils with a higher phosphorus content. However, the additional clay needed adversely affects the economic viability of the process.

The best acids found for use in the pretreatment stage have been a combination of citric and phosphoric acids. The pretreatment step is critical to the success of the process. If the contact between the treatment reagents and oil is poor then the phosphorus is not always changed into a bleachable form and therefore the bleached oil contains a high level of phosphorus.

To ensure good contact there are several approaches to be tried. Firstly it is possible to vigorously mix the oil and reagents together for a short time. Another possibility is a longer contact time and more gentle mixing. The presence of water in the pretreatment will aid in ensuring good contact between the phosphorus and acids. This contact results because the acids are located in the water phase and the polar phospholipids are attracted to the water surface. Also there is evidence that the water plays a further role in making the phospholipids bleachable.

After pretreatment the oil is then pumped to a slurry tank where clay is added. The quantity of clay needed for this process should be less than 115% of the clay used in bleaching alkali refined stocks. The oil slurry is then pumped to the bleaching vessel where it is dried. The drying temperature must be over 100°C.

To ensure good filtration the oil must be thoroughly dried. If any moisture is left in the oil, then the rate of filtration will be reduced due to plugging of the press by slimy phospholipids. If the oil is dry then the phospholipids tend to form a grainy precipitate which is easily filterable.

The bleached product should be continuously checked for color, free fatty acid, filter clay, phosphorus, and deodorized oil color, by heat test. The color should be less than a Lovibond color of 9 Red on a 5-$\frac{1}{4}$ inch tube. The heat test is performed by heating a sample of oil to 250°C under an inert atmosphere and reading the color. This gives an estimate of the deodorized oil color and should be less than a Lovibond color of 2 Red on a 5-$\frac{1}{4}$ inch cell. The clay must be fully removed from the oil as even traces that pass through can substantially reduce oil quality. The final phosphorus must be reduced to less than 5 ppm. The increase of free fatty acid during this process should be less than 0.2%.

The bleached oil hydrogenates as well as an alkali refined bleached product does. There is no reduction in the reaction rate or a need for added catalyst suggesting that all catalyst poisons are either inactivated or removed by these processes. If there is a change in hydrogenation, examination of the bleaching parameters is advised over changes in the hydrogenation conditions.

The free fatty acids, odor, and flavors are stripped from the oil during deodorization. If the deodorizer was designed for steam refining, there should be no reduction in the rate of production. However, if the steam stripping is inefficient then there might be a need to reduce production rates by as much as 25%. It is possible to make some changes to the design of the unit to maintain current production rates. It should be noted that if physical refining is to be implemented, the deodorizing temperature must be over 225°C.

The deodorized product, if properly pretreated and bleached, will be as stable and as high in quality as an alkali refined product. In some cases, a physically refined oil will show more flavor and oxidative stability than the alkali refined product. Physical refining has reduced processing losses by as much as 2%. As a result of eliminating alkali refining, many effluent problems are solved. Effluent is cleaner as a result of the elimination of soapstock production, acidulation, and reduced oil losses. The acidulation process accounts for the majority of the effluent from a refinery and is the most difficult to clean up.

V. CONCLUSIONS

These newly developed and installed processes have significantly reduced the cost of processing canola. They have improved the oil product to such an extent that canola is often easier to process than almost all other vegetable oils.

In summary, these developments have made canola processing more economical by eliminating processing steps, cutting the energy requirements by up to 25%, and reducing processing losses by 33%. It would therefore be possible today to build a canola oil processing plant which would require less capital cost than the standard plants used to process other oils and still have lower operating costs.

The improved quality of the crude oil resulting from chemical degumming and cold pressing has further improved the final products. Canola oil is now more stable and has less tendency to revert in flavor. It also has no hydrogenation problems. This means that the oil can be consumed both in hardened and liquid forms, making it a very versatile product to use.

Index

A

Acyl carrier protein, in fatty acid biosynthesis, 132–135, 358
Adrenal glands, 274–276
  accumulation of cholesterol esters in, 275
  accumulation of erucic acid in, 275
  effect of cold stress on in rats, 275–276
  effect of erucic acid on, 69, 274–276, 552
Africa, production of rapeseed, 53
Alkali-refining, 200–205
  acidulation of soapstock, 204–205
  batch, 201–202
  continuous, 202–204
Alleles
  control of eicosenoic acid in rapeseed, 151
  control of erucic acid in rapeseeed, 16, 150–152
Analides, formation of in adulterated oil, 559
Arachidonic acid
  concentration in cardiac phospholipids, 492–501, 504–505, 509
  regulation of biosynthesis, 401–406
Argentine rape, 63, 163–165

B

Behenic acid (22:0)
  cardiopathogenicity of, 249, 272
  concentration of in vegetable oils, 15, 87, 88, 91, 164, 171, 415
  formation of, 249, 272
Black mustard, see B. juncea
Bleaching, 205–208
  batch, 206

bleaching clay, 205–206
  continuous, 206–207
  for physical refining, 207–208, 566
  postbleaching, 213
Brassica
  description of, 1–13, 254
  form and cultivation, 5–6
Brassica campestris (turnip rape), 4, 62, 150, 162, see also cultivars of
  erucic acid content, 14–16, 86–95, 164, 165, 168, 171
  fiber content, 12–13, 55
  oil content, 13–14, 163, 170
  origin and distribution, 4
  Polish rape, 62, 64, 162–165
  plant and seed, 7–13
  protein content, 12–13, 163, 165, 170
  yield, 6, 64, 165, 172
Brassica carinata, 3
Brassica juncea, 5, 6, 15, 16, 18, 137, 139, 150
Brassica kaber, cv. pinnatifida, 173
Brassica napus (rape), 4, 63, 150, 163, see also cultivars of
  Argentine rape, 63, 163–165
  erucic acid content, 14–16, 86–95, 164–165, 167–168, 171
  fiber content, 55
  oil content, 13–14, 163, 170
  origin and distribution, 4–5
  plant and seed, 7–13
  protein content, 163, 165, 170
  yield, 6, 64, 165
Brassica nigra, 2–3
Brassica oleracea, 4, 153
Brassicasterol, 106–110, 441–451